INTELLECTUAL PROPERTY AND COMPETITION LAW

INTELLECTUAL PROPERTY AND COMPETITION LAW

New Frontiers

Edited by

Steven Anderman
Ariel Ezrachi

OXFORD
UNIVERSITY PRESS

This book has been printed digitally and produced in a standard specification in order to ensure its continuing availability

OXFORD
UNIVERSITY PRESS

Great Clarendon Street, Oxford OX2 6DP
United Kingdom

Oxford University Press is a department of the University of Oxford.
It furthers the University's objective of excellence in research, scholarship, and education by publishing worldwide.

Oxford is a registered trade mark of Oxford University Press in the UK and in certain other countries

Reprinted 2013

British Library Cataloguing in Publication Data
Data available

Library of Congress Cataloging in Publication Data
Data available

ISBN 978-0-19-958995-1

PREFACE

Until quite recently the interface between intellectual property rights and competition law was confined to a relatively narrow sphere. EU competition law placed limits on the exercise of intellectual property rights (IPRs) only in a few seemingly well-defined areas. There was the possibility of a compulsory licence of copyright as a competition law remedy in a limited category of exceptional circumstances under Article 102(b) TFEU. The Technology Transfer Block Exemption Regulation created a legally secure exception from Article 101 TFEU for most forms of IP licensing involving patents and know-how. Mergers were subject to scrutiny as part of the due diligence procedure to avoid a situation where a merger was refused because it involved the purchase by one technology owner of the only, or closest competing technology in the market.

Since the mid-1990s there has been a considerable expansion of competition law's reach into the sphere of intellectual property rights. The concept of an abuse of a dominant position has been stretched from refusal to license under Article 102 TFEU to cases of tying and bundling, price discrimination, unfair pricing, and abuse of the European Supplementary Protection system by a patent owner. The Microsoft case has widened the category of exceptional circumstances in which a compulsory licence can be obtained as a competition law remedy. The pharmaceutical industry has begun to raise regulatory interest in cases of abuse of dominance in preventing the entrance of generics and has demonstrated a capacity to infringe Article 101 TFEU with reverse or delay payments. The fields of standard setting and technology pools and collecting societies, involving collaboration between competitors, have attracted the attention of the competition authorities and produced cases brought under Articles 101 and 102 TFEU. Finally, the field of IP licensing in technology transfers continues to be regulated but by a more flexible system of guidelines and exemptions under a modernized Article 101 TFEU.

The main explanation for the widening of the IP/competition interface is the noticeable increase in the strategic use of patent enforcement and acquisition as part of wider commercial practice in recent decades. As corporations have developed a greater understanding of the value of their corporate intangible assets and have given a more realistic book value to their intangible assets, they have inevitably been compelled to adopt more vigorous enforcement policies. Moreover, as IPRs have become increasingly significant in overall corporate asset values, they have begun to be incorporated into the overall commercial strategies of firms resulting in more imaginative strategic uses of patent enforcement and acquisition.

While many categories of strategic patenting occur in a form that appears to be legitimate profit taking and a lawful adaptation of commercial practice to commercial realities, some do not. To a certain extent, aggressive commercialization of IPRs is compatible with competition on the merits, but when commercial practice by IP owners consists of conduct that is prohibited by the competition laws as either an exclusionary or exploitative abuse, or a restriction on competition, the competition rules can operate as a limit to the exercise of an IPR.

One aspect of these confrontations between strategic use of patent enforcement and acquisition practices is the fact that the use of the IP is often incidental to the main aim of the undertaking's commercial strategy. The 'exploitation' of the IP right may take a form that is part of a wider effort by a firm with considerable market power to raise the costs of rivals and potential competitors to compete in a particular market.[1]

This increase in the incidence of regulatory encounters raises anew the old questions about the interface. Why is the exercise of IP subject to the prohibitions in the competition rules? Why, as coexisting systems of law, are they not subject to a compromise giving appropriate recognition to each system? These cases confirm the earlier assessment that the exercise of IPRs is subject to competition rules because the IP owner exercises its IP as a private property subject to the other public laws regulating the market. There is a compromise between the two systems but it is shaped by the logic of the competition rules. That logic offers safe harbours and exceptions to the competition rules for the normal exercise of IPRs, but it determines the outer limits of normality. The 'existence' or 'specific subject matter' or 'essential function' of an IPR may coincide with these safe harbours and exceptions but these concepts do not determine the contours of the competition law prohibition. Competition law now gives explicit recognition to the precompetitive nature of IPRs and maintains that the exercise of an IPR as such is compatible with competition rules but it reserves the right to step in when an IPR is used as a means or as an instrument of a commercial strategy that involves conduct incompatible with either Article 101 and 102 of the Merger Regulation.

One important characteristic of the new interface is the increased economic sophistication of the tools used by competition authorities. These new tools have undoubtedly sharpened and refined the regulatory aim of the competition authorities. Under Article 102 TFEU, there is a more accurate measure of market power and relevant market.

[1] D Rubinfeld, and R Maness, 'The Strategic Use of Patents; Implications for Antitrust' in F Leveque and H Shelanski (eds), *Antitrust, Patents and Copyright EU and US Perspectives* (Cheltenham: Edward Elgar, 2005).

Under Article 101 TFEU after modernization, there is more economic assessment of anti-competitive harm, relevant market, market power, and the difference between agreements between competitors and those between non-competitors. Yet the actual extent of the use made of economic thinking remains contested terrain in EU competition law. The increased application of economic thinking to legal rules has certainly led to improved analysis of actual market power and more realistic assessment of threats of anti-competitive harm, based on a goal of protecting consumer welfare. However, in the EU, the goal of consumer welfare, as defined by economic thinking alone, has not completely taken over the legal norms of abuse or the assessment of exemption in Article 101(3) TFEU. Although the convergence between the EU and the USA has been substantial with respect to the new interface between competition law and IP rights, the historic preoccupations of EU competition law, based on interpretations of the Treaty, remain in place, supported by the EU Courts even where the Commission would wish to move further toward a greater role for economic analyses in the formulation of the competition rules. While antitrust law in the USA is more concerned with avoiding Type 1 errors, (where innocent firms are found to have infringed the competition rules) the EU judiciary are still more concerned with avoiding Type 2 errors (where firms who have infringed the competition rules are found innocent).

This book explores these themes in depth. It makes it a point to examine and analyse the full range of issues raised by the cases relating to the new interface. It examines the economics used in relation to the legal rules as they apply to competition law and IP rights and it considers the issues raised in both the USA and the EU in prioritizing the avoidance of Type 1 or Type 2 errors in cases and laws applicable to competition law and IP rights.

Chapters in this book are grouped under five sections which are described below.

Themes of Overlap

The first section in this book outlines the general contact points between competition law provisions and intellectual property rights. Commencing with a general, multi-sectoral overview of the interface between IP and competition laws, it provides an economic and, to the extent relevant, constitutional perspective on emergent issues in both US and EU law. This section takes a closer look at prevailing dogmas such as the incentive theory, to present a more nuanced view of the role of competition law and the economic underpinnings of both competition law and IPRs.

In *Chapter One*, Steven Anderman outlines the role of the competition regime in the IPR context by examining how the internal logic of IP laws is affected by the superimposition of the logic of competition laws. In a discussion ranging from the pharmaceutical industry to the implications of the *Microsoft* judgment, this chapter reviews the situation where the strategic use of IP rights might overstep the boundaries of competition law.

In *Chapter Two*, Rudolph Peritz argues, on the basis of the US experience, that rather than thinking of IP and competition law as two separate, and conflicting, regimes, the interpretation of the scope of patent rights by US courts suggests that competition is inherent in the notion of patent protection. Encompassing the economic justifications (including the traditional 'incentive theory') and the constitutional underpinnings of patent law, the discussion in this chapter identifies the emerging trends in US case law.

In *Chapter Three,* Cecilio Madero Villarejo and Thomas Kramler, discuss the complementarity of the IP and competition regimes, with a focus on the *Microsoft* judgment and the question of standard setting. Placing the IP-competition interface issue in the broader context of the limits of justifiable State intervention, Villarejo and Kramler argue that the ultimate twin goals of both regimes are promoting innovation and enhancing consumer welfare.

In *Chapter Four*, Mariateresa Maggiolino presents an incisive analysis of the role of economic theories in shaping competition law. This chapter discusses both orthodox neo-classical economic models as well as emerging economic theories, which offer insights that are accessible to administrators, policy makers, lawyers, and judges.

Market Power and Abuse

The second section focuses on the exercise of market power and examines key competition law concerns such as refusal to license, technological tying, and margin squeeze. Recent developments in law, including the Commission Guidance Paper on Enforcement Priorities, are explored together with an appraisal of their application in various sectors such as pharmaceuticals, information technology, and telecommunications.

In *Chapter Five*, Ariel Ezrachi considers the evolution of Article 102 TFEU jurisprudence in the European Courts and its application in cases of refusal to license. The chapter appraises the way in which the Commission Guidance Paper and economic approach to Article 102 TFEU affect the analytical framework of refusal to license and the interface between competition law and IPR.

In *Chapter Six,* John Kallaugher, examines the tension between the avowed economic approach to Article 102 TFEU and concerns over the purported misuse of IP rights, the pharmaceutical sector being one of the key areas of concern. Kallaugher questions the assumption of the 'economic approach' necessarily being an unalloyed good and suggests that when it comes to IP protection, a closer look is warranted.

In *Chapter Seven,* Ian Forrester and Katarzyna Czapracka present a practitioner's perspective on compulsory licensing. Recent decisions such as *Microsoft* are far from uncontroversial and unfortunately, the Commission's Guidance on Enforcement Priorities appears to raise more questions than it answers. The authors distil guiding principles on the basis of the case law, for the scope of intervention of competition laws.

In *Chapter Eight,* Josef Drexl discusses the interface between IP protection of trade secrets and competition law. Drexl examines *inter alia* the case law and communications from the Commission to argue that while the mere fact that protection is claimed as a 'trade secret' rather than, say, as a patent, does not automatically imply that different standards apply; however, trade secret protection has several peculiar features which ought to be borne in mind.

In *Chapter Nine*, Hedvig Schmidt reflects on competition concerns that technological integration gives rise to. As opposed to the arguably simpler case of contractual tying, a relatively sophisticated approach is required to examine the issues arising from technological integration. This chapter contrasts the differing approaches in the EU and US.

In *Chapter Ten*, Robbie Downing and Alison Jones present a comparative analysis of the US and EU position on alleged 'price squeeze' or 'margin squeeze' under section 2 of the Sherman Act and Article 102 TFEU respectively. In the telecommunications market, which until not so long ago was characterized by the presence of State-owned monopolies, drawing the line between IP law protection and competition law intervention is a particularly challenging exercise.

Pharmaceutical Sector

The third section of this book focuses on the pharmaceutical sector and examines issues arising from the interplay of multiple policy objectives, including social welfare in this sector. Pharmaceuticals have been in the spotlight of competition law, as demonstrated by the sectoral enquiry in the EU, and the three chapters in this section present a general discussion of the IP-competition interface as well as discussion of certain specific concerns, in areas such as parallel trading and out of court settlements.

In *Chapter Eleven*, Simon Priddis and Simon Constantine discuss the tensions generated by the multifarious policy objectives that impact the pharmaceutical industry. They review issues arising from competition policy and IP policy, and issues pertaining to social welfare and the scope of State intervention. The chapter explores the possible impact of the sectoral enquiry by the Commission and judicial interpretations on the pharmaceutical industry.

In *Chapter Twelve*, Pat Treacy and Sophie Lawrance draw upon their dispute settlement experience to present a realistic evaluation of the prospects of out-of-court settlements in IP litigation. Examining the EU approach in light of the US experience, the chapter proposes a framework for analysing settlements of matters involving IP and competition law issues.

In *Chapter Thirteen*, Andreas Heinemann, examines the impact of the market integration imperative under EU law on parallel trading in pharmaceuticals. There have been instances of taking a somewhat more lenient view toward market partitioning in case of pharmaceuticals (such as the *GlaxoSmithKline* judgment). This chapter examines the dynamics of the single market imperative in context of the 'innovation' argument in the pharmaceutical sector.

Standard Setting and FRAND Commitments

The fourth section focuses on standard setting and examines the US and European perspectives on the balancing exercise between the beneficial effects of standard setting activities and competition law concerns. Fair, reasonable, and non-discriminatory ('FRAND') terms are also analysed and this section includes an attempt to reconceptualize FRAND as a contractual obligation rather than an IP or competition law issue.

In *Chapter Fourteen*, Alden Abbott and Nicholas Kim discuss the US antitrust perspective to standard setting in context of the application of Section 5 of the FTC Act to deter firms from engaging in a 'patent hold-up'. The chapter *inter alia* discusses the *Rambus* litigation and the *N-Data* settlement and proposes several limiting principles to guide the application of Section 5.

In *Chapter Fifteen*, Michael Carrier evaluates the pro-competitive effects of standard setting exercises as well as potential competition concerns. The discussion includes issues related to reasonable and non-discriminatory terms for IP licences pertaining to standards and US case law, and examines the underlying principles restricting interference with standard setting activity.

In *Chapter Sixteen*, Philippe Chappatte and Paul Walter address the issues raised in the context of non-practising entities ('NPEs') who acquire patents, not for the purpose of manufacturing a product themselves, but rather with a view to obtaining royalties by licensing on the same. FRAND access is a key concern in such cases and the chapter examines the EU law position in this area.

In *Chapter Seventeen*, Roger G Brooks and Damien Geradin, propose an alternate approach toward analysis of FRAND commitments, in terms of analysing FRAND terms as contractual obligations undertaken by parties to a transaction. The chapter discusses the policies of major standard setting organizations and argues that the contract law aspect of FRAND terms should not be conflated with IP and competition law issues.

Licensing

The fifth, concluding, section of this collection presents US and EU perspectives on issues arising in relation to licensing agreements. Laws, economics, and policies underlying licensing are considered in the general context of IP licensing, as well as in relation to specific issues such as 'hard-core' restrictions in the EU block exemption for technology transfer agreements.

In *Chapter Eighteen*, Stefan Enchelmaier examines licensing issues in EU law in the context of the so-called 'hard-core' restrictions in Article 4 of Regulation 772/2004, concerning the block exemption for technology transfer agreements. The chapter examines the challenges that the self-assessment exercise presents to undertakings and critiques the complexity of Article 4.

In *Chapter Nineteen*, Kevin Arquit, examines the economics underlying the US antitrust law approach toward IP licensing. The chapter examines developments in US case law in the context of policy developments as signs of shifting away from traditional neo-classical economic theories.

In *Chapter Twenty*, Vladimir Bastidas Venegas examines theories of the economic rationales underlying licensing as well as the US and EU jurisprudence on licensing issues. The chapter argues that the IP-competition law interface in context of licensing issues has to mediate between static or allocative efficiency and innovation or dynamic efficiency concerns.

Ariel Ezrachi and Steven Anderman
Oxford and Stockholm, 2010

CONTENTS—SUMMARY

CONTENTS

IV STANDARD SETTING AND FRAND COMMITMENTS

TABLES OF CASES

DECISIONS—EUROPEAN COMMISSION

JUDGEMENTS—EUROPEAN UNION

Court of First Instance (General Court)

European Court of Justice

BELGIUM

FRANCE

UNITED KINGDOM

UNITED STATES OF AMERICA

TABLES OF NATIONAL LEGISLATION

EUROPEAN SECONDARY LEGISLATION

Directives

Regulations

TABLE OF TREATIES AND CONVENTIONS

LIST OF CONTRIBUTORS

Alden Abbott Deputy Director for Special Projects, Office of International Affairs, Federal Trade Commission.

Steven Anderman Professor of Law Emeritus University of Essex and Visiting Professor University of Stockholm.

Kevin Arquit Partner at Simpson Thacher & Bartlett LLP.

Roger G Brooks Partner at Cravath, Swaine & Moore LLP.

Michael Carrier Professor at Rutgers University School of Law—Camden.

Philippe Chappatte Partner, Slaughter and May, London and Brussels.

Simon Constantine Associate, Freshfields Bruckhaus Deringer LLP.

Katarzyna Czapracka Member of the New York Bar, JSD (Columbia Law School), White & Case, Brussels.

Robbie Downing Of Counsel, Baker & McKenzie.

Josef Drexl Director of the Max Planck Institute for Intellectual Property, Competition and Tax Law, Munich, Honorary Professor at the University of Munich.

Stefan Enchelmaier Professor of European and Comparative Law, York Law School, University of York.

Ariel Ezrachi Director, University of Oxford Centre for Competition Law and Policy. Slaughter and May Lecturer in Competition Law, Oxford University. Fellow, Pembroke College, Oxford.

Ian Forrester Queen's Counsel at the Scots Bar, Honorary Professor, University of Glasgow, LLDhc; White & Case, Brussels.

Damien Geradin Professor of Competition Law and Economics at Tilburg University, a William W Cook Global Law Professor at the University of Michigan Law School, and a partner at Covington & Burling LLP.

Andreas Heinemann Professor at the University of Zurich, Chair of Commercial, Economic and European Law.

Alison Jones Professor of Law, King's College London.

John Kallaugher Visiting Professor University College London, Partner Latham & Watkins.

Nicholas Kim Registered Patent Attorney.

Thomas Kramler Deputy Head of Unit, Directorate C, Unit C-3, IT, Internet and Consumer Electronics, DG Competition, European Commission.

Sophie Lawrance Associate, Bristows, London.

Mariateresa Maggiolino Assistant professor of Commercial Law at Bocconi University, Milan.

Rudolph Peritz Professor and Director, IProgress Project, New York Law School, Visiting Professor, Economics and Law Departments, LUISS University, Rome.

Simon Priddis Partner, Freshfields Bruckhaus Deringer LLP.

Hedvig Schmidt Lecturer, School of Law, University of Southampton.

Pat Treacy Partner, Bristows, London.

Cecilio Madero Villarejo Director, Directorate C 'Markets and cases II: Information, Communication and Media' DG Competition, European Commission.

Vladimir Bastidas Venegas Lecturer at Örebro University.

Paul Walter Associate, Slaughter and May, London and Brussels.

Part I

THEMES OF OVERLAP

1

THE IP AND COMPETITION INTERFACE: NEW DEVELOPMENTS

Steven Anderman

Introduction

EU competition law, in its newly modernized form, continues to regulate the exercise of intellectual property rights (IPRs) over an increasingly wider front. New cases in the information technology sector under Article 102 TFEU have targeted abuses of a dominant position such as price discrimination[1] and unfair pricing as well as refusals to license and tying and bundling.[2] The pharmaceutical industry has offered cases of misuse of the patent system as an abuse of dominance,[3] as well as suggesting a capacity to infringe Article 101 TFEU with reverse or delay payments, both aiming to prevent the entrance of generics into existing markets.[4] The field of standard setting and technology pools has attracted the attention of the competition authorities not only because of cases of patent ambush[5] and FRAND ambush[6] but also because of competition concerns with the process of selecting patented technologies for industrial standards. These forms of collaboration amongst competitors as well as

---

[1] See Case COMP/37.990 *Intel* Commission Decision of 13 May 2009 (D (2009) 3726 final) ('*Intel*').

[2] Case T-201/04 *Microsoft Corp v Commission* [2007] ECR II-3601 ('*Microsoft*').

[3] Case T-321/05 *Astra Zeneca v Commission* (Judgment of the General Court dated 1 July 2010) ('*Astra Zeneca*'). See also the recent *Bohringer* case being investigated by the Commission to determine whether Bohringer misused patents to exclude competitors. The Commission's charges concerned 'misuse of the patent system in order to exclude potential competition in the area of chronic obstructive pulmonary disease (COPD) drugs.

[4] European Commission Pharmaceutical Sector Inquiry Final Report, 8 July 2009.<http://ec.europa.eu/competition/sectors/pharmaceuticals/inquiry/staff_working_paper_part1.pdf> accessed 3 September 2010 ('Sector Inquiry Report').

[5] EC Notice on Acceptance of Commitments in Case IP/09/1897 *Rambus*, 9 December 2009.

[6] See 'Antitrust: Commission welcomes IPCom's public FRAND declaration' MEMO/09/549, 10 December 2009 (IPCom's s role in the UMTS pool after its belated acceptance that it was bound by its predecessor's commitment to a FRAND licence of essential patents).

that of licensing collection societies, have been examined by the European Commission under Articles 101 as well as 102 TFEU. Finally, the modernization of Article 101 TEU has produced new detailed guidelines and block exemptions for R&D and specialization agreements and IP licensing agreements in the form of technology transfers.

This pattern of competition law enforcement is first of all a reminder that the new competition rules are a development of an existing 'second tier' of regulation of the exercise of IPRs, an 'external' system of regulation that applies to anti-competitive conduct not prevented by the 'internal' system of regulation offered by IP legislation.[7] All IPRs have a system of protection that balances the exclusivity granted to 'pioneer' inventors and creators with the limits and exceptions in favour of 'follow on' innovators. However, the limits of permitted exercise of IPRs are drawn not only by the laws of IP but also by reference to the rules of competition law.

When the competition rules apply to the exercise of IPRs, they apply as a separate system according to their own logic. They treat the exercise of an IPR, once granted, as any other private tangible right, subject to the public law limits on market behaviour created by the competition rules. On closer inspection, the logic of competition law reveals a predisposition to accommodations to the exercise of IPRs. In addition to some limited special concessions such as the exceptional circumstances rule for refusals to license, the very design of the general competition rules leaves room for the 'normal' exercise of IPRs and focuses on the cases where their exercise happens to coincide with extreme forms of commercial conduct.

The reasons for this are not far to seek. In the first place, there is a natural overlap in the aims of the two fields of law. The exclusive rights created by IP laws provide an incentive to inventors to create substitute products within markets and new products which establish new markets. Similarly, IP licensing is a vehicle to enlarge exploitation of protected technologies and thereby create a wider diffusion of the new technology which either creates new markets or brings substitutes to existing markets. Finally, the internal rules of IP legislation prevent copying but in fact encourage 'competition by substitution' between 'follow on' innovators and 'pioneer' innovators.[8] They do this both by offering exceptions to exclusive rights such as 'experimental use' in patent law and 'fair use' in copyright law during the period of protection and informational benefits to innovation during the period of protection as well as after the period of protection has expired. And of course, if there is weak market power associated with an IPR, its capacity to impose a restraint on competition or a barrier to entry is limited or non-existent.

[7] See I Rahnasto, *Intellectual Property Rights, External Effects and Antitrust Law: Leveraging IPRs in the Communications Industry* (Oxford: OUP, 2003).

[8] Of course almost every so called 'pioneer' innovator is itself a 'follow on' innovator.

A second factor is that the competition rules are not meant to apply to the exercise of IPRs as such. They are only meant to apply when the IPR is used as an 'instrument' of abuse[9] or as a means of restricting competition.[10]

On the whole, the modernization of the EU competition rules has improved their accommodation to the exercise of IPRs. The modern approach to abuse of dominance is based on a much more careful assessment of the actual market power of the legal monopolies conferred by patents and industrial copyright when these are used as part of corporate strategies. These legal 'monopolies' are now assessed entirely on the basis of their *de facto* market power. There were periods in the past, in both the USA and the EU, when the legal monopoly conferred by a patent in the form of a right to exclude anyone else from exploiting the patented invention was presumed to coincide with market power. This led to judicial and administrative competition rules which paid inadequate attention to the effects of IPRs upon investment in R&D and innovation. This historical presumption has finally been replaced by a more appropriate judicial and administrative understanding that the actual market power associated with an IPR must be assessed empirically in each case.

Moreover, a new and more realistic economic approach has been introduced into the block exemption regulations and guidelines under Article 101 TFEU, reducing much of the substantive burden of regulation on IP licensing agreements between non-competitors as well as R&D Joint Ventures and standardization agreements. The new TTBER regulation provides a 'safe harbour' and the TT Guidelines offer *inter alia* a new methodology to apply Article 101 TFEU to IP licensing agreements outside the safe harbour of the TTBER. This focuses regulation more carefully upon agreements in which the IP-owning parties have or achieve real market power, and upon agreements between *competitors* thus significantly lessening the effect of the competition rules upon licensing agreements between *non-competitors*.

Despite all this, the actual number of confrontations between the owners of IP and the competition authorities has been on the increase. Why has this been so? What accounts for this increase in regulatory confrontations when the overall framework has become more nuanced? Not surprisingly, the answer can be found in both sides of the equation: the conduct of the regulated and the actions of the regulator. On the one hand, IP owners have developed new and more aggressive commercial strategies in their exploitation of IPRs. On the other hand, it can be seen that regulatory authorities have adapted their techniques for measuring market power and defining abuse in response to recent developments in commercial practice involving IPRs.

---

[9] See eg Case 102/77 *Hoffman La Roche v Centrafarm* [1978] ECR 1139 para 16.

[10] See eg Case 262/81 *Coditel v Ciné-Vog Films* [1982] ECR 3381 para 14 ('*Coditel II*').

A. The New Commercial Strategies

The first factor that has caused an increase in encounters between IPRs and the competition rules has been the noticeable rise in the strategic use of IPRs in new commercial practices. As corporations have developed a greater understanding of the value of their corporate intangible assets and have given a more realistic book value to their intangible assets, they have not only pursued more rigorous enforcement policies; they have found more imaginative uses for IPRs within their overall commercial strategies. Many of these strategies occur in a form that is legitimate profit taking and a lawful adaptation of commercial practice to commercial realities. An example of this is the strategic decision of Qualcomm to concentrate on research-only IPRs and derive its income from royalties from licensing out of its CDMA wireless technology and other patents[11] This breaks with the strategy of R&D and patenting for the purpose of manufacturing and distribution and causes demands for higher royalty rates but is a legitimate exercise of entrepreneurial imagination. Another strategy increasingly adopted by firms is to acquire wide patent portfolios. The annual number of micro-processing patents doubled during the 1990s in the USA.[12] In that sector, the emphasis on acquisition of patent portfolios has been motivated by the defensive concern to have sufficient bargaining chips when negotiating cross-licences or pre-empting claims of infringement. In other sectors, such as telecoms, patents proliferate as part of the process of creating new technology standards. Further, in pharmaceuticals, strategic patenting involves patent 'flooding' as part of a wider strategy of staving off competition from competitors during the period of patent protection as well as offering competition during the post-protection years. The aim of the portfolio is to create a defensive wall around the core protected technology.

These strategies undoubtedly have the purpose and/or effect of 'raising rivals costs'.[13] For example,

> . . . a dominant firm leading the race in research and development [may] . . . intentionally select a technology for which scale economies are substantial, knowing that the fringe firms will have to follow along. Alternatively, a firm may create an array of patents on marginal or even non-existent innovations, knowing that other firms will either have to invent around the patents or else litigate their validity.[14]

---

[11] See 'Antitrust: Commission closes formal proceedings against Qualcomm', MEMO/09/516, 24 November 2009 <http://europa.eu/rapid/pressReleasesAction.do?reference=MEMO/09/516> accessed 3 September 2010.

[12] See C Shapiro, 'Navigating the Patent Thicket: Cross Licences, Patent Power and Standard Setting' (2000) 1 Innovation Policy and the Economy 119.

[13] D Rubinfeld, and R Maness, 'The Strategic Use of Patents; Implications for Antitrust' in F Leveque and H Shelanski (eds), *Antitrust, Patents and Copyright EU and US Perspectives* (Cheltenham: Edward Elgar, 2005).

[14] H Hovenkamp, *Federal Antitrust Policy: The Law of Competition and its Practice* (West Publishing Co, 1994) 286.

It is true that a,

> . . . successful raising rivals costs strategy can serve to foreclose competition and/or to be predatory; the higher costs imposed on rivals allows the predator to profitably undercut rival's prices, to gain market share, and in some cases to drive firms from the market. Further in a dynamically competitive market . . . the abuse of intellectual property that reduces rivals' incentives to innovate may lessen innovation competition.[15]

However, not all such activity by IP owners can be viewed as anti-competitive.

For example, some patent flooding may cause 'patent thickets' and thereby deter innovation but this practice will not itself attract the attention of the competition authorities unless it is allied with some form of bundling and tying conduct or mandatory package licensing. Similarly, some attempts to concentrate on research-only patenting may be designed to raise royalty rates both within and outside technology pools but this is normally not an offence under the competition rules unless some form of patent or FRAND ambush is involved. Further, the excesses of 'patent trolling' may have occasionally resulted in widespread condemnation and alterations to the IP rules.[16] However they have not necessarily been viewed as prohibited conduct under the competition rules. Further where patent enforcement strategies move from enforcing valid patents to enforcing known invalid patents, there may be a fraud on the Patent Office but the result may not necessarily involve a breach of competition law. In the famous line of *Walker Process*[17] cases in the USA which involved attempts to enforce known invalid patents there were cases drawing a distinction between inappropriate conduct by IP holders which fell on either side of the line as far as antitrust law was concerned. For example in the 1984 case of *Brunswick Corporation v Riegel*,[18] Judge Richard Posner distinguished between a fraudulent procurement of a invalid patent in circumstances that create market power and the theft of a perfectly valid patent which only shifted the patent monopoly into different hands. He stated that the first type of fraud on the Patent Office, fraudulent procurement, could be a violation of section 2 of the Sherman Act[19] until unmasked or cancelled, whereas the second, theft of a patent, '. . . has no antitrust significance or consumer interest because antitrust law is designed to protect the process of competition and not to promote the welfare of particular competitors.' Overall, a considerable amount of aggressive and imaginative commercialization of IPRs has long

---

[15] D Rubinfeld, and R Maness, 'The Strategic Use of Patents; Implications for Antitrust' in F Leveque and H Shelanski (eds), *Antitrust, Patents and Copyright EU and US Perspectives* (n 13 above).

[16] *EBay Inc v MercExchange LLC*, 126 S Ct 1837 (2006).

[17] *Walker Process Equipment Inc v Food Machinery & Chemical Corp* 382 US 172 (1965); see also *Handguards, Inc v Ethicon Inc* 601 F2d 986, 996 (9th Cir, 1979).

[18] *Brunswick Corporation v Riegel Textile Corporation* 752 F 2d 261 (7th Cir, 1984).

[19] Judge Posner specified three conditions for such a violation: (i) the patent must dominate a real market;(ii) the invention sought to be patentable must not be patentable; and (iii) the patent must have some colourable validity, conferred by the patentee's efforts to enforce it by bringing patent infringement suits.

proved to be essentially compatible with 'competition on the merits' under the competition rules.

On the other hand, a number of new commercial practices by IP owners have been caught as prohibited exclusionary or exploitive abuses under the competition rules. The *Microsoft*[20] case highlights the sensitivity of the competition authorities to vertical integration strategies aimed at capturing 'aftermarkets' by incumbent dominant firms in the information technology sector. The *Intel*[21] case reminds of the limits to dominant firms engaging in anti-competitive discounting designed to create exclusivity in primary markets. The *Astra Zeneca*[22] case raises the issue when a misuse of the patent process amounts to a violation of Article 102 TFEU. The emergence of strategic attempts to obtain greater than FRAND royalties through patent ambush[23]and FRAND ambush[24] in the context of patent pools and industrial standards has awakened the interest of the competition authorities to the possibility of anti-competitive aspects of those processes.

This competition interest has not only spread within the field of unilateral action but has been extended to various forms of bilateral and multilateral cooperation between competitors. The reverse payment settlements in the pharmaceutical sector have raised competition issues in the conduct of the pharmaceutical firms directed against competition by generic producers. Even before the *Rambus* case, the competition authorities had shown an interest in regulating the process of standard setting and R&D joint ventures and that interest is currently intensifying. Finally there has been a renewed interest by the European Commission in collective licensing. The key feature of all these confrontations and investigations is the fact that the IP is used either as a component of market power or as a suspected instrument of the abuse in the chosen commercial strategy that leverages the dominance into abusive conduct.

B. IPRs and the Logic of the Competition Rules

As we have seen, the logic of the current rules of competition law embodies a far more nuanced understanding of the pro-competitive nature of IPRs. However, the way in which this plays out in its treatment of IPRs varies dramatically depending upon the constituent element of the competition prohibition being assessed. Thus, in the first stages of analysis of potentially anti-competitive conduct, the definition of relevant market, and the assessment of dominance, there are few if any special concessions to IPRs. As we have seen, these stages consist of a factual analysis, assisted by economic

[20] *Microsoft* (n 2 above).
[21] *Intel* (n 1 above).
[22] *Astra Zeneca* (n 3 above).
[23] EC Notice on Acceptance of Commitments in *Rambus* case IP/09/1897, 9 December 2009.
[24] See 'Antitrust: Commission welcomes IPCom's public FRAND declaration' (n 6 above).

reasoning, which is similar to the analysis of the market power of any other form of property. IPRs are assessed in terms of the extent to which they create real restrictions of competition in the form of barriers to entry or impediments to entry. At this stage IPRs are examined largely in terms of their factual contribution to market power and their innovative features are viewed almost entirely in this light. Consequently, even though there is no longer an assumption that a patent is automatically associated with market power, an IPR can be found to be associated with an asset that enjoys dominance in a market.

In contrast, in the definitions of abuse, there are more concessions to the innovative and pro-competitive features of IPRs within the logic of the norms of the competition rules. There is for example the 'exceptional circumstances' rule that applies to refusals to license an IPR. Moreover the application of general norm of Article 102(a) TFEU to the pricing of IPRs leaves room for IP asset pricing that takes into account investment risks as well as costs in the case of IPRs.

At the same time, however, there are other features of the current EU normative definition of abuse that have proved to be less receptive to sweeping IP pro-competitive arguments. The concept of abuse under Article 102 TFEU tends to rely on per se rules of abuse to implement its regulatory norms. The logic of the EU competition rules is to protect effective competition in the here and now. The first consequence of this is that Article 102 TFEU has a low level test of effects as a constituent element of its concept of abuse of dominance. The Courts are reluctant to take into account speculative arguments about future events. The second and related consequence is that the Courts have tended to confine arguments based on the pro-competitive features of prohibited acts to the category of objective justifications rather than allow benefits to offset current harms in a more balanced way.[25] This approach reduces the scope for arguments based on dynamic efficiency to operate as a defence in cases of alleged abuse in a way that might be possible in a 'rule of reason' style balancing exercise similar to that used in the USA. Thirdly, such an approach appears to make few concessions to the existence of the alternative remedies offered by the IP laws. Let us look more closely at each of these points.

C. The Definition of Relevant Market and IPRs

Introduction: narrow product markets

According to the logic of the competition rules, a preliminary step in the assessment of the market power of the owner of a product is the definition of the relevant market in which the market power is to be assessed. Market power can only be measured once the relevant market has been identified. It is this preliminary step that has

[25] *Microsoft* (n 2 above).

undoubtedly made a major contribution to the modern collisions between certain predatory commercial strategies and competition law. The competition authorities display a predilection for defining product markets narrowly and narrow product markets tend to create single product markets. This approach is not aimed solely at IP owners; it is a standard feature of EU competition policy. However it has had a direct impact upon IP owners.

In a number of earlier cases the narrow market definitions have had the effect of creating a single product market and thereby setting the foundations for findings of dominance in the case of IP owners. The classic example of this is the *Hilti* case[26] where the undertaking's powder-actuated fastening system was broken down into three separate products—the Hilti gun, Hilti gun cartridge, and nails. Both the gun and the cartridge were protected by patents

In the *Volvo v Veng* case[27] the implications of narrow market definition for dividing clusters of products owned by the same firm into primary and secondary markets were made much clearer. This placed competition law firmly in the position of acknowledging the owner of the IP-protected spare part product as a lawful monopolist in the primary market for manufacturing and selling a spare part, but having special responsibilities towards competitors in a secondary market. In the first market, the IP holding undertaking was entitled to returns on its investment in the IPR as well as protection from copying. In the second or 'after' market, the IP owner had obligations under Article 102 TFEU in 'exceptional circumstances' to supply the IP-protected product. This case introduced the concept in competition law that the exploitation of an IP in 'neighbouring' markets was not immune from the competition law rules. These rules prevented an IP owner from unlawfully leveraging its dominance in the primary market to obtain added market power in the secondary market.

'Technologies' and secondary product markets

A related feature of narrow product markets definition in recent cases has been the way in which IPRs such as patents and industrial copyright 'products' have been identified as technologies and not as end products for the purpose of 'market' definition.

In the EU, as well as in the USA, 'technology markets' have been defined as essentially 'licensing markets', consisting of markets for a licensed technology and its close substitutes.[28] This concept of 'technology markets' has been applied in the

[26] Case T-30/89 *Hilti AG v Commission* [1991] ECR II-1439.

[27] Case 238/87 *AB Volvo v Erik Veng (UK) Ltd* [1988] ECR 6211.

[28] That is, 'the technologies or products that are close enough substitutes to constrain the exercise of the market power of the licensed technology.' See US Department of Justice and FTC, Antitrust Guidelines for the Licensing of Intellectual Property (1995) para.3.2.2.

assessment of the SIEC and SLC test to mergers[29] as well as to IP licensing in technology transfers.[30]

In cases of abuse of dominance, this designation of markets allows a licensed technology to be viewed as an upstream product and the licensed product a downstream product, allowing the characterization of the licensed technology as a primary market and the product market as a secondary market.[31]

This designation can result in a finding that the IP owner is dominant in a market and even that it is an essential facility. In the *IMS* case,[32] for example, the Court of Justice thought that even if the 1860 brick structure database may not have been established as a separate market, it could nevertheless be viewed as an indispensable input. Similarly in the *Microsoft* case,[33] the CFI, now the General Court, confirmed this approach by stating that 'The fact that the indispensable product or service is not marketed separately does not exclude from the outset the possibility of identifying a separate market.'[34] Again, this treatment of technology markets can be viewed as logical under the competition rules. It applies the approach directed to tangible upstream infrastructures such as ports and ferries to IPRs when they enjoy the real market power of essential facilities. The competition rules reserve the attempt to distinguish between IPRs and physical infrastructures to the analysis of the concept of abuse when the test of 'exceptional circumstances' is applied.

Technology markets and primary markets

The demand-oriented method of defining product markets by reference to substitutes can also make its contribution to a finding of narrow product markets for IPRs. A good example is the *Astra Zeneca* case,[35] involving a finding of abuse based on a misuse of the patent process or more precisely the European supplementary protection certificate (SPC) scheme for medicinal products[36] in relation to its patented product, Losec, a proton pump inhibitor used to treat ulcers. The product market chosen by the Commission was the market for proton pump inhibitors (PPIs).

---

[29] See M Glader, *Innovation Markets and Competition Analysis: EU competition law and US antitrust law* (Cheltenham: Edward Elgar 2006).

[30] See SD Anderman and J Kallaugher, *Technology Transfer and the New EU Competition Rules: Intellectual Property Licensing after Modernisation* (Oxford: OUP, 2006).

[31] A Heinemann 'The Contestability of IP-Protected Markets' in J Drexl (ed) *Research Handbook on Intellectual Property and Competition Law* (n 8 above) 60.

[32] Case C-418/01 *IMS Health GmbH & Co OHG v NDC Health GmbH & Co KG* [2004] ECR I-5039 ('*IMS*').

[33] *Microsoft* (n 2 above).

[34] In retrospect, in Case C-241/91 P and C-242/91 *RTE and ITP v Commission* [1995] ECR I-743 ('*Magill*'), the programme listings could have been viewed as a separate product instead of a market and as such treated as an indispensable input for the programme guides.

[35] *Astra Zeneca* (n 3 above).

[36] Council Regulation No 1768/92 Concerning the creation of a Supplementary Protection Certificate for medicinal products ('SPC Regulation').

PPIs could have been viewed as a subset of anti-ulcer drugs but the Commission found that PPIs had progressively replaced the previous generation of anti-ulcer drugs—H2 Blockers—even though the PPIs were more expensive. It also found that there was no sign of substitution in the other direction—from PPIs to H2 Blockers. In addition non-price factors in consumer demand were taken into account. In the event, Astra Zeneca could be found to have considerable market power by virtue of the position of its product, Losec, in the PPI 'market'.

The technology owner can also find itself in a single product market owing to the nature of the technology fixing process. Thus, in the *Rambus* case, a case of abuse consisting of misusing market power to obtain higher than FRAND royalties for certain patented technologies in the context of a technology pool and standardization process—the four patented technologies owned by Rambus would have been viewed in technology markets containing substitutes before the selection of the technology for the standard. However, once the standard was selected and they became essential patents in the Joint Electron Device Engineering Council's (JEDEC) standards—SDRAM and DDR-SDRAM, they were treated as technology monopolies. They were now in a single product market by virtue of the facts and not as an exercise of discretion by the Commission. This treatment of technology markets is not an aberration under the competition rules. It is a method of determining the degree of dominance of an IP-protected product by assessing its actual market power.

This method of defining the relevant market is used regularly by the Commission even where no IPRs are involved in a product. It is Commission policy to look primarily at actual substitutes at the market definition stage because the Commission pursues a policy of measuring the process of competition in the here and now. Potential substitutes can be taken into account to enlarge the relevant market only where they meet a strict test of being real potential substitutes.[37] In the event, current methods of market definition have the consequence for IPRs that with narrow product markets, the IPR can be found to make a significant contribution to a finding of market power.

Finally, although generally under Article 102 TFEU, product markets have tended to be defined as separate components of complex products rather than the assembled product itself, it is theoretically possible in alleged tie-in cases to counteract these narrow product market definitions if one has a case of genuine technological integration. By and large, EU competition distinguishes between purely commercial strategic bundling and technological bundling. The first will not deter narrow product market definition. The second, if there is genuine evidence of technological advantage in integration, could and should preclude a narrow market definition because

[37] See the European Commission Notice on Defining the Relevant Market for the purposes of community competition law [1997] OJ C372/5 paras 20 et seq. See too the discussion in A Heinemann (n 31 above) 56.

the bundling has in effect produced a new single product. However, this presupposes that the IP owner can produce adequate evidence of technological benefit resulting from the product integration. In the *Microsoft* case, the Microsoft Corporation made the claim that its media player was integrated with its Windows operating system and consequently there was no unlawful tie-in but it failed to adduce adequate evidence of technological benefit produced by the bundling.[38]

D. The Concept of Dominance and IPRs

Introduction

Once a relevant market has been defined, the next step is to measure the market power of the accused firm, ie its power to set prices and otherwise act on that market independently of its competitors, its customers, and ultimately of its consumers.

The starting point for the investigation of market power is normally the market share of the IPR-protected product. Today, after the contributions of the Chicago school, this analysis of structure is complemented by an examination of the process of competition in the market, both actual and potential competition. The Commission regards the market share of the accused undertaking as only a preliminary indication of the existence of a dominant position. It looks at the market shares of any other products that may exist in the relevant market but it also looks at the dynamics of the market, product differentiation, barriers to expansion and entry, and countervailing buying power.[39] In the case of IPRs the analysis of the extent of actual and potential competition raises the issue of how the IPRs protecting a product affect the possibilities of substitution both in respect of actual substitutes and potential substitutes, ie the extent to which they constitute real barriers to entry and impediments to competition.

In cases where there are *actual* substitutes for the protected product in the market, there is still a need to investigate their real possibilities of substitution. First mover advantage, brand name, and distribution system may all impede competitors in competing with the dominant undertaking in the primary market. An IP can be associated with a dominant undertaking enjoying market power deriving from factors in addition to the IP protection against copies.

For example in the *Astra Zeneca* case, their product Losec was found to be dominant in certain national markets for proton pump inhibitors, by virtue both of its high market share and by its position of incumbent on that market because Losec was the

[38] *Microsoft* (n 2 above) para 1160.

[39] DG Competition, 'Communication from the Commission: Guidance on the Commission's Enforcement Priorities in Applying Article 82 EC Treaty [now Article 102 TFEU] to Abusive Exclusionary Conduct by Dominant Undertakings' [2009] OJ C45/7 paras 13–18.

first PPI and Astra Zeneca had considerable financial resources. Losec's market share was also protected by its patent on omeprazole and its brand name. However, its position as first mover and its financial resources were at least as important as the patent in enabling Astra Zeneca to obtain and maintain higher prices than later entrants on the market. The Commission also found that the power of monopsony buyers was considerably reduced in the case of innovative new products. As evidenced by its premium pricing, Astra Zeneca's market power was not seriously constrained by existing competition in the market. On these issues, the Commission's decision was supported by the General Court.[40] However, the General Court emphasized the contribution to market power offered by the patent protection of Losec when used as a means to put significant pressure on competitors and force their hand in entering patent settlements.[41]

If the market share of the product of the accused firm is very high, even approaching 100 per cent there will still be a need for an assessment of *potential* substitutes and the extent to which they are barred from entry by the IPRs of the incumbent. An enhanced understanding of the potential contestability of markets has drawn the regulators' attention to the role of IPRs as a barrier to entry to potential competitors.[42]

In some rare cases, the IPR itself can operate as a barrier to entry, creating a *de facto* monopoly. For example in *Volvo v Veng (UK) Ltd*,[43] the design right prevented all substitution because the design of the front wing coincided with its function, to provide a wing with the correct shape to fit in with the design of the car. Similarly in *Hilti*, the designation of the stapling gun and cartridges as separate products on separate markets meant that the legal monopoly of the patent resulted in a finding of real monopoly. Further, in the *Rambus* case, the finding of monopoly power was linked to the selection of the patented technology as an essential patent in the industrial standard. Prior to this step, the market power of the patented products was limited by the existence of substitute technologies. Once the patented products were chosen as essential to the Joint Electron Device Engineering Council's (JEDEC) standards—SDRAM and DDR-SDRAM, the market power of the Rambus patented technology became an indispensable input and it enjoyed a lawful monopoly of that input owing to its patent.

In other cases, however, the finding of 100 per cent market share and the existence of insurmountable barriers to entry is based less on the existence of the IPR and more on other factors producing the de facto monopoly. For example, in *Magill*, the ECJ found that the de facto monopoly was created by the TV companies' control over the

---

40 *Astra Zeneca* (n 3 above) para 261.

41 *Astra Zeneca* (n 3 above) para 272.

42 See eg A Heinemann (n 31 above) 54.

43 *Volvo v Veng* (n 27 above).

source of the information about programme listings as well as the existence of copyright protection. And in *Microsoft*, the Court found that the major source of market power was the applications barrier to entry and 'network effects'. These barriers to entry precluded substitutes far more effectively than the IP protection as such. The method used by the Commission is to assess the factual market power of the IP owner and not make any assumptions about the market power of the legal monopoly.

It may sometimes appear as if narrow product market definition, together with its inevitable influence on the test of market power, sometimes has the effect of bringing us full circle whereby the existence of the patent is at times equated with a finding of dominance or de facto monopoly. However, as long as the modern approach concentrates on assessing the existence of market power empirically and recognizes that not all patents are automatically monopolies because of the real possibilities of product substitution, it is difficult to criticize a finding of dominance based on a factual matrix.

A note on the degree of dominance

Finally, in the case of a finding of a primary market protected by an IPR and a derivative market, whether a downstream 'after market' or a 'neighbouring' market, the competition rules apply a filtering rule that restricts remedies for compulsory access to IPRs to extreme cases of market power. This filtering process is found in the higher standard of the test of market power and is not left solely to the concept of abuse. Unlike other abuses which often only require a prior finding of dominance as generally defined as above 40 per cent or 50 per cent or above and placing the undertaking in a position to price independently, the abuse of refusal to license, as well as refusal to supply, presupposes a special degree of dominance. In many IP cases, this degree of dominance has been factually a case of essential facilities dominance, ie a de facto monopoly over an *indispensable* input, although this is rarely described as essential facilities in IP cases.

However, in some cases it is enough for the IP owner to be the owner of a near monopoly as long as it has control over an indispensable technology.[44] The reason for this is that quasi monopolies as well as essential facilities have the capacity to create a basis for leveraging the dominance of the incumbent firm in a primary market into a derivative market and thereby to create a risk of unlawfully eliminating effective competition in a secondary market. This would follow from the Court's finding that a risk of elimination of effective competition, may be a basis for a finding of the abuse of refusal to supply. However the General Court in *Microsoft* has modified the IMS

[44] *Microsoft* (n 2 above) para 392.

test of indispensability from its earlier definition in *IMS*[45] and *Oscar Bronner*[46] by defining it as not allowing competitors in neighbouring markets to *compete viably* on the market.[47]

The legal status of dominance under Article 102 TFEU

Finally, an important feature of dominance as defined by Article 102 TFEU is that a finding that an IP-protected product is dominant in a market, or even super-dominant, does not by itself make an IP owner guilty of unlawful conduct. Article 102 jurisprudence has from its inception accepted that the achievement of dominance by a firm unilaterally using its business acumen, its R&D, and its IPRs to create successful new products is lawful. The French effort to make dominance unlawful per se in the drafting of the original EC Treaty was thwarted by the German insistence that it remain lawful and that the concept of unlawful conduct should be contained in the definition of abuse. In common with section 2 of the Sherman Act, Article 102 TFEU embodies an understanding that where a system encourages and protects legitimate competition, it should not 'punish its winners'. This philosophy applies to firms that succeed by imaginative exploitation of IPRs as an incidental consequence of a wider principle that applies equally to firms that base their success on exploiting other types of goods and services. The fact that an IP owner like Microsoft devises a business strategy of interoperability with applications makers and thereby obtains dominance in the operating systems market through its business acumen is perfectly lawful under the competition law rules. Similarly, when IP owners cooperate to select a technology that becomes a standard, the process of cooperation in R&D and standard setting, if legitimate in other respects, can itself be a lawful means of achieving dominance. It is not the acquisition of the power of an industrial standard itself that is unlawful; it is conduct that attempts to use that market power for certain ends that can be unlawful.

Moreover, in examining the acquisition of dominance by IP owners, there is an important distinction between unilateral and bilateral (or multilateral) acquisitions of market power. Whereas the unilateral acquisition of market power through internal growth is a form of competition on the merits, it can be unlawful to obtain dominance by an acquisition of the only effective competing technology in the market.[48] The acquisition of dominance solely owing to the size of the treasury of the acquiring company in this way is impermissible because it offers no guarantee of efficiency or innovation and extinguishes competition by substitution.

---

[45] *IMS* (n 32 above).

[46] Case C-7/97 *Oscar Bronner GmbH & Co KG v Mediaprint Zeitungs- und Zeitschriftenverlag GmbH* [1998] ECR I-7791.

[47] *Microsoft* (n 2 above) para 230. See discussion in I Forrester and K Czapracka, 'Compulsory Licensing in European Competition Law: The Power of the Adjective', Chapter 7 in this volume.

[48] See eg *Tetrapak I (BTG Licence)* [1988] OJ L72/27.

Furthermore, there is also a distinction to be drawn between different types of bilateral and multilateral acquisitions of market power. If a joint venture has been created to produce a new product that otherwise would not have been created and that creates a monopoly, that can be lawful because the acquisition of the monopoly is due to actual innovative efficiency. The collaboration creates a new market and thereby fosters competition by substitution; it does not repress existing competition. Similarly collaboration between competitors in joint R&D and the industrial standard setting process that results in a new industrial standard based on a new technological product can be lawful on those grounds assuming that the actual standard setting process was not a cover for a technology limiting effort or cartel to squeeze higher royalty rates from manufacturing licensees. This reasoning indicates that on this issue, the competition rules are in harmony with the IP rules. Both encourage innovative efficiency by stimulating 'competition by substitution'.

E. The Assessment of Abuse and IPRs

Introduction

The issue of strategic use of IPRs by dominant firms, just as the strategic use of tangible assets, must be based on an assessment whether a practice amounts to abuse or legitimate competition on the merits. This requires *inter alia* a weighing of whether its objects or effects are exclusionary or exploitive or justified competition on the merits.[49] What is important here is *how* the line is drawn.

In defining the line between the two, competition law has proved to be more user friendly to IPRs in some respects. That is not entirely surprising because the concept of competition on the merits embraces a theory of dynamic competition which recognizes that IPRs offer a reward/incentive to invent new products which is complementary to the pressure on firms to innovate created by effective competition. A prime example of this is the 'exceptional circumstances' test under Article 102(b) TFEU. Another example is the test of unfair pricing in Article 102(a) TFEU which takes into account the costs and risks of investing in IPRs in determining whether or not the pricing of an IPR is unfair.

There are, however, other features of the definition of abuse that have proved to be less receptive to IP pro-competitive arguments. As mentioned, the normative approach to the definition of abuse under Article 102 TFEU tends to rely on per se rules and this has resulted in a definition of abuse that incorporates a low level test of effects. The current approach of the Courts is moreover to confine arguments based

[49] Certain forms of aggressive enforcement of patents may initially appear to be exploitive conduct, but on closer inspection can be seen more accurately as exclusionary in nature because they are calculated to raise rivals' costs.

on the pro-competitive features of prohibited acts to the category of objective justifications. This approach reduces the scope for arguments based on dynamic efficiency to operate as a defence in cases of alleged abuse in a way that might be possible in a 'rule of reason' style balancing exercise similar to that used in the USA. Finally, the current approach appears to make few concessions to the existence of the alternative remedies offered by the IP laws. Let us look more closely at each of these points.

The 'exceptional circumstances' test under Article 102(b) TFEU

In cases of exceptional dominance involving both a near or full monopoly and control over an indispensable input, an IP owner can be treated differently than the owner of a tangible product. The normal competition rules of essential facility do not apply in the sense that compulsory access is used to ensure that there are sufficient competitors on the derivative market to maintain effective competition. Instead, it is accepted that the refusal to license is normally the prerogative of the IP owner. The normal exploitation of IPRs is in a sense viewed as competition on the merits because of its contribution to dynamic competition and the stream of new products. The case law under Article 102 TFEU has evolved an exceptional circumstances test that will provide a remedy of compulsory license of an IPR either where a single new entrant into a derivative market offers a new product with potential or demonstrable consumer demand or where a refusal to license will otherwise result in limiting technological development in the derivative market. The competition law rules accept that its role is to ensure that the exercise of IPRs cannot legitimately prevent competition by substitution by leveraging market power in a primary market. That is the underlying policy behind the General Court's endorsement of the Commission's decision in *Microsoft* that Article 102TFEU can be used to prevent an IP owner using its market power in one market to prevent competition by substitution in a separate but dependent market.[50] The IP rules are only respected by competition law where they are exercised to prevent copying and not when they are used to prevent substitution. In *Volvo v Veng*[51] this was set out in bold relief, its distinction between a primary market in which the IP owner could fully enjoy its monopoly[52] and a downstream market in which it had special responsibilities towards its competitors.[53]

In other cases of abuse, there is less evidence of special treatment for IPRs. The reason for this is that the IPR is not being used as part of a process of competing in an open market against other IP-protected products; its owner has engaged in a process of

---

[50] See SD Anderman, 'Microsoft v Commission and the Interoperability Issue' in Luca Rubini (ed) *Microsoft on Trial: Legal and Economic Analysis of a Transatlantic Antitrust Case* (Cheltenham: Edward Elgar, 2010).

[51] *Volvo v Veng* (n 27 above).

[52] *Volvo v Veng* (n 27 above) para 8.

[53] *Volvo v Veng* (n 27 above) para 9.

repressing legitimate competition in a market, with the IPR either an integral part of the process or a mere incidental part of the process. For example in the *Intel* case, the IP was only an incidental factor in the conduct of the undertaking. The abuses consisted of loyalty rebates meant to induce exclusive distribution agreements, tie-ins, and were designed to extinguish effective competition in the market. As it happened, effective competition in this case also entailed demonstrable competition by substitution since there were some differences in the nature of AMD's chips and Intel chips. However, at this point, the competition rules were concerned primarily to protect effective competition; the preservation of competition by substitution was simply a by-product of the protection of effective competition.

This is also true where the IPR has been more integral to the abusive commercial strategy. In such cases, the test of abusive conduct makes no special allowance for the innovative nature of the IPR. Instead, it views the IPR purely in terms of its factual contribution to market power and exclusionary conduct. In the *Rambus* case for example, the issue of dominance was dealt with by measuring the undertaking's market power before the selection of the technology and the market power of the IP owner of an essential patent in the chosen technology. The abuse was defined as one of unlawfully acquiring market power and then misusing it to impose unfairly high prices. The interpretation of Article 102(a) TFEU left no room for a concession to an IPR used in this way.

Similarly in *Astra Zeneca*, the issue was not only a case of falsely prolonged patent protection but also its knock-on effect of falsely prolonging market power and the power to exclude generics. The abuse, as defined by the Commission, consisted of an unlawful acquisition of market power as well as a misuse of that market power. One could argue that the competition rules are attempting to preserve effective competition in the market for PPIs and this includes the competition with generics envisaged under the patent law rules after the patent runs out and the technology enters the public domain.

Per se rules and effects

Although the definition of relevant market and dominance are heavily influenced by economic considerations, the definition of abuse under Article 102 TFEU is principally aimed at offering protections for the continued participation of competitors still existing in a market weakened by the existence of a dominant firm.[54] This approach is viewed as an indirect protection for consumers but the primary concern

[54] Article 102 TFEU covers practices which are likely to affect the structure of a market where, as a direct result of the presence of the undertaking in question, competition has already been weakened, and which, through recourse to methods different from those governing normal competition in products and services based on trader's performance, have the effect of hindering the maintenance or development of the level of competition still existing on the market. Case 322/81 *Michelin v Commission* [1983] ECR 3461 para 70 ('*Michelin I*').

is focused on the protection of competitors. There have been two notable features of this definition of abuse.

The first is that that certain types of conduct can be abusive per se if they consist of conduct which has as its purpose the further weakening of a market by means other than competition on the merits. The mere finding that a firm engages in such conduct will be sufficient to prove abuse; such conduct is thought to be inherently capable of having exclusionary effects.[55]

Secondly, it is well established that if abusive conduct is found to be capable of having exclusionary effects, there is no need to prove actual effects or actual harm; the law is meant to apply preventively before the damage is done, perhaps irretrievably. However, current EU law goes one step further by adding that there is no need even to prove likely effects on all occasions. For example in the *Microsoft* case, Microsoft had argued that in IP cases, following the case law, there must be a strict test of effects: it must be shown that the conduct was 'likely to eliminate' all competition or that the conduct of the dominant undertaking had a 'high probability' of elimination of all competition.

The General Court however accepted the Commission's argument that it was enough to prove that the refusal to supply gave rise to 'a risk of' elimination of competition in the second market in order to establish an abuse under Article 102 TFEU (ex-Article 82 EC). The test did not require the Commission to establish that competition had been eliminated or that its elimination was imminent.[56] Moreover, for the Community courts, the General Court said, the two tests '. . . reflect the same idea, namely that Article [102 TFEU] does not apply only when there is no more competition in the market' or that elimination is 'sufficiently imminent'.[57] This would be counter to the objective of the provision which is to maintain undistorted competition in the market and safeguard the competition that still exists.[58]

The Court went on to state that under Article 102 TFEU what matters to establish an infringement, is that the refusal *is liable to or likely to eliminate all effective competition on the market.* The fact that some competitors '. . . retain a marginal presence in some niches of the market' does not constitute the existence of such competition'.[59]

By lessening the importance of the difference between the two standards of proof, 'liable to' and 'likely to' eliminate competition, the Court emphasized that the test of Article 102 TFEU is not concerned with proof of effects but rather of proof of

55 See eg G Monti, *EC Competition Law* (Cambridge: CUP, 2007) 171.

56 *Microsoft* (n 2 above) para 457.

57 *Microsoft* (n 2 above) para 561. The General Court stated that Microsoft's complaint is purely one of terminology and wholly irrelevant.

58 The Court also noted that 'network effects' would make it harder for competitors once ejected to come back into the market.

59 *Microsoft* (n 2 above) para 563.

conduct that *could possibly* produce effects and this is in line with previous ECJ case law on Article 102 TFEU.[60]

On the other hand, *Microsoft* was a case in which there was actually strong evidence that the *likely* effect of the abuse would have been to eliminate competition because of the prior finding that access to MS Windows OS was '. . . necessary to maintain viability in the market for work servers'[61] and that such information was essential to viability on the market.[62] Indeed, insofar as the interface information was indispensable to a continued presence on the work group server market, it was in fact not only likely that elimination would occur, but also that once ejected, it would not be possible for competitors to return to the market because of the existence of the 'network effects' barrier to entry. Consequently, the language of the General Court in this particular case could be read in that light.[63]

The Commission has recently argued that it intends to take a more effects-based approach to the interpretation of Article 102 TFEU. It has indicated that proving the likelihood of harm is sufficient and that this can be inferred from the harm to the process of effective competition.[64] However, the Commission itself has not been entirely consistent in its advocacy of an effects-based test. It has stated that it will not take this approach in cases where conduct seriously restricts competition[65] because of the need to prioritize existing effective competition as the best guarantee of productive and innovative efficiencies in the long run.[66] Moreover, as we have seen, this approach has been called into question by the recent judgment of the General Court in *Microsoft*. Furthermore, in *British Airways*, the Court of Justice made it clear that it preferred a per se approach to Article 102 TFEU.

To economists as well as lawyers, one implication of these judgments is that Courts are not particularly impressed by the concept of reforming Article 102 TFEU to add a stronger test of effects as a constituent element of *this* abuse. The judgment reiterates that the test of abuse under Article 102 TFEU is conduct to which objectively abusive intent could be attributed and which creates a plausible risk of the harm of elimination of effective competition. As AG Kokott stated in *British Airways v Commission*,

---

[60] See eg Case 6/73 *Commercial Solvents v Commission* [1974] ECR 223 ('risks eliminating competition'); Case 311/84 *CBEM Telemarketing v CLT* [1985] ECR 3261('possibility of eliminating all competition').

[61] *Microsoft* (n 2 above) paras 392–3.

[62] *Microsoft* (n 2 above) paras 421–2.

[63] However, see contrary point made by Forrester and Czapracka in Chapter 7 in this volume.

[64] DG Competition, Communication from the Commission, Guidance on the Commission's Enforcement Priorities in Applying Article 82 EC Treaty [now Article 102 TFEU] to Abusive Exclusionary Conduct by Dominant Undertakings [2009] OJ C45/7 paras 19–20 (hereinafter '*Guidance on the Commission's Enforcement Priorities*'); Guidelines on horizontal cooperation agreements [2001] OJ C3/2 para 19.

[65] *Guidance on the Commission's Enforcement Priorities* (n 64 above) para 92.

[66] *Commission Discussion Paper on Exclusionary Abuses* para 91.

'. . . a line of conduct of a dominant undertaking is abusive as soon as it runs counter to the purpose of protecting competition in the internal market from distortions.'[67]

Objective justification

A second feature of the normative approach to the definition of abuse under Article 102 TFEU is its treatment of the innovative efficiency defence. In fact, this point is less of a defence and more a constituent element of the proof of abuse. The *Magill*[68] and *IMS*[69] cases indicated that an objective justification was also to be viewed in the same light.

At one point it was argued that the exercise of IPRs should enjoy a complete immunity under the competition law rules because of the property right prerogatives of IP owners.[70] More recently, an innovative efficiency justification has been raised based on the argument that if competition rules provide compulsory licences of IPRs as a remedy, they will reduce the incentives of undertakings to innovate and thereby reduce dynamic efficiency. For example, in the recent *Microsoft* case, Microsoft asserted that it would have less incentive to develop a given technology if it would be required to make that technology available to its competitors.

The Commission had attempted to tackle this assertion by Microsoft by dealing with it as a balancing exercise which balanced the arguably negative impact on Microsoft's incentives to innovate of an order to license its IPRs against the counter argument that the compulsory licence would produce improved innovative efficiency for the industry as a whole because interoperability would maintain plural sources of innovation.

The General Court clearly had other ideas about how the efficiency defence should be treated under Article 102 TFEU. It made it clear that what was called for was not a balancing exercise of the two theories of innovation but a more legally structured assessment of Microsoft's arguments. It held that Microsoft's assertion could only be raised under the head of objective justification as an integral part of the proof of abuse.[71] It stated that once the constituent elements of the abuse have been established, in this case that there were exceptional circumstances requiring an obligation to provide interface information, the burden shifts to the dominant firm to raise any plea of objective justification and support it with arguments and evidence.[72]

---

[67] *Commission Discussion Paper on Exclusionary Abuses* para 91. Opinion of AG Kokott in Case C-95/04P *British Airways v Commission* [2007] ECR I-2331 paras 68–9.

[68] *Commission Discussion Paper on Exclusionary Abuses* para 91, *Magill* (n 35 above).

[69] *IMS* (n 32 above).

[70] It has also been suggested that there should be some irreducible minimum area of non-interference with IPRs by the competition rules based on the specific subject matter or essential function of the IPR. However, the courts and competition authorities seem clear that the grant or 'existence' of an IP right may be based on public law but once granted, the IPR is exercised by its owner as a private right subject to the limits of the competition rules in the Treaty.

[71] *Microsoft* (n 2 above) para 659.

[72] *Microsoft* (n 2 above) paras 688 and 709.

In the event, Microsoft did not meet its burden. Its arguments were vague, general, and theoretical. It merely stated that '... [d]isclosure would ... eliminate future incentives to invest in the creation of more intellectual property', without specifying the technologies or products to which it thus referred.[73]

The Court concluded that the Commission's conclusion was not based on a balancing of the two conflicting views of innovation but rather upon its factual findings—that there was no basis to Microsoft's claims that its products would be cloned; that disclosure of interoperability was widespread in the industry—IBM had already committed to this course in the 1980s; and that Directive 91/250 was consistent with this policy.[74]

This indicates first of all that the innovative efficiency 'defence' in Article 102 TFEU is not viewed as a separate defence as such but rather objective justification is viewed as a constituent element of proving an abuse.

Related to this is that the test of objective justification will be limited to arguments about pro-consumer efficiencies and incentives to innovate only insofar as they are rooted in findings of fact about current conduct or information known at the time the conduct was engaged in. Predictions about the future may be too speculative to find acceptance in analysing the issue of objective justification. The lawfulness of a firm's actions must be assessed at the time when the firm acts.[75]

Finally, the innovative efficiency 'defence' will be curbed if the practice also deprives consumers of a new product or creates a risk of elimination of effective competition. There is no judicial acceptance of an approach that gives equal standing to pro-competitive and anti-competitive arguments in a balancing test.

The General Court's approach ensures that the Commission can continue to define the norms of abuse adhering to the distinctively European method of preservation of the process of competition as a guarantee of consumer benefit. It avoids an unweighted rule of reason balancing exercise of pro-competitive and anti-competitive effects similar to US antitrust law. If the Commission can prove that there is a risk of elimination of existing competition, that factor cannot be neutralized by arguments claiming greater efficiency. Nor can greater efficiency be a defence where it is merely an improvement in total welfare; there must be a fair 'pass on' of benefit to consumers. In a sense, the values and structure of Article 101(3) TFEU are built into the objective justification test. By carefully controlling excesses of dynamic efficiency defences, the European approach ensures that the test of abuse gives a priority to real effective competition in the short term and ignores less easily provable assertions about dynamic efficiencies in the long term. This feature makes its own distinctive

[73] *Microsoft* (n 2 above) para 698.
[74] *Microsoft* (n 2 above) para 710.
[75] See eg Case T-271/03 *Deutsche Telekom v Commission* [2008] ECR II-477.

contribution to cases involving findings of abuse. This is a reminder that Article 102 TFEU shares with Article 101 TFEU and the Merger Regulation a treatment of innovative efficiencies defences that is subjected to the two conditions of consumer benefit and non-elimination of effective competition.

This technique of dealing with assertions of objective justification by the General Court will not preclude defences based on genuine objective justification. In the first place, the defendant undertaking only has the burden of coming forward with the evidence to support its claim. The actual burden of proof of the abuse, and the absence of an objective justification, remains with the Commission. One could imagine successful defences based on technological innovation in the case of alleged tie-ins. Moreover, patent owners may have a strong case of objective justifications for refusing to license in cases where they have prepared plans to meet demand themselves but have delayed implementation to allow a sequencing of products.

However, the calculus will be restricted by the distinctive European approach to weighing the competing claims of innovative efficiency which prizes the existence of effective competition.

The existence of alternative IP remedies

The interface between IP and competition law has been characterized by the overlap in both system's aims to promote innovation and consumer welfare. Hence there are a number of accommodations to IPRs embodied in the logic of the competition rules. Nevertheless, the boundary between permitted and prohibited forms of exercise of IPRs under the competition rules is determined by the constituent elements of the prohibited conduct rather than the scope of the IPR. While some have argued that competition law applies mainly to cases where the IPR has been exercised outside its lawful scope the fact is that there are times when the IPR can be exercised in a perfectly lawful manner under the IP laws and nevertheless be held to infringe the competition rules.

Moreover, the process of enforcing the competition rules in those cases where the exercise of IPRs is unlawful both under the IP laws and the competition law rules show little sign of accommodation. This feature has been apparent in a number of cases. For example in *Astra Zeneca* generic competitors had a possibility of complaining under the SPC law. The practice seems to be that if the conduct of the IP owner has reached the stage of offending the competition rule, it will attract a competition law remedy. There is no evidence thus far of a policy to defer the competition law case to allow the parties to exhaust their IP remedies. In most cases so far, the existence of an alternative remedy has not prevented the immediate application of the competition rules. At the point of litigation the logic of the competition rules is straightforward. If there has been an infraction, there must not only be a fine as a penalty but the undertaking committing the abuse must discontinue the abuse. At that stage there is no room to consider a possible remedy offered under IP legislation since that possibility has not precluded the abuse. For example in *Microsoft* the argument that the competitor had a right to reverse engineer the interface codes under Article 6 of the Computer Program Directive had no effect on the finding of abuse.

Conclusion

Both features of the competition law rules—narrow product markets and exceptional circumstances—may be justified according to the internal logic of the competition law framework. However, both may also be assessed in terms of whether such rules have been successful in their impact upon the competition process. Have they contributed to a line drawn between abuse and competition on the merits that is appropriate? Do they help EU competition law to concentrate on condemning only the more extreme forms of anti-competitive behaviour? Do they help EU competition law to leave sufficient room for legitimate competition to breathe?

The Commission has expressed a concern that there is a need to move to a more economic approach to the enforcement of Article 102 TFEU to help avoid Type 1 errors (where the innocent are found to infringe the competition rules). However, while the Courts have proved to be receptive to a *more* economic approach to Articles 101 and 102 TFEU they have resisted the replacement of the political/legal philosophy expressed in their interpretations of the Treaty provisions in relation to abuse. The German/Ordoliberal concern with abuse of power seems to have become an EU concern with abuse of power and this continues to guide the Court's judgments thus far in relation to Article 102 TFEU cases in the ways set out in this chapter. The EU Courts would appear at this stage to be affirming that the values of the Treaty suggest that the avoidance of Type 2 errors (where the infringers of the competition rules are found innocent) is of greater concern than the reduction of Type 1 errors. As long as this continues to be the case Article 102 TFEU will apply as a relatively robust limit to abuses of market power.

2

COMPETITION *WITHIN* INTELLECTUAL PROPERTY REGIMES: THE INSTANCE OF PATENT RIGHTS

Rudolph J R Peritz

Introduction

This chapter describes an emergent jurisprudence and a residual economics that converge to support the reconceptualization of US patent policy as a competition regime. Its approach is inspired by an opinion that Justice Sandra Day O'Connor wrote for a unanimous Supreme Court some 20 years ago. The Court's recent patent jurisprudence sounds an echo of the opinion, which described the foundation of patent policy this way—'. . . [F]ree exploitation of ideas will be the rule, to which the protection of a federal patent is the exception.' There is, Justice O'Connor explained, a '. . . baseline of free competition upon which the patent system's incentive to creative effort depends.'

The chapter develops this proposition in three sections. The first explicates the economics of incentive theory, both its limits and its residual value. The second analyses the jurisprudence of recent decisions by the Supreme Court and Federal Circuit Court of Appeals—the speciality court for patent and trademark. The third section presents some instances of progressive change that would come of extending the reconception of the patent system as fundamentally a competition regime, an extension inspired by Justice O'Connor's image but informed by the failure of incentive theory as the economic logic for patent protection.

A. Crisis, Stalemate, and Progress

Intellectual property[1] protection and free competition have long been viewed as alternative means to encourage inventive activity and, through it, promote progress in the form of economic growth. Their relationship as means has sometimes been characterized as conflicting and other times as congruent. The same can be said about the progress they are intended to promote. In the United States, mainstream policy cabins these tensions of means and ends by treating IP protection as a domain of exclusionary rights and by removing free competition to a separate domain, to the domain of antitrust. With this bifurcation, the problem has been largely transformed into a question of adjudicating the relationship between two separate bodies of public policy.

Nonetheless, some competition doctrines linger within the IP realm. These doctrines, such as patent misuse and copyright fair use, have been characterized as intruders in the domain of exclusionary rights. Patent misuse is labelled an historical anomaly that properly belongs in antitrust, if anywhere at all, while copyright fair use is described as an interloper—either an alien article of political faith in First Amendment Speech Rights or a commercial artefact of market failures that temporarily limit the author's fundamental right to exclude.[2] Since the 1980s, IP policy makers have settled the problem of malingering doctrine by favouring exclusionary rights over free competition and by propertizing and otherwise extending IP rights.

This dynamic of bifurcation, preference, and expansion rests on the asserted superiority of exclusionary rights over open access in encouraging invention, a superiority that derives from reliance on an IP economics that holds neither in theory nor in practice.

There has long been trouble brewing in the IP economics that prevails in the United States. The trouble with IP economics recently reached boiling point with an admission by William Landes and Richard Posner, the Chicago School's dynamic duo of law and economics, that there is no ground for the dominant view of IP economics, no ground for the view that incentive theory can justify, explain, or rationalize IP rights. They made this confession in their book entitled *The Economic Structure of*

---

[1] Unless otherwise specified, intellectual property refers only to the rights granted under patent and copyright statutes enacted by Congress in accord with the Constitution's call to promote progress, though the term can plausibly be understood as referring as well to trade secret and trademark protection insofar as they have been increasingly justified in similar instrumentalist terms. The distinction is made in this chapter because the analysis of the patent domain takes account of Constitutional origin.

[2] See eg, DJ Gifford, 'Antitrust's Troubled Relations with Intellectual Property' (2003) 87 Minnesota L Rev 1695 (patent); WJ Gordon, 'Fair Use as Market Failure: A Structural and Economic Analysis of the Betamax case and its Predecessors' (1982) 82 Columbia L Rev 1600 (copyright).

IP Law.[3] The book has received wide attention and much praise. But the public confession of incentive theory's failure has been largely ignored.

At virtually the same moment, a related but separate development was bubbling to the surface of IP jurisprudence—in a recent series of surprising opinions, the US Supreme Court weakened patent protection and in the process expanded the role of free competition as an internal engine for promoting economic progress. The opinions were surprising because they run against the dominant view that pits an IP domain of exclusionary rights against an exogenous antitrust domain of free access. The recent opinions have destabilized this binary opposition between IP rights and free competition.[4]

In tandem, the failed economics and unstable jurisprudence have thrown the dominant approach to IP rights into crisis. The crisis is an emergent form of a long-term problem at the heart of both the economics and the jurisprudence, and it cannot be easily resolved.

On the economics side, informed policy makers have long recognized that economic progress is driven by the twin engines of IP monopoly and free competition. As economist Kenneth Arrow wrote in his landmark 1962 paper, the great difficulty lies in determining an optimal balance between them. Economist Joseph Schumpeter had earlier sought to merge the two engines in his vision of competition as serial monopoly, his perennial gale of creative destruction.[5]

As for the jurisprudence, the US Constitution presents a corresponding legal challenge to balance the exclusionary rights of IP protection and the open access of

---

[3] WM Landes and RA Posner, *The Economic Structure of Intellectual Property Law* (Cambridge, Massachusetts: Harvard University Press, 2003).

[4] This chapter takes up patent law as a competition regime, a theme that with respect to copyright, trade secret, and trademark as well is explored in my earlier writing, beginning with *Report to the IP Academy of Singapore* (2002–2003) (revised and published sub nom 'Competition Policy and its Implications for Intellectual Property Rights in the United States' in SD Anderman (ed) *The Interface Between Intellectual Property Rights and Competition Policy* (Cambridge: CUP, 2007) and variously investigated in other writing). The theme is a special case of the complex relationship between property rights and competition policy in American political economy, which I first developed in the domain of antitrust; see, eg, 'A Counter-History of Antitrust Law' in *Symposium: The Frontiers of Legal Thought* [1990] Duke LJ 263; 'The "Rule of Reason" in Antitrust Law: Property Logic in Restraint of Competition'(1989) 40 Hastings LJ 285, excerpted in ET Sullivan (ed) *The Political Economy of the Sherman Act: The First Hundred Years* (New York: Oxford University Press, 1990) 116, reprinted in R Graves (ed) *Competition Law* (Burlington: Dartmouth Publishing Co, 2003), reprinted in *Competition Law* (London: Ashgate Publishing, 2004). The theme was extended to other domains in *Competition Policy in America: History, Rhetoric, Law* (New York: Oxford University Press, 1996 rev edn 2001). This chapter on patent policy is part of a larger project, whose working title is *The Political Economy of Progress: IP Rights and Competition*.

[5] J Schumpeter, *Capitalism, Socialism and Democracy* (New York: Harper & Row, 1942, 3rd edn 1950); KJ Arrow, 'Economic Welfare and the Allocation of Resources for Invention' in R Nelson (ed) *The Rate and Direction of Inventive Activity: Economic and Social Factors* (Princeton: Princeton University Press, 1962) 609. It should be noted that Arrow wrote about invention while Schumpeter emphasized innovation—that is, the commercialization of invention.

free competition.[6] The Constitution empowers Congress to enact copyright and patent protection for the explicit purpose of promoting '. . . the Progress of Science and useful Arts.' So copyright and patent are not rewards, not natural rights. They are incentives—private means to a public end. But when does the private incentive of property protection promote the public benefits of progress? In both economic and jurisprudential terms, when does such protection produce more progress than would otherwise accrue with free competition? The answer to this question has proved elusive to both theorists and empirical researchers.[7]

Despite this indeterminacy, mainstream IP economics still rests on incentive theory, which holds that the scale and scope of IP rights should be determined by the degree to which they promote economic progress. Incentive theory's incapacity to guide such determinations results in an analytical stalemate between the exclusionary rights of IP protection and the open access of free competition, a stalemate because both produce economic growth but to indeterminable degrees. In this light, neither alternative deserves priority as the better means to promote economic progress.

This stalemate, this open question at the very core of IP policy, has put analysts and decision makers, including federal judges, between a rock and a hard place—on the one side, policy makers are pressed to make decisions; on the other, they are blocked from making reasoned decisions because there is no analytical methodology at hand. Policy makers have sought to extricate themselves from this predicament by taking a fall-back position, the position that maximizing the means maximizes the ends, that greater IP protection naturally leads to more invention and thus to more progress. In my view, this fall-back position explains the so-called propertization of IP rights, the normative shift to a Lockean entitlement from an instrumentalist (or means-ends) evaluation.

This fall-back into natural rights is not surprising, given the powerful ideology of private property rights in the United States. But it makes no logical sense. Nor is it supported in theory or fact. Indeed, it is well-known that too much IP protection as well as too little can stifle invention and impede economic progress. So both the economics and the law present IP policy makers with a Goldilocks problem. But there is no calculus for determining what amount of IP rights is 'just right,' particularly in a unitary system that does not discriminate among different kinds of inventions. And, of course, there is the other side of the indeterminacy coin; economic justification is equally lacking for simply eliminating IP rights entirely as a means for encouraging invention and thus promoting economic progress.

[6] United States Constitution, Art. I, Sec. 8, cl 8 states: 'Congress shall have Power: . . . To promote the Progress of Science and useful Arts, by securing for limited Times to Authors and Inventors the exclusive Right to their respective Writings and Discoveries.'

[7] For close analysis of these issues, see my essay 'Thinking about Economic Progress: Arrow and Schumpeter in Time and Space' in J Drexl (ed) *Liber Amicorum: for Hanns Ullrich* (Bruxelles: Larcier Pub, 2009).

So, what's to be done? In my view, the answer is clear—change the fall-back position. Reverse the presumption. When confronted with jurisprudential or economic indeterminacy, adopt the presumption that free competition better promotes the progress called for by constitutional directive. Given the indeterminate economic value of both free competition and IP rights in encouraging invention, policy analysis should begin with the presumption of free competition. In choosing between two rules or standards, policy makers should adopt the one that better expresses the policy of free competition.

Why adopt the presumption of free competition? In economic terms, because competition produces a tie-breaker for its indeterminacy stalemate with IP rights. The tie-breaker is competition's superior distributional outcome. When patents and other IP rights produce monopoly prices, they create welfare losses in both static and dynamic terms. In the short run, consumers pay higher prices or go to second best substitutes. In the longer run, subsequent inventors also pay higher prices or turn to second best substitutes, causing some combination of decline and path-diversion in follow-on inventive activity.[8] In this light, a rule or policy that would strengthen IP rights should first be shown to promote greater progress than would otherwise occur.

B. Patent Economics: Incentive Gap, Stalemate, Presumption of Free Competition

This part begins by examining the state of mainstream IP economics in the United States, particularly the failure of incentive theory as the economic justification for IP protection, and proceeds by sketching the IP economics that remains viable. The section closes with discussion of the IP economics of competition.

Incentive theory and its critiques

In the United States, the current economics of progress has adopted a mythical origin not unlike that of Athena, the Greek goddess of wisdom and culture who sprang fully formed from the head of Zeus. Like Athena, the economic logic of progress is seen as springing fully formed from the divine thinking of Kenneth Arrow, whose eminence was established even before his award in 1972 of a Nobel Prize in Economics. His eminence stems from his canonical 1962 paper entitled *Economic Welfare and the Allocation of Resources for Invention*.[9]

---

[8] The dynamic effects are a decrease in inventor welfare that results from the increased cost of new information or the denial of access at any price. See Peritz 'Freedom to Experiment: Toward a Concept of Inventor Welfare' (n 67 below).

[9] Arrow (n 5 above).

Of course despite such mythology, there is a substantial pre-history that posed fundamental questions and deep criticism of IP protection, much of it still pertinent today. Virtually all the questions emerged in the widespread European debates of the 19th century over patent protection; many of the criticisms were sharpened in the trenchant analysis of Sir Arnold Plant in his 1934 article entitled *The Economic Theory Concerning Patents for Inventions*[10] and in a companion piece on copyright. Plant raised many of the searching questions later addressed by American economists. The most difficult question concerned the opportunity cost of invention. Plant asked, when is use of society's resources to invent '... superior to alternative uses from which they are diverted[?]'

The opportunity cost of invention opens a wide gap in the incentive logic of IP rights, a gap between the private value and the public benefits of IP rights. There is little doubt that IP rights create a private incentive to invent—indeed, few could afford simply to give time to the enterprise of invention without remuneration. Yet the private value of IP rights has no necessary logical or economic relationship with their public benefits, benefits that depend on a wide array of factors. The opportunity cost of invention is but one powerful admonition to take account of what can be called the Incentive Gap. Ignoring it produces the category error of equating IP rights' private value with their public benefits. Taking the Incentive Gap into account transforms the question into an empirical inquiry.

None of this had noticeable impact in the United States before economist Fritz Machlup authored his 1958 Report to Congress, entitled *An Economic Review of the Patent System*. His was the most influential of 15 reports commissioned by a Congress concerned whether the costs of the patent system were justified. Here is Machlup's summary of the economic literature:

> None of the empirical evidence at our disposal and none of the theoretical arguments presented either confirms or confutes the belief that the patent system has promoted the progress of the technical arts and the productivity of the economy.[11]

---

[10] A Plant, *Selected Economic Essays and Addresses* (London: Routledge Kegan Paul, 1974 reprint [1934]) 35. Compare F Machlup and E Penrose, 'The Patent Controversy in the Nineteenth Century' (1950) 10 J Economic History 1 (chronicling the European debates).

[11] Study of the Subcommittee on Patents, Trademarks, and Copyrights of the Senate Judiciary Committee, 85th Cong., 2d Sess., An Economic Review of the Patent System, Study no. 15 (Comm. Print 1958) (written by Fritz Machlup) 79 (hereinafter 'Machlup Report'). The Machlup Report observes that '... there is no functional relation between the earnings under a patent.... and the "social usefulness" of the invention which it covers.' Machlup Report 30. In this line of analysis, the Machlup Report observes that:

> The question is no longer whether the patent system stimulates inventive talents to use more of their time and energy than they otherwise would for the development of new technology, but rather whether it stimulates business corporations to hire more of these talents than they otherwise would for this task. If this is affirmatively answered, the second question arises whether this use of the talents is superior to the alternative uses from which they are diverted. (Machlup Report 36)

What's to be done? 'Muddle through', wrote Machlup. Why? Because there has been '. . . a patent system for a long time', he declared, '. . . it would be irresponsible, on the basis of our present knowledge, to abolish it.'[12] There could be no weaker rationale for keeping patent protection.

It was under this cloud of indeterminacy that Kenneth Arrow published his landmark paper four years later. Like so many other sacred texts, Arrow's paper has become the touchstone for theorists and others who identify themselves with the orthodox approach as well as those who oppose it.

Arrow questioned the impact of competition on incentives to invent. For economists, perfect competition is the Holy Grail. Its miraculous power produces allocative efficiency by taking society's resources and putting them to their highest and best uses. But Arrow argued that perfectly competitive markets fail. They fail by discouraging inventors from inventing.

Arrow's story has become a commonplace—without patent protection, inventions are easily copied or imitated. Free access to their ideas discourages inventors from inventing and, thus, harms society. Patent rights correct this market failure by allowing inventors to profit and society to benefit from increased invention. Patent protection and the profits it generates are the means to an end. Patents are private rights that produce the public benefits of technological advancement and economic progress.

The dominant camp relies on the following quotation to support their call for stronger patent protection—'[Invention that is]. . .. available free of charge. . .. provides no incentive for investment in research.'[13]

Those who call for more access and thus more competition rely on this quotation—'. . .. the incentive to invent is less under monopolistic than under competitive conditions.'[14]

Both statements are accurate quotations from Arrow's landmark article. So it turns out both camps are right; and both are wrong. Why? Because reliance on one or the other quotation ignores Arrow's recognition that an incentive theory of patent protection creates a dilemma for welfare economics. The dilemma is that both patent rights and competition promote economic progress. And both impede it. Here is how Arrow described the dilemma,

---

Note that this is a modern form of the question posed in the 19th century European debates about what we would term the opportunity costs of diverting scarce resources. Compare Peritz, 'Patents and Progress: The Incentive Conundrum' in A Kur (ed) *Intellectual Property Rights: Does one Size Fit All?* (Aldershot: *Edward Elgar Publishing,* 2009) (selected papers, 2008 *ATRIP Annual Conference*, Max-Planck Institute, Munich, Germany, 21 July 2008).

[12] Machlup Report (n 11 above) 80.

[13] Arrow (n 5 above) 609.

[14] Arrow (n 5 above) 619.

> In a free enterprise economy, inventive activity is supported by using the invention to create property rights; precisely to the extent that it is successful, there is an underutilization of the information.[15]

In short, patent protection is the private incentive necessary to spur invention and at the same time the social cost that prevents its optimal use. Arrow transformed this dilemma of IP rights and competition into a trade-off over time—pay more now for better products in the future.[16]

Let's call this Arrow's Trade-Off. Arrow posed the social welfare question as a trade-off over time insofar as the current costs of patent monopoly pay for the future benefits of increased invention.

But in the end, Arrow's Trade-Off encompassed only part of the problem of IP's social value. It addressed the narrow question of IP's private value and its relationship to direct public costs and benefits but not its indirect effects, including opportunity costs. In consequence, the analysis did not speak to the Incentive Gap between the private value and the overall public costs and benefits of IP protection. The subsequent economic literature continued to pursue the broader question. But the theoretical scholarship largely rehearsed the European debates and Arnold Plant's economic analysis.

Empirical investigation

Ultimately the theoretical impasse resolved into empirical inquiry. What of the empirical literature that followed?[17]

A wide array of studies, almost all involving patents, developed various data sets to investigate different proxies for economic progress. Researchers have interviewed corporate decision makers; they have measured research and development expenditures and patenting activity on the input side, and productivity gains and economic growth on the output side. Studies have looked at single sectors, individual countries, and across countries.

---

[15] Arrow (n 5 above) 617.

[16] '[A]n incentive to invent can exist even under perfect competition in the product markets, though not, of course in the "market" for the information contained in the invention.' Arrow (n 5 above) 619. In his hypothetical world, Arrow does even better than transform a dilemma into a trade-off. He creates a model that neatly eliminates the present cost of the trade-off. He posits perfectly monopolistic markets for invention that provide inventors the greatest profit incentive and buyers in perfectly competitive markets for goods provide consumers the widest distribution at the lowest price. It's the best of all possible worlds though it is not the real world. In the real world, monopoly prices do not dissolve into the thin air of economic models. The hypothetical is perfected by Arrow's assumption that the invention is a new process that provides cost savings in the goods market that equal the monopoly prices paid for the new invention. Thus price and output in the goods market are not affected. Neat and tidy. But unlikely and perhaps economically illogical. See Peritz, 'Thinking about Economic Progress: Arrow and Schumpeter in Time and Space' (n 7 above).

[17] A working paper that takes a closer look at the literature is available from the author: 'Patents and Progress: The Incentive Conundrum' (2008).

The longest series of studies developed interview data from senior executives in the research and development departments of commercial firms. Five studies between 1959 and 2001 all reached the same conclusion: The prospect of patent protection was typically a factor of third or fourth order importance to research and development decisions, with the exception of the drug industry and perhaps chemicals. Still, it must be understood that these studies investigated only the private value of patents; neither public benefits nor public costs were addressed.

Other recent studies *have* inquired into the public benefits by looking at the relationship between changes in patent protection and changes in research and development expenditures. Japanese and US studies found the data inconclusive. One study across 29 countries found a mild positive correlation and another across 60 countries found a weak negative one. Moreover, the findings have been mixed in studies of statistical correlation between patent protection and the ultimate economic goal of increasing growth.

A rare statistical study of copyright protection has just been published. Relying on data from 1870 to 2006, the authors conclude, 'Despite the logic of the theory that increasing copyright protection will increase the number of copyrighted works, the data do not support it.'[18]

In sum, the empirical literature on the public benefits of patent and copyright is at best inconclusive.[19] This brings us full circle back to the theoretical impasse that preceded it. Small wonder, then, that so many policy makers in the United States have taken the fall-back position, the mistaken focus on the means itself—on maximizing IP protection in the erroneous belief that progress will be maximized as a natural result.

[18] RSR Ku, J Sun, Y Fan, 'Does Copyright Law Promote Creativity? An Empirical Analysis of Copyright's Bounty' (2009) 62 Vanderbilt L Rev 1669.

[19] Regardless of findings, all the empirical work confronts methodological difficulties. Here are two. First, the variables used are controversial. The uses of patent counts, citations, or renewal rates as measures of technological progress have all been criticized, as has the use of research and development expenditure data. Simply counting patents, or copyright registrations for that matter, does not take into account differences in their importance and social value. And more R&D spending does not necessary lead to more or better inventions.

There is a second methodological difficulty—the intractable problem of disentangling patent or R&D data from other sources of economic growth, sources including trade secrets, improved technical education, or increased production, to name a few. A noted American legal scholar put the general methodological problem this way, 'If a state of affairs is the product of *n* variables, and you have knowledge of or control over less than *n* variables, if you think you know what's going to happen when you vary "your" variables, you're a booby.' A Leff, 'Economic Analysis of Law' (1974) 60 Virginia L Rev 451 (comparing the first edition of Richard Posner's *Economic Analysis of Law* to Cervantes' *Don Quixote*). 'Booby' denotes a stupid person.

Patent economics: the residue

Where does that leave IP economics? Some alternatives to the mainstream approach have emerged, alternatives ranging from conservative incrementalism to radical repeal. Landes and Posner sit at the conservative end of the spectrum, where they argue that we should try to optimize the system and do the best we can with what we have. Economists Michele Boldrin and David Levine have been the latest to lay claim to the radical end, where they argue that IP rights are not necessary because free competition produces adequate profits to attract the invention necessary to promote economic progress. Of course these positions as well as those between them are not new.[20]

A moment's tarry at the Landes and Posner position is worthwhile, in my view, because it is likely to become the mainstream position, once the shock of incentive theory's demise has subsided. Landes and Posner take up the view espoused in Machlup's 1958 Report to Congress, the view that while the patent regime per se cannot be rationalized, changes can be evaluated for their effectiveness.[21] For this, Machlup developed a nine-step analysis and provided an example. The example is an increase in the patent term. As the author pointed out, the analysis requires quantitative and qualitative assumptions at every step and, even then, it cannot take into account the opportunity cost of more investment in research and development. In short, even though the more confined analysis of changes in rules or standards benefits from having a defined baseline of current invention levels that is lacking in an analysis of the patent regime per se, other problems of experimental design and measurement remain. Machlup concluded that the analysis of whether an increase in the patent term increases economic growth depends on '. . . a complex set of probabilities, the magnitudes of which depend [on]. . . many unknown variables.'[22]

Following Plant and Machlup, Landes and Posner reject incentive theory. In its place they adopt a series of more specific goals emphasizing reductions in, for example, transaction costs, rent seeking, and congestion externalities.[23] In their chapter on patent law, the authors proceed from the general point that patent protection '. . . makes economic sense because it curbs certain inefficiencies unavoidably created by trade secrecy.'[24] In their view, those inefficiencies derive from a number of sources,

---

[20] Compare Peritz, 'Patents and Progress: The Incentive Conundrum' (n 11 above).

[21] Machlup (n 11 above) 64–7, discussed in Peritz, 'Patents and Progress: The Incentive Conundrum' (n 11 above) and a 2008 working paper (n 17 above).

[22] Machlup (n 11 above) 64–7.

[23] Congestion externalities reflect a questionable reintroduction of tragedy-of-the-commons logic to public goods. The issue is of questionable importance for two reasons. First, because use of information (or invention, as Arrow called it) does not deplete its supply or quality; in that sense, there cannot be over-use. Second, because privatization presents an analogous problem, if there is one at all, in the form of the anti-commons—often called patent thickets. In sum, congestion is either a two-sided problem that does not resolve the question of propertization or it is no problem at all.

[24] Landes and Posner (n 3 above) 294.

including the following—first, from the very nature of trade secrecy, which keeps information out of the market. Second, from the higher costs of trade secret licensing. But these assertions turn out to be controversial. As to the first, Landes and Posner themselves develop an elegant analysis of the relationship between patent and trade secret that belies the impact on information. In the chapter on trade secrets, they assert patents are preferable only to the extent an invention is self-disclosing or likely to be invented independently.[25] That is, patents tend to disclose information that has lowest public value. As to the higher cost of trade secret licensing, the authors identify a number of higher costs associated with patents that seem to offset the advantage in licensing costs. Most telling, patent disclosure may lower the time to invent around or, perhaps worse, enable infringement that triggers expensive litigation with a substantial risk of finding patent invalidity.[26]

Landes and Posner take the substantive patent regime as a given and seek to optimize its implementation, an enterprise that seems likely to become the mainstream approach even though it is rife with the indeterminacy that devastates incentive theory. In my view, sound economics calls for change in patent policy more severe than fine-tuning.

Surprisingly, there might be a place in patent policy for a more limited conception of incentive theory, a conception that takes account of its limitations as well as the primacy of the free competition baseline. In this view, incentive theory becomes a sharp instrument of focused industrial policy, one applied to target particular goals. Patent rules might be changed to channel specific inventive activity toward green technology, cancer research, equality-inducing business methods, or other specific goals. Such targeting would introduce a qualitative dimension to economic progress. These judgments would place bets on particular social welfare consequences, political economic judgments that do not purport to serve the quantitative goal of economic growth. Thus, neither large scale nor narrow gauge cost-benefit analysis would ensue. The question would be whether the added incentive would increase the targeted inventive activity beyond the current rate. But targeted incentives would be bets and would raise difficulties of evaluation discussed above. Still, as Arrow recognized in his landmark paper, governments both here and abroad have long made these sorts of bets.

In addition to the risks of unsuccessful research and development, such judgments bring the danger of unintended consequences. One current example is the unintended anti-competitive impact of the Hatch-Waxman Act, the 1984 amendment to the Food and Drug Act that was intended to increase the incentive to produce patented drugs by extending the patent term and at the same time increase competition by opening the door to early market entry by generics manufacturers who claimed

[25] Landes and Posner (n 3 above) 355–6.

[26] Landes and Posner (n 3 above) 357.

their generics did not infringe valid patents. In actual experience, such generic filings are quickly answered with patent infringement cases filed by branded manufacturers. These cases have often produced settlements that include reverse payments of large sums from plaintiff branded pharmaceutical companies to defendant generics manufacturers in exchange for promises to keep their generic drugs off the market. Courts have approved the settlements and rejected antitrust claims of agreements in restraint of competition, finding them not only consistent with the general law that encourages settlements but also within the exclusionary rights of the contested patents. In consequence, consumers pay billions of dollars in higher prices and follow-on inventors are given the perverse incentive to invest resources that position them to litigate and settle rather than develop and commercialize generic drugs.[27]

Certainly, patents can serve as a more focused instrument for targeted industrial policy. But even there, risks of failure and unintended consequences call for careful analysis to overcome the presumption that free competition better serves the goal of promoting progress.[28] While the dynamic efficiency effects of both free competition and patent rights are indeterminate, distributional effects provide a tie-breaker. Patents that actually have economic value produce monopoly prices and, with them, welfare losses in both static and dynamic terms. Not only consumers but subsequent inventors are worse off. In this light, a rule or policy that would strengthen patent rights should first be shown to promote greater progress than would otherwise occur.

C. Patent Jurisprudence: Ends, Means, Emergent Emphasis on Competition

The frailties of patent economics leave policy makers in a quandary. On the one hand, there is no economic justification for patent protection as the primary means for promoting economic growth. Indeed, the residual economics points to free competition as the presumptive means. On the other hand, the constitutional instruction remains—Congress and the judiciary must formulate patent policy to promote progress. How can policy makers advance the constitutional purpose of patent protection in light of the economics?

---

[27] See eg, RJR Peritz, 'Three Statutory Regimes at Impasse: "Reverse Payments" in "Pay-for-Delay" Settlement Agreements between Brand-Name and Generic Drug Companies' in J Drexl, W Grimes, RJR Peritz, and E Swaine (eds) *More Common Ground for International Competition* Law? (Aldershot: *Edward Elgar Publishing*, forthcoming 2010); CS Hemphill, 'Paying for Delay' (2006) 81 New York U L Rev 1553; C Shapiro, 'Antitrust Limits to Patent Settlements'(2003) 34 RAND J Economics 31; M O'Rourke and JF Brodley, 'An Incentives Approach to Patent Settlements' (2003) 87 Minnesota L Rev 1767.

[28] In some circumstances, including the FDA example, there is no free competition to presume; there, the question becomes one of betting that one targeted incentive is better than its alternatives.

This section takes up patent jurisprudence, whose constitutional quandary is reflected in a pair of tensions in means and ends. The tension in means has been expressed in dominant and emergent strains of the jurisprudence. While the dominant strain continues to treat patents as the primary engine for promoting progress, an emergent alternative has recognized competition as the primary engine or, at the very least, an instrumentality that deserves more recognition for its value in promoting progress. These strains parallel the tension in means earlier seen in Arrow's Trade-Off. At the same time, a second tension, this one in ends, lies entirely within the dominant approach. It is a tension between the goals of more public knowledge or more material benefits. After unravelling these tensions, the section closes by organizing the jurisprudence according to what can be called the patent life cycle. Patents are shown to move through a life cycle in three stages, each one characterized by its own mix of means and ends, and all of them driven by a fundamental commitment to competition.

The dominant approach: an internal tension in ends

The Supreme Court has long declared that patent policy is founded on an incentive theory, 'Since the primary aim of the patent laws is to promote the progress of science and useful arts, an arrangement which diminishes the incentive is said to be against the public interest.'[29] Last year the Federal Circuit Court of Appeals characterized an amendment to the Patent Act as a '. . . legislative effort to reinforce the value of the patent statute as an innovation incentive.'[30] Although these pronouncements might seem to be synonymous statements of patent policy, they are not. There is a subtle but significant difference between them.[31] While the Supreme Court addressed the general enterprise of promoting progress, the Federal Circuit focused on innovation, which reflects only one aspect of progress. Innovation is not invention but rather its commercialization. The distinction between invention and innovation is important in two respects. First, because attracting investment to innovation can draw investment away from invention. Second, because a focus on innovation defines the primary form of progress as material advancement of day-to-day life through commercial development of extant invention.

Indeed, the Federal Circuit's focus on material advancement diverges from numerous statements by the Supreme Court that '. . . [t]he primary purpose of our patent system. . .. is directed to disclosure of advances in knowledge which will be beneficial

[29] *Transparent-Wrap Mach Corp v Stokes & Smith Co*, 329 US 637, 646 (1947) (Douglas, J).

[30] *Cardiac Pacemakers, Inc v St Jude Medical, Inc*, 576 F3d 1348, 1371 (Fed Cir 2009).

[31] There is a second subtle difference as well—note that the Supreme Court writes that '. . . an arrangement which diminishes the incentive *is said to be* against the public interest.' The Court is careful to avoid the implication that it adopted this view. This is consistent with the scepticism expressed in the text accompanying the next footnote.

to society; it is . . . an incentive to disclosure.'[32] In other words, patents make public new knowledge which would otherwise be hidden under the blanket of trade secrecy. The public value of new knowledge goes beyond the Enlightenment virtue of edification. It has use value for follow-on inventors—disclosure reduces the costs of competition by invention. Moreover, it accelerates the learning curve. In short, the public benefit of disclosure is the free competition that results from free riding on the patented efforts of prior inventors.[33]

This divergence is embedded in the Constitution's language of promoting the 'Progress of *Science* and *useful Arts*.' A twenty-first century restatement of the constitutional language calls for promoting the progress of *knowledge* and *industrial technology*. Yet courts have seldom been asked to adjudicate the relationship between advancing knowledge and advancing the material conditions of everyday life. Here are two examples of court decisions whose outcomes turn on the choice of primary public benefit.

The first example involves a dispute between two researchers who filed patent applications for the same pharmaceutical compound. The first to file was a biochemist for a Japanese company but the first to invent was a professor at Cornell Medical School. As a general rule, patents in the United States are awarded to the first to invent. The time of invention dates back to the moment of conception. In the US patent system, the first to conceive the idea is supposed to win. It matters not who files first.

The professor should have won. But he lost. Why? The court refused to apply the standard US rule because it determined the professor did not proceed with 'reasonable

[32] 'The primary purpose of our patent system is not reward of the individual but the advancement of the arts and sciences. Its inducement is directed to disclosure of advances in knowledge which will be beneficial to society; it is not a certificate of merit, but an incentive to disclosure.' *Sinclair & Carroll Co v Interchemical Corp*, 325 US 327, 330 (1945). 'The basic quid pro quo contemplated by the Constitution and the Congress for granting a patent monopoly is the benefit derived by the public from an invention with substantial utility.' *Brenner v Manson*, 383 US 519, 534 (1966). '. . . the public may have the full benefit thereof, after the expiration of the patent term.' *Bonito Boats, Inc v Thunder Craft Boats, Inc*, 489 US 141, 147 (1989).

Yet the Supreme Court has long expressed some scepticism about the incentive value of IP rights. For example, in the *Marconi Wireless* case of 1943, the Chief Justice remarked—'For all I know the basic assumption of our patent law may be false, and inventors and their financial backers do not need the incentive of a limited monopoly to stimulate invention. But whatever revamping our patent laws may need, it is the business of Congress to do the revamping.' *Marconi Wireless T Co of America v US*, 320 US 1, 63–4 (1943).

Here is an economic analysis of disclosure: Patent right and its disclosure obligation present the right strategy for those inventions not adequately protected as trade secrets. In this light, patented inventions are those most likely to be disclosed anyway and so the public really gains very little if anything. Note the tension between this account and traditional norms and incentives to disclose in the scientific community, tensions exacerbated with increased propertization and thus increased incentive to withhold disclosure until the patent application is filed. Note also the patent regime's disincentives to read patents, especially intentional infringement liability for multiple damages.

[33] Of course the incentive problem re-emerges. Should patent rights be shaped to encourage publication of new knowledge or encourage internalization of pecuniary benefits?

diligence' from the time he conceived the new idea to the time he reduced it to a practical invention. In the court's view, there was unreasonable delay in his waiting for outside research funding and for his chosen graduate student to enter the programme. Confronted by a conflict between '... the interest in rewarding and encouraging invention [and] the public's interest in the earliest possible disclosure',[34] the court chose 'earlier disclosure over earlier invention' because it saw 'early public disclosure [as] the "linchpin of the patent system".'[35] The outcome appears very European or very Japanese insofar as the first to file was awarded the patent. But the rationale reflects a uniquely American issue characterized as a conflict between the goals of advancing knowledge and advancing industrial technology.[36]

The second example of patent jurisprudence that seeks to adjudicate this conflict of ends is even more dramatic. It is more dramatic because it involves the entire relationship between the patent and trade secret regimes, between the federal requirement of patent disclosure and the state trade secret requirement of secrecy. In the United States, the Constitution expresses a general principle of harmonization in what is called the Supremacy Clause. When a state law conflicts with federal law, the federal law prevails; the state law is unconstitutional and thus unenforceable.

In practical terms, the Supreme Court had to find a way to harmonize two regimes that had co-existed for more than 100 years; throwing out the trade secret laws of 50 states was unacceptable.[37] What to do? The Court characterized patent and trade secret protection as harmonious because they both encourage technological advancement. In this view, they share the same goal. As for the conflict in disclosure, the Court reduced its importance to triviality by asserting without foundation that an inventor would always choose stronger patent over weaker trade secret protection. But in some circumstances trade secrecy can provide stronger protection.

[34] *Griffith v Kanamaru*, 816 F2d 624 (Fed Cir 1987). More specifically, the date of invention generally relates back to the date of conception. The first inventor is the first to conceive the idea. But here, there was the unacceptable delay between the professor's conceiving the idea and his reducing it to practice. And so the reduction date was treated as the date of invention. As for the commercial researcher, there was no evidence to support either a date of conception or a date of reduction—both of them irrelevant in the Japanese as well as EU patent regimes. And so the patent filing date was used as the date of invention. The Japanese filing date preceded the professor's reduction date, each of them proxies for the date of invention.

[35] *Horwath v Lee*, 564 F2d 948, 950 (Crt Customs & Patent App 1977).

[36] Before going to my second example of the conflict between encouraging invention and encouraging its disclosure, between advancing technology and advancing knowledge, I want to take a quick look at the court's treatment of reasonable diligence. The professor's seeking necessary outside funding and awaiting the return of his graduate student do not seem unreasonable per se. Still, the court refused these typical academic reasons, stating that only personal reasons such as family illness, personal finances, or vacation time would excuse delay, even though arguments were made that outside funding was a university research policy to validate projects through outside competitions for funds. In that sense, there was no delay but rather another kind of evaluation or even competition that was going on. Without explanation, the court seemed much more approving of the commercial research environment.

[37] *Kewanee Oil Co v Bicron Corp*, 416 US 470 (1974).

Indeed, market studies have found that many inventors prefer trade secret protection for reasons IP economics, including the work of Landes and Posner, has made clear.[38]

And so the orthodox patent jurisprudence has sought to mediate a tension within the constitutional incentive logic of promoting progress, the tension in ends between advancing knowledge and improving the material conditions of life. The mediation has affected adjudication of questions both narrow and broad—our two examples, the narrow question of whom to deem the inventor of a particular product and the broad question of how to characterize the relationship between the patent and trade secret regimes. In each case, the choice of end affected the outcome, changed the circumstances of inventive enterprise, and defined the conditions of competitive activity—first in prosecuting patent applications and second in making strategic choices between patent or trade secret protection.

The emergent approach: three recent patent cases and their countenance

While the dominant approach has grappled with a conflict in ends, the constitutional logic for promoting progress has produced a second tension as well. Recent Supreme Court jurisprudence has shown signs of an approach different from the orthodoxy, an emergent approach that raises questions about patent protection as the presumptive means for promoting progress, questions that parallel those raised in IP economics. This emergent approach is more properly termed a re-emergent strain of IP jurisprudence insofar as the recent decisions summon the policy stated in Justice O'Connor's opinion some 30 years ago for a unanimous Court, the statement that there is a '. . . baseline of free competition upon which the patent system's incentive to creative effort depends.' In short, '. . . free exploitation of ideas will be the rule, to which the protection of a federal patent is the exception.'[39]

This line between patent monopoly and free competition is drawn by the statutory requirements for patentability, most notably the requirement that a patented invention be non-obvious in light of prior art.[40] Beginning in the 1980s, the non-obviousness requirement was increasingly trivialized. For example, in 1999, the Federal Circuit Court ordered that the US Patent Office issue a patent to an applicant who decorated large black plastic garbage bags with orange pumpkin faces. The Federal Circuit declared that this combination of garbage bags and Halloween decoration, each element itself obvious, was a non-obvious combination that merited a patent.[41]

[38] See discussion accompanying n 48 below for the economic analysis of Landes and Posner on strategic choices between trade secrecy and patent protection.

[39] *Bonito Boats, Inc v Thunder Craft Boats, Inc*, 489 US 141, 145–7 (1989). Note that Justice O'Connor poses the baseline imagery within the orthodox view of incentive theory. But, as the first section demonstrates, the baseline metaphor itself has an economic logic that does not depend on the orthodoxy.

[40] Patent Act § 103(a).

[41] *Re Dembiczak*, 175 F3d 994 (Fed Cir 1999).

In 2003, the Federal Trade Commission issued a widely praised Report criticizing patent protection's descent into triviality.[42]

Then, in 2007, the Supreme Court published the *KSR* decision, which elevated the non-obviousness requirement for the largest category of patents, those like the Halloween garbage bag that involve combinations of prior art. The decision instructed the Patent Office to reject applications for combination that show only 'ordinary creativity'.[43] The Patent Office has since rejected on the ground of obviousness a number of applications for combination patents, and the courts have regularly upheld those rejections.

The Court in *KSR* took issue with the Federal Circuit's '. . . transform[ation of a] general principle into a rigid rule that limits the obviousness inquiry.' The patent principle holds that a combination is obvious to '. . . a person of ordinary skill in the relevant field' when the prior art '. . . demonstrate[es] a teaching, suggestion, or motivation to combine known elements' into that combination.[44] The Federal Circuit rigidified the principle '. . . by overemphasis on the importance of published articles and the explicit content of issued patents.' This approach failed to take account of 'common knowledge and common sense', which consider a larger body of public knowledge, including 'design need and market pressure', knowledge which seldom finds its way into the literature of prior art. Justice Kennedy observed that a '. . . person of ordinary skill is also a person of ordinary creativity, not an automaton'.[45] This observation brings to the fore the difficulty of separating ordinary creativity from the non-obvious type because '. . . inventions in most, if not all, instances rely upon building blocks long since uncovered, and claimed discoveries almost of necessity will be combinations of what, in some sense, is already known.'[46]

The practical question, then, is what to do about the great bulk of inventions that lie in the bandwidth between the obviously ordinary and the obviously non-obvious. In expanding the range of references for determining prior art, the Court in *KSR* raised the level of non-obviousness required for patentability. Now, a combination may be found obvious even without a reference in the prior art to 'teaching, suggestion or motivation to combine known elements'.

---

[42] See Federal Trade Commission, 'To Promote Innovation: The Proper Balance of Competition and Patent Law and Policy' (October 2003) <http://www.ftc.gov/os/2003/10/innovationrpt.pdf> accessed 2 September 2010; see also, JH Barton, 'Non-Obviousness' (2003) 43 IDEAJ L and Technology 475; Working Group on the New Economy, American Antitrust Institute, 'Antitrust and the New Economy: Comments to the Antitrust Modernization Commission, Washington, D.C.' (July 2005).<http://govinfo.library.unt.edu/amc/public_studies_fr28902/new_economy_pdf/050715_AAI-New_Economy.pdf> accessed 2 September, 2010.

[43] *KSR Int'l Co v Teleflex Inc*, 127 S Ct 1727, 1743 (2007) (hereinafter '*KSR*').

[44] *KSR* (n 43 above) 1741, 1742.

[45] *KSR* (n43 above) 1743 (citing *DyStar Textilfarben GmbH & Co Deutschland KG v CH Patrick Co*, 464 F3d 1356, 1367 (Fed Cir 2006)) (internal quotation marks omitted).

[46] *KSR* (n 43 above) 1741.

It should be noted that the very process of determining non-obviousness in the course of patent application is in effect a contest in ideas, a competition between prior art and the prosecuted invention. The standard is whether the invention embodies an advance in ideas that is not obvious in the light of prior art. *KSR* raises the level of difficulty for the new arrival to win this competition in ideas.[47]

Given that the heightened standard will exclude a class of combination inventions that met the old standard for non-obviousness but fail the new one, what are the likely effects? Some of the newly obvious combinations, especially those involving processes, can be hidden from public view and, thus, their owners can seek trade secret protection. In this instance, public information about such combinations will be lost until the secrets are discovered. Owners of newly obvious combinations which are self-disclosing on sale or use will proceed in reliance on licensing provisions, first-mover advantages, or simply the benefits of the new combination when they outweigh the competitive costs of imitation by others. The resulting mix of secret and public combinations is an empirical question.

Moreover, there is a strategic question that sheds some light on the matter. As Landes and Posner have observed, the choice between patent and trade secret protection depends on their relative value.[48] On the cost side, patents are more expensive to obtain. And patent disclosure provides the very information rivals need to invent around more cheaply or simply infringe when that makes strategic sense. On the benefit side, a patent becomes more valuable than trade secret protection as the risk of disclosure, reverse engineering, or independent invention increases. As a general matter, the inventor is more likely to seek patent protection for inventions that are more likely to become public knowledge or otherwise legally available to rivals. Inventions whose secrecy is more readily maintained are less likely to be patented.[49]

In sum, *KSR*'s heightened standard for non-obviousness increases the play of competition, either immediately by direct imitation or eventually by investigation, independent discovery, or reverse engineering. The Court has denied patent protection for inventions that reflect only 'ordinary creativity' and, in consequence, expanded access to inventions that were protected under the old rule.[50] The result is that

[47] The statutory requirement of usefulness assures that the invention is not a disembodied idea: 35 USC § 101. For further discussion of this point, see Peritz, 'Patents and Progress: The Incentive Conundrum' (n 11 above).

[48] Landes and Posner (n 3 above) 354–71. Note as well that patents are less expensive to maintain and license.

[49] In this light, the loss of public knowledge from patent publication should not be overestimated. Nor should a decline in public knowledge resulting from the heightened standard of patentability.

[50] The extent of access to competitors under the new approach deserves further comment insofar as it depends on the character of prior art embodied in the combined elements. If no elements are protected by patents still in force, then access to the new combination is entirely free and competition is simply extended. If, however, any element is still protected, then use of the new combination requires a licence from each patent holder. But no patent licence is required to practice the combination. The net effect in either case is free access to the combination and, with it, lower bargaining and licensing costs.

competitors now have free access to make, use, and sell inventions that would have been protected by combination patents under the lower level of creativity.

In a second recent patent decision, a unanimous Supreme Court in *eBay* tightened the requirement for obtaining an injunction against a patent infringer. A more stringent requirement means that infringing competitors are not so easily restrained from making, using, or selling patented inventions; instead, the remedy of compulsory licences opens competition to patent infringers who would otherwise have been excluded from the market.[51]

The unified Court in *eBay* once again rejected an instance of the Federal Circuit's rigid jurisprudence of expansive patent rights, this time its '... general rule that courts will issue permanent injunctions against patent infringement absent exceptional circumstances.' In rejecting this general rule, the Court held that issuance of permanent injunctions summons '... familiar principles [of equity that] apply with equal force to disputes arising under the Patent Act.'[52] Justice Thomas' opinion for the Court provides a clear and unembellished basis for a more flexible approach,

> As this Court has long recognized, 'a major departure from the long tradition of equity practice should not be lightly implied.' Nothing in the Patent Act indicates that Congress intended such a departure. To the contrary, the Patent Act expressly provides that injunctions 'may' issue 'in accordance with the principles of equity'.[53]

While the opinion for the Court does not venture beyond the statutory text and equity doctrine to make plain the outcome, two concurring opinions offer differing policy analysis for support. Both address an issue raised in the opinion by Justice Thomas, particularly in a passage that rejected the Federal Circuit's reasoning for its general rule for issuing permanent injunctions. The Federal Circuit had concluded that the Patent Act's explicit definition of a patent as '... having the attributes of personal property', particularly '... the right to exclude others from making, using, offering for sale, or selling the invention... alone justifies its general rule'. Justice Thomas quoted specific statutory language that provides for the judicial discretion associated with traditional equity practice, observing that '... the creation of a right is distinct from the provision of remedies for violations of that right.'[54] The two

In all circumstances, however, the intervention of trade secret protection must be taken into account, with consequences as described in the text accompanying this footnote.

[51] *eBay Inc v MercExchange, LLC,* 126 SCt 1837 (2006) (hereinafter '*eBay*'). Of course the infringing user must bear a reasonable royalty as determined by the court. This can be understood as shifting from the patent holder to the court the power to determine royalties. In consequence, the patent holder cannot hold up would-be competitors in what is typically a one-sided monopoly bargaining scenario that does not promise the efficient solution generally attributed to settlements and bargain contracts more generally, per the Coase Theorem.

[52] *eBay* (n 51 above) 1839 (both quotations in the paragraph).

[53] *eBay* (n 51 above) 1839 (citations omitted).

[54] *eBay* (n 51 above) 1840 (citing 35 USC. §§ 261, 154(a)(1)).

concurring opinions assert sharply different rationales for treating the distinction between the exclusionary nature of property rights and the exclusionary remedy of injunction.

Chief Justice Roberts understood the relationship between right and remedy reflected in the statutory provisions to be reflected in a 'long tradition of equity practice' to grant injunctions 'upon a finding of infringement in the vast majority of patent cases' on account of 'the difficulty of protecting a right to *exclude* through monetary damages that allow an infringer to *use* the invention against the patentee's wishes.'[55] In sum, Justice Roberts was instructing federal judges not to stray from that 'long tradition' of recognizing patents as fundamentally property rights to exclude, rights to empower individual choice about how to practice the invention, or whether to practice it at all.

Justice Kennedy in his concurring opinion gave a diametrically opposed rationale for the Court's declaration that the statutory definition of patent as property right does not necessarily define the remedy for its violation. At the outset, Justice Kennedy rejected the Chief Justice's view that the difficulty of fully protecting patent rights with monetary damages underlies a 'long tradition' that calls for judges to conserve the property rights in patents. In sharp contrast, Kennedy's opinion looks forward rather than back. It invests the equitable nature of injunctive relief with a progressive ability to adjust to change, '[I]n many instances the nature of the patent being enforced and the economic function of the patent holder present considerations quite unlike earlier cases.' Two examples are given—first, 'industries in which firms use patents not as a basis for producing and selling goods but, instead, primarily for obtaining licensing fees'; second, 'patents over business methods', which raise significant questions of 'vagueness and suspect validity'.[56] Both examples reflect concerns that patent rights to exclude can be questionable barriers to the market entry needed for competition to flourish.

It is no accident that Justice Kennedy's source for both examples is the Federal Trade Commission report entitled 'To Promote Innovation: The Proper Balance of Competition and Patent Law and Policy'. A balance between access and exclusion is required because experience and economics tell us that both competition and patent rights can promote innovation as well as the invention that precedes it. In this light, injunctive relief for patent infringement should not be granted, particularly to patent trolls or business patent holders, when it results in less progress than competition or compulsory licensing.[57] Justice Kennedy cautions against the dangers of excessive

[55] *eBay* (n 51 above) 1840 (joined by Justices Ruth Bader Ginsberg and Antonin Scalia) (emphasis in original).

[56] *eBay* (n 51 above) 1842 (Justice Kennedy joined by Justices Stevens, Souter and Breyer) (for all quotations in the paragraph).

[57] *eBay* (n 51 above) 1842 (based on the FTC Report).

patent protection and, with it, inadequate regard for competition as a powerful means to promote progress through innovation.

In the third recent case, the competition logic driving Justice Kennedy's concurrence emerges even more emphatically in Justice Stephen Breyer's dissent from the *Metabolite* decision. Justice Breyer's opinion questions the wisdom of dismissing the writ earlier granted in a case that addresses the fundamental patent imperative to '[e]xclude from . . . patent protection . . . laws of nature, natural phenomena, and abstract ideas.'[58]

What is so important about this issue? In Justice Breyer's view, granting a 'monopoly over a basic scientific relationship' upsets a careful balance embodied in patent rights, '. . . [S]ometimes too much patent protection can impede rather than "promote the Progress of Science and useful Arts," the constitutional objective of patent and copyright protection.'[59]

Justice Breyer was concerned about public access to '. . . the basic tools of scientific and technological work' and, as such, to '. . . part of the storehouse of knowledge and manifestations of laws of nature as free to all men and reserved exclusively to none.' The rationale for free access lies in the public policy to promote progress by encouraging 'development and the further spread of useful knowledge itself.'[60]

What exactly is this careful balance embodied in patent rights? Justice Breyer incorporates it by reference to Justice O'Connor's *Bonito Boats* opinion for a unanimous Court,

> The Patent Clause itself reflects a balance between the need to encourage innovation and the avoidance of monopolies which stifle competition without any concomitant advance in the 'Progress of Science and useful Arts.' . . . [T]he stringent . . . novelty and nonobviousness requirements express a congressional determination that the purposes behind the Patent Clause are best served by free competition and exploitation of either that which is already available to the public or that which may be readily discerned from publicly available material.[61]

Justice Breyer was reminding readers that the patent regime begins, as Justice O'Connor put it, with 'the baseline of free competition . . . [from] which the protection of a federal patent is the exception.'[62] And so Justice Breyer concluded his opinion in *Metabolite* with references to competition policy—not only the *Bonito Boats*

[58] *Lab Corp of Am Holdings v Metabolite Labs, Inc*, 126 SCt 2921, 2922 (2006) (hereinafter '*Metabolite*') (Justice Breyer, joined by Justices John Paul Stevens and David Souter, dissenting from opinion to dismiss writ of certiorari as improvidently granted).

[59] *Metabolite* (n 58 above) 2925 (citations omitted).

[60] *Metabolite* (n 58 above) 2923 (citations and internal quotation marks omitted).

[61] *Bonito Boats, Inc v Thunder Craft Boats, Inc*, 489 US 141, 146, 150 (1989) (hereinafter '*Bonito Boats*'); *Metabolite* (n 58 above) 2926 (citing *Bonito Boats* 146).

[62] *Bonito Boats* (n 61 above) 156, 151.

decision but also the FTC Report and former FTC Commissioner Robert Pitofsky's article on antitrust and intellectual property rights.[63]

Each of these patent cases expresses an aspect of an emergent jurisprudence—first, granting rights to exclude competitors only with respect to non-obvious inventions; second, determining the propriety of exclusionary remedies by equitable principles rather than by the property logic of patent ownership; and finally, maintaining public access to 'laws of nature, natural phenomena, and abstract ideas'. Every one of these propositions limits the exclusionary power of patent protection. Each one widens public access to inventions or to the knowledge embodied in those inventions. The result is increased weight attributed to the patent regime's internal policy of free competition as an engine to promote progress.

These recent calls to competition are not exceptional.[64] Patent monopoly has long been disfavoured in the United States. As Thomas Jefferson put it over 200 years ago, the patent system must draw '. . . a line between the things which are worth to the public the embarrassment of an exclusive patent, and those which are not.'[65] Indeed, the Supreme Court is currently considering a closely-watched case that presents questions posed by Justices O'Connor, Kennedy, and Breyer, questions about the scale and scope of patentable subject matter and thus the reach of exclusionary rights in information technology.[66]

The patent life cycle: three stages of competition

The twin tensions reflected in the patent jurisprudence, the tensions in means and ends, resolve differently in the course of what can be called the patent life cycle. Patents move through three stages, each one comprising a technological and a legal component. Schumpeter, FM Scherer, and other economists have characterized technological change as the well-known steps of invention, innovation, and imitation or diffusion.[67] The legal component of the patent life cycle runs through the stages of patent prosecution, patent term, and patent expiry.

63 *Metabolite* (n 58 above) 2929.

64 For a recent trade dress decision that echoes Justice O'Connor's call to a baseline of competition, see *Wal-Mart v Samara*, 529 US 205, 213–14 (2000). The opinion was written by Justice Antonin Scalia, usually the Court champion of property rights and freedom of contract.

65 13 Writings of Thomas Jefferson (Memorial ed 1904) 335, cited in, eg, *Bonito Boats* (n 61 above) 148. This passage and others suggest the possibility that for the 18th century founding fathers, property rights had a natural incentive effect. In this view, there was no fundamental distinction between property as natural rights and as incentives.

66 *Re Bilski*, 545 F3d 943 (Fed Cir 2008), *cert. granted sub nom Bilski v Doll*, 129 SCt 2735 (2009).

67 'Invention to [Schumpeter] was the act of conceiving a new product or process and solving the purely technical problems associated with its application. Innovation involved the entrepreneurial functions required to carry a new technical possibility into economic practice for the first time—identifying the market, raising the necessary funds, building a new organization, cultivating the market, etc. Imitation or diffusion is the stage at which a new product or process comes into widespread use as one producer after another follows the

During the patent prosecution stage, the claimant must persuade the patent examiner that there is an invention and that it merits protection.[68] The applicant is free to engage in innovation and further invention during prosecution. The process is confidential.[69] Diffusion of knowledge is delayed until the application and file folder are published by the Patent Office—until the patent is issued though even sooner in many cases.[70] Patent prosecution can be understood as a competition in ideas pitting the invention against the body of prior art in a contest refereed by a patent examiner according to a strict set of rules and guidelines. If the invention is useful and proper subject matter, if it is clearly described, and, finally, if it is not anticipated by the prior art, if it is not obvious, then it embodies new knowledge whose embodiment is worthy of patent protection.[71] The competition in this phase of the life cycle produces the private right to exclude and what can be called the patent's public knowledge benefit.

The knowledge benefit's crucial importance to the prosecution stage can be seen in the strict requirement that the description of the invention in the patent application be clear and complete, and that it enable those reasonably skilled in the art to make and use it. The applicant must also include any additional knowledge concerning the best mode of making and using the invention. The description and enablement requirements provide two kinds of public knowledge benefit. First, the description requirement separates the idea from its embodiment, the public benefit from the private property by assuring that the applicant has reduced the idea to practice. Without a strict description requirement, there would be the danger of patenting the idea, of turning the public benefit into private property. Second, as the Supreme Court stated long ago, if the description is so vague and uncertain that no one can

innovating firm's lead.' F Scherer, *Industrial Market Structure and Market Performance* (1970) 350.

The three steps are not mutually exclusive. Indeed, they are best understood as overlapping and intertwined. Compare Peritz, 'Freedom to Experiment: Toward a Concept of Inventor Welfare' (2008) 90 J Patent and Trademark Office Society 245.

[68] Though the statute states that patent '. . . will be granted unless', in practical terms most applications are rejected initially and thus the burden falls on the applicant.

[69] In some circumstances, provisional rights to damages are available for third party use during the prosecution stage, but only after the patent has been issued. Patent Act § 154(d).

[70] 'Publication of patent applications is required by the American Inventors Protection Act of 1999 for most plant and utility patent applications filed on or after 29 November 2000. On filing of a plant or utility application on or after 29 November 2000, an applicant may request that the application not be published, but only if the invention has not been and will not be the subject of an application filed in a foreign country that requires publication 18 months after filing (or earlier claimed priority date) or under the Patent Cooperation Treaty. Publication occurs after the expiration of an 18-month period following the earliest effective filing date or priority date claimed by an application. Following publication, the application for patent is no longer held in confidence by the Office and any member of the public may request access to the entire file history of the application.' USPTO, 'General Information Concerning Patents' <http://www.uspto.gov/web/offices/pac/doc/general/#pub> accessed 2 September, 2010. Patent Act § 122.

[71] Of course, in addition to non-obviousness, requirements of utility, novelty, and proper subject matter must be met.

tell, except by independent experiment, how to construct the patented device, the patent is void.[72] In other words, the information must be sufficient to enable subsequent inventors to learn from the description. With less stringent requirements of description and enablement, the patent prosecution phase would produce the worst of all possible outcomes—private commercial rights to an idea and public knowledge without use value. The domain of ideas would shrivel while monopoly in commercial markets would expand.

The life cycle's second stage begins when the patent is issued. During the patent term, the owner holds a right to exclude others from using the invention for any purpose including innovation and further invention.[73] This right to exclude is the condition underlying the patent holder's power to license the technology and in consequence take advantage of downstream efficiencies of development, production, and distribution. Licensing the technology or otherwise using the invention is of course subject to general legal requirements and restrictions. A private right to the commercial benefits of the invention promises a public benefit—the material benefit of improved conditions of everyday life. In this stage of the life cycle, two kinds of competition are anticipated. First, the patent holder is encouraged to commercialize the invention and offer it on the market. Nonetheless, a material benefit from commercial competition is not guaranteed because the patent holder has no obligation to work the patent and even if she does, consumers might not buy it. Second, competitors have access to the new knowledge and often the invention itself, and can make practical use of it by improving or inventing around it. However, this competition by experimental use is severely restrained in the United States.[74]

The third stage of the patent life cycle begins with the grant's expiry. The patent's limited term creates a further material benefit when, after 20 years in the case of a utility patent, the invention itself falls into the public domain. This reversion to public use[75] triggers a general privilege to use the invention and, in so doing, invites commercial competition by imitation that promises to lower prices and, thus, to disseminate more widely the invention's material benefits. Moreover, to the extent invention follows imitation, there is further competition in both ideas and commerce.

---

[72] The Incandescent Lamp Patent, 159 US 465 (1895). The patenting of computer software raises important questions about the knowledge benefit. The description element is satisfied by language of general means that does not require publication of source code. The result is patents that are too broad and information that is too vague to be useful. My approach would not permit the current approach to software patents.

[73] In the United States, there is virtually an absolute ban on unlicensed experimental use. Peritz, 'Freedom to Experiment: Toward a Concept of Inventor Welfare' (n 67 above).

[74] Peritz, 'Freedom to Experiment: Toward a Concept of Inventor Welfare' (n 67 above).

[75] The public's future interest is a reversion to the transferor in the constitutional sense that patents are not common law property rights but rather statutory grants for a term of years that reserve a reversionary interest in the public.

Each stage of the patent life cycle reflects a different sort of competition, each one conditioned by the regime's rules and policies, which express resolutions of the tensions in means and ends. As discussion of the *Griffith* and *Kewanee Oil* decisions has shown, resolving the tension in ends by (not) privileging the knowledge benefit over the material benefit can have dramatic consequences. In similar fashion, taking competition as the presumptive starting point for adjudication of patent rights can also have powerful effects. For instance, in making it more difficult to obtain an injunction remedy for patent infringement, the Supreme Court in *eBay* weakened the patent holder's property right to exclude and thereby opened the market to increased competition in the second stage of the patent life cycle. Similarly, when the *KSR* decision raised the standard of non-obviousness for combination patents in the patent prosecution phase to change the conditions of competition in ideas, it effectively cut back the scope of exclusionary rights and increased commercial competition in the second stage. And when Justice Breyer in his *Metabolite* dissent explicitly called for protection of the public knowledge benefit by limiting the subject matter of patent rights, the intended effect was to change the prosecution stage's conditions for competition in ideas and thereby extend the reach of public access to subject matter that would otherwise fall under the private control of patent holders.

These recent cases present not only a common dynamic of tensions, but a common resolution. They reflect an underlying commitment to competition policy expressed in the patent regime. Their approach resonates with the political economy described in Justice O'Connor's *Bonito Boats* opinion for a unanimous Court. That is not to say this sample of opinions provides enough data to infer a new orthodoxy in patent jurisprudence, one that recognizes the illogic of an IP economics founded on incentive theory. Indeed, Justice O'Connor's opinion itself presents the baseline of free competition as the necessary condition for an incentive theory of patents. But the sample is enough to say there is an emergent strain of patent jurisprudence that reveals a preference for competition policy, an emergent strain of patent jurisprudence that converges with a residual IP economics of competition, an economics that does not depend on an unfounded incentive theory as the logic for privileging exclusionary rights to promote economic progress.

D. Patents as a Competition Regime: Some Consequences

This section concludes the chapter by suggesting some additional changes that would result from extending the patent regime's emergent jurisprudence and the residual economics of competition.

The recent Supreme Court decisions discussed offer examples of what complexity theorists call the 'butterfly effect'—a small change in initial conditions that produces a radical change in system behaviour. It takes its name from the familiar image of a butterfly in New York City's Central Park, a butterfly whose fluttering wings alter the course of an entire weather system in the Amazon rain forest.[76] A presumptive shift in patent jurisprudence to competition policy is a small change, a change well within the traditional view that both exclusionary rights and free competition drive economic progress. The same can be said for privileging the public knowledge benefit over the material benefits anticipated from the patent regime. These changes in initial conditions are incremental, not radical. They are small but, as recent decisions demonstrate, they can effect sharp and surprising turns in patent policy.[77]

Further changes, large and small, could come of this shift in initial conditions, a procedural shift in patent jurisprudence to the presumption that free competition promotes progress, a shift supported by the residual economics. Each stage of the patent life cycle would be understood as reflecting a baseline of competition, a primary commitment to the public knowledge benefit,[78] a narrowly targeted version of incentive theory, and in sum a patent policy that serves the constitutional purpose of promoting progress. Here are a three further instances of such changes, the last one extended into a specific example derived from the EU *Microsoft* case.

The first involves the 'experimental use' defence to patent infringement. Almost 25 years ago, the Federal Circuit transformed unauthorized experimental use of another's patented invention into patent infringement. The rationale lay in a questionable extension of the already questionable logic of incentive theory. The court determined that a patent holder's power over the invention should extend beyond commercial profit to control of its every use. Why? The court began by attributing a 'business interest' to everyone from garage tinkerers to research scientists, a business interest that was itself seen as endangering the incentive value of patents. An unlicensed researcher could overcome this powerful presumption of a business interest only when the purpose was literally the 'idle curiosity' of a

---

[76] This effect can be called radical incrementalism, meaning that small differences in the initial condition of a dynamic system may produce large variations in its long-term behaviour. The concept of sensitive dependence on initial conditions was developed by French mathematician R Thom, *Structural Stability and Morphogenesis: An Essay on the General Theory of Models* (Reading, Mass: Benjamin, 1975) and was popularized later as Catastrophe Theory in EC Zeeman, *Catastrophe Theory,* (April, 1976) Scientific American 65. It was a precursor to Chaos Theory and Complexity Theory. On Complexity Theory and dynamic efficiency, see Peritz, 'Dynamic Efficiency' in A Cuccinota, R Pardolesi, & R Van den Bergh (eds) *Post-Chicago Developments in Antitrust Law* (Aldershot: Edward Elgar Publishing, 2002) 108 fn 30 and accompanying text.

[77] The shift in IP economics, however, would not be perceived as small. This effect is an extreme form of the tipping phenomenon derived from mathematician René Thom's Catastrophe Theory: a sudden and irreversible change in direction from a preceding course that appeared steady and reversible. Examples include stock market volatility, fight and flight reactions to danger, and the last straw.

[78] Of course a primary commitment to the public knowledge benefit would call for reconsideration of the *Kewanee Oil* decision. For discussion, see Peritz, 'Patents and Competition: Toward a Knowledge Theory of Progress' in LM Genovesi (ed) *Intellectual Property Rights and Market Power* (2009) (selected papers from ATRIP annual conference, Parma, Italy, September 2006).

'dilettante affair'.[79] Since the doctrine's announcement, not one published decision has reported a successful 'experimental use' defence to patent infringement.

The demise of the traditional privilege to engage in unauthorized experimental use of a patented invention is another instance of the propertization trend that has been expanding IP protection in the United States. It is a particularly harmful instance because experimental use is perhaps the most important form of competition during the patent term. If unauthorized experimentation were seen instead as presumptively competitive conduct, then the patent holder would be required to prove actual commercial injury and public harm, all of which results not from imagined intentions but from actual commercial conduct, from making, using, and selling. In short, a viable 'experimental use' defence would not harm patent holders' legitimate interests in exclusive rights to commercial profit during the patent term.

Moreover, the current stranglehold on unlicensed experiment disserves the public interest in three ways. First, it in effect extends the 20-year patent monopoly by the time necessary for rivals to engage in research and development of products for offer on the market.[80] Second, the current regime empowers patent holders to control too much of follow-on research, a power inconsistent with the unlimited availability of improvement patents to all who meet the statutory requirements. Patent's open door policy for follow-on research is in sharp contrast to the Copyright Act's treatment of derivative works, whose protection is available only to the holder of the underlying copyright.[81] Third, the patent holder's control over research also channels and restrains the production of new knowledge intended to replenish the public domain. More widespread competition and cooperation in research during the patent term would produce public benefits by lowering the costs, expanding the field of improvement patents, opening the production of new knowledge, limiting the patent term to its statutory boundary, and, if relevant, serving the national interest by bringing the United States in line with most of the rest of the world, to which unlicensed research activities likely immigrate to escape the harsh US regime.

My second example involves purified forms of naturally occurring substances. Product patents have been granted for them regularly since an early 20th-century decision, which affirmed a grant for the purified hormone adrenalin on the ground that it was 'a new thing commercially and therapeutically'.[82] While this rationale emerged from a focus on commercial markets, the actual effects were much broader

[79] *Roche Prods v Bolar Pharma Co*, 733 F2d 858 (Fed Cir 1984) (citing as most persuasive precedent *Pitcairn v US*, 547 F2d 1106 (Ct Cl 1976)). This expanded view of patent rights is a natural result of viewing them through the prism of property logic.

[80] Compare *Brulotte v Thys Co*, 379 US 29, 32 (1964) (licensing agreement extending beyond patent term per se violation of federal patent law); *Pitney Bowes, Inc v Mestre*, 701 F2d 1365 (11th Cir 1983) (same).

[81] The current treatment of derivative works also reflects overprotection, in this author's view.

[82] *Parke-Davis v Mulford*, 189 F2d 95 (SDNY 1911).

because of the standard scope of protection afforded product patents in the United States—the product patent practically encompassed the very idea of purified adrenalin insofar as it included not only the product but its equivalents for all uses not only known at the time but also discovered later.[83]

An approach beginning with the presumption of competition as patent's baseline would begin by confining the scope of protection to what was actually invented—the new process of purification and the method of using purified adrenalin. Beyond the specific process and method of use, open competition would prevail. The product and with it the idea of purified adrenalin would be freely available in the public domain.[84]

In a very recent decision that has attracted attention, a federal court in New York City ruled that isolated and purified DNA was not patentable because it lacked 'markedly different characteristics' from native DNA.[85] The plaintiffs referred to the *Adrenalin* case in arguing, 'Isolated DNA molecules should be treated no differently than other chemical compounds for patent eligibility.' But the court rejected the reference by distinguishing DNA from other chemical compounds in the body—while adrenalin and other compounds necessarily convey information, DNA encodes an entirely different kind of information, not about its own molecular structure involving its own biological function but rather about its biological function of directing the synthesis of other molecules in the body. This distinction was dispositive because, in the court's view, the isolated and purified DNA carried precisely the same information as the native DNA and thus lacked 'markedly different characteristics'.

The court recognized the importance of the case in stating,

> The widespread use of gene sequence information as the foundation for biomedical research means that resolution of these issues will have far-reaching implications, not only for gene-based health care and the health of millions of women facing the specter of breast cancer, but also for the future course of biomedical research.

In short, competition and cooperation in gene research would not be controlled by patent holders. Despite the court's special treatment of DNA, the same could be said

[83] Only patents for improved or new production processes or methods of use were possible. The result would be blocking patents.

[84] With a targeted incentive theory, the question might be whether it would be good industrial policy to support the exclusionary regime of a product patent for adrenalin and a suitably defined category of naturally occurring substances in order to channel research and development in a direction that is currently neglected under what would otherwise be a regime of open access required by free competition. Since the question would call for a judgment about industrial policy, it would be for Congress to legislate some combination of general standards and specific rules for the Patent Office, which would promulgate guidelines for its examiners, who would provide technological expertise, as they do now, according to guidelines in the prosecution stage of the patent life cycle.

[85] *Association for Molecular Pathology v US Patent and Trademark Office*, SDNY, No 09 Civ 4515, 29 March 2010.

for generally denying product patents for purified forms of all naturally occurring substances.

The last example involves the description and enablement requirement already discussed in the jurisprudence section—here, the requirement as it applies to computer software. Ten years ago the Federal Circuit Court declared that a general functional description satisfies the requirement[86] for software patents. The practical consequence of these cases is lack of adequate description and enablement. The description is insufficient to assure that the claimant actually 'has possession' of the invention rather than simply a general idea about its function. Moreover, enabling a skilled programmer to make or use the software would require flow charts, source code, and the detailed descriptions that annotate modules, descriptions that computer programmers customarily include as documentation for others who subsequently need to understand, change, or fix the source code. Both protocols[87] and programmer comments are embedded in source code listings, while protocols also appear in software documentation. Why is a general functional description enough for the Federal Circuit Court and, thus, for the Patent Office though it does not meet industry standards? Because, according to the court, conversion of functional description into source code is 'a mere clerical function to a skilled programmer'.[88]

The court's rationale rings hollow for anyone who has actually designed or written operating systems or complex applications software. Indeed, no judge sitting on the Federal Circuit could have taken the stated view after having any actual experience in the field. The author of this chapter spent some years designing and writing such software, and the experience evidences the reality that conversion of systems design specifications to source code is often challenging work that involves much more than mere clerical function. Nonetheless, the Federal Circuit in its nescience requires only a general description of the software process.

While general information about software function has some limited value, its satisfaction of the patent disclosure requirement creates two problems. First, general claims and descriptions produce software patents that are too broad and, as a result, foreclose too much competition as functional equivalents. This problem includes treatment of business method inventions, which are typically embodied in software. Second, there is insufficient information flow for subsequent inventors. The combination is deadly—broad patent rights and little public information about them. This situation is exacerbated by the acknowledged difficulty in locating and identifying prior art in the category of computer software.

[86] Patent Act § 112. The court acknowledged that more might be required in special cases.

[87] A protocol is a standard procedure and format that two computers or other devices must understand, accept, and use in order to communicate with one another. Examples include network log on procedures and html format.

[88] *Northern Telecom, Inc v Datapoint Corp*, 908 F2d 931, 942 (Fed Cir 1990) (citing *Re Sherwood*, 613 F2d 809, 817 fn 6 (CCPA 1980)).

If the patent regime is intended to encourage learning from prior art and thereby foster competition by invention, the level and quality of information must be improved. The current requirement of a general process description requires only a low level of information and, in consequence, erects a barrier to further invention, a barrier that benefits the patent holder by keeping rivals out, rather than an information flow that benefits society by enabling others to improve and surpass the invention.

Moreover, a surprising anomaly arises—despite the patent requirement of disclosure, specific code modules in patented software can be protected as trade secrets. The patent requirement of disclosure and anomaly of trade secrecy in its midst can co-exist because of the Federal Circuit's general description requirement. Recognizing the public knowledge benefit of competition in ideas during the prosecution stage of the patent life cycle would call for a more demanding description and enablement standard for computer software. The change would call for disclosure of the source code and system documentation that industry practices recognize as needed to enable subsequent work on the software.

Moreover, in the broader ambit of innovation policy, a proper patent requirement to disclose would obviate the need for antitrust litigation to resolve some issues of interoperability and disclosure through compulsory licensing of patented software. The EU *Microsoft* antitrust case provides a handy example because it involved computer software for which Microsoft asserted patent and trade secret protection as defences to antitrust liability for refusals to disclose information rivals needed for the continued interoperability of their software with Microsoft WINDOWS for PCs. How would the analysis of IP claims proceed if antitrust were no longer seen as the sole source of a conflicting competition policy? And if patent policy were no longer seen as driven exclusively by property rights to exclude competitors? Instead, let's look at policies of exclusion and access not in opposition to one another, not in antithetical domains of patent and antitrust, but rather in a joint venture to set the conditions for relationships of competition and cooperation.[89]

[89] Wesley Hohfeld made the fundamental point that property rights can be best understood as relations between persons with respect to a thing rather than between a person and a thing. W Hohfeld, 'Some Fundamental Legal Conceptions as Applied in Judicial Reasoning' (1923) 23 Yale LJ 16; JW Singer, 'The Legal Rights Debate in Analytical Jurisprudence from Bentham to Hohfeld' (1982) Wisconsin L Rev 975. This chapter takes up this well-known relational conception of property rights and extends it to competition and cooperation. The extension is indebted to the voluminous literature about the interplay between competition and cooperation—whether literature relating directly to innovation and efficiencies or more broadly throughout the social and management sciences. Much of the literature is informed by game theory, from the simple prisoners' dilemma to complex multi-layered iterative games. For a brief introduction to a game theory approach to parallel commercial conduct, see Peritz, 'Doctrinal cross-dressing in derivative aftermarkets: Kodak, Xerox and the copycat game' (2006) 51 Antitrust Bulletin 287 and sources cited therein.

In the actual *Microsoft* case, the Court of First Instance (CFI), now the General Court, affirmed the Commission judgment that Microsoft abused its dominant position in the market for PC operating systems by leveraging WINDOWS' dominance into the market for work group server operating systems. The CFI concluded that Microsoft wrongfully applied this leverage in refusing to disclose to rivals in the server market information they needed for continued interoperation with work group PCs running WINDOWS.[90] The information included interface protocols and an 'Active Directory' ('Directory'), which organized the protocols in an arguably original way that allowed Microsoft's server software to interoperate smoothly and efficiently with WINDOWS. Microsoft claimed that patent and trade secret protection allowed them to deny access to this information.[91]

The CFI began its analysis by resolving what it treated as a conflict between competition policy and intellectual property rights. The conflict was resolved as follows—first, the Court simply assumed that Microsoft had patent and trade secret protection of the protocols and the Directory, despite some hesitation over the strength of the claims. Second, the Court affirmed the Commission's determination that, under exceptional circumstances, competition policy can trump patent and trade rights. The exceptional circumstances turned on the question of access to an indispensable asset controlled by a dominant firm, here WINDOWS protocols controlled by Microsoft. The protocols were deemed an essential facility for competition in the market for server operating system software. The CFI concluded that in the special circumstances competition policy trumped Microsoft's exclusionary rights in the absence of objective justification. It followed that the proper remedy was a decree compelling Microsoft to disclose the information to their competitors.

The effect was a sharp change in the competitive and cooperative relationships between Microsoft and its rivals. Until the decision, both competition and cooperation were restrained by Microsoft's asserted property rights to refuse disclosure of interoperability information, an exclusionary right asserted under the aegis of trade secret and patent ownership. As is often the case, competition on the merits was not possible without some cooperation between participants. The CFI decision applied the competition

---

[90] The case also involved distribution of WINDOWS MEDIA PLAYER. As Professor Steven Anderman put it, four threads run through the CFI analysis:

> '(1) The significance of the findings of "indispensability" of the interface protocols to interoperability in the "second market." (2) The significance of the finding that there was a "risk" of elimination of competition in the second market. (3) The "exceptional circumstances" in which competition law will find that a refusal to license an IPR will be an infringement of Art. 82 [now Article 102 TFEU]. (4) The finding of an absence of objective justification.' Anderman, Pro-Consumer Efficiencies in Antitrust Law and Practice (26 October 2007) LUISS University, Rome.

[91] Case T-201/04 *Microsoft v Commission* [2007] ECR II-3601 (interoperability protocols for some front-end server software) ('*Microsoft*'); see also, *Lockwood v American Airlines, Inc*, 107 F3d 1565 (Fed Cir 1997).

policy of the then Article 82 EC (now Article 102 TFEU) to compel Microsoft to cooperate with rivals in order to allow competition on the merits of the server software rather than on the advantage derived from Microsoft's ownership of exclusive access to an essential component in PC networks comprising numerous components.

So much for EU competition policy and exclusionary rights. In the US, the outcome would have been in doubt. First, the Supreme Court has in effect gutted essential facility doctrine as a basis for antitrust liability. Second, turning on its head the EU view, patent rights trump US antitrust policy. And third, as a general matter, US courts are indisposed toward granting compulsory licences, seeing them as insults to the institution of private property. In sum, US antitrust is a weak voice for expressing competition policy.[92]

Now comes the emergent view of the US patent domain as a distinct competition regime and, in consequence, a more rigorous requirement of description and enablement. How would this play out on the bare-bone facts of the EU *Microsoft* antitrust case?[93]

As already discussed, the current description and enablement requirement for computer software calls only for a general description of the process. And so Microsoft is not currently required to specify the protocols or the Directory. In short, the information would likely be secret. But if the requirement were reformulated in consonance with the emergent view, then both the protocols and the Directory would be disclosed—the protocols as necessary to enable skilled practitioners to use the software and the Directory as reflecting the protocols' best mode of use.[94]

A more demanding description and enablement would have two effects. First, it would improve the information flow during the patent term, the public knowledge benefit expected from the prosecution stage's competition in ideas. Second, it would define more clearly and more narrowly the metes and bounds of the patent monopoly. In relational terms, it would expand the patent holder's obligation to cooperate with rivals and other interested parties, a third party obligation enforced during the prosecution stage of the patent life cycle. As a result, it would change the conditions of competition during the patent term by having given rivals the information needed

[92] See, eg, RJR Peritz, 'The Microsoft Chronicles' in L Rubini (ed) *Microsoft on Trial: Legal and Economic Analysis of a Transatlantic Antitrust Case* (Aldershot: *Edward Elgar Publishing,* forthcoming 2010); Peritz, 'Microsoft e il flusso di informazioni' (2007) 9 Mercato, Concorrenza, Regole 523 (Italian translation by Andrea Giannaccari).

[93] For an expansive introduction to patent, copyright, trade secret, and trademark as competition regimes, see Peritz 'Competition Policy and its Implications for Intellectual Property Rights in the United States' (2006) (n 4 above).

[94] For a demanding approach to the description requirement, see *The Gentry Gallery v The Berkline Corp*, 134 F3d 1473 (Fed Cir 1998). For criticism, see *Moba v Diamond Automation, Inc*, 325 F3d 1300, 1323 (Fed Cir 2003) (Rader, J, dissenting).

to compete on the merits. In sum, the relational changes would track those of the CFI decision in the EU *Microsoft* decision.

But the relational changes would not be identical. First, the patent resolution would require no litigation for disclosure. Second, it would involve no licensing, no judicial oversight, and thus no licensing fee, no bargaining or other transaction costs. Third, however, the use value of the information would depend on the scope of experimental use permitted. In the United States, the use under current law is for all practical purposes forbidden. But with a viable experimental use doctrine properly understood as competition during the patent term, experimentation short of commercial use would be permitted. Finally, unlike the compulsory licence in the EU case, patent disclosure of the Directory as the best mode of organizing the protocols would not necessarily permit its commercial use. Certainly second comers could use the protocols commercially because they lack invention; but if the Directory is a non-obvious invention, patent rights would block its commercial use. Here, the patent regime's compelled cooperation between Microsoft and its rivals should not include commercial use of the Directory, if determined a non-obvious invention, because commercial competition on the merits is possible without access. Indeed competition in the development of more efficient or otherwise superior protocol organization in other directories holds the promise of technological progress.

This final example has shown how the patent regime can be understood as an instrument of economic progress that shapes relationships of competition and cooperation. The analysis begins with the presumption that free competition promotes economic progress. Any policy or adjudication that would expand the scale or scope of patent rights requires evidence of its progressive value. What justifies this shift from the current patent regime's presumption that exclusionary rights promote progress? It begins with recognition of a policy stalemate that derives from the indeterminacy of incentive theory as the basis for preferring either patent protection or free competition as the superior engine of progress. This stalemate is broken by free competition's superior distributional effects, superior because competition generates more allocatively efficient outcomes and, with them, conditions more conducive to future inventive activity. At the same time, recent Supreme Court jurisprudence reflects an emergent strain of patent doctrine that recognizes an internal competition policy—what an earlier decision by a unanimous Court called 'a baseline of free competition'. More broadly, adopting a baseline of competition would change the current US view of IP and antitrust as antithetical regimes, as a binary opposition between monopoly and competition, between exclusion and access. What would emerge is a more progressive and more functional view of IP and antitrust as two intertwined regimes comprising policies of both exclusion and access, two sets of rules and policies that set the conditions for relationships of competition and cooperation to promote the progress of knowledge and industrial technology.

Conclusion

This chapter has sketched the orthodox view of the patent regime as founded on exclusionary rights to promote progress, its jurisprudence, and its dependence on a failed incentive theory, as well as an emergent view expressed in the jurisprudence and in a residual economics that converge to support the reconceptualization of patent protection as a competition regime. The emergent jurisprudence echoes an opinion by Justice Sandra Day O'Connor written some 20 years ago for a unanimous Supreme Court, an opinion that described the foundation of patent policy as a 'baseline of free competition'.

While the chapter adopts the baseline of free competition, it otherwise diverges from Justice O'Connor's opinion insofar as the chapter's ensuing analysis reflects the failure of incentive theory as the economic logic for patent protection, a failure that is not acknowledged in the opinion or in today's mainstream jurisprudence. Still there is a residual economic logic that is surprisingly straightforward despite its absence in the mainstream literature: while the dynamic efficiency effects of both free competition and patent rights are indeterminate, their distributional effects point toward free competition. Why? Because patents that actually have economic value produce monopoly prices and, with them, welfare losses in both static and dynamic terms. In consequence not only consumers but subsequent inventors are better off under a regime of free competition because it gives inventors open access to new information. The result is improved conditions for subsequent invention. In this light, a rule or policy that would strengthen patent rights should first be shown to promote greater progress than would otherwise occur. Yet patent protection can serve the public interest as a sharp instrument for targeted industrial policy though it fails as a magic potion for promoting economic progress. But even with patents as sharp tools of industrial policy, risks of failure and unintended consequences call for careful analysis to overcome the presumption that free competition better serves the public interest.

3

INTELLECTUAL PROPERTY RIGHTS AND COMPETITION RULES, A COMPLEX BUT INDISPENSABLE COEXISTENCE

*Cecilio Madero Villarejo and Thomas Kramler**

Introduction

The debate on the relationship between intellectual property rights (IPRs) and competition rules and their respective roles for innovation is intrinsically linked to broader economic discussion about how markets work and the role that government should take. The long-running debate on how monopoly power interacts with innovation is part of this larger controversy on the right level of state intervention.

We believe that there should be no antagonism between the two sets of rules as both serve the same purpose, namely to ensure that true innovation is rewarded and to facilitate the innovation process.[1]

Competition rules have to be seen as a safety net that only intervenes in cases where the market does not deliver the innovative products or services one would expect. This, however, leaves the question of when to make use of this safety net in terms of striking the right balance between cases that merit intervention by a competition authority and cases that do not.

---

* The views expressed do not necessarily reflect those of the European Commission.

[1] See also Commission Guidelines on the applicability of Article 81 of the EC Treaty (now Article 101 TFEU) to horizontal cooperation agreements [2001] OJ C3/2 para 7:

> Indeed, both bodies of law share the same basic objective of promoting consumer welfare and an efficient allocation of resources. Innovation constitutes an essential and dynamic component of an open and competitive market economy. Intellectual property rights promote dynamic competition by encouraging undertakings to invest in developing new or improved products and processes. So does competition by putting pressure on undertakings to innovate. Therefore, both intellectual property rights and competition are necessary to promote innovation and ensure a competitive exploitation thereof.

Competition usually induces companies in a market with a given technology to offer the best products at the lowest prices, ie to realize what economists call static efficiency. However, it is innovation which causes product markets to change as improved products and production processes are introduced. This leads to greater consumer satisfaction and lower production costs.

Some economists, in the footsteps of Joseph Schumpeter, claim that innovation is spurred by monopoly as monopoly profits may fund further research and development. The 'Schumpeterians' argue that by attacking monopoly and preventing market power from arising, competition policy may have positive effects on static efficiency but at the same time undermines dynamic efficiency and innovation.

The 'Schumpeterian' view has been contradicted by Nobel prize winner Kenneth Arrow and other economists, who have put forward a number of reasons why competition may provide more incentives for innovation than monopoly. Arrow contends that in favouring monopolies, Schumpeter underestimated the incentives for innovation that competition can offer, for example by not recognizing that firms under competitive pressure are less complacent and know that inventing a new product is their best strategy for taking over their competitors' market share. In a competitive market new entrants know that innovation can help them to succeed in the market.[2] Or as the American analyst David Isenberg put it 'The milk of disruptive innovation does not flow from cash cows.'

It is against this theoretical economic background that the Commission develops its case-practice and seeks to strike the right balance between the protection of intellectual property and the protection of the competitive process to the benefit of the consumer. It should be stressed that an IPR does not *ipso facto* convey market power on its holder. However, under certain circumstances an intellectual property right may enable those possessing such an exclusive right to charge monopoly prices or to otherwise restrict competition.

It is particularly in the area of enforcement of Article 102 TFEU (ex Article 82 EC) where the Commission is regularly faced with the question of how to strike the aforementioned balance correctly. This chapter focuses on IPRs and market power and explores some of the challenges the Commission faces by reference to our policy with regard to refusals to supply and to competition issues related to standard setting.

---

2 For a summary of the debate see RJ Gilbert, 'Competition and Innovation', <http://repositories.cdlib.org/iber/cpc/CPC07-069> Competition Policy Center, Institute of Business and Economic Research, UC Berkeley, 27 January 2007, accessed 2 September 2010.

A. The Commission's Policy with Regard Refusal to Supply

The Commission has in 2009 adopted a Communication providing 'Guidance on the Commission's enforcement priorities in applying Article 82 of the EC Treaty (now Article 102 TFEU) to abusive exclusionary conduct by dominant undertakings'.[3] This Communication makes clear that when setting its enforcement priorities, the Commission starts from the position that any undertaking, whether dominant or not, should have the right to choose its trading partners and to dispose freely of its property. However, it may be an abuse of a dominant position to refuse to supply another company with an input this company needs in order to compete. This 'input' may be access to an infrastructure, for instance a port, but may also be access to interface information or intellectual property. The standard of proof for finding an abuse in such a situation is high, precisely to protect investment and innovation incentives. An obligation to supply may undermine the incentive to invest or innovate by the dominant company.

It is for these reasons that the Commission has only found a refusal to supply to be an abuse of a dominant position in exceptional circumstances. The *Microsoft* case, in which the relationship between intellectual property protection, innovation, and competition featured prominently, is a good example in this respect.

The *Microsoft* case

The 2004 Decision

On 24 March 2004 the Commission adopted a decision pursuant to Article 102 TFEU concluding that Microsoft had abused its dominant position in the PC operating system market by (i) refusing to provide interoperability information necessary for competitors to be able to effectively compete in the work group server operating system market and (ii) tying its Windows Media Player with the Windows PC operating system.[4] This chapter will deal exclusively with the refusal to supply interoperability information.

The Commission's findings in the 2004 Decision rest on Microsoft's exceptional, super-dominant[5] market position on the PC operating system market. Microsoft's market share in this market is between 90 and 95 per cent, and it has enjoyed these high market shares for many years. Microsoft's market position is protected by the applications barrier to entry in the market which derives from the fact that the vast

---

[3] [2009] OJ C45/7.

[4] *Microsoft* (COMP/C-3/37.792) [2007] OJ L32/23, (hereinafter 'the 2004 Decision').

[5] Advocate General Fennelly refers to 'superdominance' with respect to undertakings with an 'overwhelming dominance verging on monopoly' in his opinion in Joined Cases C-395/96P & C-396/96P *Compagnie Maritime Belge Transports and others v Commission* [2000] ECR I-1365 ('*Cewal*') [2000] ECR I-1365, opinion of Advocate General Fennelly, para 137.

majority of PC applications are written for the Windows PC operating system platform.

The 2004 Decision found that Microsoft abusively leveraged its PC operating system dominance onto the work group server operating system market through withholding essential interoperability information necessary for competing work group server operating systems to 'communicate' with the Windows PC and server operating system.

The 2004 Decision established that access to the interoperability information was indispensable to compete viably in the work group server operating system market and that Microsoft's refusal risked eliminating competition in the work group server operating system market to the detriment of innovation and consumer choice.

The 2007 judgment

On 17 September 2007, the Court of First Instance (now General Court) rendered judgment with regard to Microsoft's application for annulment against the 2004 Decision.[6] The Court upheld the Commission's findings with regard to Microsoft's refusal to supply interoperability information and the tying of Windows Media Player.[7] The Court's reasoning will be summarized in this section.

In its application for annulment with regard to the Commission's findings on the refusal to supply interoperability information, Microsoft relied essentially on the argument that it would illegally be required to grant a licence to its intellectual property rights.[8]

Although the 2004 Decision did not take any position as to whether the interoperability information was indeed covered by IPRs, the Court proceeded on the presumption that the interoperability information was covered by IPRs or constituted trade secrets.[9]

The Court reiterated well-established case-law[10] according to which,

> . . . [T]he refusal by an undertaking holding a dominant position to license a third party to use a product covered by an intellectual property right cannot in itself constitute an abuse of a dominant position within the meaning of Article 82 EC

---

[6] Case T-201/04 *Microsoft v Commission* [2007] ECR II-3601 (hereinafter 'the judgment').

[7] The Court, however, annulled Article 7 of the 2004 Decision which foresees the establishment of a monitoring mechanism, including a monitoring trustee, to oversee Microsoft's compliance with the 2004 Decision, insofar as Article 7 entails the delegation of powers of investigation to the monitoring trustee and orders Microsoft to bear the costs of the monitoring trustee.

[8] Microsoft also argued that it had not in fact refused to supply the interoperability information and that the 2004 Decision was incompatible with the TRIPS Agreement. Both pleas were rejected by the Court (see paras 713–76 and 777–813 of the judgment).

[9] Para 289 of the judgment.

[10] Case 238/87 *Volvo v Veng* [1988] ECR 6211; Joined Cases C 241/91 P and C 242/91 P *RTE and ITP v Commission* [1995] ECR I-743 ('*Magill*'); Case C 418/01 *IMS Health* [2004] ECR I-5039.

> [now Article 102 TFEU]. It is only in exceptional circumstances that the exercise of the exclusive right by the owner of the intellectual property right may give rise to such an abuse.[11]

The Court noted that,

> ... [T]he following circumstances, in particular, must be considered to be exceptional:
>
> 1. in the first place, the refusal relates to a product or service indispensable to the exercise of a particular activity on a neighbouring market;
> 2. in the second place, the refusal is of such a kind as to exclude any effective competition on that neighbouring market;
> 3. in the third place, the refusal prevents the appearance of a new product for which there is potential consumer demand.[12]

The Court then went on to analyse whether these exceptional circumstances were present in the *Microsoft* case.

Indispensability

With regard to the indispensability of the interoperability information, the Court agreed with the Commission that Microsoft was able to impose Windows as the *de facto* standard for work group computing.[13]

The Court therefore concluded that non-Microsoft work group server operating systems had to be capable of interoperating with Windows PC and server operating systems on an equal footing if they were to be marketed viably[14] and that there were no viable solutions to achieve interoperability other than disclosures[15] from Microsoft.

Elimination of competition

As regards the impact of Microsoft's refusal to supply on the competitive situation in the work group server operating system market, the Court noted at the outset that the Commission does not have to wait until there is no competition left in the relevant market before acting under Article 102 TFEU.[16] The Court also emphasized that because of the significant network effects in place in the concerned market the elimination of competition brought about by Microsoft's behaviour would be difficult to reverse.[17]

---

[11] Para 331 of the judgment.
[12] Para 332 of the judgment.
[13] Para 392 of the judgment.
[14] Para 422 of the judgment.
[15] Para 435 of the judgment.
[16] Para 561of the judgment.
[17] Para 562 of the judgment.

The Court fully confirmed the Commission's definition of the relevant product markets as well as its analysis of market data and the competitive situation.[18] The Commission collected a very significant amount of customer evidence showing that it was the artificial 'interoperability advantage' that Microsoft reserved for its products via the refusal to supply that drove Microsoft's rapid gain of market share and prevented other vendors of work group server operating systems from viably competing on the market.

The Court concluded that,

> . . . Microsoft's refusal has the consequence that its competitors' products are confined to marginal positions or even made unprofitable. The fact that there may be marginal competition between operators on the market cannot therefore invalidate the Commission's argument that all effective competition was at risk of being eliminated on that market.[19]

New product/consumer welfare

The Court noted that whether Microsoft's:

> . . . conduct prevents the appearance of a new product on the market falls to be considered under Article 82(b) EC [now Article 102(b) TFEU], which prohibits abusive practices which consist in 'limiting production, markets or technical developments to the . . . prejudice of consumers'.[20]

The Court confirmed the Commission's analysis that the insufficient degree of interoperability with Windows discouraged Microsoft's competitors from developing and marketing work group server operating systems with innovative features to the prejudice of consumers.[21]

The Court also confirmed the Commission's analysis that,

> . . . consumers consider non-Microsoft work group server operating systems to be better than Windows work group server operating systems on a number of features to which they attach great importance.[22]

In the same vein, the Court emphasized that Microsoft's competitors would not be able to clone or reproduce its products solely by having access to the interoperability information.[23]

18 Paras 479–620 of the judgment.
19 Para 593 of the judgment.
20 Para 643 of the judgment.
21 Para 653 of the judgment.
22 Para 661 of the judgment.
23 Para 657 of the judgment.

Therefore, the Court agreed with the Commission's findings that Microsoft's refusal to supply interoperability information limited technical development to the prejudice of consumers within the meaning of Article 102 (b) TFEU.[24]

Objective justification

The Court accepted that even when the above-mentioned circumstances are present, the company that refused to supply a product could objectively justify its conduct. The Court noted that the burden of proof lies with the company invoking the objective justification and it is for the Commission to rebut the company's arguments.[25]

As already pointed out Microsoft's first claimed objective justification for its refusal to supply interoperability information was the fact that the technology concerned was covered by intellectual property rights.[26]

However, the Court rejected this argument as,

> . . . inconsistent with the raison d'être of the exception which that case-law thus recognises in favour of free competition, since if the mere fact of holding intellectual property rights could in itself constitute objective justification for the refusal to grant a licence, the exception established by the case-law could never apply.[27]

The Court also rejected Microsoft's argument that the disclosure of the interoperability information would significantly reduce or eliminate Microsoft's incentives to innovate.[28]

In particular, the Court pointed out that it was,

> . . . normal practice for operators in the industry to disclose to third parties the information which will facilitate interoperability with their products and Microsoft itself had followed that practice until it was sufficiently established on the work group server operating systems market.[29]

The Court also noted,

> . . . that there was ample scope for differentiation and innovation beyond the design of interface specifications. In other words, the same specification can be implemented in numerous different and innovative ways by software designers.[30]

Conclusion on the *Microsoft* case

As evidenced by the quotes from the judgment in the previous section the *Microsoft* case was very fact-intensive and was based on a great deal of analysis of market

[24] Para 665 of the judgment.
[25] Para 688 of the judgment.
[26] Para 689 of the judgment.
[27] Para 690 of the judgment.
[28] Para 701of the judgment.
[29] Para 702 of the judgment.
[30] Para 655 of the judgment.

conditions, effects on consumers, and incentives to innovate. The Commission found that by abusing its dominant position in the PC operating system market Microsoft was able to impose its less performant technology as *de facto* standard. This in turn led to a lock-in of consumers and a denial of consumer choice.

The Commission was very careful to establish that Microsoft's competitors would not be able to duplicate Microsoft's operating system and that the disclosure of the interoperability information would not negatively impact on Microsoft's incentives to innovate in the work group server operating system market. On the other hand, the Commission also established that Microsoft's conduct prevented innovation in the work group server market as products which were valued by consumers were prevented from viably competing with Microsoft's products.

Proof that the Commission's analysis with regard to innovation incentives in this case was correct can be found in Microsoft's own actions and statements following the judgment.

In February 2008 Microsoft announced that it would make interoperability information, including the information it had to disclose pursuant to the 2004 Decision, available on its website free of charge.[31] At the same time Microsoft's CEO Steven Ballmer confirmed the benefits of interoperability disclosures,

> . . . what we are permitting is more innovation around our products, more interoperability, maybe also more potential for third parties to cannibalize what could have been Microsoft business . . . But it is a path we have committed ourselves to because we think it is good for customers and is consistent with our legal obligations.[32]

Following intensive discussions with the Commission, Microsoft published on 16 December 2009 an interoperability undertaking,[33] in which it committed to disclose a large amount of interoperability information related to its main products Windows, Windows server, and Office free of charge on its website. Pursuant to the undertaking Microsoft shall ensure that the information is complete and accurate and shall provide a warranty to that effect. Access to and use of the information shall be subject to no more than a nominal upfront fee and licensing terms which are compatible with open source licences. The warranty agreements which Microsoft offers as part of the undertaking foresee private enforcement including a fast-track

[31] See 'Interoperability Principles: Open Connections, Standards Support, Data Portability' <http://www.microsoft.com/interop/principles/default.mspx> 21 February 2008, accessed 2 September 2010.

[32] As quoted in KJ O'Brien, 'Microsoft declares its troubles with EU over' <http://www.nytimes.com/2008/03/03/business/worldbusiness/03iht-msft.4.10661050.html> International Herald Tribune, New York, 3 March 2008, accessed 2 September 2010.

[33] See 'Microsoft Statement on European Commission Decision' <http://www.microsoft.com/Presspass/press/2009/dec09/12-16Statement.mspx> 16 December 2009, accessed 2 September 2010.

dispute resolution mechanism and a right to liquidated damages in case the warranties are breached.

Microsoft's informal interoperability undertaking thus relies on some elements familiar from the post-judgment implementation of the disclosure remedy imposed by the 2004 Decision.[34] As pointed out by Commissioner Kroes[35] this is a very welcome initiative by Microsoft as enhanced interoperability is crucial for competition in the software industry. The Commission will now carefully monitor the impact of Microsoft's undertaking on the market.

It appears that the Commission's enforcement actions have induced Microsoft to announce a different, more open, policy with regard to interoperability. But as 'actions speak louder than words' it remains to be seen whether these announcements result in a real change of course.

Finally we should stress that in view of the specific circumstances of the *Microsoft* case it is highly misleading to derive from this case, as some commentators have done, that the 'floodgates will open after *Microsoft*' and that refusal to supply cases will become the norm rather than the exception in the Commission's enforcement practice.

We are sure that antitrust intervention in refusal to supply cases will remain exceptional as cases will be decided on the basis of rigorous assessment of specific facts and in view of their impact on consumers and innovation.

B. Standard Setting

Another area where innovation or hindrance to innovation and the relationship between IPRs and competition rules is at stake is standard setting. The Commission as an antitrust regulator is confronted with an increasing number of cases on alleged anti-competitive effects of IT standards.

Agreements to set standards may be either concluded between private undertakings or set under the aegis of public bodies.

Under competition rules, insofar as a standard is set by agreement among competitors on a given market to choose one solution over another, consumer choice and other competitive restraints are foregone. In light of this, in order to achieve the

[34] See 'Microsoft case: Implementation of the Decision' <http://ec.europa.eu/competition/antitrust/cases/microsoft/implementation.html> accessed 2 September 2010.

[35] See 'Neelie Kroes, European Commissioner for Competition Policy, Your Internet, Your Choice: Microsoft web browsers decision, Opening remarks at press conference Brussels 16 December 2009'. <http://europa.eu/rapid/pressReleasesAction.do?reference=SPEECH/09/582&format=HTML&aged=0&language=EN&guiLanguage=en> accessed 2 September 2010.

benefits that derive from standards, standard setting must be subject to a number of principles.

For example, access to the standard must be given on fair, reasonable, and non-discriminatory terms to any company which wishes to use it for the manufacturing of its goods or the provision of its services. Competition authorities must also remain vigilant in making sure that the standardization process is not instrumentalized by single companies with market power for anti-competitive purposes. The transparency and openness of the standard-setting process is therefore of utmost importance.[36]

Standards may confer market power on a company which it may otherwise not have possessed, in particular where the implementation of a standard requires access to IPR or know-how which only one company controls.[37] This is true for both formal and *de facto* standards.

For example, the owner of a proprietary technology essential to a standard might hide the fact that it holds essential IPRs over the standard being developed, and only start asserting such IPRs after the standard has been agreed and other companies are 'locked in' to using it.

In such a case switching to a new standard might theoretically be possible. However, in practice, it is likely to entail significant costs. Once path-dependent investments have been made in the technologies chosen for the standard, it becomes extremely difficult for technologies which had not been chosen to compete.[38]

This 'patent ambush' allows the company to extract an artificially inflated *ex post* value for its patented technology. It is a lucrative trick, but one that is hard to admire and that is objectively bad for consumers. There is therefore an important pro-competitive rationale behind requiring disclosure of patents and patent applications in the framework of standard setting before a standard is set in order to avoid 'patent ambushes'.

Many standard-setting organizations therefore insist that all IPR holders commit to licensing on fair, reasonable, and non-discriminatory (FRAND) terms and disclose relevant IPR before a standard is adopted.

However, some IPRs still might not be disclosed during the standard-setting process and the IPR holders might subsequently charge unreasonable royalties, or one person's FRAND terms might become another person's alleged exploitative abuse of a dominant position.

[36] See Commission Guidelines on the applicability of Article 81 of the EC Treaty (now Article 101 TFEU) to horizontal cooperation agreements [2001] OJ C3/2 paras 159 *et seq.*

[37] See C Madero and N Banasevic, 'Standards and Market Power' <https://www.competitionpolicyinternational.com/standards-and-market-power/> Global Competition Policy (now Competition Policy International Antitrust Chronicle) 16 May 2008, accessed 2 September 2010.

[38] See *Rambus* (COMP/38.636) [2009] OJ C133/16.

Both scenarios, ie the non disclosure of essential IPRs during the standard-setting process ('patent ambush') and a failure to respect a FRAND commitment could constitute a breach of competition rules, notably Article 102 TFEU, under the assumption that the holder of the IPR possesses market power by virtue of his technology being included in the standard.

An essential question in both scenarios is to determine the 'but for world'. That is to say to establish what the company would have been able to charge absent the inclusion of its technology in the standard. Such an *ex post* assessment of the reasonableness of royalty rates is, however, difficult to carry out.

It is therefore sensible for standard-setting bodies, with input from competition authorities where appropriate, to look at ways that markets can be made to operate better, avoiding the need for *ex post* control.

One way *ex post* interference of competition authorities in standard-setting processes and pricing of IPRs can be avoided are *ex ante* disclosure mechanisms.

What is meant by *ex ante* disclosures? Let us give an example. If one would imagine that at the point at which a standard is under discussion, there are four viable, roughly comparable alternative technologies each competing to be selected as the essential technology for the standard. If the selection is made on the basis of both the price and quality of the relevant technologies, then the price is competed down to the market level *ex ante*.

Broadly speaking, such mechanism could work as follows. One potential essential technology holder declares his rights before the standard is set, and at the same time, specifies the maximum royalty he would charge if his technology is selected for the standard. Then, another potential essential technology holder may declare his rights and the associated maximum royalty that he would charge. There is therefore competition on the basis of technology and price before the standard is selected where the price is competed down to the market level.

There have been increasing calls for standard-setting bodies to allow and implement *ex ante* selection of this kind. There is no reason, as some claim, that this is in itself anti-competitive because it risks leading to some kind of inappropriate price collusion.

Of course, there is a need for vigilance if such *ex ante* schemes are merely some kind of artificial front for illegal price-fixing. But *in and of themselves*, such disclosure schemes are not anti-competitive, but rather have the potential to bring strong pro-competitive benefits by competing the price down to the market level before the standard is set.

All this being said, we do not believe that it is the role of a competition regulator to prescribe actively what rules standard-setting bodies must adopt. Different rules may be appropriate for different bodies and sectors, and industry will generally know best what works. However, we can give guidance as to what may or may not be problematic from an antitrust perspective to ensure that industry can make the most informed choices.

Conclusion

To conclude we should reiterate that we do not see an antagonism between IPR and competition rules. Competition rules should serve as a safety net in cases where the IPR system does not deliver the desired results in terms of innovation and consumer welfare.

If applied with the necessary restraint, competition policy can contribute to stimulate innovation in certain markets and cases.

In the Commission we recognize that we need truly to understand markets before we can make sound decisions about them. We have to ground our conclusions in strong economic analysis and that depends on capturing the true state of a market. But once a real competition problem has been identified we should not shy away from acting swiftly in order to remove obstacles to innovation.

4

THE ECONOMICS OF ANTITRUST AND INTELLECTUAL PROPERTY RIGHTS

Mariateresa Maggiolino

Introduction: The Relationship between Economics, Law, Facts, and Policies

Economics studies the market conduct of agents, such as individuals, firms, and institutions; it describes the consequences of these behaviours and, if its explanations are successful, it may even provide forecasts and policy suggestions as to the impact that these practices produce on markets.[1] Therefore legal scholars, who wish to offer *well-founded* and *reliable* policy suggestions as to the proper way of shaping, interpreting, and enforcing statutes that rule over business conduct, may find it useful to employ economics to 'bridge'[2] these legal rules with the empirical phenomena that they address.[3]

[1] See G Kirchgassner, 'Homo oeconomicus: the economic model of behavior and its applications' in *Economics and other Social Sciences* (New York: Springer, 2008) 6–7.

[2] The use of economics to describe empirical phenomena, which eventually underpin some legal rules, may introduce economic elements into the legal structure of the offence/felony in question, especially when its wording is poor and loose. See I Lianos, 'Categorical thinking in Competition Law and the "Effects-Based" Approach in Article 82 EC' in A Ezrachi (ed) *Article 82 EC: Reflections on its Recent Evolution* (Oxford: Hart Publishing, 2009) 19. This approach has nothing to do with the so-called 'law and economics approach', whose aim is, instead, to assess the effects that legal rules produce in terms of costs and benefits. See, eg, CK Rowley, 'Public Choice and the Economic Analysis of the Law' in N Mercuro (ed) *Law and Economics* (Boston: Kluwer Academic, 1989) 125.

[3] It is by using economics to 'bridge' legal rules with empirical phenomena that one can push the antitrust and IP worlds to overlap. In other words, by using such a methodology, these two fields of law succeed in speaking the same language, because the same economic principles and concepts are called upon to shape them.

Competition and innovation are, as a matter of fact,[4] the two empirical phenomena that antitrust, patent, and copyright provisions address.[5] Therefore, legal scholars, who want to elaborate well-founded and reliable policy suggestions about antitrust and IP statutes, may profitably employ economics to understand how competition and innovation develop.

Like any other social science, economics encompasses both a 'hard core' of well-established principles and concepts, and a 'fringe' of new models and insights,[6] which introduce new research questions and explore some of the issues that hard core theories cannot solve.

Usually, scholars that use economics as a bridge between legal provisions and empirical phenomena rely on these hard core theories, because they already belong to common knowledge and, hence, are more *administrable* for lawyers, judges, and jurors. In fact, textbook economics of IP is quite basic: although the Chicago school has described it only recently,[7] it is rooted in the instrumental/utilitarian approach and in the cost-benefit analysis of neo-classic economics,[8] as well as in the transaction costs insights that derive from the Coase theorem.[9] Likewise, the current textbook

[4] Indeed, the existence of a material connection between competition and innovation is not a theoretical issue: it is an issue of fact. These two phenomena are actually interdependent. What is a policy choice, instead, is the decision to consider innovation among the aims of antitrust law, and competition among the aims of IP laws. In this last regard see, eg, RJ Gilbert, 'Looking for Mr Schumpeter: Where Are We in the Competition—Innovation Debate?' in AB Jaffe, J Lerner and S Stern (eds) 6 *Innovation Policy and the Economy* (Cambridge: The MIT Press, 2006) 159; and G Ghidini, *Intellectual Property and Competition Law: the Innovation Nexus* (Cheltenham: Edward Elgar Publishing, 2006).

[5] Hereinafter, the word 'innovation' is not used in a technical way: it addresses both inventions and works of art, that is to say, the subject matter of patents and copyrights. Indeed, in light of what the following chapters will explain, this chapter does not discuss trademarks, trade secrets, and other IPRs.

[6] The term 'hard core' is freely taken from I Lakatos, *The Methodology of Scientific Research Programmes* (Cambridge: Cambridge University Press, 1978). The term 'fringe', instead, is not to be connected to the Lakatos' 'belt of auxiliary hypothesis' intended to shield 'hard core theories'. It addresses the opposite phenomenon—the existence of new models and frameworks that, if accepted by the scientific community, could substitute the existing hard core theories.

[7] See, eg, WM Landes and RA Posner, *The Economic Structure of Intellectual Property Law* (Harvard University Press, 2003).

[8] For brevity's sake, it can be stated that five simple ideas lie at the heart of neo-classical economics (also named orthodox or mainstream economics) and combine to create the so-called neo-classical theoretical system; namely: (i) the scarcity assumption, according to which resources are available only in limited quantities; (ii) the *Homo Oeconomicus* paradigm, which prescribes that (a) economic agents are self-interested individuals (and not political bodies or social classes) with Olympian rationality, exogenous preferences, complete information, and homogeneous motivations; and (b) the human good consists in nothing other than economic welfare; (iii) the utilitarian approach, according to which any human behaviour is exclusively reducible to a rational calculation aimed at the maximization of welfare; (iv) the subjective theory of value, which states that something has a value as long as an economic agent considers it valuable; and (v) the substitution principle, according to which economic agents pursue their goals deciding the optimal choice among alternative possibilities. See, in this regard, E Screpanti and S Zamagni, *An Outline of the History of Economic Thought* (Oxford: OUP, 2005) 165–7. Economic neo-classicism arose in the 1870s and 1880s, but it established itself in the 1920s, after about 30 years' of struggle against the old-classicist school.

[9] See RH Coase, 'The Problem of Social Cost' (1960) 3 J L and Economics 1. The Coase theorem explains that, with low transaction costs, the existence of property rights allows economic agents to

economics of antitrust law consists of various industrial organization models,[10] which originate from the application of the above-mentioned principles of neo-classic economics to imperfect market realities.[11]

Yet, competition and innovation are kaleidoscopic phenomena that, sometimes, extend beyond the reach of these hard core economic theories. For instance, the industrial organization models that nowadays shape antitrust law[12] are so focused on short-run price and output effects that they cannot fully explain how innovation causes economic growth, by changing the long-run performance of firms and markets.[13] This is an issue, especially when antitrust enforcers confront business conduct that involves IPRs—whether agreements, monopolists' conduct, or mergers—because this conduct produces both short-run and long-run effects on both competition and innovation. Likewise, despite what orthodox economics suggests, some empirical studies show that in specific industries there is no linear relationship between the scope of IPRs and innovation.[14] This is an issue too, because if broader and stronger IPRs do not boost innovation, the recent propertization phenomenon[15] may turn out to be counterproductive, that is, wider and stronger IPRs may actually slow down or hinder innovation.

As a consequence, in order to provide a better 'bridge' between the antitrust and IP rules on the one side, and competition and innovation phenomena on the other side, some scholars have recently begun to look at 'fringe' economic theories.

While not attempting to provide an exhaustive discussion of past and current economic theories, this chapter explores the mainstream economic backbone of present

maximize allocative efficiency, to reach Pareto efficiency, by assigning (material as well as intellectual) resources to whom value them the most highly.

[10] For a description of what industrial organization is, see Part A of this chapter. Moreover, for a brief history of the industrial organization thinking, see R Schmalensee, 'The New Industrial Organization and the Economic Analysis of Modern Markets' in W Hildenbrand (ed) *Advances in Economic Theory* (Cambridge : Cambridge University Press, 1982) 253.

[11] In economics, markets are said to be imperfect when the hypothesis underpinning the perfect competition model does not apply, that is to say, when: economic agents do not act independently one from the other but collude; when resources are immobile and products are differentiated; when information is incomplete; and when there are economies of scale, barriers to entry, and externalities.

[12] Nowadays, US and EU antitrust authorities (judges and officers) look at what economic models suggest. Often, they swing between the simple, clear cut insights of the Chicago school and the very sophisticated models of Post-Chicago economics. By and large, whereas US courts are prone to follow the Chicago approach, in order to limit the risk of prohibiting practices that could turn out to be harmless, US agencies and EU institutions are more willing to prohibit practices that the post-Chicago models deem harmful, perhaps because many economists work in these institutions.

[13] See Part A of this chapter.

[14] See Part B of this chapter.

[15] With the term 'propertization' scholars address the fact that recently IP laws have been subject to subsequent reforms that have not only created new IPRs and included new items in the subject matter of traditional IPRs, but have also enlarged the duration and the scope of the traditional IPRs. See, eg, MA Carrier, 'Cabining Intellectual Property through a Property Paradigm' (2004) 54 Duke L J 1, 4–7.

antitrust and IP laws and what future developments these legal rules could undergo because of the influence of some fringe economic theories.

Part A explains that the present antitrust discourse mirrors the 'normative meaning' of the perfect competition model, but employs industrial organization models—sometimes, far too sophisticated and complex models—in order to explain firms' actual behaviours. Part B discusses how neo-classic economics justifies the existence of patents and copyrights by using both the appropriability theory and the Coase theorem. Given that industrial organization models cannot fully explain dynamic efficiency and long run phenomena, Part C looks at some heterodox economic theories that '. . . place innovation and change at the heart of the competitive process'.[16] Similarly, given that arguably the neo-classic justification for IPRs does not always support a proper set of incentives for follow-on inventions and derivative creations, Part D describes how one of those heterodox economic theories, that is to say, evolutionary economics,[17] seems to be capable of striking a better equilibrium between present and future innovation, by elaborating on the ideas of 'knowledge' and 'learning processes'. The chapter concludes with the finding that at the moment no drastic change can be required in order to apply these fringe economic theories to an understanding of present antitrust and IP rules.

A. The Economics of Antitrust Law: An Orthodox Approach

In order to explain why protecting competitive markets is important, many antitrust law textbooks employ the 'perfect competition model'[18]—a neo-classic economic model that shows that perfect competition is a state of affairs that guarantees the achievement of allocative efficiency, productive efficiency,[19] maximum consumer

[16] See J Ellig and D Lin, 'Introduction' in J Ellig (ed) *Dynamic Competition and Public Policy. Technology, Innovation, and Antitrust issues* (Cambridge: Cambridge University Press, 2001) 1, 4.

[17] See Part C of this chapter.

[18] It was not until the 1920s that, thanks to F Knight, *Risk, Uncertainty and Profit* (1921), the conception of perfect competition, born with AA Cournot, *Researches into the Mathematical Principles of the Theory of Wealth* (1838) and FY Edgeworth, *Mathematical Psychics* (1881), was finally received into the corpus of mainstream economics. See GJ Stigler, 'Perfect Competition. Historically Contemplated' (1957) 65 J Political Economy 1. Yet, the real consecration of the model of perfect competition happened with J Robinson, *The Economics of Imperfect Competition* (1933) and E Chamberlin, *The Theory of Monopolistic Competition* (1933), with Hicks-Samuelson general equilibrium theory, and with the new welfare economics. See, in this regard, I Kirzner, 'Competition and the Market Process: Some Doctrinal Milestones, in J Krafft (ed) *The process of competition* (Cheltenham: Edward Elgar, 2000) 11; M Blaug, 'Competition as an End-State and Competition as a Process' in BC Eaton and RG Harris (eds) *Trade, Technology and Economics: Essays in honor of Richard Lipsey* (Cheltenham: Elgar, 1997) 241–242; NJ Foss, 'The Biological Analogy and the Theory of the Firm: Marshall and Monopolistic Competition' (1994) 28 J Economic Issues 1115, 1124; and S Moss, 'The History of the Theory of the Firm from Marshall to Robinson and Chamberlin: The Source of Positivism in Economics' (1984) 51 Economica 307.

[19] Allocative efficiency (also named pareto Efficiency) is reached when the existing stock of (final and intermediate) goods is assigned via the system of market prices to those purchasers who value them most

welfare, maximum producer welfare, and maximum total welfare.[20] In this way, antitrust law textbooks justify why policy makers look with suspicion at firms that acquire or strengthen their market power by agreements, monopolization, and mergers, and, thus, push markets away from perfect competition. Indeed, away from perfect competition, firms with substantial market power (also cartelized and merged firms):

(1) produce the so-called 'deadweight loss', that is, a net reduction of total welfare, by decreasing market output and raising market price profitably;
(2) benefit from a welfare transfer from consumers—a welfare transfer that, according to the orthodox approach, they are likely to waste in 'rent seeking activities' aimed at protecting their market power;[21] (See Figure 1, page 92)

and, because of the lack of competitive challenges coming from the monopolized market,[22] are

(1) not only prone to degrade the quality and variety of their own products and, hence, deprive consumers of choice;[23] but also
(2) not willing to innovate.[24]

---

in terms of willingness to pay or willingness to forego other consumption possibilities. When allocative efficiency is achieved, market prices are equal to the marginal costs of producing and supplying those goods; no more mutually advantageous bargains or contracts can be made as none of the players, sellers, or buyers, could be made better off without someone being made worse off. Productive efficiency (also named operating efficiency) is instead a narrower concept: it is achieved when a particular firm is able either to produce the maximum output possible for any given level of inputs cost, or to employ the cheapest combination of inputs possible for any given level of output. When operating efficiency is achieved, firms produce at the minimum average cost permitted by the existing input prices and technology. Indeed, whereas input prices determine which combination of inputs is the one that entails the lowest production costs for any given level of output, technology conditions determine what alternative combinations of inputs can produce a given amount of output. Allocative and productive efficiencies are both static concepts that address the performances that markets and firms achieve at a given point in time. For these definitions see, eg, RJ Van den Bergh and PD Camesasca, *European Competition Law and Economics: A Comparative Perspective* (2nd ed, London: Sweet & Maxwell, 2006); and M Motta, *Competition Policy—Theory and Practice* (Cambridge: Cambridge University Press, 2004).

[20] The welfare (or surplus) of an individual consumer results from the difference between the consumer's willingness to pay for a specific good—that is her valuation of the good—and the price she really has to pay for it. Thus, consumer welfare is nothing more than the sum of the individual surplus of all the consumers. In contrast, the welfare of an individual producer results from the difference between the price at which she sells the good and the costs she has to sustain in order to produce it. Thus, producer welfare is (again) nothing more than the sum of all the profits that producers make in an industry. Total welfare results from the sum of consumer surplus and producer surplus. It is the standard concept that economics uses to measure how well a market performs.

[21] See, eg, AO Krueger, 'The Political Economy of the Rent-Seeking Society' (1974) 64 American Economic Rev 291.

[22] According to the orthodox approach, indeed, whereas competitive pressure sharpens management's incentives to avoid sloth and slack, managers in monopolies may resist change and innovations that would threaten the existing organizational structure. This insight can be traced back to Adam Smith, *An Inquiry into the Nature and Causes of the Wealth of Nations.*

[23] See, eg, H Leibenstein, 'Allocative Efficiency vs. "X-Efficiency"' (1966) 56 American Economic Rev 392.

[24] According to the neo-classical approach, there are further reasons why firms with market power are not interested in innovating. First, they may have less to gain from innovation than their rivals

Yet, such a description of the rationales of both market functioning and antitrust law is good for novices, but is over-simplified for two reasons, at least. First, since real markets do not verify the hypothesis underpinning the perfect competition model,[25] to enforce antitrust law in order to reproduce perfect competition would mean to morph competition law into a piece of regulation—a very intrusive one. Second, since several empirical studies show that perfect markets and firms without market power are not always very conducive to innovation,[26] to protect perfect competition without any further caution or clarification could be counterproductive.

Industrial organization models offer a kind of reply to the first of these critiques to the perfect competition model. These models study imperfect markets, that is to say,

because, unlike them, they already have most of the business that there is to get. Further, whereas the price competition that occurs in competitive markets may induce firms to search for competitive advantages to escape price competition, the monopolist's quiet life could induce dominant firms to economize on research and development by imitating rather than innovating. See, eg, KJ Arrow, 'Economic Welfare and the Allocation of Resources for Innovation' in National Bureau Committee for Economic Research (ed) *The Rate and Direction of Inventive Activity* (Princeton: Princeton University Press, 1962) 609–25.

[25] Stated briefly, the conditions that must hold for having perfect competition are that: (i) consumers and producers do maximize their welfare; (ii) sellers and buyers act independently of one another; (iii) all the capital and human input can move freely among markets, also because there are no barriers to entry or exit; (iv) there are no externalities; (v) sellers and buyers are so numerous that no one has the power to condition market price via her production or consumption decisions—ie, all the economic agents are price takers; (vi) all the economic agents have perfect knowledge about all the relevant market variables.

[26] Virtually all of the evidence suggests that market structure is not obviously the most important driver of innovation. See, eg, J Gual, 'Time to Rethink Merger Policy' <http://ssrn.com/abstract=987093> (2007) 3 Competition Policy International, Spring 2007, accessed 3 September 2010, 29; and WM Cohen and RC Levin, 'Empirical Studies of Innovation and Market Structure' in R Schmalensee and RD Willig (eds) 2 *Handbook of Industrial Organization* (Amsterdam: North-Holland, 1989) 1059. Empirical studies on innovation and firm size are somewhat more definitive, finding that smaller firms (beyond a minimum threshold) are as research-intensive as larger firms and are more productive. However, the contribution of small firms to innovation is known to vary greatly by industry. See, eg, DL Rubinfeld and J Haven, 'Innovation and Antitrust Enforcement' in TM Jorde and DJ Teece (eds) *Antitrust, Innovation, and Competitiveness* (New York: OUP, 1992) 75. There is also a substantial body of evidence that leading incumbents prefer a different path of innovation than challengers. See NS Dorfman, *Innovation and Market Structure: Lessons from the Computer and Semiconductor Industries* (Cambridge: Ballinger, 1987) (observing that whereas leading companies generally use technology as a means of reinforcing their position for the already mentioned substitution effect, small firms and major firms working outside their main area of specialization are more conducive to the introduction of path-breaking technological breakthroughs). Furthermore, it has been argued from both a theoretical and empirical standpoint that some intermediate levels of competition might be optimal, as if there was an inverted U relationship between competition and innovation. See, eg, FM Scherer, 'Market Structure and the Employment of Scientists and Engineers' (1967) 57 American Economic Rev 524 (assuming that the relationship between market structure and innovation follows an inverted-U pattern: innovation is observed to be low at high levels of competition, reach its peak at intermediate levels of oligopoly (where the four leading firms control roughly half the market), and then fall off as market structure approaches monopoly). Several studies replicated and confirmed Scherer's results. See, eg, P Aghion et al, 'Competition and Innovation: An Inverted-U Relationship' (2005) 120 Q J Economics 701, 714; and RC Levin et al, 'R&D Appropriability, Opportunity and Market Structure: New Evidence on Some Schumpeterian Hypotheses' (1985) 75 American Economic Rev 20.

markets where economic agents may collude; where resources may be immobile and products may be differentiated; where information is incomplete; and where there are economies of scale, externalities, and barriers to entry. In particular, they focus on the strategies of firms (whether cooperative, or non-cooperative) aimed at acquiring and strengthening market power, in a scenario where economic agents do not passively adapt to market realities anymore.[27] Therefore, industrial organization models are well-suited to shape a 'more realistic' rationale for antitrust law and market functioning: given that real markets are imperfect, the sole aim of antitrust law must be that of preventing firms from worsening markets' actual performances via some specific types of conduct and industrial organization models allow recognition of these types of conduct, by describing the effective consequences of business practices upon efficiency and welfare (that,[28] however, are two economic variables that root in the perfect competition model).

For instance, in respect of the non-price conduct of dominant firms, the usual concern of the IP-antitrust interface, such as refusals to deal, exclusive dealing, and tie-ins, current industrial organization models—named 'post-Chicagoan' models[29]—on the one hand, acknowledge what the 'Chicago' and the 'NIE' doctrines[30] showed

[27] Therefore, industrial organization models are premised upon a new model of firm—a firm that interacts with its rivals, consumers, and markets. Indeed, as explained in DW Carlton and JM Perloff, *Modern Industrial Organization* (Boston: Pearson Addison Wesley, 2005) 350–1:

> . . . a strategic behavior is a set of actions a firm takes to influence. . . all factors that influence the market outcome (prices, quantities, profits, welfare), including the beliefs of customers and of rivals, the number of actual and potential rivals, the production technology of each firm, and the costs or speed with which a rival can enter the market. By manipulating [all these factors], a firm may be able to increase its profits.

Therefore, given that as noticed by PJ McNulty, 'Economic Theory and the Meaning of Competition' (1968) 82 Q J Economics 639, 645, '. . . in a free enterprise economy. . . there are two primary institutions through which resources are organized, transformed, and channeled for ultimate consumption as goods or services: the private business firm and the market', with industrial organization doctrines the focus of analysis shifted in part from the market (and, thus, from the issues connected to exchange, determination of the economic value, prices) to the firm (and, hence, to the matters related to production, determination of physical form, and quality).

[28] O Budzinsky, 'Modern Industrial Economics and Competition Policy: Open Problems and Possible Limits' (16 June 2009) <http://ssrn.com/abstract=1420855> accessed 3 September 2010, 2.

[29] See, eg, SC Salop (ed) *Strategy, Predation, and Antitrust Analysis* (Washington DC: Federal Trade Commission,1981) and, for the relationship between post-Chicagoan and other industrial organization doctrines, see JB Baker, 'Recent Developments in Economics That Challenge Chicago School Views' (1989) 58 Antitrust L J 645, 646 (characterizing post-Chicago's work as a refinement of Chicago thinking, that is to say, as an effort to provide decision makers with more accurate pictures of marketplaces and more sensitive tools for detecting inefficient behaviours). In particular, in order to question Chicagoan outcomes, post-Chicago scholars used game theory, which focuses on firms' strategic interactions. See, eg J Von Neumann and O Morgenstern, *Theory of Games and Economic Behavior* (Princeton: Princeton University Press, 1944).

[30] The Chicago school used 'price theory' to show that many types of conduct such as predatory pricing, vertical agreements, vertical unilateral practices, and vertical mergers are driven by firms' desire to achieve greater efficiencies. See, eg, RA Posner, 'The Chicago School of Antitrust Analysis' (1979) 127 U Pennsylvania L Rev 925. Likewise, the analysis of New Institutional Economics (NIE), by focusing on the costs that a firm sustains by trading with other parties, made it possible to identify

about the efficiencies that these behaviours entail; and, on the other hand, discover the very specific conditions under which these practices strengthen market power and might harm welfare and innovation. For instance, whereas according to the conventional Chicagoan wisdom, dominant firms do not engage in vertical conduct in order to leverage their market power across markets,[31] post-Chicago scholars have found particular circumstances when vertical conduct works for strengthening and prolonging market power, by deterring entry into primary and secondary markets.[32]

To be sure, since these models are quite sophisticated, even the 'more realistic' economics that they provide to antitrust law has to face a difficult policy choice as to the level of complexity that antitrust authorities should embrace in making their economic analysis. The problem is that the more such analysis is complicated the less it is administrable by lawyers, judges, and jurors.

Furthermore—and moving to the second of the above-mentioned critiques to the perfect competition model—apart from a few exceptions,[33] industrial organization models remain focused on short-run price and output effects—they share, by and large, the same static perspective that both underpins the perfect competition model and, as said before, cannot fully explain innovation. In other words, although it is acknowledged that innovation boosts economic growth far more than does the elimination of static inefficiencies,[34] the theoretical framework of perfect competition and industrial organization models rests with the neo-classical perspective, which is '. . . at best a caricature of how innovation takes place . . . [n]or does it give any credit

---

the efficiency reasons for which firms use various contracts and, in particular, vertical agreements and vertical mergers. See, eg, R Coase, 'The Nature of the Firm' (1937) 4 Economica 386; O Williamson, 'Transaction Costs Economics: The Governance of Contractual Relations' (1979) 22 Journal of Law and Economics 233; and GR Carroll and DJ Teece (eds) *Markets and Hierarchies. The Transaction Cost Economics Perspective* (Oxford: Oxford University Press, 1999).

[31] Indeed, the Chicago school showed that a dominant firm that monopolizes secondary markets is not able to reap additional monopoly profits from them, because it can earn the total maximum amount of monopoly profits just once. See, eg, RH Bork, *The Antitrust Paradox: a Policy at War with Itself* (New York: Basic Books, 1978).

[32] See, eg, DW Carlton and M Waldman, 'The Strategic Use of Tying to Preserve and Create Market Power in Evolving Industries' (2002) 33 Rand J Economics 194; RC Feldman, 'Defensive Leveraging in Antitrust' (1999) 87 Georgetown L J 2079; MD Whinston, 'Tying, Foreclosure, and Exclusion' (1990) 80 American Economic Rev 837.

[33] Some industrial organization models approach the idea of change by employing game theory in order to look at entry and exit mechanisms in the light of the interaction between scale economies and irreversible investments. The model of contestable markets, for instance, describes an easy entry-exit mechanism by connecting it to business investments that can be easily liquidated. See WJ Baumol, JC Panzar and RD Willig, *Contestable Markets and the Theory of Industry Structure* (New York: Harcourt Brace Jovanovich, 1982).

[34] See, eg, TF Bresnahan, 'The Mechanisms of Information Technology's Contribution to Economic Growth' in JP Touffut (ed) *Institutions, Innovation and Growth: Selected Economic Papers* (Cheltenham: Edward Elgar, 2003) 135–7; and CI Jones and JC Williams, 'Measuring the Social Return to R&D' (1998) 113 Q J Economics 1119.

to entrepreneurship.'[35] Think, for instance, that in neo-classical models entrepreneurs/innovators are assumed to act like any other kind of Olympian agent—they know from the beginning the full range of innovative choices available to them and have an accurate understanding of the consequences that each innovative choice entails. As a consequence, in such an ideal scenario, entrepreneurs/innovators are presumed to pick up—in case, after elaborating a quite complicated sequential strategy—the innovative option that maximizes their expected profits or,[36] at least, that protects at best their market power—a very strong presumption.[37]

B. The Economics of IPRs: Still an Orthodox Approach

According to neo-classic economics, intellectual goods are not excludable and not rivalrous.[38] In other words, inventors and creators cannot reap all the profits generated by their investments in inventions and works of art;[39] and one's consumption of a specific intellectual good does not deplete it of its value. As a consequence, competitive markets cannot price intellectual goods in order to give self-interested agents, on the one hand, sufficient incentives to produce the socially optimal amount of intellectual goods and,[40] on the other hand, the ability to efficiently allocate intellectual goods to whoever values them the most highly.

Orthodox IP systems suggest a remedy to this twofold market failure by morphing intellectual goods in the subject matter of specific property rights and, thus, in scarce resources, they allow inventors and creators, first, to gain a sufficient portion of the value of their innovations and, second, to efficiently bargain them. In other words,

[35] DJ Teece, 'Dosi's Technological Paradigms and Trajectories: Insights for Economics and Management' (2007) 17 Industrial and Corporate Change 507, 509

[36] RP Merges and RR Nelson, 'On limiting or encouraging rivalry in technical progress: the effects of patent scope decisions' (1994) 25 J Economic Behavior and Organization 1, 3.

[37] For sure, as long as firms are not conceived as passive agents, it must be admitted that they can use innovation and R&D investments to change market structure and competition. In other words, the relationship of structure/competition-innovation cannot be deemed unilateral. Indeed, there are industrial organization models that study how firms' strategies that pivot around R&D investments and innovations do have an effect by creating barriers to entry and making rivals exit so as to gain and increase market power.

[38] Arrow (n 24 above) 614–16.

[39] In particular, see JA Ordover, 'Economic Foundations and Considerations in Protecting Industrial and Intellectual Property' (1985) 53 Antitrust L J 503, 505–6 (arguing that the greater the gap between the costs sustained for producing intellectual goods and the benefits that their producers cannot appropriate: (i) the less likely it is that producers will be able to recoup their production fixed costs through higher revenues; (ii) the more likely it is that a significant post-creation rivalry will occur lowering the producers' expected profits and, hence, their *ex ante* incentives to produce; and (iii) the more likely it is that each producer will strategically wait with its producing activities so as to be the one who takes a free ride on the producing efforts of others).

[40] Indeed, within the neo-classic framework, economic agents respond to incentives. Therefore, it is understood that social planners can, on the one hand, predict their behaviours by knowing the existing pattern of incentives and, on the other hand, influence their behaviours by providing them a new set of incentives.

for mainstream economics IPRs are both appropriability[41] and allocative tools; whereas their allocative function guarantees the spread of innovation, the stronger is the appropriability power of IPRs, the higher is the level of innovation that the whole society can reach.[42]

Recently, this last insight has somehow urged IP legislators to strengthen IPRs appropriability power by enlarging IPR holders' right to control complementary products,[43] follow-on inventions, and derivative works of art,[44] that is to say, by increasing IPR holders' ability to collect benefits that flow from future innovative and creative spillovers, as well as prior positions in related markets.

Yet, by focusing on industries characterized by cumulative knowledge,[45] such as high-tech industries, several empirical studies refute this kind of linear relationship between the scope of IPRs and the incentives to produce intellectual goods.[46] In fact,

---

[41] Broadly speaking, the term 'appropriability' addresses one's ability to earn the net-value of those things in which one has invested. In other words, wondering about appropriability means discussing whether economic agents that have invested in intellectual goods and dominant positions can actually enjoy the benefits that flow from them. As explained above, this concept is typical of IP law. As to its use in antitrust law see, eg, RM Brunell, 'Appropriability in Antitrust: How Much is Enough?' (2001) 69 Antitrust L J 1.

[42] See, eg, M Sakakibara and L Branstetter, 'Do Stronger Patents Induce More Innovation? Evidence from the 1988 Japanese Patent Law Reforms' (2001) 32 Rand J Economics 77; V Denicolò, 'Patent Races and Optimal Patent Breadth and Length' (1996) 44 J Industrial Economics 249; R Gilbert and C Shapiro, 'Optimal Patent Length and Breadth' (1990) 21 Rand J Economics 106; P Klemperer, 'How Broad Should the Scope of Patent Protection Be?' (1990) 21 Rand J Economics 113; FM Scherer, 'Nordhaus's Theory of Optimal Patent Life: A Geometric Reinterpretation' (1972) 62 American Economic Rev 422; and WD Nordhaus, *Invention, Growth, and Economic Welfare: a Theoretical Treatment of Technological Change* (MIT Press, 1969).

[43] DJ Teece, 'Profiting from Technological Innovation: Implications for Integration, Collaboration, Licensing and Public Policy' (1986) 15 Research Policy 285 (arguing that successful exploitation and commercialization of an innovation requires it to be utilized in conjunction with other complementary capabilities and assets).

[44] In particular, see EW Kitch, 'The Nature and Function of the Patent System' (1977) 20 J Land Economics 265, 271–5 (arguing that granting a patent early in the development process allows the inventor to invest in development without fear that another firm will steal his or her work, thus encouraging the inventor to improve the innovation and to coordinate activities with other firms). Kitch stated that this 'prospect function' of patents necessarily implies broad patents in favor of pioneers. Instead, looking at copyright see, eg, OG Hatch and TR Lee, 'To Promote the Progress of Science: The Copyright Clause and Congress's Power to Extend Copyrights' (2002) 16 Harvard J L and Technology 1, 20–1 (commenting on the Sonny Bono Copyright Term extension Act by stating that the 20-year extension of copyright protection was needed to incentivize the production and dissemination of copies that require intermediaries' investments that otherwise would not have occurred).

[45] For a description of the different types of innovative regimes see, eg TM Jorde and DJ Teece, 'Introduction' in TM Jorde and DJ Teece (eds) *Antitrust, Innovation, and Competitiveness* (n 26 above) 4–5 (distinguishing between regimes of incremental innovation, which occurs when new products are introduced in rapid succession, each one as an improvement on the previous product so that the new drives out the old; and the regime of radical innovation, where each innovation represents a clearly displacing factor for the old products).

[46] See, eg, RM Hunt 'When Do More Patents Reduce R&D?' (2006) <http://ideas.repec.org/e/phu40.html> accessed 3 September 2010; N Gallini 'The Economics of Patents: Lessons from Recent U.S. Patent Reform' (2002) 16 J Economic Perspectives 131; RH Ziedonis and BH Hall

when the probability and the costs of creating new intellectual goods depend on the extent to which innovators may build upon earlier works, broad IPRs make it harder for new innovators to develop their innovative projects and unbalance the relationship between present and future innovations.

In other words, anyone who questions the universal neo-classic ideal that innovation is a sequence of discrete inventions/creations that do not relate to each other, can understand that the policy choices aimed at strengthening pioneer patents and copyrights and, hence, at both protecting initial innovators from the competitive effects of subsequent innovations, and giving initial innovators the option to monopolize subsequent innovations, can have surprisingly depressing effects upon future innovative and creative activities.

C. Fringe Economic Theories and Antitrust Law: Competition as a Dynamic Process Pivoting around Innovation

Over time, scholars have developed numerous theories to explain innovation in connection to the functioning of markets—theories that represent competition as a dynamic and on-going process of development, where firms' innovations trigger changes in markets.[47] For instance, the Austrian doctrine describes competition as a constant process of disclosure where economic agents innovate when they discover (or uncover) pieces of information and knowledge previously ignored. Firms that innovate enjoy a competitive advantage in comparison to their rivals because they become the best at coping with the uncertainty that ignorance generates. Yet, since information and knowledge are supposed to be always incomplete, at any time any firm can develop new products, new techniques, new management methods, new corporate

'The Effects of Strengthening Patent Rights on Firms Engaged in Cumulative Innovation: Insights from the Semiconductor Industry' (2001) 32 Rand J Economics 101; J Bessen and E Maskin 'Sequential Innovation, Patents, and Imitation' (2000) <http://papers.ssrn.com/paper.taf?abstract_id=206189> accessed 3 September 2010; C Shapiro 'Navigating the Patent Thicket: Cross Licenses, Patent Pools, and Standard-Setting' in AB Jaffe, J Lerner and S Stern (eds) 1 Innovation Policy & the Economy (Cambridge: The MIT Press, 2000) 119; AB Jaffe, 'The U.S. Patent System in Transition: Policy Innovation and the Innovation Process' (2000) 29 Research Policy 531; M Heller and R Eisenberg, 'Can Patents Deter Innovation? The Anticommons in Biomedical Research' (1998) 280 Science 698; and S Scotchmer 'Standing on the Shoulders of Giants: Cumulative Research and the Patent Law' (1991) 5 J Economic Perspectives 29, 32, 35.

[47] Stigler was one of the first to have drawn attention to the distinction between competition as an end-state and competition as a process. See, Stigler (n 18 above) 11. His reading of the classical economists as those that saw competition as a state of affairs but were not able to articulate it correctly and precisely has been questioned by PJ McNulty, 'A Note on the History of Perfect Competition' (1967) 75 J Political Economy 395; and FM Machovec, *Perfect Competition and the Transformation of Economics* (London: Routledge, 1995). On the conceptions of competition, see also J Vickers, 'The Concepts of Competition' (1995) 47 Oxford Economic Papers 1; and I Kirzner, *Competition and Entrepreneurship* (Chicago: University of Chicago Press, 1973).

culture, and new organizational processes to keep on competing with its rivals. In this way, therefore, competition becomes an open-ended race between rivals.[48]

Analogously, although it has been somehow misunderstood,[49] the Schumpeterian doctrine describes competition as a continuous fight for innovations—a continuous fight that leads the economy to superior equilibria. When an innovation occurs, on the one hand, the innovative firm jumps from the state of equilibrium where it stays with its competitors to another state of affairs, where it enjoys its competitive advantage and earns extra profits. On the other hand, its competitors begin trying to imitate the innovation in order to reduce the gap between them and the innovative firm. When they succeed in imitating fully the innovation, the economy will reach a new and better state of equilibrium, where no firm will enjoy a competitive advantage and earn extra profits, and this process will start again when a new innovation occurs.[50]

A similar concept of competition as an open-ended process of innovation and experimentation that consist in new 'manners of doing things' is at the core of the evolutionary theory as well. As it arose in the context of Schumpeterian competition, this concept combines the behavioural theory of the firm with the evolutionary theory of technological change.[51] Indeed, evolutionary economics assumes that firms enjoy only 'bounded rationality', that is to say, they can take their decisions only by using a limited amount of time and information and by suffering from significant cognitive limitations. Therefore, they are not able to detect the behaviours that maximize their profits: they can only use 'routines', that is to say, sets of 'learned behaviours', to guide their responses to unforeseen market conditions. Firms learn routines by doing, by using, by searching, by interacting, and by monitoring new developments. Therefore, developing new routines by putting together old ones, by copying some from rivals, and by developing brand new ways of doing things, is the way of surviving

[48] See, eg, D Kallay, *The Law and Economics of Antitrust and Intellectual Property. An Austrian Approach* (Cheltenham: Edward Elgar, 2004); LM Lachmann, *The Market as an Economic Process* (Oxford: Blackwell, 1986); IM Kirzner (ed), *Methods, Process and Austrian Economics: Essays in honor of Ludwig von Mises* (Lexington Books, 1982); L von Mises, *Human Action: A Treatise on Economics* (London: W Hodge, 1949); and FA Hayek, *Individualism and Economic Order* (London: Routledge and K Paul, 1949).

[49] See, eg, RR Nelson, *The sources of economic growth* (Harvard University Press, 1996) 87 (arguing that much economic research urged by the Schumpeterian hypothesis is based on '. . . a misreading of Schumpeter or, at least, a failure to think through what was basic in Schumpeter's arguments'. The simple-minded search for correlations between market structure and innovation neglects the dynamism and disequilibrium that was the essential element of Schumpeter's vision of capitalism). The development of mainstream economic research does not have much in common with the original spirit of Schumpeter's work. See also F Malerba, 'Innovation and the Evolution of Industries' in U Cantner and F Malerba (eds) *Innovation, Industrial Dynamics and Structural Transformation. Schumpeterian Legacies* (Heidelberg: Springer, 2006) 8.

[50] See, eg, J Schumpeter, *Capitalism, Socialism, and Democracy* (London : Allen & Unwin, 1943).

[51] See RR Nelson, 'Understanding Economic Growth as the Central Task of Economic Analysis' in F Malerba and S Brusoni (eds) *Perspectives on innovation* (Cambridge: Cambridge University Press, 2007) 27, 30.

the selection of the market. Each new routine that the market selects as a good one represents, hence, a competitive advantage for the firm and, therefore, a source of extra profits. Other firms can fill in such a competitive gap by imitating the winning routines.[52]

Now, antitrust law can gain some guidance from these 'heterodox' streams of economic thinking. For instance, antitrust authorities could conclude that what may appear inefficient today, such as monopolies and barriers to entry, may actually turn out to be a necessary passage toward achieving efficiency tomorrow. In other words, they could recognize in dominant positions and entry barriers that harm allocative efficiency the same appropriability function that neo-classic economics gives to IPRs, which impair today's allocative efficiency in order to guarantee tomorrow's innovation. Likewise, competition authorities could realize that, on the one hand, market power, barriers to entry, and over-competitive prices are evidence that the firm in question has made a substantial contribution to economic growth, by developing new products, processes, and organization models that better satisfy consumers. And, on the other hand, they could acknowledge that firms unable to develop new products, production techniques, or models of organization should not be 'saved'. In this perspective, antitrust authorities could further acknowledge that concentrated markets have to be tolerated.

Yet, unfortunately, these policy recommendations are not easy to reconcile with the insights that underpin the static analysis proper of industrial organization models, nor with the teaching of the so-called path-dependence theory, which antitrust scholars have recently looked at because it focuses mainly on new economy markets.[53]

Path dependency proposes that antitrust authorities should intervene in infant markets, helping deconcentration before lock-in occurs, to prevent any firm from gaining too much market power and, hence, allow consumers to reach superior alternatives that they do not otherwise enjoy. Indeed, the path-dependence doctrine introduces

[52] See, eg, G Dosi et al, *Technical Change and Economic Theory* (London: Pinter, 1988); RN Langlois, *Economics as a Process: Essays in the New Institutional Economics* (Cambridge: Cambridge University Press, 1986); and RR Nelson and SG Winter, *An Evolutionary Theory of Economic Change* (Belknap Press, 1982).

[53] See, eg, JT Soma and KB Davis, 'Network Effects in Technology Markets: Applying the Lessons of Intel and Microsoft to Future Clashes between Antitrust and Intellectual Property' (2000) 8 J Intellectual Property L 1; ML Katz and C Shapiro, 'Antitrust in Software Markets' in JA Eisenach and TM Lenard (eds) *Competition, Innovation and the Microsoft Monopoly: Antitrust in the Digital Marketplace* (Boston: Kluwer Academic Publishers, 1999); ML Katz and C Shapiro, 'Systems Competition and Network Effects' (1994) 5 J Economic Perspectives 93; ML Katz and C Shapiro, 'Network Externalities, Competition, and Compatibility (1986) 75 American Economic Rev 93; ML Katz and C Shapiro, 'Technology Adoption in the Presence of Network Externalities' (1986) 94 J Political Economy 822; and SM Besen and J Farrell, 'Choosing How to Compete: Strategies and Tactics in Standardization' (1994) J Economic Perspectives 117.

the idea—questioned from an empirical standpoint[54]—that in particular scenarios, such as those of the so-called new economy, competition might not select the most efficient firm.[55] More exactly, this theory maintains that those who succeed in being dominant might just have benefited from small events unrelated to the superiority of their product or processes, but connected to some market features, such as increasing network effects,[56] and tipping effects,[57] that are completely unrelated to the particular firm's merits.[58] Therefore this theory, trusting governments more than markets in picking up winning technologies,[59] suggests to antitrust authorities the need to carry

[54] SJ Liebowitz and SE Margolis, 'Are Network Externalities a New Source of Market Failure?' in RO Zerbe, Jr and W Kovacic (eds) 17 Research in Law and Economics (New York: Elsevier, 1995) 1; SJ Liebowitz and SE Margolis, 'Path Dependence, Lock-In and History' (1995) 11 J L Economics, and Organization 205; SJ Liebowitz and SE Margolis, 'Network Externality: An Uncommon Tragedy' (1994) 8 J Economic Perspectives 133; SJ Liebowitz and SE Margolis, 'The Fable of the Keys' (1990) 33 J L and Economics 1.

[55] In other words, according to this theory a product could lose out to an objectively inferior alternative if it enjoys peculiar market advantages. See, eg, WB Arthur, 'Competing Technologies, Increasing Returns, and Lock-In by Historical Events' (1989) 99 Economic J 116; PA David, 'Clio and the Economics of QWERTY' (1985) 75 American Economic Rev 332; and J Farrell and G Saloner, 'Standardization, Compatibility, and Innovation (1985) 16 Rand J Economics 70.

[56] There are two kinds of network effects; direct and indirect network effects. Direct network effects occur when the benefit to one person from using a product increases as more people use the product; in other words, when average consumer benefit rises as output increases. For instance, direct network effects spring off with telephones and computers as each telephone and computer acquires a larger value as more people are able to communicate one with the other, in this case exchanging compatible files. Because of direct network effects, without some offsetting advantage a potential entrant has no prospect of success, so that an incumbent may employ some strategies just in order to reach a number of consumers to enjoy direct network effects. For a similar discussion in the specific context of PC software, see DS Evans and RL Schmalensee, 'Be Nice to Your Rivals: How the Government Is Selling an Antitrust Case without Consumer Harm in United States v Microsoft' in DS Evans et al (eds) *Did Microsoft Harm Consumers? Two Opposing Views* (AEI Press, 2000) 45, 65–6. Instead, the term 'indirect network effects' serves to address the consequences that the existence of a network has on the production of complements. See, eg, N Gandal, 'A Selective Survey of the Literature on Indirect Network Externalities: A Discussion of Liebowitz and Margolis' in RO Zerbe, Jr and W Kovacic (eds) 17 Research in Law and Economics (New York: Elsevier, 1995) 23. With indirect network effects, the potential entrant's disadvantage relates to the availability of complements of various sorts. To the extent that the complements are incompatible in some way, an entrant might have to establish its own complementary network.

[57] Tipping effects occur when one network has taken such a large portion of the market that competing networks no longer have enough members to be viable. Even though an individual consumer may prefer a competing network's technology, the benefits (due to network effects) from joining a dominant network may swamp these considerations. Thus, marketing techniques, or anti-competitive practices that gain large market share for a network, may 'tip' the market in favour of that network. See, in this regard, M Schanzenbach, 'Network Effects and Antitrust Law: Predation, Affirmative Defenses, and the Case of U.S. v Microsoft' [2002] Stanford Technology L Re 4 <http://stlr.stanford.edu/STLR/Articles/02_STLR_4>accessed 3 September 2010, 6 and WB Arthur, 'Positive Feedback in the Economy' (1990) 262 Scientific American 92.

[58] See, eg, GJ Werden, 'Network Effects and Conditions of Entry, Lessons from the Microsoft Case' (2001) 69 Antitrust L J 87.

[59] See, eg, S Liebowitz and SE Margolis, 'Network Effects and the Microsoft Case' in J Ellig (ed) *Dynamic Competition and Public Policy: Technology, Innovation, and Antitrust Issues* (Cambridge: Cambridge University Press, 2001) 190–1 (arguing that network markets, such as software products and operating systems, are highly competitive even though some firms have large market shares); RA Cass and KN Hylton, 'Preserving Competition: Economic Analysis, Legal Standards and Microsoft'

on such a duty rather than to leave markets to operate freely and tolerate dominant firms and concentrated markets.

In sum, although the fringe economic theories here described offer precious insights, and possible new approaches, there is currently no consensus on a final microeconomic theory that explains change, economic growth, the relationship between competition and innovation, and, in the end, whether and how long monopolies should last in order to promote competition and innovation. In particular, what seems difficult to establish is which models, static and dynamic, are best suited to analyse firms' conduct, given that their application can lead to different outcomes.[60]

D. Fringe Economic Theories and IPRs: Two Different Conceptions

The previously mentioned fringe economic theories have not always elaborated an explicit conception of what IPRs are and should be. However, whereas the Schumpeterian and the Austrian doctrines stress the appropriability function of IPRs, by emphasizing—as the present orthodox approach does—that they serve to incentivize innovators across time periods,[61] the evolutionary doctrine points in the opposite direction.

For evolutionary scholars, intellectual goods stem from knowledge. Knowledge—which they deem the engine of change and economic growth—originates by

(1999) 8 George Mason L Rev 1, 36–7 (noting while entry barriers may be higher in network markets, payoffs to successful entry are higher as well); and RA Levy, 'Microsoft and the Browser Wars' (1999) 31 Connecticut L Rev 1321, 1352–8 (disputing the significance of network effects in the Microsoft case).

[60] Consider, however, that, once dynamic theories are well developed, Occam's razor suggests that dynamic theories are better only where they are needed. Where a case can be solved with the simpler, less realistic, static theories, these last remain preferable. On the other hand, the search for simplicity should not lead to misleading analysis. In this regard, see TJ Brennan, Should Innovation Rationalize Supra-Competitive Prices? A Skeptical Speculation, prepared for the Pros and Cons of High Prices Seminar, Swedish Competition Authority/Konkurrensverket (Stockholm, Sweden, 9 November 2007) <http://papers.ssrn.com/sol3/papers.cfm?abstract_id=1030115> accessed 3 September 2010, 15–17 speaking about Microsoft and arguing that:

> ... reduction of a difficult dynamic case to an easy static case had at least three consequences. First is the foregone opportunity to clarify how monopolization or abuse of dominance law should work in dynamic or 'new-economy' industries characterized by Schumpeterian competition. Second, avoiding the hard work of a dynamic case undercut the basis for consequential relief. ... Third is that allegations involving control over distribution of other forms of software, eg media players, are held (by some commentators) to an inappropriate dynamic standard, when they should be brought only as straightforward static exclusion cases.

[61] As to the Schumpeterian approach, see, eg, J Schumpeter, *Capitalism, Socialism, and Democracy* (n 50 above) 88 (arguing patents are protecting devices). As to the Austrian approach see, eg, Kallay (n 48 above) 24–5, 104 (however observing that Austrian economists would not recommend too broad IPRs, fearing both a block to the process of competition and a too intrusive precence of the government in the economic life).

applying to sheer information cognitive and procedural 'routines', that is to say, sets of learned behaviours, which are to a large extent tacit and embedded in human organizations. Thus, knowledge is something more than information: it is specific to each agent that develops it; it is strongly dependent on the unique features of each applied routine;[62] and—because of these two factors—it does not spill over easily to other agents as information.[63]

Therefore, for evolutionary economics intellectual goods are more excludable and more rivalrous than is suggested by orthodox economics.[64] As long as knowledge is complex, cumulative, tacit, and difficult to replicate,[65] intellectual goods are not pieces of pure information and IPRs should not be conceptualized as appropriability and allocative tools, but as means that contribute to explain and disseminate such a kind of knowledge among people, in hope that this would help to foster economic and cultural progress.

In particular, as to the appropriability issue that is so dear to neo-classic economics, evolutionary economics does not only agree with the fact, which some orthodox scholars and empirical studies however noticed, that in some industries too much appropriability blocks further innovation, it also highlights that appropriability does not hinge on the scope and enforceability of IPRs. Rather, it argues for limited, 'quite open' IPRs, because it believes that appropriability rests on team-embodied knowledge, that is to say, on the ability of firms to replicate effectively those routines that lead to successful innovations as well as to leverage them in other markets.[66]

In the extreme, evolutionary economics considers that what affects the pace and direction of innovation is not the possibility to fully benefit from the profits that flow from inventions and works of art, but the opportunities that each body of

[62] G Dosi, L Marengo and C Pasquali, 'How much should society fuel the greed of innovators?: On the relations between appropriability, opportunities and rates of innovation' (2006) <http://ideas.repec.org/p/ssa/lemwps/2006-17.html> accessed 3 September 2010, 4–5.

[63] See Michael Polanyi, *Personal knowledge. Towards a post critical philosophy* (London: Routledge and Kegan Paul, 1958) (noticing that in some industries knowledge is not easy to codify and transmit but is instead sticky and 'tacit').

[64] See, eg, G Dosi, L Marengo, and C Pasquali, 'Knowledge, Competition and Innovation: Is Strong IPR Protection really needed for More and Better Innovations?' (2007) 13 Michigan Telecommunications and Technology L Rev 471, 479 (arguing that '. . . knowledge can be misleadingly identified with information. When this occurs, there is a deletion of any reference to cognitive and procedural devices, whose role is to transform raw information into "useful knowledge".').

[65] See, eg, G Dosi, L Marengo, C Pasquali and M Valente, 'Appropriability, Patents, and Rates of Innovation in Complex Products Industries' (2009) LEM Papers Series 2009/05 (Laboratory of Economics and Management (LEM), Sant'Anna School of Advanced Studies, Pisa, Italy) <http://www.lem.sssup.it/WPLem/files/2009-05.pdf> accessed 3 September 2010, 5–6.

[66] See, eg, G Szulanski, 'Appropriability and the challenge of scope: banc one routinizes replication', in G Dosi, RR Nelson, and SG Winter (eds) *The nature and dynamics of organizational capabilities* (Oxford: OUP, 2000) 69, 69.

knowledge—also named technological paradigms[67]—provides.[68] Indeed, opportunities are the fundamental determinants of innovation rate and trajectories, since they represent '. . . the width, depth and richness of the sea in which incumbents and entrants go fishing for innovation'.[69] In other words, these initial chances contribute to establish how far and how fast an industry will develop. Entrepreneurs/innovators have to discover them, or at least those that are specific to their firms, and choose among them which one to exploit.

Yet, in opposition to what neo-classic economics believes, evolutionary models believe that entrepreneurs/innovators, who 'go fishing for innovation', are unable to see clearly all the innovative alternatives actually available to them (ie they enjoy limited information). Further, when they choose what opportunity to develop, they cannot calculate the option that actually is the best for them (ie they enjoy bounded rationality). Therefore, for evolutionary economics, different economic agents do different things, because they see and rank opportunities differently.[70] Each entrepreneur/innovator affects the pace and direction of innovation differently,[71] sometimes in a bad way.

In such a scenario, IPRs are no longer neutral to the rate and direction of change as they are for neoclassic economics: by denying access to intellectual goods, ie to some pieces of knowledge, they limit the opportunities available to entrepreneurs/innovators and, hence, affect innovation's speed and course.

In sum, by shifting the focus of the analysis on knowledge and the characteristics of learning processes,[72] evolutionary economics calls attention to the scope and distribution of IPRs. As to the scope of IPRs, it highlights that broad IPRs do not only produce the static inefficiencies, such as higher prices, predicted by neo-classic economics, they also produce dynamic inefficiencies in terms of constraints on future

[67] G Dosi, 'Technological Paradigms and Technological Trajectories: A Suggested Interpretation of the Determinant and Direction of Technological Change', (1982) 11 Research Policy 147, 147–8 (arguing that a technological paradigm concerns problem-solving activities). Broadly speaking, according to evolutionary economics a paradigm encompasses: (i) the definition of the relevant problems that are to be solved; (ii) the definition of the specific knowledge necessary to the solution of those problems; (iii) a set of procedures to reach points (i) and (ii); (iv) frames of shared knowledge about 'how to do things'; and (v) basic templates of artifacts and systems, which over time are progressively modified and improved.

[68] Dosi, Marengo and Pasquali (n 62 above) 7.

[69] Dosi, Marengo and Pasquali (n 62 above) 16.

[70] Merges and Nelson (n 36 above) 5.

[71] See Nelson and Winter (n 52 above) ch 6.

[72] L Marengo and S Vezzoso, 'Dynamic inefficiencies of intellectual property rights from an evolutionary/problem-solving perspective: some insights on computer software and reverse engineering' (2006) <http://ssrn.com/abstract=1358920 > accessed 3 September 2010, 2, and SX Shi, 'Knowledge and Its Inventors/Carriers in Evolutionary Economics: Towards A Relational Theory of Authorship and Copyright' (2008) <http://ssrn.com/abstract=1311662> accessed 3 September 2010, 2.

innovation,[73] especially when it comes to complex industries and system innovations.[74] As to the distribution of IPRs, evolutionary economics considers that it can seriously determine the total and sub-optimal level of innovation, because once an entrepreneur/innovator took an innovative path, to abandon it is very difficult,[75] especially when the complexity of products is high.[76] Therefore, evolutionary economics invites scholars to reevaluate the role that IPRs are called to play in incentivizing the production of inventions and creations. It claims, first, that 'weaker', 'quite open' IPRs, would be good at encouraging knowledge dissemination across individuals and time periods,[77] and would not prevent the development of complementary products, follow-on innovations, and derivative works. Second, evolutionary economics seems to ask for a more reasonable distribution of IPRs, arguing that their allocation can affect future industries' development. Third, by asserting that knowledge, opportunities, paths, and paradigms are specific to each industry, evolutionary economics questions the universal nature of IPRs, even claiming that they represent an '. . . unfortunate juridical category for the regulation of an issue that should be essentially one of finely tuned industrial policy'.[78]

Conclusion

This chapter is built upon the following reasoning. As long as it is accepted that economics is good at explaining the empirical phenomena that IP and antitrust rules try to govern, ie, innovation and competition, scholars have to interpret legal rules in accordance with what economics suggests about innovation and competition as processes. Therefore, when economics provides new explanations about innovation and competition, scholars have to modify their approaches to IP and antitrust rules accordingly. There are two exceptional cases in which this may not be true: (i) when new economic explanations are not well-founded and reliable; and (ii) when new explanations are too sophisticated to be administered by lawyers, judges, and jurors.

Leaving to economists both the duty to explore whether or not the new theories herein briefly described are well-founded and reliable, and if the opportunity offers

[73] R Mazzoleni and RR Nelson, 'The benefits and costs of strong patent protection: a contribution to the current debate' (1998) 27 Research Policy 273, 280.

[74] Dosi, Marengo, Pasquali, Valente (n 65 above) 6.

[75] Dosi, Marengo, Pasquali (n 64 above) 477–8.

[76] Dosi, Marengo, Pasquali, Valente (n 65 above) 22–3.

[77] P David, 'Intellectual Property Institutions and the Panda's Thumb: Patents, Copyrights, and Trade Secrets in Economic Theory and History', in MB Wallerstein, ME Mogee and R Schoen (eds) *Global dimensions of Intellectual Property Rights in Science & Technology* (Washington DC: National Academy Press, 1993) 19, 54–7; P David, 'Does the Economy Need All the Old IPR Institutions? Digital Information Goods and Access to Knowledge for Economic Development', Address at the Wider Conference on The New Economy in Development (11 May 2002).

[78] Dosi, Marengo, Pasquali, Valente (n 65 above) 22–3.

viable and economically sustainable business models,[79] legal scholars have to establish which of these new theories is administrable enough to became a tool of lawyers, judges, and jurors.

Fortunately, fringe economic theories provide insights less sophisticated than those that some streams of orthodox economics, such as game theory, have already supplied. Better still, since they are not rooted in extreme abstractions or in complex mathematics, but in qualitative descriptions of how firms and human beings behave, they can be better understood by laymen and by people who do not study economics.

Although especially evolutionary economics could urge a radical change in the IP and antitrust systems, more realistically it invites IP scholars to rethink the propertization phenomenon[80] and antitrust scholars to look with less scepticism at firms that are going to acquire and strengthen their market power[81]—a suggestion that the Schumpeterian and Austrian theories share in connection to dominant firms.

To be sure, even fringe economic theories cannot clarify—or, at least, have not clarified yet—the boundaries of such a change. They cannot establish how much appropriability is enough, either in the antitrust realm, where they cannot indicate the proper scope of market monopolies, or in the IP realm, where they cannot clarify the proper scope of IPRs. In other words, up to now 'open access'[82] and 'no fear of bigness' could be at most the slogan of a new stream of policy suggestions, more than new passwords for solving old problems.

---

[79] See, eg, MW Carroll, 'Creative Commons and the New Intermediaries' (2006) 45 Michigan State L Rev 45; and U Gasser et al, 'Content and Control: Assessing the Impact of Policy Choices on Potential Online Business Models in the Music and Film Industries' (2005) <http://papers.ssrn.com/sol3/papers.cfm?abstract_id=654602> accessed 3 September 2010.

[80] See, eg, N Elkin-Koren and NW Netanel (eds), *The commodification of information* (Kluwer Law International, 2002).

[81] *Verizon Communication, Inc v Law Offices of Curtis V Trinko, LLP*, 540 US 398, 407 (2004).

[82] GB Ramello, 'Intellectual property, social justice, and economic efficiency: insights from law and economics', in A Flanagan, ML Montagnani (eds) *Intellectual Property And Social Justice: A Law And Economics Approach* (forthcoming, 2010).

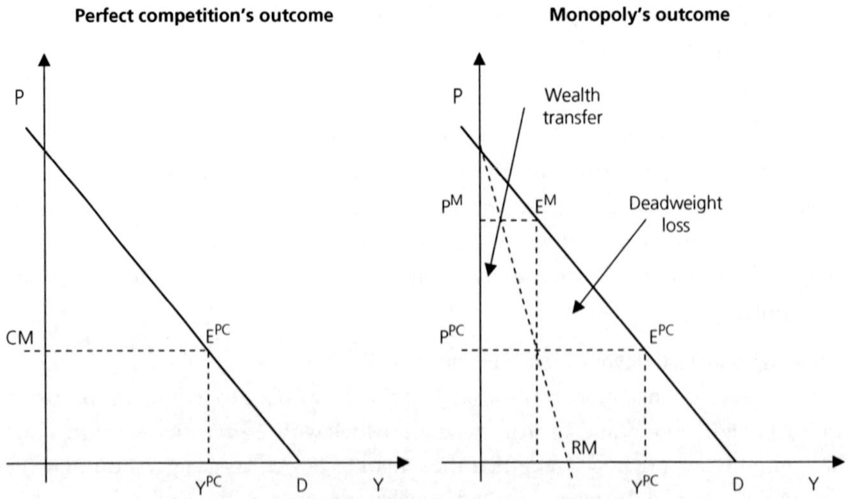

Figure 1 Graphs of perfect competition and of monopoly

Part II

MARKET POWER AND ABUSE

5

COMPETITION LAW ENFORCEMENT AND REFUSAL TO LICENSE: THE CHANGING BOUNDARIES OF ARTICLE 102 TFEU

Ariel Ezrachi

Introduction

The landscape of European competition law has seen significant changes in the past decades, both in terms of enforcement and substantive application. Albeit belated, the winds of reform have also reached the shores of Article 102 TFEU (ex Article 82 EC). Recent years have been characterized by a lively debate on the introduction of an economic-based approach and the way such approach may affect Article 102's nature and scope.

The origins of the public debate may be traced to the publication of the 2005 EAGCP Consultation Paper[1] which was followed by the release of the 2005 Discussion Paper.[2] The more recent publication of the Guidance Paper on enforcement priorities has further sparked the public debate on the soul of Article 102 TFEU.[3] This debate is interesting, not least as it has changed the proposed analytical framework of abuse and affected the boundaries of Article 102 TFEU.

[1] EAGCP Report on 'An Economic approach to Article 82 EC' (July 2005) <http://ec.europa.eu/competition/publications/studies/eagcp_july_21_05.pdf> accessed 3 September 2010 (hereinafter 'EAGCP Report').

[2] DG Competition discussion paper on the application of Article 82 of the Treaty [now Article 102 TFEU] to exclusionary abuses (December 2005) <http://ec.europa.eu/competition/antitrust/art82/discpaper2005.pdf> accessed 3 September 2010 (hereinafter 'Discussion Paper').

[3] Communication from the Commission, Guidance on the Commission's Enforcement Priorities in Applying Article 82 EC Treaty to Abusive Exclusionary Conduct by Dominant Undertakings (February 2009) [2009] OJ C45/7 <http://ec.europa.eu/competition/antitrust/art82/index.html> accessed 3 September 2010 (hereinafter 'Guidance Paper').

This chapter sets to explore the ways in which changes to the ambit of Article 102 TFEU have affected the meeting point between competition law and intellectual property rights (IPR). More specifically it is concerned with the way in which the Guidance Paper has altered the analysis of refusal to license and the effects of that change. To position the discussion in context, the chapter begins with a brief review of the evolution of the European Court's approach to refusal to supply and license. Following this, the emergence of the effects-based approach to Article 102 TFEU is considered. The discussion then centres on the treatment of refusal to license in the Guidance Paper and highlights the imperfect overlap between the Guidance Paper and the European Court judicature. Subsequently, it considers how changes to the analysis of abuse may affect legal and business certainty at Union and Member States levels.

A. Refusal to License: The Evolution of the European Court Judicature

In the area of refusal to license one may identify a gradual trend which has widened the scope of Article 102 TFEU on account of the protection awarded to IPRs. The case law in this area is well known and a brief overview of the landmark cases on refusal to supply and license should suffice to set the background for the discussion which follows.

Following the European Court of Justice's judgment in *Commercial Solvents Corporation v Commission* the principal possibility of condemning a dominant company for refusing to supply an undertaking at the downstream level, and subsequently eliminating competition in that market, was firmly embedded in the European judicature.[4] The doctrine of refusal to supply was subsequently used in a number of cases, concerning tangible and intellectual property, where the dominant company abused its upstream market power and leveraged it to improve its position at the downstream market.[5] The doctrine has also been used in instances in which the refusal to supply was intended to protect the upstream market where the company enjoyed a dominant position.[6]

[4] Case 6/73 *Instituto Chemioterapico SpA and Commercial Solvents Corporation v Commission* [1974] ECR 223, [1974] 1 CMLR 309.

[5] See for example: Case 238/87 *AB Volvo v Erik Veng (UK) Ltd* [1988] ECR 6211; Case 77/77 *BP and others v Commission* [1978] ECR 1511; Case 53/87 *Conzorzio Italiano della Componentistica di Ricambio per Autovericoli and Maxicar v Regie National des Usines Renault* [1988] ECR 6039, [1990] 4 CMLR 265; Case IV/34.689 *Sea Containers Ltd/Stena Sealink* (Interim measures) [1994] OJ L15/8; *Sealink/B&I Holyhead* (Interim measures), [1992] 5 CMLR 255; Joined cases C-468/06 to C-478/06 *Sot Lélos kai Sia EE and Others v GlaxoSmithKline AEVE* [2008] ECR I-7139; Case T-301/04 *Clearstream v Commission* (CFI 9 September 2009) [2009] 5 CMLR 24.

[6] Case IV/32.279 *BBI and others v Boosey & Hawkes plc* (Interim measures) OJ [1987] L286/36, [1988] 4 CMLR 67.

The application of the doctrine to IPRs in the *Magill* case provided a significant landmark in the debate on the interface between the two legal regimes.[7] While the Court held that the mere ownership of an intellectual property right does not necessarily establish the existence of a dominant position, it outlined the instances in which refusal to license would amount to an abuse. It held that the exercise of an exclusive right by the proprietor may, in *exceptional circumstances*, involve abusive conduct.

The special circumstances in *Magill* which swung the balance in favour of an obligation to license were later highlighted, and limited, in *Oscar Bronner*.[8] Accordingly, to amount to an abuse, a refusal to license/supply would (i) concern a product the supply of which was indispensable for carrying on the business in question, (ii) such refusal prevents the appearance of a new product for which there was a potential consumer demand, (iii) the refusal is likely to exclude all competition in the secondary market, and (iv) it is not justified by objective considerations.

By clarifying the notion of *exceptional circumstances*, the *Oscar Bronner* judgment seemed to have limited the enforcement boundaries of Article 102 TFEU. As such it echoed the need for careful balancing of conflicting considerations prior to interfering with a dominant undertaking's freedom to contract. As highlighted by Advocate General Jacobs in his opinion,

> . . . in assessing such conflicting interests, particular care is required where the goods or services or facilities to which access is demanded, represent the fruit of substantial investment. That may be true in particular in relation to refusal to license intellectual property rights.[9]

The *Oscar Bronner* balancing formula was short lived, and has been eroded in subsequent cases.

In *Tierce Ladbroke* the Court of First Instance (CFI, now General Court) distinguished the circumstances of that case from those of *Magill*.[10] In doing so it noted that the refusal to license could not fall within the Article 102 TFEU prohibition unless it concerned a product or service which was *either* essential for the exercise of the activity in question, *or* was a new product whose introduction might be prevented, despite specific, constant, and regular potential demand from consumers. In doing so the Court treated conditions (i) and (ii) in *Magill* as alternatives rather than cumulative, thus widening the instances in which the test could be applied.

---

[7] Joined Cases C-241, 242/91P *RTE & ITP v Commission* (The Magill Case) [1995] ECR I-743, [1995] 4 CMLR 718.

[8] Case C-7/97 *Oscar Bronner GmbH Co KG v Mediaprint* [1998] ECR I-7791, [1999] 4 CMLR 112.

[9] AG opinion para 62.

[10] Case T-504/93 *Tierce Ladbroke SA v Commission* [1995] ECR II-923, [1997] 5 CMLR 309.

Further erosion was introduced in the *IMS Health* judgment.[11] There, the CFI reiterated the formula and conditions stated in *Oscar Bronner*, but in relation to the third condition, notably that the refusal must be likely to exclude all competition in the secondary market, the Court held that the finding of a 'potential or even hypothetical market' would suffice. The finding of a 'fictional' downstream market is significant as it may result in compulsory licensing from a dominant undertaking to its competitor.[12] Arguably, this widening of the breadth of the doctrine, which departed from the upstream downstream model, risks undermining companies' incentive to innovate.[13] Note however that this widening was later reaffirmed in the *Microsoft* judgment where the CFI noted that 'The fact that the indispensable product or service is not marketed separately does not exclude from the outset the possibility of identifying a separate market.'[14]

The *Microsoft* judgment is interesting as it provides the latest signal from the European Court on the interface between competition law and IPRs. It concerned an appeal to the CFI on the Commission decision which found Microsoft to have abused its dominant position, among other things, by refusing to grant licences to interoperability information. While the Commission accepted that some level of interoperability was present on the market, it argued nonetheless that this was not sufficient to provide for viable competition.[15] On appeal, the CFI upheld the Commission finding that Microsoft abused its dominant position by refusing to provide interoperability information. At three distinct levels, the judgment contributed to further widening of Article 102 TFEU.

First, with respect to the requirement of the appearance of a new product, the court held that the circumstances as envisaged in *Magill* and *IMS Health*,

> . . . cannot be the only parameter which determines whether a refusal to license an intellectual property right is capable of causing prejudice to consumers within the meaning of [Article 102(b) TFEU]. As that provision states, such prejudice may arise

[11] Case C-418/01 *IMS Health GmbH & Co OHG v NDC Health GmbH & Co KG* [2000] ECR I-5039, [2004] 4 CMLR 28.

[12] Contrast this approach with the US approach as evident in *Verizon Communications Inc v Trinko LLP*, 540 US 398. There the US Supreme Court considered the absence of two distinct markets as a relevant factor in its decision not to condemn a refusal to supply (para 5); on this point see further: J Drexl 'Intellectual Property and Antitrust law—*IMS Health* and *Trinko*—Antitrust Placebo for Consumers instead of Sound Economics in Refusal-to-Deal cases' (2004) 35 Intl Rev Industrial Property and Copyright L 788.

[13] Note however that, similar to the circumstances which surrounded the *Magill* case, it is possible to question the justification of the copyright protection in this case. Since IMS developed the brick structure in collaboration with pharmaceutical companies and it became the de facto industry standard, one might argue that competition law was used here to remedy unjustified over protection by the intellectual property laws.

[14] Case T-201/04 *Microsoft Corp v Commission* [2007] ECR II-3601, [2007] 5 CMLR 11 para 335.

[15] Commission Decision of 24 March in Case COMP/C-3/37.792 *Microsoft* [2004] OJ L 32/23 paras 666–701.

> where there is a limitation not only of production or markets, but also of technical development.[16]

The CFI used Article 102(b) TFEU to 'import' innovation into the '*Magill* equation'. This 'internal' expansion tipped the balance of the carefully crafted list of exceptional circumstances in the *Magill* and *IMS Health* judgments and lowered the threshold for finding an abusive refusal to license. The reference to innovation not only lowered the threshold for intervention, but has arguably also introduced uncertainty to the test, as one may find it difficult to determine the level of innovation which may trigger intervention.

Second, the court widened the notion of *indispensability*. Earlier cases such as *Commercial Solvents, Magill*, and *IMS Health* were based on the absence of a potential or actual substitute which was objectively indispensible. Yet, in *Microsoft* the CFI broadened the concept to cover 'economic indispensability'.[17] In line with that it held that although access to the market might have been technically possible, Microsoft's refusal to license the interoperability information eliminated the economic viability of such entry.[18] The Court added in this context that the test for indispensability is satisfied also when competition is eliminated gradually rather than immediately.[19]

Third, the Court lowered the threshold for intervention by eroding the requirement for elimination of all competition in the secondary market, which formed part of the conditions in earlier cases. It held that it is not,

> necessary to demonstrate that all competition on the market would be eliminated. What matters, for the purpose of establishing an infringement of [Article 102 TFEU], is that the refusal at issue is liable to, or is likely to, eliminate all effective competition on the market.[20]

Reflecting on the developments of the European Court's judicature, from *Magill* to *Microsoft*, the widening trend of Article 102 TFEU is evident. Case by case application and balancing resulted in (at times inconsistent) modification of the test on account of IPRs. Nonetheless, the evolution of the application of Article 102 TFEU to refusal to license has been restrictive to the extent that it has been, and still is, underlined by the consideration of *exceptional circumstances*. Naturally, that in itself does not make the current balancing point immune from criticism as one may argue that it reflects an overly interventionist approach which hampers innovation and investment. This discussion, however, concerns the more general debate on the

---

[16] *Ibid.*

[17] B Vesterdorf 'Article 82 EC: Where do we stand after the *Microsoft* judgement?' (2008) 1 Global Antitrust Rev 1 available online <http://www.icc.qmul.ac.uk/GAR/GAR2008/Vesterdorf.pdf> accessed 3 September 2010.

[18] *Ibid.*

[19] Case T-201/04 *Microsoft* (n 14 above) para 428.

[20] Case T-201/04 *Microsoft* (n 14 above) para 563.

interface between competition law and IPR and is beyond the scope of this chapter.[21] In what follows, the paper considers the changes fostered by the introduction of an economic approach to Article 102 TFEU and the Guidance Paper.

B. The Rise of the Economic Approach to Article 102 TFEU

As mentioned above, year 2005 has been characterized by the launch of the public debate on the analysis of abuse and the economic approach to Article 102 TFEU. In its Consultation Paper, titled 'An economic approach to Article 82',[22] the Economic Advisory Group for Competition Policy (EAGCP) questioned the merits of the traditional formalistic approach to Article 102 TFEU. The paper advocated an effects-based approach which focused on competitive harm and refrained from treating different categories of behaviour as abusive per se. It stemmed from the understanding that similar actions may result in different effects on the market, some abusive and others not. Likewise, different actions may lead to the same anti-competitive effect.

Shortly after the publication of the EAGCP Consultation Paper, the Commission released its 2005 Discussion Paper on the application of Article 102 TFEU to exclusionary abuses.[23] While more restrictive in its approach than the EAGCP Consultation Paper, the Discussion Paper echoed some of its economic spirit and embedded an effects-based analysis into Article 102 TFEU. Noticeable, in this respect, were the Commission's treatment of rebates and the proposed treatment of efficiencies.[24] However, as clearly indicated by the Commission, the Discussion Paper did not create any legitimate interest and was not meant to be relied upon to provide guidance on Commission enforcement policy.[25]

Both papers generated wide-ranging debate on the scope and nature of analysis in Article 102 TFEU cases.[26] Commenting on the review of Article 102 TFEU, then

[21] See generally: ML Katz, 'Intellectual Property Rights and Antitrust Policy: Four principles for a complex world' (2002) 1 J Telecommunications & High Technology 325; J Langenfeld, 'Intellectual Property and Antitrust: Steps Towards Striking a Balance', (2001) 52 Case Western Reserve L Rev 91; L Peeperkorn, 'IP Licenses and Competition Rules: Striking the Right Balance' (2003) 26(4) World Competition 527; OECD, 'Policy Roundtables: Competition, Patents and Innovation' (2007) DAF/COMP(2007)40.

[22] EAGCP Report (n 1 above).

[23] DG Competition, Discussion Paper (n 2 above).

[24] Guidance Paper (n 3 above) paras 84–92; see also Discussion Paper (n 2 above) para 8.

[25] Discussion Paper (n 2 above) para 7.

[26] See for example: G Niels, 'The Article 82 Discussion Paper: A Comment on the Economic Principles' (2006), available at <http://www.oxera.com> accessed 28 June 2010; J Padilla, 'The Law and Economics of Article 82' (2007) 44 (4) Common Market L Rev 1185–6; P Papandropoulos, 'Implementing an effects-based approach under Article 82' (2008) 1 Concurrences 1–5; G Monti, 'Article 82 EC: What future for the effects based Approach?' (2010) 1 J European Competition L & Practice 1; A Ezrachi (ed), *Article 82 EC—Reflections on its Recent Evolution* (Oxford: Hart Publishing, 2009).

Commissioner Neelie Kroes stated that the Commission's intention was not to propose a radical shift in enforcement policy but rather to 'develop and explain theories of harm on the basis of a sound economic assessment'.[27] While many agree on the clear merit of introducing more carefully balanced effects-based analysis when establishing abuse, the practicalities of such an approach have been difficult to agree upon.

In December 2008, the Commission published the Guidance Paper on its Enforcement Priorities in Applying Article 102 TFEU (ex Article 82 EC) to Abusive Exclusionary Conduct by Dominant Undertakings. Alongside the Commission's specific enforcement decisions, the Guidance Paper was,

> ... intended to provide greater clarity and predictability on the general framework of analysis which the Commission employs in determining whether it should pursue cases concerning various forms of exclusionary conduct and to help undertakings better assess whether a certain behaviour is likely to result in intervention by the Commission under Article 82.[28]

The Guidance Paper lays down the Commission's modern approach to Article 102 TFEU and delivers several of the innovative elements which were aired in the preceding papers.

In an attempt to weave an economic approach into the analysis of Article 102 TFEU, the Commission departed at times from the European Court's case law. Noteworthy, however, is the fact that the institutional division in Europe limits the Commission's ability to widen the analytical framework of Article 102 TFEU and depart from the European Court case law. In line with this limitation, the Guidance Paper stipulates that '... it is not intended to constitute a statement of the law and is without prejudice to the interpretation of [Article 102 TFEU] by the European Court of Justice and the Court of First Instance'.[29] Embedded, therefore, in the Guidance Paper, is the tension between the Commission's desire to promote a new approach to Article 102 TFEU and its subrogation to the European Court.

In principle, the introduction of additional economic-based variants to the Commission's analysis of abuse is to be welcomed. It reduces the risk of over regulation and of chilling competition which may result from formalistic analysis. It aims to distinguish between competition on the merits and anti-competitive unilateral action. In the context of refusal to license, such approach may provide a valuable fine-tuning instrument. An effects-oriented analysis may assist the dominant company when contesting a finding of abuse and establishing that a refusal to license does not generate anti-competitive effects. At the same time, it may also assist establishing

[27] Neelie Kroes, 'Preliminary Thoughts on Policy Review of Article 82' (23 September 2005) Speech at the Fordham Corporate Law Institute.

[28] Guidance Paper (n 3 above) para 2.

[29] Guidance Paper (n 3 above) para 3.

abuse in cases which do not amount to outright refusal to supply but involve weaker forms of exclusivity such as constructive refusal to license.[30] Such may be the case when the dominant undertaking unduly delays or undermines the licensing or charges unreasonable conditions in return.[31]

Yet, the practicality of the effect-based analysis is challenging. The move from a formalistic to an economic-based approach creates uncertainty as to the relevant benchmarks used to establish abuse.[32] In this respect, it has been argued that the Guidance Paper, being a relatively short document of enforcement priorities, provides limited account of the new concepts and puts forward a suboptimal framework.[33] New effects-based variants are considered but are generally left with no clear details as to their application and limiting principles. These variants, to the extent that they depart from the European Court case law, create uncertainty as to their application and scope.

C. The Guidance Paper

In the context of refusal to license, it is interesting to consider the way in which the analytical framework, as manifested in the Guidance Paper, may have affected the protection of IPRs. While the Guidance paper did not openly set to change the interface between competition law and IPRs, it seems to have nonetheless modified it. As will be illustrated below, these modifications affect the level of intervention and subsequently the protection awarded to IPR holders.

Refusal to supply and license

At first glance, the discussion of refusal to supply and license in the Guidance Paper reflects in broad lines the European Court judicature. The Commission's approach as manifested in the Guidance Paper is anchored in the principles developed in case law. Accordingly, the Commission states that intervention in such cases is the exception and will only take place following careful consideration of the effects it may have on innovation and investments.[34]

---

[30] The term is used by the Commission in the Guidance Paper (n 3 above) para 79.

[31] See generally, Guidance Paper (n 3 above) para 79.

[32] The notion of 'an equally efficient competitor' is heavily used in the Guidance to underpin the effects-based approach. See eg Guidance Paper (n 3 above) paras 42 ff and 59 ff.

[33] See for example: J Temple Lang 'Article 82 EC—The Problems and the Solution' (2009) FEEM Working Paper No. 65.2009 <http://papers.ssrn.com/sol3/papers.cfm?abstract_id=1467747##> accessed 3 September 2010; 'The Commission's Guidance on Article 82 EC and the Effects Based Approach—Legal and Practical Challenges' in A Ezrachi (ed) *Article 82 EC—Reflections on its Recent Evolution* (n 26 above).

[34] Guidance Paper (n 3 above) para 75.

A closer look at the Guidance Paper reveals, however, several areas in which the Commission rhetoric departs from existing case law. As will be illustrated below, the consequences of this departure are yet unclear.

Consumer harm

An apparent change may be found with respect to the benchmark for intervention. Here, the Commission modified the overall test as established in the case law and shifted emphasis to the finding of consumer harm. The Guidance Paper stipulates that the Commission will consider a refusal to supply as an enforcement priority and will intervene when three cumulative conditions are met:(i) the product is objectively necessary, (ii) the refusal would lead to elimination of effective competition in the downstream market, and (iii) the refusal would lead to consumer harm.[35]

This change of rhetoric signals an emphasis on effect. In an attempt to advance an effects-oriented approach, the Commission introduced *consumer welfare* as a main variant in the analysis. Accordingly, the Commission will balance the likely negative consequences of the refusal to supply on consumer welfare, against the negative consequences of imposing an obligation to supply.[36] The Commission considers prevention of innovation as an example of such harm, but leaves the condition open ended. One would expect the Commission to pour additional content into this benchmark as its case law develops. Until then, it is unclear what the new condition might encompass.

Exceptional circumstances

The role played by 'exceptional circumstances' in limiting the application of Article 102 TFEU has also been affected by the Guidance Paper. As illustrated in section A of this chapter, the Court has held that a duty to contract arises only in 'exceptional circumstances'.[37] This instrument serves as an important safety valve which embodies the need for proportionality, and restrains the application of competition law. It served as a balancing instrument even in instances in which the Court widened the realm of competition law, such as in the *Microsoft* case. Commenting on that case, Bo Vesterdorf, former president of the CFI, noted its significance in ensuring restrictive application and interpretation of Article 102 TFEU.[38]

Surprisingly, despite the central role of this instrument in the Union's judicature, and earlier references to it in the 2005 Discussion Paper,[39] the Commission omitted it

---

[35] Guidance Paper (n 3 above) para 81.

[36] Guidance Paper (n 3 above) para 86.

[37] On this point see: Lang, 'Article 82 EC—The Problems and the Solution' (n 33 above).

[38] B Vesterdorf 'Article 82 EC: Where do we stand after the *Microsoft* judgement?' (n 17 above) 9.

[39] In the Discussion Paper (n 2 above) para 239, the Commission noted that 'Only under exceptional circumstances can the refusal to license an IPR be considered an abuse' citing Case 238/87 *Volvo* (n 5 above), Joined Cases C-241/91 P and C-242/P *Magill* (n 7 above), and Case C-418/01 *IMS Health* (n 11 above).

from the Guidance Paper. While the Commission states that intervention requires careful consideration, it does not state clearly that a duty to contract would arise only in 'exceptional circumstances'.[40] As such it appears to have removed from its tool box, or at least push aside, a significant limiting principle.[41]

Follow-on innovation

Another example which illustrates a departure from case law, concerns follow-on innovation. The Guidance Paper stipulates on that point that,

> The Commission considers that consumer harm may, for instance, arise where the competitors that the dominant undertaking forecloses are, as a result of the refusal, prevented from bringing innovative goods or services to market and/or *where follow-on innovation is likely to be stifled* [Case T-201/04 *Microsoft v Commission*]. This *may be particularly the case* if the undertaking which requests supply does not intend to limit itself essentially to duplicating the goods or services already offered by the dominant undertaking on the downstream market, but intends to produce new or improved goods or services for which there is a potential consumer demand or is likely to contribute to technical development [Case C-418/01 *IMS Health v NDC Health*].[42]

This statement illustrates the imperfect overlap between the broad concepts described in the Guidance, and as developed by the European Court. With respect to *follow-on innovation*, the Guidance Paper includes a reference to the *Microsoft* and *IMS Health* cases and suggests that such innovation is likely to be stifled *in particular* when the undertaking which requests supply does not intend to duplicate the goods. Yet, in its *IMS Health* judgment the ECJ held that a refusal by a dominant undertaking to license an indispensable product '... may be regarded as abusive *only* where the undertaking which requested the licence does not intend to limit itself essentially to duplicating the goods or services already offered on the secondary market.'[43] The Commission favoured a less restrictive approach and has put forward a broad statement which is not accompanied by the factual matrix surrounding the cases from which the benchmark was imported. The result is a likely widening of the instances in which abuse may be found. [44]

---

[40] On this point see: Lang, 'Article 82 EC—The Problems and the Solution' (n 33 above).

[41] *Ibid.*

[42] Guidance Paper (n 3 above) para 87 (emphasis added).

[43] C-418/01 *IMS Health* (n 11 above) para 49.

[44] The Commission's statement in the Guidance Paper may also be contrasted with its own statement in the 2005 Discussion Paper where it noted that a refusal by a dominant company to license access to the IPR could be considered abusive when '... the refusal to grant a licence prevents the development of the market for which the licence is an indispensable input, to the detriment of consumers.' This the commission added, '... may *only* be the case if the undertaking which requests the licence does not intend to limit itself essentially to duplicating the goods or services already offered on this market'. (Discussion Paper (n 2 above) para 239).

Establishing foreclosure

More generally the Guidance Paper presents the Commission's approach for establishing foreclosure. The term anti-competitive foreclosure is used in the Guidance Paper to describe,

> . . . a situation where effective access of actual or potential competitors to supplies or markets is hampered or eliminated as a result of the conduct of the dominant undertaking whereby the dominant undertaking is likely to be in a position to profitably increase prices to the detriment of consumers.[45]

The Commission stipulates the factors which would be assessed when considering intervention.[46] Among these, the Commission notes the evidence which may be used to establish actual foreclosure,

> . . . [I]f the conduct has been in place for a sufficient period of time, the market performance of the dominant undertaking and its competitors may provide direct evidence of anti-competitive foreclosure. For reasons attributable to the allegedly abusive conduct, the market share of the dominant undertaking may have risen or a decline in market share may have been slowed. For similar reasons, actual competitors may have been marginalised or may have exited, or potential competitors may have tried to enter and failed . . .[47]

The direct, simplified, link between structural changes in the market and abuse is surprising. It stands at odds with an economic approach, as it treats market share and market structure as direct evidence when establishing abuse. Needless to say, changes in market share may be brought about by research and development, innovation, and changes in consumer preferences, without abusive conduct of a dominant company. Despite this, the Commission's approach may further widen the instances in which a company may be found to abuse its dominant position by refusing to supply or license.

Efficiency defence

The introduction of an efficiency defence in the Guidance Paper is welcome as it may help inject an economic appraisal into the analysis of abuse. It introduces fine tuning which moves away from a formalistic approach and explores the real welfare consequences of the conduct. As such it could potentially counterbalance any widening trends as illustrated above. Such defence could in principle allow for the dominant company to justify its action when it can show that efficiencies brought about by the refusal to license outweigh the likely negative effects stemming from the alleged anti-competitive conduct.

---

[45] Guidance Paper (n 3 above) para 19.
[46] Guidance Paper (n 3 above) para 20.
[47] Guidance Paper (n 3 above) para 20.

In the Guidance Paper the Commission notes two instances which may give rise to such claim by an IPR holder, both focusing on the incentive to invest and innovate. The first is justified by the need to allow the dominant undertaking to realize an adequate return on the investment required to develop its input business. The second concerns the effect on the dominant undertaking's innovation and the need to generate incentives for it to continue to take risks and invest in innovation. With respect to the latter, the Commission will consider the effects that an obligation to supply and follow-on innovation by competitors will have on the dominant undertaking.[48] This is in line with the 2005 Discussion Paper where in the context of licensing, the Commission noted the need to consider whether the intellectual property rights may be the result of substantial investment entailing significant risks.[49] This may particularly be the case when the investment would not have been made if the investor had known that it would have a duty to supply.[50]

Note however that these considerations have traditionally formed part of the initial assessment of abuse by the Commission and European Court. As indicated by the Commission in the Guidance Paper—intervention in cases of refusal to supply is the exception and will only take place following careful consideration of the effects it may have on innovation and investments.[51] One may therefore question the extent to which the novel efficiency defence adds and refines the analysis in the context of IPRs. It is worth noting, however, that it does change the balance of proof. Whereas before it was an element of establishing abuse by the Commission, under the Commission Guidance Paper it would be a defence that the defendant company would need to establish itself.

In its 2005 Discussion Paper the Commission noted that if the conduct of a dominant undertaking generates efficiencies and satisfies the other conditions of Article 101(3) TFEU, such conduct should not be classified as an abuse under Article 102 TFEU.[52] Accordingly the proposed efficiency defence under Article 102 TFEU was based on four conditions:

(1) that efficiencies are realized or likely to be realized as a result of the conduct concerned;
(2) that the conduct concerned is indispensable to realize these efficiencies;

[48] Guidance Paper (n 3 above) para 89. Note Case T-201/04 *Microsoft* (n 14 above) in which a similar claim made by Microsoft was rejected (paras 688–711).

[49] Discussion Paper (n 2 above) para 235.

[50] Discussion paper (n 2 above) para 236.

[51] Guidance Paper (n 3 above) para 75.

[52] Discussion Paper (n 2 above) para 8. The Commission substantiates this approach by reference to the CFI judgment in Case T-193/02 *Laurent Piau v Commission* (CFI, 26 January 2005) [2005] ECR II-209 para 119.

(3) that the efficiencies benefit consumers;
(4) that competition in respect of a substantial part of the products concerned is not eliminated.[53]

The Guidance Paper retains, by and large, this approach to efficiencies.[54] With respect to the fourth condition it stipulates that,

> ... where there is no residual competition and no foreseeable threat to entry, the protection of rivalry and the competitive process outweigh possible efficiency gains.[55]

The Commission then adds that in its view,

> Exclusionary conduct which maintains, creates or strengthens a market position approaching that of a monopoly can normally not be justified on the grounds that it also creates efficiency gains. [56]

To satisfy the fourth condition the dominant undertaking should show that '... competition in respect of a substantial part of the products concerned is not eliminated'. Note however that this condition is (an imperfect) mirror image of the refusal to license test in its narrow interpretation. Recall, that when applied narrowly the finding of abuse will be substantiated due to the refusal being found '...likely to exclude all competition in the secondary market'.[57] In such case the elimination of all competition would exclude the use of the efficiency defence. This is not surprising, considering the general view, as expressed by the Commission, according to which '... the protection of rivalry and the competitive process is given priority over possible pro-competitive efficiency gains.'[58]

The defence may however play a role when the refusal to supply or license was established on a more liberal interpretation than that established in the *Magill* judgment, as seems to be favoured by the Commission. In cases where the Commission established abuse on the basis of negative effect on competition in the downstream market[59] or partial elimination of competition, the defence could provide a (limited) balancing tool. It would, however, be for the undertaking to bear the burden of proof and establish the conditions of this yet unexplored vehicle.

Overall effect

As illustrated above, while the Commission's Guidance Paper retains the overall framework of analysis of refusal to supply and license, it strips it from some of its vital

[53] Discussion Paper (n 2 above) para 84.
[54] Guidance Paper (n 3above) paras 30–1.
[55] Guidance Paper (n 3 above) para 30.
[56] Guidance Paper (n 3 above) para 30.
[57] See the *Magill* and *Oscar Bronner* judgments (n 7 and n 8 above).
[58] Discussion Paper (n 2 above) para 91.
[59] Discussion paper (n 2 above) para 231.

limiting principles. This not only raises fundamental questions on the interface between competition law and IPR, but also stands at odds with the European Court judicature. The introduction of the efficiency defence, while valuable, provides limited and yet uncertain relief.

To better understand the intricacy presented by the Guidance Paper, one may recall comments made by William Blumenthal in his work on the US Merger Guidelines. There he identified a number of variables which contributed to their success.[60] He noted that the main strength of the 1982 US Merger Guidelines, and the revisions which followed, was in the detail of the document and the fact that it left the basic analytical framework intact by adhering to pre-existing law. In addition he pointed to the balance between simplicity and flexibility and the practicality of the merger guidelines, as some of the other variants which contributed to their success.

Quite to the contrary, the Guidance Paper seems to depart, at times, from the case law and fails to deliver the requisite level of detail. Principles are detached from the factual matrix which accompanied them in each case—resulting in more flexible and vague benchmarks. New concepts, such as the efficiency defence and consumer harm are relatively abstract and provide limited relief.

As hinted above, some of these deficiencies stem from the difficult position in which the Commission finds itself. The European institutional framework subrogates the Commission's creative leeway to the Community judicature. The Commission is therefore limited in its ability to modify the law. Accordingly, the Guidance Paper clarifies that it is not intended to constitute a statement of the law.[61] The practical impact of this was highlighted, in *British Airways v Commission,* when Advocate General Kokott noted that even if the Commission's administrative practice was to change, it '. . . would still have to act within the framework prescribed for it by [Article 102 TFEU] as interpreted by the Court of Justice.'[62] Arguably, this limitation led the Commission to a creative approach in treating the document as guidance on enforcement priorities and not guidelines. The title echoes the hybrid nature of the document and the potential tension between some of the new concepts and the case law of the European Courts. Whilst its positioning as an enforcement priority document might have been useful in avoiding direct conflict with the courts, it undermines the relative certainty which other guidelines have provided in the past.

[60] W Blumenthal 'Clear Agency Guidelines: Lessons from 1982' (2000) 68 Antitrust LJ 5; W Blumenthal 'Ambiguity and discretion in the new guidelines: some implications for practitioners' (1993) 61 Antitrust LJ 469.

[61] Guidance Paper (n 3 above) para 3.

[62] Case C-95/04 *British Airways v Commission* [2007] ECR I-2331 para 28.

D. Application at Union and Member States Levels

Some may argue that the new balancing point advanced in the Guidance Paper provides for a clear and adequate benchmark for intervention. Note for example Commission officials, Luc Peeperkorn and Katja Viertiö, who have commented on the three conditions for intervention as set in the Guidance Paper and described them as setting a 'stricter standard for intervention.'[63] Accordingly, one would argue that the Guidance Paper provides a better standard for intervention which would foster investment and innovation.[64]

Others may disagree and consider the changes described above to undermine the essence of IPR. It seems that the Commission has fostered an evolution which aims to widen the instances in which it may intervene. Assuming the Commission has carefully considered the wording in the Guidance Paper (any other assumption would be worrying in itself), the omission of some of the limiting principles, and the use of a broad brush to define the boundaries of abuse, were used to lower the threshold for intervention. New balancing principles such as consumer harm and efficiency defence lack, at least at present, the necessary detail to effectively counter balance this widening trend.

Whichever view one takes on this issue, there is little doubt that the Guidance Paper presents a change in the rhetoric and as such affects the perceived benchmark for intervention. Holders of IPRs face the difficult challenge of decoding this recent evolution. Faced with a blurred contact point between Article 102 TFEU and IPRs, they will have to establish their view on the possible consequences of refusal to license and the likelihood for intervention from the Commission.

One organ which may assist in clarifying the position is the European Court which has the power to limit any departure from existing case law, or clarify its scope. Note however that the limited number of cases on refusal to license and the slow turnover of cases on appeal, hinder the courts' ability to react in a timely fashion to the recent expansion in the Guidance Paper. Also recall that the Court has traditionally restricted its appraisal of complex economic and technical matters.[65] True, the court may, if it so desired, examine the Commission's appraisal of economic data and engage in complex analysis.[66] Yet the *Microsoft* judgment illustrated that it doesn't often have the appetite to do so. Subsequently, time may pass before the Court instils clarity.

---

[63] L Peeperkorn and K Viertiö 'Implementing an effects-based approach to Article 82' (2009) 1 Competition Policy Newsletter 20 <http://ec.europa.eu/competition/publications/cpn/2009_1_5.pdf> accessed 3 September 2010.

[64] On the treatment of IPR see generally: C Ritter, 'Refusal to Deal and "Essential Facilities": Does Intellectual Property Require Special Deference Compared to Tangible Property?' [2005] 28(3) World Competition 281.

[65] Case T-201/04 *Microsoft* (n 14 above) paras 87–8.

[66] Case T-201/04 *Microsoft* (n 14 above) para 482.

It is important to emphasize that the debate extends beyond the Commission's reading of the case law and its enforcement priorities. The reason for this is the parallel competence shared by the Commission, National Courts, and National Competition Agencies in applying Article 102 TFEU.[67] The multitude of players which may refer to the Guidance Paper when determining the adequate level of intervention, adds further complexity. Evidently, the Guidance Paper, being soft law, functions outside the precedent system and has no binding effect on National Courts. However, it is well accepted that,

> ... national courts may find guidance in Commission regulations and decisions which present elements of analogy with the case they are dealing with, as well as in Commission notices and guidelines relating to the application of [Articles 101 and 102 TFEU] and in the annual report on competition policy.[68]

This position is also reflected in national legislation, for example in the provisions of the UK Competition Act 1998 (the Act). Section 60(2) of the Act stipulates that the national court should avoid inconsistencies with decisions of the European Courts. Section 60(3) establishes that the national court '... must have regard to any relevant decision or statement of the Commission'.[69]

The role of the Guidance Paper in the development of competition law at Member States' level is thus more significant than that which its informal non-binding nature may suggest.[70] Accordingly, despite its focus on enforcement priorities and its lack of binding effect, the Guidance Paper can potentially play a paramount role at Member States' level.[71] Here, possible inconsistency may emerge as different National Courts and Competition Agencies across Europe endorse, in an inconsistent manner, the Commission's modified approach to refusal to license. Note that to the extent that these national organs believe the Guidance Paper to reflect the Union judicature or to provide a natural continuation of the widening trends in the case law, they may favour adopting its stance.[72] The potential for inconsistency is clear. The large group

[67] Article 5, Regulation 1/2003; see also Commission Notice on the cooperation between the Commission and the courts of the EU Member States in the application of Articles 81 and 82 EC [2004] OJ C 101/54.

[68] *Ibid* para 8.

[69] Note more generally Council Regulation (EC) 139/2004 on the control of concentrations between undertakings (hereinafter the 'Merger Regulation') [2004] OJ L24/1 art 16; *Inntrepreneur Pub Company (CPC) and others v Crehan,* [2006] UKHL 38 (House of Lords (English Court)).

[70] *Ibid.*

[71] See for example the ECJ judgment in C-322/88 *Grimaldi* [1989] ECR I-4407 para 18.

[72] On the possible wide interpretation of the European case law (in the pre-Guidance era) note for example the English Court judgment in *Intel Corporation v Via Technologies Inc* and its open ended approach toward 'exceptionality'. In a summary judgment the Court held that a patent holder may be restricted from exercising its IP rights in circumstances which differ from those identified in *Magill* and *IMS:* 'Magill and IMS indicate the circumstances which the Court of Justice and the President of the Court of First Instance respectively regarded as exceptional in the cases before them. It does not follow that other circumstances in other cases will not be regarded as exceptional.' It therefore concluded that it is at least arguable that there could be a breach of Article 102 TFEU

of 'users' which may put their faith in the document increases the 'uncertainty multiplier' and highlights the possible negative effect that even marginal inconsistencies may generate.[73]

Conclusion

While not setting to revolutionize the law, the Guidance Paper has shifted the balancing point between competition law and IPR. In doing so it modified the position established in the case law.

While the Guidance Paper heralds change, this change is only partially driven by an economically oriented approach. On this point it seems that the focus on consumer harm and the use of efficiency defence can be viewed as valuable effects-driven changes. Although these concepts are yet abstract they may provide an important contribution once they mature in future years.

On the other hand, other changes in the Guidance Paper seem to pull in the other direction and are underlined by a widening trend. This is predominantly the result of the 'broad brush' applied by the Commission when describing its approach to refusal to license. On this point note that unlike judgments of the European Court, modifications to the analysis of abuse in the Guidance Paper are not confined by the factual matrix of each case. Their inherent nature, subsequently, provides for wider scope for flexibility.

The result is a perceived lowering of the threshold for intervention and subsequently further erosion to the protection awarded by IPRs. As illustrated above, these trends are not confined to the Commission and may also find their way to the Member States' level.

The consequence of these processes is yet unclear. One may question the extent to which the Commission set to lower the benchmark for intervention. Was the perceived lowering of the threshold for intervention an unintentional evolution?

without the exclusion of a wholly new product or all competition. Case No A3/2002/1380, A3/2002/1381 *Intel Corporation v Via Technologies Inc* [2002] EWCA Civ 1905 (English Court of Appeal) para 47.

[73] In this respect the Guidance's 'uncertainty multiplier' differs from previous Commission Guidelines on merger control where the Commission has sole jurisdiction to take decisions (see Article 21 of Council Regulation 139/2004, the Merger Regulation (n 69 above)). The impact of any possible ambiguity in merger guidelines is easily contained and resolved through development of case law by the European Commission and the courts on appeal. It also differs from other guidelines which are addressed to a 'large group of users' but are more in line with the community judicature. See for example the Guidelines on the application of Article 81(3) of the Treaty (2004) which reflects the case law as developed over the years by the Commission and Community courts.

And even if so, can IPR holders risk such assumptions? Can they assume that all stakeholders would share it?

The lack of clarity and fear of litigation may affect the dominant undertaking's behaviour when considering refusal to license and royalty charges.

While the widening trends described above may only be perceived, the costs they carry are real.

6

EXISTENCE, EXERCISE, AND EXCEPTIONAL CIRCUMSTANCES: THE LIMITED SCOPE FOR A MORE ECONOMIC APPROACH TO IP ISSUES UNDER ARTICLE 102 TFEU

*John Kallaugher**

Introduction

The shift to a 'more economic approach' to Article 102 TFEU enforcement reflected in the Commission's Article 102 Enforcement Guidance[1] has been welcomed by many academic commentators and practitioners. Focusing on economic analysis can, however, lead to impatience on the part of regulators with 'formalistic' legal limitations on the use of competition law powers once economic harm has been identified. A principal target for such impatience is the limitation on application of competition law in the context of IP rights. Such frustration with formal limits on economic application of competition rules is evident in the Commission's Pharmaceutical Sector Inquiry. The final report in the Inquiry[2] (the 'Report') is a

* The author would like to record his particular thanks to his colleague Alana Tart, for her assistance in researching and revising this paper.

[1] Communication from Commission, Guidance on the Commission's Enforcement Priorities in Applying Article 82 of the Treaty to abusive exclusionary conduct by dominant undertakings [2009] OJ C 45/02 ('Article 102 Enforcement Guidance').

[2] Pharmaceutical Sector Inquiry Final Report (DG Competition Staff Working Document) dated 8 July 2009, accompanied with a Commission Communication (executive summary). These documents, and other public materials related to the pharmaceutical sector inquiry are available at the Commission's website: <http://ec.europa.eu/competition/sectors/pharmaceuticals/inquiry/index.html> accessed 8 September 2010. A good summary of the legal and policy issues that would arise if the Commission attempts to use Article 102 TFEU to address the issues identified in the Report can be found in L Kjolbe, 'Article 82 EC as Remedy for Patent System Imperfections: Fighting Fire with Fire?' (2009) 2 World Competition 163.

500-page description of serious economic harm attributable to dominant firms. Although the authors of the Report have generally been careful to avoid formal legal conclusions, the strong implication is that much of this conduct ought to be subject to control under Article 102 TFEU. The barrier to this economically sensible application of Article 102 TFEU is, of course, the special position of IP rights. The Report does recognize that IP rights play an important role in encouraging and rewarding innovation. The Report concedes that where patents fulfil this role they should be protected under competition law. The Report identifies, however, examples of cases where it suggests that patents are not fulfilling their proper economic role. The Report questions whether follow-on patents for production processes or chemical formulations represent a significant addition to the creative value of the original innovation.[3] The Report is also critical of the situation where a dominant firm obtains patents that it does not intend to practice itself, but which prevent a rival from introducing its own competing products. Here the Report observes that the patents do not reward innovation because the benefits to the patentee do not derive from either licensing or manufacture using the patented innovation.[4] A more economic approach that allowed a case-by-case assessment of whether IP rights were performing their economic function would allow the Commission to challenge this behaviour.

The Pharmaceutical Sector Inquiry raises the broader question of whether the general shift to a more economic approach under Article 102 TFEU calls for a new look at the interface between Article 102 enforcement and IP rights. Perhaps the scope of the legal exception to Article 102 TFEU created by IP rights should be defined on the basis of an economic analysis of both the benefits of IP rights and the full economic costs of their exercise in a specific case. Alternatively, the 'exceptional circumstances test' could be adapted to provide a basis for a full economic analysis.

This paper addresses these questions from a theoretical and practical perspective. The first part provides a brief discussion of whether a more economic approach is really the best way to resolve the tensions inherent in the relationship between IP rights and competition law enforcement. The second part reviews the established structure for resolving conflicts between competition policy and IP rights—the existence/exercise paradigm, as well as the role of intent evidence in Article 102 cases. This review suggests that there is little scope for introducing economic analysis to analysis of core IP rights other than through the 'exceptional circumstances' rule. The third part considers the extent to which 'exceptional circumstances' test provides a basis for introducing a case-specific economic analysis.

[3] Pharmaceutical Sector Inquiry Final Report (n 2 above), sections C 2.1.2 through C 2.1.4 on 'patent clusters' and 'divisionals' and section 3.2.1 of Commission Communication (executive summary).

[4] Pharmaceutical Sector Inquiry Final Report (n 2 above), section C 3.1.2 on 'defensive patenting' and section 3.3.1 of Commission Communication (executive summary).

A. Does a 'More Economic Approach' to IP-Competition Law Issues Make Policy Sense?

The 'more economic approach' to Article 102 enforcement is based on the premise that the primary goal of competition policy should be to enhance consumer welfare. The Commission's Article 102 Enforcement Guidance[5] works from this principle to assess the consumer welfare effects of dominant firm conduct in a specific case or category of cases, limiting enforcement to cases where conduct is likely to lead to competitive harm. This approach follows the case-specific economics-based approach applied to restrictive agreements in the Commission's *Horizontal Guidelines*[6] and *Vertical Guidelines*[7] and in the Commission's well developed case law under the Merger Regulation.[8] It represents a shift from formalistic categories of permitted or prohibited behaviour, similar to that achieved in the context of restrictive agreements with the abolition of the Regulation 17[9] notification system in favour of self assessment under Regulation 1/2003.[10]

An economic approach is already the basis for applying competition law to 'non-core' exercise of IP rights[11] and there is a certain logic to the idea that the resolution of all IP-competition law conflicts would benefit from a more economic approach.[12] After all (this argument runs), the traditional limits on competition enforcement in IP cases are a clear example of legal formalism. Indeed, the creation of separate rules for a category of 'core rights' is itself an example of formal analysis at odds with a full economic approach. Defining those limits on the basis of legal categories may have been sensible when competition law itself relied on formal legal categories. Now that

---

5 Article 102 Enforcement Guidance (n 1 above).

6 Commission guidelines on the applicability of Article 81 of the EC Treaty to horizontal cooperation agreements OJ 2001 C3/2.

7 Commission Notice—Guidelines on vertical restraints [2000] OJ C 291/1.

8 Council Regulation (EC) No 139/2004 of 20 January 2004 on the control of concentrations between undertakings [2004] OJ L24/1.

9 Council Regulation No 17: First Regulation implementing Articles 85 and 86 of the Treaty [1962] OJ 294/62, English Special Edition [1959–1962] OJ 87.

10 Council Regulation (EC) No 1/2003 of 16 December 2002 on the implementation of the rules on competition laid down in articles 81 and 82 of the Treaty [2003] OJ L 1/1.

11 Commission Notice—Guidelines on the application of Article 81 to technology transfer agreements [2004] OJ C101/2. As explained in Part B of this chapter, contractual restrictions imposed by a licensor or licensee have traditionally been viewed as outside the specific subject matter of IP rights and therefore subject to control by competition law.

12 For a considered argument for a more economic approach, see J Drexl, 'Is there a "more economic approach" to intellectual property and competition law?' in J Drexl (ed), *Research Handbook on Intellectual Property and Competition Law* (Cheltenham: Edward Elgar, 2008) 27, where the author concludes:

> European competition law is in need of a more economic approach to intellectual property and competition law. The case law of the European Court of Justice on refusal to license proves that a normative approach to striking a balance between the exclusivity of the IPR and competition law intervention cannot provide a test that adequately mirrors the goals and functions of the specific IPR in the context of the relevant market.

competition law has moved to a more case-specific economics-based approach, should IP law-based defences not also look beyond formal status to economic reality? Should deference to IP rights not therefore be limited to the cases where IP rights promote their legitimate economic objectives?

While the policy argument for a more economic approach may be superficially attractive, it does not take full account of the nature of the conflict between IP law and competition law. This conflict does not simply involve the definition of an 'economic justification' defence to a competition law claim, but rather the resolution of different policy objectives that are achieved through different legal structures. Although the policy goals are broadly complementary and the legal structures overlap at critical points, a case-specific economics-based approach may not provide the best resolution of the conflict where core IP rights are concerned.

It is broadly understood that the policy goals of the IP law system and the competition law system are complementary, in that both systems contribute to economic welfare in a broad sense. The policy conflict arises because each system contributes to consumer welfare in a fundamentally different way. The IP system contributes to consumer welfare primarily by creating dynamic efficiencies.[13] Competition law, while acknowledging the importance of dynamic efficiency at various points,[14] primarily contributes to consumer welfare by encouraging static or allocative efficiencies.[15] The focus on static or allocative efficiency reflects, in part, the different economic analysis applicable to dynamic efficiency claims. The well-developed economic analysis of static effects in competition cases means that in case-by-case assessment, short-term harm will always be easier to quantify. The legal test in competition cases

[13] Dynamic efficiency refers to the efficient allocation of resources between different investment projects to develop new products and services. In many dynamic markets, firms compete for the market rather than in the market. Intellectual property rights can encourage innovations in products and services by permitting successful innovators to recoup the benefits of their investments through such rights. It is noted, however, that trade mark rights may contribute more to allocative rather than dynamic efficiency as trade marks can influence the value consumers place on a product (allocative efficiency relates to the difference between the cost of producing the marginal product and the valuation of that product by consumers). As a result trade mark rights may be easier to incorporate into competition law analysis than other rights such as patents and copyright.

[14] See for example, possible efficiency defences in merger cases: Guidelines on the assessment of horizontal mergers under the Council Regulation on the control of concentrations between undertakings [2004] OJ C31/5 ('Horizontal Merger Guidelines') paras 15, 76–8; the Article 102 Enforcement Guidance (n 1 above) paras 89–90; and efficiency justifications open to undertakings under Article 101(3) TFEU: Guidelines on the application of Article 81(3) of the EC Treaty [2004] OJ C101/97, section 3.2.2.2 and para 105, which distinguish between cost efficiencies and efficiencies of a qualitative nature which create value in the form of new or improved products.

[15] Static efficiency models focus on prices and quantities and take the range of products available and methods of producing these products as a given. Static efficiency is attained when the price of a good is set as equal to the true opportunity cost of producing it. Economic and legal analysis in competition policy and regulation tend to take more of a static view and ignore dynamic aspects of the market, see eg the Horizontal Merger Guidelines (n 14 above) and the Commission Guidelines on the Assessment of Non-Horizontal Mergers under the Council Regulation on the control of concentrations between undertakings [2008] OJ C265/6.

is also biased against justifications based on dynamic efficiencies, since defendants face a high burden of proof where a short-term reduction in allocative efficiency has been shown.[16]

The tension between the role of the IP system in promoting dynamic efficiency and the focus of the competition law system on creating and preserving static efficiency is accentuated by a fundamental difference in the way that each system contributes to achieving efficiency goals. IP law is largely 'rules-based', while competition law enforcement is largely outcome-based. As a general matter, obtaining or enforcing an IP right requires an applicant to satisfy legal requirements that are reasonably well-defined by the relevant statutes and case law. Some of these requirements do tend to contribute to dynamic efficiency (such as the requirement that a patent reflect an advancement on prior art). The examiner or the enforcing court will not generally look beyond the formal legal requirements, however, to determine whether a patentee has in fact engaged in costly research and development or whether copyrighted material reflects culturally valuable creative effort. Furthermore, while the goal of the IP rights system may be to provide incentives for innovation and creative activity, the grant of an IP right does not create an automatic right to an economic return, which will depend on demand for products protected by the rights. Thus IP rights contribute to dynamic efficiency as a 'system'—they do not require investigation of whether the grant of a right in a specific case rewards genuine creative effort. The economics-based approach to competition law, in contrast, focuses on the assessment of individual cases against the goal of maximizing consumer welfare. Rules and guidelines provide a structure for analysis, but the focus is on determining the optimal economic result in a specific case.

These fundamental differences have important consequences for the interaction of IP law and competition policy. The contribution of IP law to dynamic efficiency depends on the operation of a set of rules that over time will tend to create incentives for conduct leading to improvements in dynamic efficiency. Changing those rules on a case-by-case basis could affect the incentives offered by the IP system with results that cannot be easily predicted. Simply 'asking the question' of whether rights in a specific case are economically justifiable creates a potential new legal issue in every infringement case and thereby reduces the value of IP rights across the board. Furthermore, since the competition law system functions on the basis of case-specific assessment, the comparison of the immediate and demonstrable impact of conduct on static efficiency will almost always appear more important than the largely theoretical impact of intervention on the efficiency benefits of the IP system as a whole.

[16] In merger cases, for example, efficiencies must benefit consumers, be merger specific and be verifiable: Horizontal Merger Guidelines (n14 above) paras 78–88; Under Article 2 of Regulation 1/2003 the undertaking or association of undertakings claiming the benefit of Article 101(3) TFEU shall bear the burden of proving that the conditions of that paragraph are fulfilled. The conditions of Article 101(3) are cumulative and therefore undertaking(s) must fulfil all four.

This creates an entirely understandable tendency for competition law enforcement authorities to expand the envelope of competition law enforcement by limiting the scope of IP rules.[17]

Taken together, the difference in policy objectives and the different manner in which these objectives are achieved suggest that IP/competition law conflicts may not be best resolved using economic analysis on a case-by-case basis. A form-based approach may indeed be the optimal way to define the boundaries of competition law enforcement where core IP rights are concerned. The existence/exercise paradigm, as developed in the case law of the European Courts over the last 40 years may achieve a balance of the competing interests that is preferable to open-ended economic analysis, particularly with the safety valve of 'exceptional circumstances'.[18]

B. Specific Subject Matter, Essential Function, and Anti-competitive Intent: The Limits on Article 102 Enforcement

Commentators have identified a number of different ways to implement a 'more economic assessment' of IP issues in competition cases. This could involve an assessment of whether the IP rights in question fulfil their economic function of fostering dynamic efficiency; such an assessment might require an evaluation of whether the rights go further (in terms of scope or duration) than required to serve their legitimate economic function. It could involve assessment of the economic context in which rights are exercised.[19] It could involve a balance of the economic harm

[17] The tendency to expand to competition law competence receives an additional impetus in the context of EU competition law because of the division of competence between the European Commission, which has primary responsibility for enforcement of competition law, and the Member States, which retain direct responsibility for IP rights (see eg Case C-24/67 *Parke Davis& Co v Probel & Centrafarm* [1968] ECR 55 para 71). Many observers perceive the IP system as too generous, both in the criteria for granting IP protection and in the application of those criteria. Since the Commission cannot act directly to remedy these deficiencies by changing the IP system, there is an understandable interest in testing its ability to deal with these deficiencies using its competition law tools.

[18] The purpose of this paper is not to reach a final conclusion on the optimal basis for resolving IP-Competition Law conflicts from an economic perspective. The purpose of the discussion in the text is rather to suggest that it cannot be *assumed* that in this context more economic analysis equates to better decision-making. The possibility that a formal 'per se' legality rule or a strict exceptional circumstances test might be preferable from a welfare perspective to extensive economic analysis is discussed in Y Katsoulacos, 'Optimal Legal Standards for Refusals to License Intellectual Property, A Welfare-Based Analysis' (2008) 5 J Competition L and Economics 269.

[19] For example, Professors Drexl and Heinemann have argued that a mandatory licence requirement might be justified where a market is characterized by high barriers to entry, so that the market is not 'contestable': A Heinemann, 'The contestability of IP-Protected Markets' in J Drexl (ed), *Research Handbook on Intellectual Property and Competition Law* (n 12 above) 54; J Drexl, 'Abuse of Dominance in licensing and Refusal to License: "A more economic approach" to competition by imitation and competition by substitution' in CD Ehlermann and I Atanisiu (eds), *European Competition Law Annual 2005: the Relationship Between Competition Law and Intellectual Property* (Oxford: Hart, 2007) 647. Professor Schweitzer has accepted this suggestion while otherwise arguing that the conditions for mandatory licensing should be limited; H Schweitzer, 'Controlling the Unilateral Exercise of Intellectual

resulting from a refusal to license in a specific case against the effects on 'incentives to innovate'. This balancing exercise could be limited to the impact on dynamic efficiency or could include the impact of the conduct on static efficiency.[20] A more limited assessment could focus on the 'intent' of IP owners so that conduct could be controlled by competition law where it pursued a goal outside the scope of the relevant rights.[21]

In developing models for achieving a more economic approach, academic commentary has understandably focused on identifying the optimal policy balance between IP protection and competition law enforcement. A move to more economic analysis of the treatment of core IP rights must also, however, be accommodated with the existing law. This is not necessarily straightforward, particularly since a change in the analytical basis for assessing IP-related Article 102 claims could affect the rules applicable under Article 101 and could also have a broader impact on the rules involving free movement.

The limitation on application of Article 102 TFEU to core IP rights originates in the case law of the ECJ regarding the distinction between existence and exercise in free movement cases. To appreciate the legal constraints on introduction of a 'more economic approach' it is important to understand the basis for this rule in the free movement context and its application first in Article 101 and then in Article 102 TFEU to limit the application of competition rules. The proponents of a 'more economic approach' are not writing on a 'clean slate'. The idea that core IP rights are subject to Article 102 TFEU where they do not serve their economic function is not new.

Property Rights, A Multitude of Approaches but no Clear Way Ahead' in I Govaere and H Ullrich (eds), *Intellectual Property, Market Power and the Public Interest* (Brussels: Peter Long, 2008) 59.

[20] See, eg, in addition to the authors mentioned in the previous note, B Conde Gallego, 'Unilateral refusal to license indispensable intellectual property rights—US and EU approaches' in J Drexl (ed) (n 12 above) 215; M Dolmans, R O'Donoghue, and J Loewenthal, 'Article 82 EC and Intellectual Property: the State of the Law Pending the Judgment in Microsoft v. Commission' (2007) 3 Competition Policy Intl 108 ('*State of the Law*'); W Kerber and C Schmidt, 'Microsoft, Refusals to Deal and the Incentive Balance Test of the EC Commission' (2008) <http://ssrn.com/abstract=1297939> accessed 8 September 2010.

[21] While this would not involve a full economic assessment, it would still require case-specific consideration. The Preliminary Report does acknowledge that intent is irrelevant under patent law, but notes (for example at footnotes 375 and 376) that 'intention can . . . be taken into account in competition law assessments'. (European Commission, DG Competition Staff Working Paper, 'Pharmaceutical Sector Inquiry Preliminary of likely effects and identification of proper scope of the right in question' ('Preliminary Report'), 28 November 2008,<http://ec.europa.eu/competition/sectors/pharmaceuticals/inquiry/preliminary_report.pdf> accessed 8 September 2010). See also the discussion of the Commission Decision in *AstraZeneca* (COMP/A.37.507/F3), 15 June 2005 in Annex A to the Report at para 13. This approach was criticized by the European Patent Office in its response to the Preliminary Report, since the intent of applicants did not form part of the assessment of patent claims (<http://ec.europa.eu/competition/consultations/2009_pharma/index.html> accessed 8 September 2010).

It formed the basis for the CFI judgment in *Magill*[22] and some commentators suggest that 'essential function' remains a legal basis for limiting the exercise of core IP rights.[23] The idea that the absence of 'incentives to innovate' justifies interference with core rights is also not new and was rejected by the ECJ in *Renault*[24] and implicitly in *Magill.* The idea that an anti-competitive object might render IP-related conduct subject to Article 102 TFEU was raised by the CFI in *Magill*, soundly criticized by the Advocate General, and ignored by the ECJ. This means that a more economic approach can probably only be introduced through development of the 'exceptional circumstances' test.

Specific subject matter and essential function in free movement cases[25]

Starting with *Consten and Grundig*[26] in 1966, the European courts have asserted on numerous occasions that the 'existence' of IP rights granted under national law is not affected by the European treaties. In cases involving the free movement of goods, this exclusion of IP rights is based directly on Article 36 TFEU.[27] In cases involving the free movement of services or the application of competition law, this exclusion is based on the policy established in Article 36, together with the general principle that Member State systems of property law must be respected (Article 345 TFEU).[28] These cases distinguish, however, between the existence of such rights and their 'exercise', which in principle may be subject to regulation. This distinction cannot be taken at face value, because it is clear that an IP right has no value if it cannot be 'exercised', for example by the threat of litigation against infringers. The real distinction has therefore been between forms of exercise that involve the 'core' of the IP right, and other conduct that may be regarded as a mere exercise of the IP right. The courts have described these core legal rights as the 'specific subject matter' of the IP right.

---

[22] Case T-69/89 *RTE v Commission* [1991] ECR II-485; Case T-70/89 *British Broadcasting Corporation and BBC Enterprises Ltd v Commission* [1991] ECR II-535. Case T-76/89 *ITP v Commission* [1991] ECR II-575.

[23] Dolmans et al argue for the continued validity of this analysis, *State of the Law* (n 20 above) 113 and 119. Some passages in the Report, which focus on whether IP rights perform their essential function in a specific case, also appear to point in this direction. See in particular section of F of the Report. The Preliminary Report puts greater emphasis on the function of patents being a reward for the creative effort of the inventor.

[24] Case 53/87 *Consorzio italiano della componentistica di ricambio per autoveicoli and Maxicar v Régie nationale des usines Renault* [1988] ECR 6039. See discussion below at n 51.

[25] This paper does not purport to re-examine the origins or development of these concepts in any detail or to defend the policy basis for these concepts in IP law. The focus here is on the use of these concepts to define the respective exclusive competence of Member State law (in terms of IP) and EU law (in terms of competition or free movement). A good critical overview of the broader issues from the IP perspective can be found in DT Keeling, *Intellectual Property Rights in EU Law,* Vol I (Free Movement and Competition Law) (Oxford: OUP, 2003).

[26] Joined Cases 56/64 and 58/64 *Consten and Grundig v Commission* [1966] ECR 299.

[27] See eg, Case 16/74 *Centrafarm v Winthrop* [1974] ECR 1183.

[28] See eg, Case 62/79 *Coditel v Ciné Vog Films* [1980] ECR 881 ('*Coditel I*').

It is important to appreciate that the limitation on application of EU law to core rights does not reflect an initial policy judgment regarding the respective merits of competition law (or free movement rules) and IP rights in fostering economic efficiency. The initial basis for the limitation is rather a jurisdictional division between the competence of the Commission or courts to apply EU law and the competence of the Member States to create systems of intellectual property. Since the existence/exercise distinction is in essence a device for determining regulatory competence, it is not surprising that the assessment involves some degree of legal 'formalism'. The policy objectives that underlie this division of competence and which might inform the exercise of such competence become important only where the allocation of competence in a specific case is unclear.

The ECJ used the term 'specific subject matter' of a right to describe the basic content of an intellectual property right for the first time in its 1971 *Deutsche Grammophon*[29] judgment. In that case, the Court ruled that a copyright owner could not assert its rights to prevent imports of recordings from other Member States because its rights were 'exhausted' where the products had been sold in a Member State with its consent.[30] The ECJ *defined* 'specific subject matter' for the first time in its 1974 judgments in *Centrafarm v Sterling Drug*[31] and *Centrafarm v Winthrop*.[32] These cases applied the *Deutsche Grammophon* exhaustion principle to patent and trademark rights respectively. Thus the Court in *Sterling Drug* described the specific subject matter of a patent as:

> . . . the guarantee that the patentee, to reward the creative effort of the inventor, has the exclusive right to use an invention with a view to manufacturing industrial products and putting them in circulation for their first time, either directly or by the grant of licenses to third parties, as well as the right to oppose infringements.[33]

It should be noted that in *Sterling Drug*[34] the passage relevant to the case before the Court was the reference to '. . . putting industrial products manufactured under a patent in circulation *for the first time*'. This allowed the Court to conclude that control on goods after first sale in the Community was not a core right and therefore

[29] Case 78/70 *Deutsche Grammophon v Metro* [1971] ECR 487 para 11.

[30] The Court did not, however, define the specific subject matter of copyright in *Deutsche Grammophon* (n 29 above) (and would not do so until 15 years later in the *Coditel* cases: *Coditel I* (n 28 above) paras 12 and 14; Case 262/81 *Coditel v Ciné Vog Films* [1982] ECR 3381 para 17 ('*Coditel II*'). Note that neither of the Coditel cases uses the phrase 'specific subject matter'. A more complete definition of the specific subject matter of copyright is found in Joined cases C-92/92 and C-326/92 *Phil Collins v Imtrat Handelsgesellschaft mbH; Patricia Im-und Export Verwaltungsgesellschaft mbH, Kraul, Leif Emmanuel v EMI Electrola GmbH* [1993] ECR I-5145 para 20.

[31] Case 15/74 *Centrafarm v Sterling Drug* [1974] ECR 1147.

[32] *Winthrop* (n 27 above).

[33] *Sterling Drug* (n 31 above) para 9. The same formulation for trademarks was provided in *Winthrop* (n 27 above) para 8.

[34] As well as in *Winthrop* (n 27 above) para 8 and, by implication, in *Deutsche Grammophon* (n 29 above) (where the specific subject matter of copyright was not defined).

could be controlled through the rules on free movement. The passage describing the 'purpose' of the right ('to reward the creative effort of the inventor') explains the policy basis for the right. There is no suggestion, however, that this phrase is intended to limit the scope of IP rights protected under Article 36. Indeed, in the context of *Sterling Drug* or the other early cases there was no reason for the Court to do so.[35]

The 'purposive' language in the specific subject matter formulation has been incorporated by the Court in a separate concept—the 'essential function of the right'.[36] In the 1978 judgment in *Centrafarm v American Home Products*, the Court stated that this concept became applicable '. . . in order to establish in exceptional circumstances the precise scope of that exclusive right granted to the proprietor'.[37] In *American Home Products* the Court invoked the essential function of a trademark to support its ruling that a proprietor could prohibit a reseller of its goods from repackaging the products and then applying the proprietor's trademark to the new package. Thus in that case the essential function concept served to expand the specific subject matter beyond the core rights previously identified.

In subsequent free movement cases, the Court has consistently used the 'essential function' concept as the criterion for expanding (or not expanding) the core rights that constitute the specific subject matter of patents, copyrights, trademarks, and other rights. Many of these cases have involved the extent to which pharmaceutical companies can invoke trademarks to prevent repackaging of products.[38] In cases

[35] Some commentators (eg, Keeling (n 25 above) 63) have suggested that the 'specific subject matter' concept serves a dual function. It describes the case rights granted to an IP owner, but it also defines the purpose of these rights. These commentators observe that the French term 'objet specifique' which is used in the French uniformly in French texts has a double meaning connoting both subject matter and purpose. This linguistic argument is unpersuasive, given that the relevant German and Dutch texts using words 'Gegenstand' or 'Voorwerp' do not support the small double meaning.

[36] As a general matter, the formulation of the essential function concept relates back to the purposive language in the definition of specific subject matter. Thus the essential function of a patent is to reward the creative effort of the inventor, following the Court's statement in *Sterling Drug* (n 31 above). This approach has not been strictly followed, however. In the case of trademarks, the Court identifies the essential function in Case 3/78 *Centrafarm v American Home Products Corporation* [1978] ECR 1823 para 12 as '. . . to guarantee the identity of the origin of the trademarked product to the consumer or ultimate user.' In *Winthrop* (n 27 above), in contrast, the Court had described the purpose of the trademark as allowing the holder to prevent rivals cashing-in the reputation of the product. Thus the scope of a right's essential function is fixed from a legal perspective, which for evolution in the Court's understanding of the role played by the IP right provides a further basis for caution in using this purposive language to limit basic core rights.

[37] *American Home Products* (n 36 above) para 12.

[38] See, eg Case 102/77 *Hoffman-La Roche v Centrafarm* [1978] ECR 1139 and Case 1/81 *Pfizer v Eurim-Pharm* [1981] ECR 2913 (both decided prior to Council Directive 89/104/EEC to approximate laws relating to trade marks [1989] OJ L40/1); see also, eg Joined Cases C-427/93, C-429/93 and C-436/93 *Bristol-Myers Squibb and others v Paranova* [1996] ECR I-3457; Case 443/99 *Merck, Sharp & Dohme v Paranova* [2002] ECR I-3703; Case 143/00 *Boehringer v Swingward* [2002] ECR I-3759; and Case 433/00 *Aventis Pharma Deutschland v Kohlpharma* [2002] ECR I-7761. These cases do not involve the core specific subject matter of trademarks as defined by the Court because they do involve the use of the trademark in connection with the first sale of the product in the EU. They have sometimes required a factual assessment to determine whether an extension of core rights is justified

where the rights themselves are novel the Court has enquired further into the economic purpose of the right created in national law.[39] In none of these cases has the case-specific enquiry extended to assessing whether a copyright in fact involved creative effort, whether a patentee required an incentive to create its innovation or whether consumers actually care about the origin of a product protected by a trademark.

Existence/exercise in competition cases: Article 101 TFEU

A rich body of law and regulation defines the relationship between IP rights and competition law in the context of Article 101 TFEU. Starting with the 'Christmas Message' in 1962,[40] the Commission has tried to define the circumstances where restrictions that might in other circumstances be prohibited by Article 101 TFEU are permissible in the context of an IP licence. To a large extent contractual control over exploitation of IP rights by licensees has been regarded as a form of 'exercise' subject to Article 101(1) TFEU so that restrictive arrangements require exemption under Article 101(3) TFEU.[41] Territorial limits on the sale of products covered by IP rights, in particular, have been regarded as a restrictive form of IP exercise capable of exemption under appropriate conditions. Commission practice and the various block exemptions have recognized, however, that some common provisions in licence agreements fall outside Article 101(1) altogether, even where on their face they restricted competitive activity by the licensee.[42] The ECJ confirmed in its 1986 *Windsurfing* judgment that the basis for excluding these restrictions from Article 101 assessment was that they were necessary to protect the specific subject matter of the

in a particular case (usually because protecting the essential function of a trademark would justify a limitation of repackaging where it could affect the quality of the product protected by a trademark). In none of these cases, however, has the case-specific assessment focused on the quality or function of the trademark.

[39] A good example is the exclusive rental right reviewed in Case 158/86 *Warner Brothers and Metronome v Christiansen* [1988] ECR 2605, where the Court carefully considered the essential function of a copyright for video lending in the context of the specific needs of the video lending industry.

[40] Commission Notice of 24 December 1962 concerning patent licensing agreements [1962] OJ 2922/62.

[41] The old procedural regulation, Council Regulation No 17 (EEC): First Regulation implementing Articles 85 and 86 of the Treaty [1962] OJ 13, 204/62, had a special provision applicable to simple vertical licensing agreements that made prior notification unnecessary for application of Article 101(3) TFEU by the Commission. Starting in 1984, the Commission also issued block exemption regulations covering licensing of patents and know-how, that were finally incorporated in a regulation covering 'technology transfer' more generally (see (n 11 above) and Council Regulation (EC) No. 2349/84 of 23 July 1984 on the application of Article 85(3) of the Treaty to certain categories of patent licensing agreement and [1984] OJ L 219/15 and Commission Regulation (EC) No. 556/89 of 30 November 1988 on the application of 85(3) of the Treaty to certain categories of know-how licensing agreements, [1989] OJ L 61/1).

[42] It was not fully clear whether the basis for this exclusion was 'ancillarity' (ie the restriction was objectively necessary for exploitation of the IP right in a specific case) or an application of the existence/exercise paradigm.

licensed IP rights.[43] Both the *Windsurfing* judgment and the various block exemptions follow the judgments in the free movement cases in defining the core rights contained in the specific subject matter by reference to a specific type of IP. There is no suggestion that the scope of core rights might be contracted on the basis of the features of the grant in a specific case.

The ECJ has also followed the approach developed in the free movement cases by invoking the 'essential function' concept to expand the scope of the IP exception to Article 101 TFEU recognized in *Windsurfing*. *Coditel II* involved application of Article 101 to an agreement granting exclusive rights to show a film on Belgian television. The defendant in national proceedings was a cable television operator that retransmitted to subscribers in Belgium programming from a German terrestrial television broadcaster. The cable operator was sued for infringement in connection with the retransmission in Belgium of a film over which the plaintiff was the exclusive Belgian licensee. In a previous case (*Coditel I*)[44] the Court had rejected the cable operator's argument that the copyright holder's rights had been 'exhausted' by the initial German broadcast, even though the German broadcaster had been fully licensed for broadcast in Germany by the same licensor. The defendant now argued that the grant to the Belgian licensee of absolute territorial protection was a clear restriction of competition prohibited under Article 101(1) (and therefore void) on the basis of the practice of the Commission and the Courts regarding patents and similar IP rights.[45]

The ECJ ruled that the Article 101 prohibition on absolute territorial protection in patent licence cases did not extend to licences of film copyright. The Court stated that in light of the 'characteristics of the cinematographic industry', the grant of exclusive rights to a sole licensee did not by itself restrict competition.[46] Although the Court does not explicitly refer to either specific subject matter or essential function of copyright in this judgment, the earlier analysis in *Coditel I* (which addressed in

[43] Case 193/83 *Windsurfing International v Commission* [1986] ECR 611 paras 45, 72, 80, 85 and 92. This case involved a licence agreement for patents covering the rig for the leading brand of windsurfing boards. The Commission had found that various provisions in the licence infringed Article 101, observing at each point that the relevant provisions, including a no challenge clause, were not necessary to protect the specific subject matter of the licensed patents. On appeal the Court followed the Commission's analytical approach, assessing each restrictive provision to determine whether it served to protect the specific subject matter of the IP rights. The assessment in *Windsurfing* was driven in part by the conclusion of the Commission and the Court that the licensor did not hold a valid patent in Germany. As a result, provisions restricting the licensee's commercial freedom in Germany could not, according to the Court, be justified by reference to the specific subject matter of the licensor's patent.

[44] The Court ruled that exhaustion was inconsistent with the essential function of film copyrights.

[45] Earlier in the same year, in Case 258/78 *Nungesser v Commission* [1982] ECR 2015, the Court had established that a grant of absolute territorial exclusivity (including protection from passive sales by licensees in other Member States) would constitute an infringement of Article 101.

[46] *Coditel II* (n 30 above) para 16.

more detail the characteristics of the industry referred to in *Coditel II*) suggests that this finding was based on the need to preserve the essential function of the copyright in this specific industrial context.[47]

Taken together, the ECJ judgments and Commission regulations applying Article 101 are entirely consistent with the approach to existence/exercise developed in the free movement cases. The specific subject matter of a right is defined generically for a 'type' of right and not on the basis of circumstances in a specific case. This is demonstrated in *Coditel II*, where the Court applied an essential function analysis to expand the scope of the IP exclusion from Article 102 based on the circumstances of licensing in the film industry generally, without considering the merits of a specific film. There is no support in these cases for using 'essential function' as a tool for limiting core rights on a case-by-case basis.

Existence/exercise in competition cases: Article 102

The ECJ firmly endorsed the central role of the existence/exercise paradigm for Article 102 analysis in its 1988 judgments in *CICCRA/Renault*[48] and *Volvo/Veng.*[49] These cases involved references from national courts concerning cases where a car manufacturer had sued independent suppliers of body panels to repair shops for infringement of its IP rights. The panel suppliers had invoked the 'Euro defence' that enforcement of the relevant IP rights by the manufacturer in each case infringed Article 102 TFEU. The Advocate General supported the suppliers, submitting that IP owners should be subject to Article 102 where IP rights in a specific case did not perform their function as recognized under Community law.[50] The Court, however, referring to the distinction between existence and exercise, ruled that the enforcement of IP rights validly granted under Member State law was not in itself an infringement of Article 102. The Court then observed that the *exercise* of the IP rights could be subject to Article 102 and provided three examples of situations where Article 102 could be applicable. Each of these examples involve situations where the IP owner itself is selling (or refusing to sell) products protected by an IP right.[51] In the Article 101 context, restrictions on sale of products protected by IP rights had always been regarded as outside the specific subject matter of those rights. It had never been

[47] *Coditel II* (n 30 above) paras 17 and 19. Interestingly, the Court goes on to observe that Article 101 could be applicable to the exercise *by the licensee* of its exclusive rights This underscores the distinction between the specific subject matter (expanded in this case to ensure that the essential function of the right can be protected) and the exercise of these rights which is subject in principle to the rule of the Treaty.

[48] *CICCRA/Renault* (n 24 above).

[49] Case 238/87 *AB Volvo v Erik Veng* [1988] ECR 6211.

[50] Opinion AG Mischo in Case 53/87 *Consorzio italiano della componentistica di ricambio per autoveicoli and Maxicar v Régie nationale des usines Renault* [1988] ECR 6039 paras 21–32.

[51] The three examples were (i) excessive pricing of the body parts; (ii) refusal to supply independent repair shops; and (iii) failure to continue production of parts.

seriously argued that excessive pricing, bundling, discrimination or exclusionary abuse fell outside Article 102 where the products were protected by intellectual property.[52] It was not therefore particularly surprising that the Court should regard these examples as involving conduct outside the specific subject matter of the rights in question and therefore subject to Article 102 and the judgments make no suggestion that an interference with core IP rights is intended.[53] Thus *Volvo* and *Renault* appeared to fit neatly with the established existence/exercise paradigm as applied in previous cases involving either free movement or competition.

The first (and so far the last) serious attempt to use the 'essential function' concept as a limitation on exercise of core IP rights came in *Magill*.[54] The *Magill* judgments followed a Commission ruling that the separate refusal by three TV broadcasters to allow an independent publisher to include their schedules in its weekly TV guide constituted an abuse under Article 102 TFEU.[55] The CFI rejected the broadcasters' appeals, stating that the conduct '. . . clearly goes beyond what is necessary to fulfil the essential function of copyright under Community law.'[56] The CFI did not base this finding, however, on an assessment of whether the copyrights in question protected real creative efforts or moral rights attributable to the broadcasters. Instead, the CFI concluded that the broadcasters' conduct was outside the essential function of copyright because the rights were '. . . exercised to pursue an aim manifestly contrary to the objectives of Article [102].'[57] This reflected the CFI's novel restatement of the essential function of copyright—'. . . to protect the moral rights in the work and ensure a reward for creative effort, while respecting the aims of, in particular, of Article [102].'[58] Thus while the CFI did invoke the 'essential function' concept to limit the scope of protected activity under Article 102, it did so on the basis of a competition policy limitation on essential function that does not appear in any other case.

---

[52] Indeed, subsequent to *Volvo* and *Renault* the Court ruled that royalty rates for licensing IP rights themselves were in principle subject to Article 102(a) TFEU, Case 397/87 *Ministère public v Tournier* [1989] ECR 2521.

[53] In retrospect, the issue of interference with specific subject matter arose with respect to the third example: failure of the dominant firm to continue supplies of the parts protected by the design rights. As the Advocate General would observe in *Magill* (AG Opinion Gulmann Joined Cases C-241/91 P and C-242/91 P *Radio Telefis Eireann (RTE) and Independent Television Publications Ltd (ITP) v Commission* [1995] ECR I-743 paras 100–1), the decision by an IP owner regarding whether or when it will produce and sell products protected by IP is generally considered part of the specific subject matter of the right. He suggested that this might be abusive due to the 'special circumstances' that the manufacturer had created a dependency on the part of consumers by selling the cars.

[54] See the judgments referred to in n 22 above.

[55] Case IV/31.851 *Magill TV Guides/ITP, BBC, and RTE* [1989] OJ L78/43. The Commission decision stated that the broadcasters were using their copyrights as an instrument of abuse in a manner which falls outside the scope of the specific subject matter of that intellectual property right.

[56] Case T-70/89 *British Broadcasting Corporation* (n 22 above) para 74.

[57] *Ibid* para 58.

[58] *Ibid* para 58.

On further appeal to the ECJ, Advocate General Gulmann was critical of the CFI's attempt to graft a competition policy objective on the essential function concept.[59] While he accepted that in principle conduct that did not implement the essential function of the right could be subject to Article 102 TFEU,[60] he apparently viewed conduct that *did* implement the essential function as immune from Article 102 control.[61] He urged the Court to overturn the CFI judgment, because in his view the conduct of the broadcasters in *Magill* did implement the essential function of the rights by allowing them to profit from creative effort.[62] He acknowledged that the creative work in *Magill* did not require any incentive (since it was a by-product of television broadcasting).[63] But argued that this factor should not affect application of Article 102, since establishing the criteria for IP protection was the proper role of Member State authorities.[64]

The ECJ did not deal with the issues of specific subject matter and essential function in its own judgment.[65] Some commentators have suggested that this means that the CFI judgment and/or the Advocate General's Opinion remain authority for the proposition that essential function provides an alternative basis for determining the scope of Article 102 in IP cases.[66] It should be recalled, however, that the Advocate General convincingly pointed out that CFI analysis was based on an incorrect understanding of the essential function concept. The ECJ judgment in turn, by affirming the Commission's finding of abuse, rejected the Advocate General's own central contention that conduct implementing the essential function of a right was always outside Article 102.

[59] AG Opinion Gulmann (n 53 above) para 70. He observed that the essential function concept is based on the objectives of IP rights in national law and that it 'does not make sense' to incorporate competition law aims derived from EU law in this definition.

[60] The Advocate General's conclusion in this regard appears to have been based on a misreading of the free movement cases, in particular those involving repackaging. He describes these cases as limiting the scope of specific subject matter, rather than limiting the *extension* of specific subject matter as explained earlier in this paper. AG Opinion Gulmann (n 53 above) para 85 (citing cases at n 38 above).

[61] Even where conduct did not implement the essential function, he still thought that interference with core IP rights required 'particularly substantial and weighty competition grounds.' AG Opinion Gulmann (n 53 above) para 78.

[62] AG Opinion Gulmann (n 53 above) paras 88–117. He suggested that this might not be the case if the licence were sought for a truly 'new product' that did not compete at all with products sold by the broadcasters, because in that situation there would be no impact on the revenue associated with the products protected by the IP rights.

[63] AG Opinion Gulmann (n 53 above) para 123.

[64] AG Opinion Gulmann (n 53 above) para 126. AG Gulmann cited in support of this conclusion the approach taken by the Court in *Renault* and *Volvo*. AG Gulmann argued that it was not appropriate to assess the strength of the IP rights in question '. . . even though it may be tempting and appear actually reasonable in the present case.'

[65] Since the ECJ found that the circumstances taken into account by the CFI were sufficient to support a finding of abuse it ruled that it was '. . . unnecessary to examine the reasoning of the contested judgments in so far as it is based on Article 36' (para 58 of the *Magill* ECJ judgment, Joined Cases C-241/91 P and C-242/91 *P Radio Telefis Eireann (RTE) and Independent Television Publications Ltd (ITP) v Commission* [1995] ECR I-743).

[66] See eg, Dolmans et al, *State of the Law* (n 20 above) 119.

Thus while the ECJ did not formally reject the argument that essential function provided the basis for determining where conduct was abusive, the analytical structures for applying that concept proposed by the CFI and by the Advocate General were both discredited. Furthermore, by citing *Volvo* for the proposition that the conditions for grant of intellectual property were a matter for national rules, the Court rejected the suggestion that the presence or absence of incentives to innovate affected application of Article 102.[67]

Following *Magill* the legal position can be summarized as follows. The 'rule' is that exercise of core IP rights, as defined by national law, is not subject to Article 101 or 102. Where the extent of core rights is uncertain, the courts may consider the 'essential function' of a right in order to assimilate some borderline conduct to that right. This determination does not involve a case-specific economic analysis, but may require consideration of the requirements of a specific 'type' of right in a specific industry. There is an 'exception' to this rule in certain 'circumstances'—the 'exceptional circumstances' identified in *Magill.* This approach has been applied consistently in subsequent cases.[68] The basic principle, that exercise of core IP rights are only subject to Article 102 in exceptional circumstances, has been repeated in the subsequent ECJ judgment in *IMS Health,*[69] as well as in the CFI's own judgment in *Microsoft.*[70] There is no suggestion in these judgments that the essential function of an IP right provides an independent basis for assessing the compatibility of its exercise in a specific case with Article 102. Indeed, when in *IMS Health* the Commission

[67] *Magill* ECJ judgment (n 65 above) para 49. In effect, the ECJ accepted the warning of the Advocate General on this point, adopting the argument that he made based on *Volvo.*

[68] The 'essential function' concept has also been invoked as a possible limitation on remedies in competition cases. This issue has arisen in the long-running '*DSD (Grune Punkt)*' litigation. Case COMP D3/34493 *DSD* [2001] OJ L166/1; Case T-151/01 *Der Grüne Punkt-Duales System Deutschland v Commission* [2007] ECR II-1607 and Case C-385/07 P *Der Grüne Punkt-Duales System Deutschland v Commission* (judgment of the ECJ dated 16 July 2009). In the CFI, DSD argued that the remedy imposed by the Commission impermissibly interfered with its rights to the 'Grune Punkt' trademark. DSD observed that the essential function of a trademark is to inform the customer of the origin of the product or service. Since packaging bearing the mark might be recycled in rival recycling centres following the Commission's decision, this essential function was, according to DSD, no longer fulfilled. Thus DSD argued that the remedy chosen by the Commission encroached on the specific subject matter of DSD's rights, which the Commission could not do without applying the *Magill* criteria. The CFI rejected DSD's argument on the basis that the essential function of the trademark in this case was to inform consumers that the product could be recycled in a DSD facility since the Commission remedy did not interfere with this more limited 'essential function'.

When the ECJ considered the case four years later, the 'essential function' point no longer featured as one of DSD's principal arguments. The ECJ, however, took substantial care to establish that the Commission remedy did not involve a compulsory licence of the trademark. The Court (at paras 156 to 162) emphasized that the remedy did not require DSD to license its mark to a new customer, but rather only altered the terms of licences concluded by DSD with existing customers. Thus the Court took a line fully consistent with the previous existence/exercise cases. The exercise fell outside the specific subject matter and preserving the essential function did not require extension of the rights.

[69] Case C-418/01 *IMS Health GmbH & Co OHG v NDC Health GmbH & Co KG* [2004] ECR I-5039.

[70] Case T-201/04 *Microsoft v Commission* [2007] ECR II-3601.

as intervener argued that the Court need not apply the *Magill* exceptional circumstances test because the rights in question were not serving their essential function, this submission was simply ignored by the Court.[71] Thus it would appear that, barring a significant change in the legal analysis (which would inevitably impact analysis in free movement cases) any move to a more economic approach to core IP rights must be accommodated in the 'exceptional circumstances' test. Before considering the scope for economic analysis in the 'exceptional circumstances' test, however, the possibility of introducing a case-by-case 'intent' analysis should be considered.

Existence/exercise and 'intent'

As noted earlier in this paper, the Final Report in the Pharmaceutical Inquiry includes a general suggestion that anti-competitive intent may be relevant to application of Article 102 to the conduct it describes.[72] The CFI judgment in *Magill* also focuses on anti-competitive intent even if it mischaracterized intent as an element in the analysis of 'essential function'. It might be argued that Article 102 should be applicable where the exercise of IP rights involves a manifestly anti-competitive aim even where it involves core rights. Even though the attempt to introduce intent as an element in assessing 'essential function' failed in *Magill*, an intent analysis might be introduced as part of a proper Article 102 analysis. This could open the door to economic evidence. If the general statement of the law 'post *Magill*' posited above is accepted, then intent evidence should not be relevant outside the context of 'exceptional circumstances.' Consideration of the cases involving intent evidence in the context of both Article 102 and the free movement rules suggest, however, that it would not be appropriate in any case to apply an intent test to expand the application of Article 102 to core IP rights.

Evidence of intent plays a limited role in Article 102 analysis. It is well established that the possibilities incorporated in Articles 101 and 102 are objective concepts. Even the concept of 'restriction by object' in Article101(1) is applicable on the basis of an objective assessment of the allegedly restrictive conduct, not by reference to

[71] The Commission's argument is summarized in the Opinion of AG Tizzano in C-418/01 *IMS Health* [2004] ECR I-5039 para 45.

[72] See n 21 above. The Commission referred in that passage to its decision in *AstraZeneca*. The recent judgment of the General Court affirming that decision does not, however, support the suggestion that intent evidence can provide a stand-alone basis for finding abuse under Article 102. On the appeal, the Commission argued that evidence showing that the dominant firm intended to restrict competition was relevant to the finding that the conduct was liable to have such effects, Case T-321/05, *AstraZeneca AB v Commission* (judgment of 1 July 2010, not yet officially reported) para. 334. The Court accepted this submission, observing that while abuse is an objective concept,

> . . . intention can still be taken into account to support the conclusion that the undertaking concerned abused a dominant position, even if that conclusion should primarily be based on an objective finding that the abusive conduct actually took place (para 359).

subjective intent.[73] Evidence of subjective intent does play a role, however, in resolving the application of Article 101[74] or Article 102 where either the classification of conduct in a specific case as 'competition on the merits' or the likelihood of competitive harm is unclear. The most prominent example of this approach is the AKZO[75] test for predatory pricing. Under this test, prices that do not recover the average variable costs of the product are presumed to be abusive, while prices that cover costs on a fully allocated cost basis constitutes competition on the merits and cannot be abusive. Pricing in the grey area (covering variable costs but not total costs) is deemed abusive only where it forms part of an anti-competitive scheme or plan to exclude rivals.

The role of intent evidence as a tool for determining the application of objective competition law concepts in 'grey areas' is analogous to the role of the 'essential function' concept in identifying the specific subject matter of an IP right.[76] It is at least superficially attractive to consider the possibility of using intent evidence to supplement an essential function analysis in defining the forms of conduct that can properly be appropriate to the specific subject matter category. There is, however, a major difference between use of intent evidence to define the boundaries of competition law enforcement powers (where the alternative is simply non-action) and the use of such evidence to circumscribe the application of a separate principle of Community Law.[77] If intent in a specific case is to become part of the analysis of essential function, this analysis will need to shift in free movement cases as well. The current case law leaves little space, however, for such a shift in analysis.[78]

---

[73] In Case C-209/07 *Competition Authority v Beef Industry Development Society and Barry Brothers* [2008] ECR I-8637 ('*Irish Beef*'), the ECJ ruled that cooperation between competitors was restrictive by object where the alternative to cooperation was further rivalry between the cooperating parties.

[74] See Case 35/83 *BAT v Commission* [1985] ECR 363 where intent was relevant to the assessment of a delimitation agreement under Article 101.

[75] Case C-62/86 *Akzo Chemie BV v Commission* [1991] ECR I-3359. *AstraZeneca* (n 72 above) provides an example of use of intent to assist in determining the existence of anti-competitive effects.

[76] See, eg *American Home Products Corporation* (n 36 above) paras 10–17; *Coditel I* (n 28 above) paras 14–19 and *Coditel II* (n 30 above) paras 17–19. Note that neither of the *Coditel* cases uses the phrase 'specific subject matter'.

[77] The cases applying specific subject matter/essential function analysis in the context of competition law all start with the assumption that a competition law infringement would exist if specific subject matter/essential function are not applicable. The issue is always therefore the extent of the specific subject matter/essential function exception, not the substantive scope of competition law.

[78] The principal reason that the established analysis of specific subject matter and essential function in free movement cases does not take account of evidence regarding anti-competitive intent is that Article 36 contains its own more limited intent provision. In the context of the free movement rules, even conduct normally falling within the specific subject matter of an IP right may be prohibited if it constitutes a 'disguised restriction on trade between Member States'. Thus registration of different trademarks in different Member States, while falling within the specific subject matter of a trademark, may nonetheless render enforcement of the trademark against parallel imports a disguised restriction on free movement. Case 119/75 *Terrapin v Terranova* [1976] ECR 1039. While it might be arguable that conduct designed to partition national markets should be subject to competition rules, this is not an issue in practice, because IP-related conduct designed to partition markets involves either licensing

Apart from the formal legal difficulty of introducing an analysis of anti-competitive intent to resolve IP issues in competition cases, this approach would also pose a significant practical problem. As the Advocate General observed in *Magill*, it is usual that a party invoking IP rights has the subjective intention of eliminating competition from a rival.[79] Evidence of an 'anti-competitive scheme or plan' such as that relied on in *AKZO* would therefore have little value in distinguishing between legitimate enforcement of IP rights and abuse. It would mean that, in practice, almost any exercise of core IP rights would be subject to competition law review.

The law regarding the relevance of intent evidence thus does not provide a basis for using such evidence to resolve a conflict between core IP rights and competition law enforcement. Intent evidence can play a useful role in competition law cases involving conduct not directly related to the specific subject matter of the right.[80] It cannot assist, however, in defining the core area where competition law is not applicable. These conflicts can only be resolved by recourse to the 'exceptional circumstances' test discussed in the next part of this paper.

C. Core IP Rights, Exceptional Circumstances, and Economic Analysis

The simplest resolution of conflicts between intellectual property and competition law would be non-interference—competition law should not apply to exercise of the core legal rights associated with intellectual property. Most IP practitioners would favour this approach, which is the approach used in the United States. It is also attractive from a policy perspective because it is easy to administer and avoids the need to balance interests that cannot readily be balanced. This approach has, however, been rejected for purposes of EU law since the ECJ in *Magill* established that competition law can be applicable to exercise of core IP rights under 'exceptional circumstances'. It is therefore necessary to consider the nature of these exceptional

terms or refusals to sell products protected by IP rights, forms of 'exercise' outside the scope of the specific subject matter and therefore subject to competition law.

[79] Advocate General Gulmann observed that the CFI's focus on the aims of Article 102 appeared to give unjustified priority to the aims of Article102 over those on IP law (AG Opinion Gulmann (n 53 above) paras 66–8).

[80] Thus in *BAT*, the fact that the owner of a trademark did not intend to use that trademark in the territory reserved to itself in a delimitation agreement meant that the agreement could not be regarded as the resolution of a bona fide trademark dispute. Similarly, in Case C-9/93 *IHT Internationale Heiztechnik v Ideal Standard* [1994] ECR I-2789 evidence of subjective intent was relevant to determine whether a network of exclusive licences created a restrictive agreement among licensees subject to Article 101. Both of those examples involve conduct (litigation settlements and licence terms) that is a classic form of exercise not generally assimilated to the specific subject matter of the right. See also Case 65/86 *Bayer AG and Maschinenfabrik Hennecke GmbH v Heinz Süllhöfer I* [1988] ECR 5249.

circumstances and the extent to which economic analysis and policy considerations assist in identifying where such circumstances exist.

The exceptional circumstances test

The ECJ judgment in *Magill* is not itself particularly helpful in defining the exceptional circumstances where competition law will prevail over IP rights. Apart from a cryptic reference to its *Volvo* judgment, the Court did not provide any legal or policy reasons for identifying the circumstances in *Magill* as exceptional.[81] The nature of the essential circumstances rule must therefore be derived from consideration of the circumstances identified in *Magill*, as they have been interpreted and applied in subsequent cases.

Although the ECJ in *Magill* referred to *three* exceptional circumstances, the subsequent ECJ judgment in *Oscar Bronner*[82] identified *four* relevant circumstances in *Magill*:

- the refusal in question concerned a product the supply of which was indispensable for carrying on the business in question
- the refusal prevented the appearance of a new product for which there was a potential consumer demand
- the refusal was not justified by objective considerations
- refusal was likely to exclude all competition in the secondary market.

The CFI in *Microsoft* adopted this four step assessment, but changed the order in which the criteria would be addressed by considering the 'exclude all competition' criterion together in tandem with the 'indispensability' criterion. This allowed the CFI to put the *Magill* criteria in a more developed legal context. The *Microsoft* judgment treats the first and fourth factors as an application of the established case law involving refusals to supply and monopoly leveraging found in cases such as *Commercial Solvents*,[83] *Télémarketing*,[84] and *Bronner* itself.[85] The third criterion, lack of objective justification, is likewise treated as the application of a general

[81] The reference to *Volvo* appears to relate to the discussion in the ECJ proceedings of the failure to manufacture abuse situation identified in *Volvo* as conduct involving the specific subject matter of the relevant rights. See discussion (n 54 above).

[82] Case C-7/97 *Oscar Bronner GmbH & Co. KG v Mediaprint* [1998] ECR I-7791 para 40. In *Magill* the ECJ had combined the indispensability of the input with the result of refusal to grant access—blocking the emergence of a new product.

[83] Cases 6/73 and 7/73 *Istituto Chemioterapico Italiano SpA and Commercial Solvents v Commission* [1974] ECR 223.

[84] Case 311/84 *Centre Belge d'Etudes du Marche-Telemarketing (CBEM) v SA Compagnie luxembourgeoise de télédiffusion (CLT) and Information publicité Benelux (IPB)* [1985] ECR 3261.

[85] This analysis was consistent with the approach taken by Advocate General Tizzano in *IMS Health* (n 71 above). In his analysis of the question of whether he relied on the general precedents regarding refusal to supply and leveraging to support his conclusion that an abusive refusal to license did not require an established upstream market for licensing of the rights in question.

Article 102 principle.[86] In effect, the CFI's approach means that the other three circumstances are not really 'exceptional' at all.[87] Since any application of Article 102 must start from the premise that a *prima facie* showing of abuse can be made, the special circumstance in *Magill* that is unique to an IP case is the 'new product' requirement.[88] This reflected the observation of the ECJ in *IMS Health*, that the 'new product' requirement was present in order to ensure the appropriate balance '. . . of the interest in protection of the intellectual property right against the interest in protection of free competition.'[89]

The CFI in *Microsoft* went beyond *Magill* and *IMS Health*, however, by re-examining the legal basis for the 'new product' exception to the rule that refusals to license are not abusive. Having observed that the ECJ *Magill* judgment had relied in this context on Article 102(b) TFEU the CFI ruled that,

> The circumstance relating to the appearance of a new product, as envisaged in *Magill* and *IMS Health*, cannot be the only parameter which determines whether a refusal to license an intellectual property right is capable of causing prejudice to consumers within the meaning of Article [102(b)]. As that provision states, such prejudice may arise where there is a limitation not only of production or markets, but also of technical development.

---

[86] As a general matter, the ability of a dominant firm to defend conduct on the basis that it is justified to defend the firms' legitimate interests is well established (Case 27/76 *United Brands v Commission* [1978] ECR 207 para 184; Commission's Enforcement Priorities in Applying Article 82 EC Treaty to Abusive Exclusionary Conduct by Dominant Undertakings [2009] OJ C45/7, section III D). Such conduct falls outside the definition of abuse in Article 102 because it is a method of competition that characterizes the performance of 'commercial operators' (see Case 322/81 *Michelin v Commission* [1983] ECR 3461 ('*Michelin I*') paras 70–1 citing Case 85/76 *Hoffmann-La Roche & Co AG v Commission* [1979] ECR 461 para 91). Some commentators have treated the discussion of objective justification in the Commission decision or the CFI judgment as articulating a 'test' for determining when a refusal to license is abuse. See, eg Lianos (n 104 below) and Kerber and Schmidt (n 20 above). This overstates its significance. There is no limitation on the grounds that may be put forward as a justification other than that these grounds are lawful and not in themselves anti-competitive, that the evidence is objective and that the measures are 'proportionate' to the interests asserted. The possibility that an otherwise abusive refusal to license may be objectively justified because it results in a reduction in incentives for the dominant firm to invest is therefore well within the general Article 102 analytical framework.

[87] The CFI in *Microsoft* (n 70 above) rejected the argument that the elimination of downstream competition requirement should be higher in a refusal to license case (elimination of all competition rather than elimination of effective competition).

[88] As the CFI observed in *Microsoft* (n 70 above) para 354.

[89] *IMS Health* (n 69 above) para 48. The CFI endorsed the view expressed by Advocate General Tizzano:

> Even where those circumstances [elimination of all competition on a secondary market] obtain, in weighing the balance between the interest in protection of the intellectual property right and the economic freedom of its owner, on the one hand, and the interest in the protection of free competition on the other, the balance may in my view come down in favour of the latter interest only if the refusal to grant the licence prevents the development of the secondary market to the detriment of consumers. AG Opinion (n 71 above) para 61.

Many distinguished commentators have suggested that this test will be easy to satisfy, often referring in particular to the reference at para 656 in the judgment to products that will be distinguished from Microsoft products on the basis of 'parameters which consumers consider important'.[90] In considering this aspect of the CFI judgment in *Microsoft*, it is important, however, to distinguish between the legal test articulated by the Court and the factual findings in the Commission decision. The legal test applied by both the Commission and the CFI was whether the refusal to license 'limited technical development'.[91] The Commission decision devoted nine paragraphs to showing that the refusal by Microsoft to provide interface information did indeed limit technical development. The CFI ruled that the Commission's findings in this regard were 'not manifestly incorrect'. While the CFI did specifically observe in respect of specific paragraphs in the decision that the Commission's observations or general approach were 'correct',[92] the Court appears to have been primarily concerned with refuting Microsoft's submission that the Commission decision would lead to 'cloning' of Microsoft products. On this basis, it would be a mistake to focus on the discussion by the CFI of an individual finding in the Commission decision as an indication of what will constitute a limitation on technical development in future cases.[93]

While the CFI judgment does not provide direct guidance on what could constitute an abusive limitation on technical progress, the Commission decision is actually more helpful. The nine paragraphs setting out the Commission findings are not drafted as a list of individual instances where technical development was limited, but rather as a series of factors that, taken together, support the ultimate conclusion that the requirements of Article 102(b) are satisfied. The emphasis is not on specific examples of product differentiation, but rather on the aggregate effect of Microsoft's refusal

[90] See, eg, Kjolbe (n 2 above) 175; D Beard, 'Microsoft: What Sort of Landmark?' (2008) 4 Competition Policy Intl 33, 36; J Langer, 'The Court of First Instance's Microsoft Decision: Just an Orthodox Ruling in an Unorthodox Case?' (2008) 35 Legal Issues of Economic Integration 183, 192; A Andreangeli, 'Case Note: Case T201/04, Microsoft v. Commission' (2008) 45 Common Market L Rev 863, 884.

[91] One odd aspect of the *Microsoft* judgment (n 70 above) is the ruling (at para 664) that the Commission did not need to show that the refusal to license had prejudiced consumers. The CFI relied in this regard on the 'settled case law' under the general Article 102 abuse test establishing that finding of abuse does not require a direct harm to consumers, but can also result from impairing an effective competitive structure. While this principle may be correct with regard to abuse generally, it is not clear the requirement of consumer harm can be so easily avoided where the abuse depends on application of Article 102(b), where the wording of the Treaty would appear to make prejudice to consumers a formal requirement.

[92] See, for example, *Microsoft* (n 70 above) paras 650 and 653.

[93] As regards the broader question of whether the *Microsoft* test creates a low threshold for creating an obligation to license, it is worth noting the observation of Bo Vesterdorf, who was President of the CFI when the judgment was issued: 'Since the conditions under which a refusal to license a product must remain *exceptional*, this definition or broadening of the "new product" requirement must, like any other exception, be interpreted restrictively.' B Vesterdorf, 'Article 82 EC: Where do we stand after the Microsoft judgement?' (2008) 1 Global Antitrust Rev 1, 9.

to license on a range of features in different products from a range of competitors. The Commission could therefore conclude that competition by rivals in the work station server market on the basis of product differentiation was '. . . currently artificially relegated to a secondary position'[94] and that in the longer term, '. . . there will be little scope for innovation—except possibly for innovation coming from Microsoft'.[95] The freeze on innovation, both short-term and long-term, in a market where significant innovation was possible, provided the basis for the Commission's conclusion that the refusal to license would limit technical progress.

Exceptional circumstances, policy, and economic analysis

If the CFI in *Microsoft* was correct in identifying the new product-limit on technical development requirement as the real 'exceptional circumstance' that determines whether a refusal to license is abusive, it has still not answered *why* this criterion should be decisive. This rule could be seen as an example of legal formalism, so that the specific requirements of Article 102(b) control in circumstances where the general abuse clause in the first paragraph of Article 102 does not.[96] The suggestion in *IMS Health* that the new product requirement was present to ensure the appropriate balance 'of the interest in protection of the intellectual property right against the interest in protection of free competition' suggests, however, that this requirement has some more fundamental policy basis.

The policy justification for making the Article 102(b) criterion decisive in determining where Article 102 can interfere with core IP rights may relate to the nature of the conflict between IP rights and competition law identified in the first part of this paper. Insofar as competition law enforcement focuses on harm to short-term static or allocative efficiency it will always conflict with IP rights, which by their nature envisage some reduction in static or allocative efficiency in return for a longer-term improvement in dynamic efficiency. Preventing the emergence of a genuine new product or limiting technical progress does not, however, primarily concern static efficiency. These forms of economic harm involve an impact on dynamic efficiency. Thus whether by accident or design, basing the exceptional circumstances test on Article 102(b), avoids the central conflict between IP rights and competition law.

---

[94] *Microsoft* (n 70 above) para 699.

[95] *Microsoft* (n 70 above) para 700.

[96] Alternatively, it might be argued that the Article 102(b) infringement is significant because it means that there are *two* abuses at once—refusal to deal *and* limit on development to prejudice of consumers. On this basis the identification of multiple abuses, such as a disproportionate termination of existing relationship (abusive under *United Brands*) or existence of an overall 'anti-competitive plan' might be deemed to constitute exceptional circumstances, even if there were no Article 102(b) infringement. The alternative approach proposed by the Commission in *Microsoft* of looking to the 'entirety of circumstances' in assessing a refusal to license appears to follow this line in part, although it also appears to rely on the severity of the consequences of the refusal in the market in question (Case COMP/C-3/37.792 *Microsoft* [2004] OJ L 32/23 paras 551–8; CFI judgment (n 70 above) para 303).

In effect, Article 102 becomes applicable only where the IP system has failed on its own terms. This analysis has significant implications for both the legal application of the exceptional circumstances test in future cases and for the role of economic analysis.

Two issues regarding the legal test merit discussion. The first concerns whether the list of 'exceptional circumstances' applied in *Magill*, *IMS Health*, and *Microsoft* is exhaustive. The Commission argued to the CFI in *Microsoft*, the language of the *Magill* judgment suggests that these are only an example of 'exceptional circumstances'.[97] The CFI did not reach this argument in its judgment, since it found that the Commission decision satisfied the requirements of the existing case law.[98] If the core of 'exceptional circumstances' is the application of Article 102(b), however, it follows that the other circumstances might change, as long as that core requirement remains in place.[99] In other words, the theory of abuse might not involve refusal to supply an essential input, but rather some other abusive conduct such as the termination of an existing supply relationship. It would still, however, be necessary to show that the putative abusive conduct either prevented the emergence of a genuine new product or led to a significant limitation on technical progress. Establishing a 'really big' impact on allocative or static efficiency would not be sufficient to establish exceptional circumstances.[100]

The second legal issue concerns the application of the exceptional circumstances test to conduct other than a refusal to license. If the purpose of the test is indeed to ensure the appropriate balance 'of the interest in protection of the intellectual property right against the interest in protection of free competition' then it should also be applicable to other conduct that constitutes the exercise of a core IP right.[101] On this basis,

[97] *Microsoft* (n 70 above) paras 316–17.

[98] *Microsoft* (n 70 above) para 336.

[99] This analysis will be different of course for cases like those identified in *Volvo* where the exercise of a core IP right reinforces the effects of conduct that would be abusive in any case, which the *Magill* court apparently regarded as also involving 'exceptional circumstances'.

[100] It may be noted that in *IMS Health* the Commission argued that the existence of a significant effect on a downstream market was sufficient to render a refusal to license abusive (see Opinion of Advocate General Tizzano (n 71 above) paras 42–5). This argument was ignored by both the Advocate General and the Court.

The Commission's Article 102 Enforcement Guidance (n 1 above) para 87 focuses on harms to dynamic efficiency in its discussion of refusal to license abuses:

> The Commission considers that consumer harm may, for instance, arise where the competitors that the dominant undertaking forecloses are, as a result of the refusal, prevented from bringing innovative goods or services to market and/or where follow-on innovation is likely to be stifled. This may be particularly the case if the undertaking which requests supply does not intend to limit itself essentially to duplicating the goods or services already offered by the dominant undertaking on the downstream market, but intends to produce new or improved goods or services for which there is a potential consumer demand or is likely to contribute to technical development.

[101] It may be noted in this regard that in both *Volvo* and *Magill* the starting point for the analysis was the observation that 'the exclusive right of reproduction' forms part of the IP right (Case T-69/89 (n 22 above) para 49, *Volvo* para 7). The rule that a simple refusal to license is not in itself abusive

enforcing an IP right in court (or threatening to do so) is clearly part of the specific subject matter of the right, as long as the rights owner has a good faith basis.[102] Likewise obtaining an IP right where the legal requirements are met would be presumptively lawful since those legal requirements constitute part of the core package of rights created by IP law.[103] This would be consistent with the evolution of the legal standard prior to the *Magill* case, as outlined in section B of this paper.

As regards economics, the exceptional circumstances test as described above provides scope for analysis of whether conduct does lead to a significant limitation of technical progress in a specific case. The impact of IP licensing practices on dynamic efficiency, particularly in markets characterized by sequential innovation, is already the subject of economic literature and it would not be surprising if the information generated in future litigation does not lead to refinements in the theoretical understanding of this issue.[104] There is also scope for developing economics-based objective justifications, including arguments that refusals to license preserve necessary incentives to invest in productive innovation. The exceptional circumstances test does not, however, provide a basis for an open-ended economic analysis incorporating an assessment of static efficiency costs. The existence of exceptional circumstances also cannot depend on analysis of the strength or benefits of IP rights in a specific case, although these factors may be relevant to an objective justification argument where a *prima facie* harm to dynamic efficiency has been demonstrated.

---

applies because a refusal to license is an exercise of the right to exclusive reproduction. Exercise of non-core rights, such as excessive or discriminatory licence fees or refusals to supply products that are protected by IP rights would remain generally subject to competition law.

[102] Case T-111/96 *ITT Promedia NV v Commission* [1988] ECR II-2937.

[103] This analysis presumes that the IP owner has a good faith basis for its infringement action and that a party obtaining an IP right has not misled the relevant national authority in respect of significant aspects of its application. Thus the conduct identified by the Commission in its Decision in Case COMP/A.37.507/F3 *AstraZeneca* (15 June 2005), if proven, would not be affected by this analysis. As the General Court emphasized in its recent judgment affirming that decision, the abuse involved obtaining IP rights to which the applicant either was not entitled or was entitled to for a shorter period (Case T-321/05 (n 72 above) para 355). The suggestion in the Commission decision, however, that the 'acquisition' of IP rights is clearly subject to competition law on the basis of the Court judgment in Case T-51/89 *Tetra Pak Rausing v Commission* [1990] ECR II-309 ('*Tetra Pak I*') is, however, seriously flawed. There is an enormous difference between a commercial transaction between two private parties for assignment of IP rights and the grant of IP rights by a national authority on application by a private party. The first situation involves a clear case of exercise of IP rights that does not involve the core rights constituting the specific subject matter of those rights. The second situation involves obtaining a legal right from a Member State to which the applicant is legally entitled and which is the prerequisite for any exercise of such rights. The General Court did not reach this issue in its recent judgment, but did remark in another context that the facts in *Tetra Pak* and those in *AstraZeneca* involved 'different situations' (*AstraZeneca* (n 72 above) para. 365).

[104] A recent paper by John Vickers outlines some of the economic models that support a possible negative impact of licensing practices on dynamic efficiency, J Vickers 'Competition Policy and Property Rights' (2010) 120 The Economic Journal 375–92. For a contrary view, see F Etro, 'Antitrust Issues in Dynamic Markets' in I Lianos and I Kokkoris (eds), *The Reform of EC Competition Law: New Challenges* (Kluwer Law International, 2010) 335 ('. . . no clear economic argument that weakening IPRs could ever strengthen innovation in the long run, even where innovation is sequential.').

Conclusion: Economic Assessment, IP Law, and the Pharmaceutical Inquiry

The extensive case law defining the application of EU law to exercise of IP rights granted by national law constrains the ability of the Commission to apply case-specific economic analysis in competition cases. Where the exercise of recognized core IP rights is concerned (such as the exclusive right of reproduction), the short-term effects of IP rights on static or allocative efficiency are simply not relevant to determination of whether conduct constitutes an abuse of a dominant position under Article 102 TFEU. The case law also provides no basis for limiting core rights on a functional basis in a specific case (ie, on the basis that the exercise of a core right in a specific case is not justified by relation to the policy objectives of that right.)

The practical implication of this analysis may be illustrated by reference to the final report in the Pharmaceutical Inquiry. The Report (which was prepared pursuant to the Commission's competition law powers under Regulation 1/2003) identifies a range of practices by major pharmaceutical companies that in the view of the Commission lead to significant competitive harm. The competitive harm identified in the Report almost entirely involves effects on static efficiency (higher prices due to delayed entry by generics manufacturers or constraints on ability of other manufactures to develop alternatives that will provide price competition for existing products). Much of the conduct challenged in the Report appears to involve the exercise of core IP rights (such as the development of exclusionary litigation strategies, evergreening, or defensive patenting). The scope of the Commission to apply Article 102 to this conduct would therefore be limited to cases where there is a significant impact on dynamic efficiency.[105]

[105] This might be the case, for example, where 'defensive patenting' prevented the development of a rival product that would constitute a significant improvement on the incumbent product in terms of efficacy or application.

The recent judgment of the General Court in *AstraZeneca* (n 72 above) does not affect this conclusion. In that judgment the Court ruled *inter alia* that a dominant firm '. . . cannot use regulatory procedures in such a way as to prevent or make more difficult the entry of competitors on the market, in the absence of grounds relating to the defence of the legitimate interests of an undertaking involved in competition on the merits or in the absence of an objective justification.' (para 672) The regulatory procedures in that part of the *AstraZeneca* case did not, however, involve the grant of IP rights. As explained in this paper, the basis for excluding exercise of core IP rights from application of Article 102 is not that such exercise constitutes 'competition on the merits' or provides an 'objective justification' for otherwise abusive conduct. Although obtaining IP rights may often constitute a form of 'performance-based competition', the cases limiting application of Article 102 where core IP rights are involved have not linked those limits to competition on the merits. Had they done so, some cases (notably *Volvo* (n 49 above) and *Renault* (n 24 above)) would have been decided differently. Thus since the special treatment of core IP exercise from Article 102 (outside exceptional circumstances) derives from allocation of competence under the Treaties, the application of Article 102 to use of regulatory procedures that do not involve IP should not impact application of Article 102 to conduct involving core IP rights.

This result may be frustrating for advocates of an economics-based approach to competition law enforcement, since it applies a formal 'legalistic' determination to issues that could be resolved by a sophisticated economic assessment tailored to an individual case. From a policy perspective, this result may nonetheless be sensible. It limits the conflict between competition law enforcement and the IP system by interfering only where the IP system has failed on its own terms—by limiting improvements in dynamic efficiency. This approach thus gives proper deference to the goals of the IP system and to the way that system operates in practice.

7

COMPULSORY LICENSING IN EUROPEAN COMPETITION LAW: THE POWER OF THE ADJECTIVE

Ian S Forrester and Katarzyna A Czapracka

Introduction

There are very few formal decisions by the European Commission and European courts on the desirability of compelling an allegedly dominant player to deal against its will. The law on refusals to deal has developed in a jerky, uneven fashion. In twenty years, we have moved from anything being surprising to everything being possible.

This difficulty of predicting outcomes is reinforced by the difference between how the law is promulgated by the European Court of Justice to courts making references (principled, rather cautious), and how the European Courts perform their appellate function (deferential, confined to a review of the legality of the act). The Courts are much more cautious in the former cases and much less critical in the latter.

Having been involved in several of these cases inevitably creates a certain distortion, and an intellectual desire to rationalize or criticize according to the posture advanced in court on a client's behalf. (One of the authors was counsel for the Commission in *Magill*[1] and for NDC in *IMS*[2] and for the company in

---

[1] Commission Decision of 21 December 1988 *Magill TV Guide/ITP, BBC and RTE* [1989] OJ L78/43 ('Magill Decision'), upheld by the General Court in Case T-69/89 *Radio Telefis Eireann v Commission* [1991] ECR II-485 ('General Court Magill Judgment'), upheld by the Court of Justice in Joined Cases C-241/91 P and C-242/91 *Radio Telefis Eireann v Commission* [1995] ECR I-743 ('Court of Justice Magill Judgment').

[2] Case C-418/01 *IMS Health GmbH & Co OHG v NDC Health GmbH & Co KG* [2004] ECR I-5039 ('IMS'). See also Commission Decision of 3 July 2001 in Case COMP D3/38.044 *NDC Health v IMS Health: Interim measures* [2002] OJ L59/18 and Commission Decision of 13 August 2003 in Case COMP D3/38.044 *NDC Health v IMS Health: Interim measures* [2003] OJ L268/69 and Case T-184/01 R *IMS Health Inc v Commission* [2001] ECR II-2349.

Microsoft,[3] and is therefore familiar with a broad range of possible theories on the topic; and maybe 'blinded' to alternative visions of the cases). But even allowing for that distortion, it is hard to deny the difficulty of stating what the law is today. When is a dominant player obliged to share its IP rights against its will?

One feature of the debate is a tendency to polarization, over-simplification, and adjectival thought—by labelling the conduct one defines its legitimacy or otherwise. Commission decisions and European Courts judgments are written in terms which are confident and give one side of the story so emphatically that the 'loser' appears to have had a hopeless case. A delicately balanced dispute over whether to encroach on a private right is presented as the righting of a wrong, the prevailing of fairness over abusiveness. Arguments to the contrary are sometimes disparaged rather than fairly advanced and rejected. The reader may understandably be confused.

These comments are substantive, not merely linguistic or aesthetic. Many scholarly articles about compulsory dealing take many hotly-disputed points for granted. Is the 'bad conduct' intrinsically bad? Bad in the particular context? Bad because profitable? Bad because it made life difficult for a competitor? What if it is highly lucrative? Would that constitute 'reasonable justification'? Do words like 'leveraging' or 'discriminatory' or 'anti-competitive' or 'excessive' have a literal meaning, or do they make manifest a verdict or moral comment?

The 1988 *Volvo/Veng*[4] case was the first EU compulsory licensing decision. It involved a classic battle—short-term compulsory sharing of property rights versus long-term respect for those rights. Should competition rules be applied to give consumers access to cheaper spare parts and after-sales services at the expense of doing some violence to intellectual property rights? Volvo made a profit from selling the car, so why did it need a monopoly over the replacement body parts? There was no creativity deserving of honour. Prices would fall. Consumers would benefit. Why should a mere design right block wholesome competition? But passionate oratory could be confidently deployed on both sides. What special ground of competition law existed which would distinguish this case from other cases? Who could deny that the right existed? Is there a good pro-competitive way to assert an IP right and a bad anti-competitive way? The right holder had been given a right, a monopoly created by the state. Why should he not use it? And what were the circumstances that indicated the unlawfulness of the action? The Court of Justice gave a balanced answer—generally speaking refusals to license are legitimate, but a refusal to license a valid IP right may violate Article 102 TFEU (ex Article 82 EC Treaty) if it involves 'certain abusive conduct', of which a few examples were given,

[3] Commission Decision of 24 March 2004 in Case COMP/C-3/37.792 *Microsoft* [2004] OJ L 32/23('Microsoft Decision'), upheld on appeal by the General Court in Case T-201/04 *Microsoft Corp v Commission* ('Microsoft Judgment') [2007] ECR II-3601.

[4] Case C-238/87 *AB Volvo v Erik Veng (UK) Ltd* [1988] ECR 6211.

> 8. It must also be emphasized that the right of the proprietor of a protected design to prevent third parties from manufacturing and selling or importing, without its consent, products incorporating the design constitutes the very subject-matter of his exclusive right. It follows that an obligation imposed upon the proprietor of a protected design to grant to third parties, even in return for a reasonable royalty, a licence for the supply of products incorporating the design would lead to the proprietor thereof being deprived of the substance of his exclusive right, and that *a refusal to grant such a licence cannot in itself constitute an abuse of a dominant position.*
>
> 9. It must *however* be noted that *the exercise of an exclusive right* by the proprietor of a registered design in respect of car body panels *may be prohibited by Article 86 [now Article 102 TFEU] if it involves,* on the part of an undertaking holding a dominant position, *certain abusive conduct such* as the arbitrary refusal to supply spare parts to independent repairers, the fixing of prices for spare parts at an unfair level or a decision no longer to produce spare parts for a particular model even though many cars of that model are still in circulation, provided that such conduct is liable to affect trade between Member States.[5]

These two famous paragraphs confirmed that the right holder was in principle free to say 'no', but could not do so with a wholly clear conscience. Thus the Court satisfied many points of view. IP conservatives were reassured, but competition enthusiasts noted that the door was not closed entirely.

Courts and agencies have wrestled ever since to define adequately what would justify imposing an obligation to license on a dominant company. In doing so, the EU Courts and the Commission were not only faced with the question of the proper balance between the need to protect competition and the need to safeguard IP rights, but also fundamental questions on the scope and purpose of competition rules.

Below we analyse the relevant case law, concluding that the results of these efforts have so far been lamentably unclear. Though the outcomes of individual compulsory licensing decisions arguably make sense, the reasoning behind these decisions is somewhat obscure, particularly if analysed in the broader context of other compulsory licensing cases. Dominant companies are left without clarity as to when saying 'no' to a competitor asking for their crown jewels could be condemned as 'arbitrary' or 'abusive,' or 'unjustified.' The Commission's long awaited Abuse of Dominance Guidance[6] not only did not make the EU law on refusals to license clearer—it made the situation even worse.

Still, we think that the precedents do offer some guidance and, if rightly read, could be used to provide some comfort to dominant IP holders. In the last section of this paper, we propose some principles which we think could be helpful in identifying obviously baseless claims and in assessing when antitrust intervention may be necessary.

---

[5] *Volvo/Veng* (n 4 above) (emphasis added).

[6] DG Competition, Communication from the Commission, Guidance on the Commission's Enforcement Priorities in Applying Article 82 EC Treaty to Abusive Exclusionary Conduct by Dominant Undertakings [2009] OJ C45/7.

A. IP as an Essential Facility?

Magill, the first case where the Commission found that the circumstances warranted granting a compulsory licence, shed some light on the notion of 'certain abusive conduct.' There were in the Republic of Ireland three weekly TV guides each devoted to the programmes of a single broadcaster. 'Listings' of the forthcoming week's TV programmes were sent by each television channel to every newspaper and magazine in the country, with a view to encouraging the widest possible publicity for the programmes. Newspapers reproduced the list of programmes every day, free of charge, with the encouragement of the broadcasters. But the newspapers were forbidden to reproduce them on a weekly basis, since each of the three broadcasters desired to prevent competition for its own weekly guide. Thus Irish TV addicts needed three magazines to plan their week's viewing, since each weekly guide described only one broadcaster's programming. Magill wanted to publish a weekly TV guide giving details of all available TV channels. In joint litigation before the Irish courts, the broadcasters invoked their copyright to enjoin Magill's publication and metaphorically stifled it at birth (after one issue had been published).

In April 1986, Magill lodged a complaint with the Commission, alleging that the TV companies' actions to close down the Magill TV Guide and their refusal to license constituted an abuse of their respective dominant positions. Magill also complained under Article 85 of the EEC Treaty (now Article 101 TFEU), claiming that the concerted action of the broadcasters in court revealed an unlawful agreement. After protracted internal debate the Commission decided to condemn the three broadcasters. In December 1988, it adopted a Decision finding that there had been a breach of Article 86 of the EEC Treaty (now Article 102 TFEU) and ordering the broadcasters to license, on reasonable and non-discriminatory terms, their advance weekly programme listings. The Commission expressed doubt as to whether copyright existed in the controversial material. The President of the Court of Justice showed considerable scepticism about the Commission's case and in May 1989, granted interim measures, ordering the suspension of the compulsory licence. In July 1989, the Irish High Court found that the claimed copyright, contrary to the Commission's doubts, did exist.[7] So the outcome was anything but certain.

[7] Mr Justice Lardner reasoned as follows:

> I am satisfied by the evidence that the BBC's weekly TV programme schedules as published in Radio Times are the end product of a long process of planning, preparation, arrangement and revision which involves a great deal of work and experience and the exercise of skill and judgment. They are the creation of the BBC and in my judgment they constitute an original literary work in the sense of a compilation within sections 2 and 8 of the Copyright Act 1963 in which the BBC and BBC Enterprises Limited have shown that they are entitled to copyright in the Republic of Ireland. [1990] ILRM 534, 550.

The Court of First Instance (now the General Court) upheld the Commission's Decision, reasoning that the circumstances of the case warranted subjecting the refusal to license to scrutiny under EU competition rules,

> Conduct of that type—characterized by preventing the production and marketing of a new product, for which there is potential consumer demand, on the ancillary market of television magazines and thereby excluding all competition from that market solely in order to secure the applicant's monopoly—*clearly goes beyond what is necessary to fulfil the essential function of the copyright as permitted in Community law.* The applicant's refusal to authorize third parties to publish its weekly listings was, in this case, arbitrary in so far as it was not justified either by the specific needs of the broadcasting sector, with which the present case is not concerned, or by those peculiar to the activity of publishing television magazines. It was thus possible for the applicant to adapt to the conditions of a television magazine market which was open to competition in order to ensure the commercial viability of its weekly publication, the RTE Guide. The applicant's conduct cannot, therefore, be covered in Community law by the protection conferred by its copyright in the programme listings.[8]

This approach suggested that the problem was invoking a right in circumstances not contemplated by the creators of the right; but it neglected the right of first publication.

The Court of Justice confirmed the lower court's decision but did so in more scientific language which could more comfortably be applied in future. It stated uncontroversially that '. . . the exercise of an exclusive right by the proprietor may, in exceptional circumstances, involve abusive conduct.'[9] The Court reasoned that such circumstances were present in the case before it because,

(1) the refusal to license concerned a product (the weekly programme listings) the supply of which was *indispensable* to the exercise of the activity in question (the publication of a complete weekly television guide);[10]
(2) the refusal prevented the appearance of a new product, a comprehensive weekly guide to television programmes, which the television companies in question did not offer and for which there was a potential consumer demand;[11]
(3) the refusal was not justified;[12] and
(4) the refusal allowed the broadcasters to reserve for themselves a secondary market, the market for weekly television guides, by excluding all competition on that market: the right was being used to prevent the emergence of something desired by the public which the broadcasters could have made but did not.[13]

---

[8] General Court Magill Judgment (n 1 above) para 73 (emphasis added).
[9] Court of Justice Magill Judgment (n 1 above) para 50.
[10] *Ibid* para 53.
[11] *Ibid* para 54.
[12] *Ibid* para 55.
[13] *Ibid* para 56.

The second test seems to us the most creative and helpful, as it potentially offered a way to distinguish bad refusals from good refusals, and it rooted the abuse in consumer harm.

Though the Court has never mentioned the essential facilities doctrine, some commentators suggested that in *Magill* the doctrine was applied to IP rights in Europe.[14] Without agreeing that *Magill* was an essential facility case, we can agree that the situation in *Magill* presented some analogies with essential facility cases. The protected subject matter was essential—Magill could not publish its weekly TV guide without TV listings drawn up each week by the broadcasters. Practical alternatives were not available. A compulsory licence substantially improved competition in the market place by increasing innovation—a new product for which there was substantial unmet consumer demand.

Forcing companies to share their assets with competitors has always been controversial. It is undisputed that such remedy has to be used with caution. The difficulty lies in defining the limiting principles and applying them consistently. Advocates of a restrained approach to competition policy have called for strict conditions limiting the application of the doctrine. Without such principles, as in the words of the late Professor Areeda, who was a prominent critic of the doctrine, the doctrine becomes a 'catch-phrase', '... indicating some exception to the right to keep one's creations to oneself, but not telling us what those exceptions are.'[15] In his 1990 essay, he suggested that the doctrine has been expanded '... to the limits of its language, with little regard to policy' and predicted that the 'cutting back' process would soon begin.[16] He advocated defining the doctrine in a rigorous manner—limit its use to cases where the plaintiff cannot compete effectively without the facility, where duplication or practical alternatives are not available, and where forced sharing '... is likely substantially to improve competition in the marketplace by reducing price or by increasing output or innovation.'[17] The last condition is unlikely to be satisfied '... when the plaintiff merely substitutes itself for the monopolist or shares the monopolist's gains.'[18] Last but not least, Professor Areeda pointed to the difficulties in designing remedies. In his view, imposing a duty to deal would only be practical if an admission to a consortium were at stake, if it could be achieved by a divestiture,

[14] See, eg, J Temple Lang, 'Defining Legitimate Competition: Companies' Duties to Supply Competitors and Access to Essential Facilities' (1994) 18 Fordham Intl LJ 437, 440; see also Opinion of AG Jacobs Case C-7/97 *Oscar Bronner GmbH & Co KG v Mediaprint Zeitungs-und Zeitschriftenverlag GmbH* [1998] ECR I-7791.

[15] PE Areeda, 'Essential Facilities: An Epithet in Need of Limiting Principles' (1990) 58 Antitrust LJ 841, 841. The article has been cited in numerous lower court opinions as well as in the United States Supreme Court ruling in *Verizon Communications, Inc v Law Offices of Curtis v Trinko, LLP*, 540 US 398 (2004).

[16] *Ibid* 841.

[17] *Ibid* 852.

[18] *Ibid* 852.

or if there is a regulatory agency that could supervise the terms of dealing.[19] Professor Areeda is not alone in promoting this narrow view on the doctrine of essential facilities. Even the proponents of the essential facilities doctrine note that the doctrine has been successfully invoked in very few cases and advocate limiting its application to very specific circumstances.[20]

The doctrine of essential facility seems to us most convincingly applied to physical infrastructure such as the port owner who also owns a ferry and who must/may/should allow a competing ferry to use the port.[21] We are doubtful of the utility of the notion in the field of IP rights. First, as IP rights are legal creations which were invented for a purpose and are defined narrowly to strike a balance between the long-term objective of promoting innovation and the short-term costs related to limiting competition, their scope is limited as compared to the scope of exclusivity granted by property law in tangible objects. IP rights typically cover only certain aspects of exploitation of the protected subject-matter. They are subject to numerous exceptions and limitations. Copyrights cover only one specific way in which information is expressed, not the idea itself revealed by the words as expressed. They are subject to a broad fair use exception. They cannot be used to prevent independent creation; and they are limited in time. The sharpest controversies relate to copyright over utilitarian materials such as maps, databases, protocols, interfaces, timetables, and lists. These genuine limitations on the market power conferred by IP rights ought to make them stronger (not immune) against competition rules. Second, the immaterial nature of IP means that any encroachment will be more severe than in the case of a physical asset. The owner of the railway bridge who has to allow a competing railway to cross his bridge still owns the bridge notwithstanding the irritations of competition law. In the case of the patentee or copyright holder, there will be much less left after the copyright has been nibbled away.

B. Applying the *Magill* Criteria

Ladbroke,[22] a case decided by the Commission a few years after *Magill*, suggested that the criteria coined in Magill would be interpreted narrowly (or confirmed that the Court is very reluctant to interfere with Commission decision-making).

Ladbroke, a betting shop chain, complained to the Commission that its competitor, the French Tote, violated Article 102 TFEU by refusing to supply televised picture

[19] *Ibid* 853.

[20] See, eg, SW Waller, 'Areeda, Epithets, and Essential Facilities' (2008) Wisconsin L Rev 360 (advocating the use of infrastructure theory to limit the essential facilities doctrine).

[21] The concept of an essential facility is based on the necessity to provide third parties, even competitors, access to an indispensable asset controlled by a monopolist. It was first developed in the US to analyse refusals to deal with competitors by vertically integrated dominant companies. See Waller (n 20 above) 361.

[22] Case T-504/93 *Tiercé Ladbroke v Commission* [1997] ECR II-923.

images and sound commentaries of French horse races to Ladbroke betting shops. Ladbroke relied on *Magill* to argue that without access to the televised pictures and sound commentaries it could not compete on the horse-race betting market. The Commission refused to act upon Ladbroke's complaint. The Court of First Instance upheld the Commission's decision, stating that the refusal to license,

> . . . could not fall within the prohibition laid down by Article [102 TFEU] unless it concerned a product or service which was either essential for the exercise of the activity in question, in that there was no real or potential substitute, or was a new product whose introduction might be prevented, despite specific, constant and regular potential demand on the part of consumers.[23]

Despite Ladbroke's claims, the live transmission of horse races did not seem essential for Ladbroke's business. Betting shops do not need live images in the shop in order to survive. The word 'essential' was used, but it was doubtful if the Court had decided to embrace the essential facilities doctrine. *Ladbroke* (together with *Bronner*,[24] not an IP case, decided a few years later) suggests that access is not indispensable if a dominant company's competitors can develop or market their products without access, even if it would be more convenient for them to have it.

Finally came the *IMS*[25] case. As in *Magill*, the Commission was asked to intervene against the otherwise fatal consequences of a dominant player successfully invoking at an interlocutory stage an improbable national IP right. IMS is the world leader in gathering and supplying data on deliveries to pharmacies by wholesalers of pharmaceuticals. Pharmaceutical companies use this data to measure the effectiveness of their promotional efforts in each town and district. On the German market a geographic format for presenting this data had been developed by IMS and its customers and had become the *de facto* industry standard. The format divides Germany into 1,860 geographic zones, based on postal codes, each containing at least three pharmacies, by reference to which marketing data describing deliveries, prices, and volumes in those zones is compiled and analysed. The precise frontier of each zone was agreed between IMS and its pharmaceutical customers, who drew the lines taking account of the location of bridges, hospitals, pharmacies, doctor's premises and the like. When a breakaway competitor NDC appeared on the German market, IMS relied on copyright to preclude it from presenting sales data, independently captured by NDC, using the same map of Germany as had been used for over a decade by IMS. Customers said to NDC that presenting the marketing data in any other way was not acceptable. NDC challenged the use by IMS of copyright before the German courts, and also complained to the Commission.

---

[23] *Ladbroke* (n 22 above) para 131.

[24] Case C-7/97 *Oscar Bronner GmbH & Co KG v Mediaprint Zeitungs-und Zeitschriftenverlag GmbH* [1998] ECR I-7791.

[25] Case C-418/01 *IMS Health GmbH & Co OHG v NDC Health GmbH & Co. KG* [2004] ECR I-5039 ('*IMS*').

The requested input was supposedly 'indispensable' to remain in the market. The Commission found that there was no practical possibility for companies wishing to compete with IMS to offer pharmaceutical sales data in any other format than that used by IMS. There were no actual or potential substitutes or alternatives to reporting the sales of pharmaceutical products in Germany other than the geographic lines delineating 1,860 regions following postal codes, which IMS was successfully claiming constituted a breach of its copyright. In the Commission's view, bringing an enforcement action in such situation was an abuse of dominance. It considered that the litigation was likely to eliminate all competition, and that the refusal to license lacked objective justification.[26]

Unlike in *Magill*, a compulsory licence would not necessarily result in the creation of a wholly new product. Though the complainant asserted that its product would be delivered by a different technology and captured more data, it could be argued that its product would not be a fundamentally different product than that already offered by IMS. Another important difference from the *Magill* case was that the 1,860 zone map was a *de facto* industry standard. IMS had developed the map together with its customers to match their needs and then invoked its copyrights to prevent NDC from offering its services in that format, which had in the meantime become the industry standard, as customers preferred it (at least in the immediate term). Thus, the Commission could have based its case on the theory that IMS was claiming IP rights over a standard developed with the client industry for the abusive purpose of excluding competition by preventing its competitors from using the standard.[27] The Commission, however, never pursued this avenue.

The President of the Court of First Instance suspended the Commission Decision.[28] The Commission did not ultimately rule on the merits, so there was no appeal. Instead, the Court of Justice gave a preliminary reference ruling in response to questions from a German court before which the national case was pending.[29] The Court of Justice, building on and referring to *Magill*, held that a refusal to license will be deemed abusive if four conditions are satisfied, namely that: (1) the protected product or service is *indispensable* for carrying on a particular business; (2) the refusal is '*such as to exclude any competition on a secondary market*'; (3) the refusal prevents the

[26] Commission Decision of 3 July 2001 *NDC Health/IMS Health: Interim Measures* [2002] OJ L59/18; withdrawn by the Commission Decision of 13 August 2003 *NDC Health/IMS Health: Interim Measures* [2003] OJ L268/69. To the best of our knowledge, the parallel copyright dispute has not yet been resolved: cases relating to the scope and ownership of copyrights in the IMS cases are still pending before the German courts.

[27] On this point see J Killick, 'IMS and Microsoft Judged in the Cold Light of IMS' (2004) 1 Competition L Rev 23, 30–2.

[28] Case T-184/01 *IMS Health v Commission* [2001] ECR II-3193.

[29] Case C-418/01 *IMS* (n 25 above).

emergence of a new product for which there is a potential consumer demand; and (4) the refusal is not justified by any '*objective considerations*'.[30]

The adjectives used to describe the conditions under which a compulsory licence is warranted are noteworthy. The access to the IP in question had to be *indispensable* for carrying out business on a particular market. It must be, in other words, absolutely necessary. As the Court clarified, the indispensability condition would not be satisfied if there are alternative solutions to the use of the requested input,

> It is clear from paragraphs 43 and 44 of *Bronner* that, in order to determine whether a product or service is indispensable for enabling an undertaking to carry on business in a particular market, it must be determined *whether there are products or services which constitute alternative solutions, even if they are less advantageous, and whether there are technical, legal or economic obstacles capable of making it impossible or at least unreasonably difficult for any undertaking seeking to operate in the market to create, possibly in cooperation with other operators, the alternative products or services.* According to paragraph 46 of *Bronner*, in order to accept the existence of economic obstacles, it must be established, at the very least, that the creation of those products or services is not economically viable for production on a scale comparable to that of the undertaking which controls the existing product or service.[31]

Access to IP should be granted only if a competitor as efficient as the dominant company could not duplicate it. The necessity to gain access to the IP in question must be of a general nature and not just limited to the competitor requesting the input. The refusal to provide access must be such as to 'reserve' the secondary market to the right holder '*by eliminating all competition on that market.*'[32]

Thus a licensee who wished to bring out a copy of the original product with let us say blue stripes or with added chlorophyll or at a lower price would not be making something new. The party requesting the licence would need to launch something wholly different from the product offered by the owner of the right,

> 48. [. . .] in the balancing of the interest in protection of the intellectual property right and the economic freedom of its owner against the interest in protection of free competition, the latter can prevail only where refusal to grant a licence prevents the development of the secondary market to the detriment of consumers.
>
> 49. Therefore, the refusal by an undertaking in a dominant position to allow access to a product protected by an intellectual property right, where that product is indispensable for operating on a secondary market, may be regarded as abusive only where the undertaking which requested the licence does not intend to limit itself essentially to duplicating the goods or services already offered on the secondary market by the owner of the intellectual property right, but intends to produce new goods or services not offered by the owner of the right and for which there is a potential consumer demand.

---

[30] *Ibid* paras 37–8.
[31] *Ibid* para 28 (emphasis added).
[32] *Ibid* para 52.

It would be for the national court to decide if the service being rendered was novel enough or a mere imitation.

The *IMS* judgment calls for restraint in applying Article 102 TFEU to refusals to license. That being said, the Court of Justice did not require itself to decide how the newly promulgated test would apply to the conduct of IMS, NDC, and the others. The nature of a preliminary ruling requires that the Court addresses the abstract principles and leaves it to the national court to apply these principles to the facts. The debate still goes on as to the nature and extent of the rights claimed by IMS.

The *IMS* test was applied by the Court of First Instance (now General Court) in *Microsoft*. Unlike the Court of Justice in *IMS*, the lower EU court was not asked to respond to hypothetical questions. It had to decide whether the Commission Decision imposing a compulsory licence should stand.

The *Microsoft* case dates back to 1998, when Sun Microsystems, a supplier of server hardware and software, requested Microsoft to provide it *inter alia* with '... complete information required to allow [Sun] to provide native support for the complete set of Active Directory technologies on Solaris [Sun's server software].'[33]

The underlying technological issues were complex and hotly debated. The marketplace involved the operating systems for computer servers, which coordinate families of personal computers and ensure that the family members will, for example, cooperate efficiently with each other with respect to sharing printers and authentication of users. Microsoft was developing a new generation of server operating system which would be launched in 2000, and one new feature of that operating system was to be its supposedly unique directory functionality.

Active Directory, the commercial name for the directory service incorporated in Microsoft's server software, allowed individual servers performing directory functions to function as exact replicas of each other. If a problem arose, each directory server was programmed to make its own decision on how to react without 'consulting' other servicers. Thus the system collectively delivered a single outcome. In practice this meant that a computer user could log into the network and get the same user experience—access the same confidential documents and restricted services—on any computer around the world. A number of patents covered the technology, which according to Microsoft was superior to that used by its rivals in various respects. It was undeniably different from them.

[33] Microsoft Judgment (n 3 above) para 4. For a detailed description of the facts of the case see IS Forrester, 'Victa Placet Mihi Causa: The Compulsory Licensing Part of the Microsoft Case' in L Rubini (ed), *Microsoft on Trial: Legal and Economic Analysis of a Transatlantic Antitrust Case* (Cheltenham: Edward Elgar, 2010).

Sun requested the information that would allow its servers to perform the Active Directory functions in the same manner as servers running Microsoft software. Sun believed that adding this feature to its server software was essential for its competitive position on the market. It asserted that without better interoperability than was currently available its future was bleak.

After a lengthy investigation, the Commission found that Microsoft abused its dominant position by refusing to provide Sun and other rivals with 'interoperability information' to enable them to develop their own technology for making operating systems for server computers performing so-called work-group functions. The Commission's theory was that by refusing to share information with its rivals, Microsoft preserved privileged connections between Windows and its work group server operating system. This allowed Microsoft to leverage its dominant position in the market for client PC operating systems in the market for server operating systems, reinforcing its position in the market for PC operating systems.

Microsoft argued that labelling the information requested by Sun as 'interoperability information' was misleading. According to Microsoft, Sun's server products were indeed able to interoperate with Microsoft products. In fact, different server operating systems interoperated perfectly well in practice in many customers' computer networks. Heterogeneous networks were routine—it was customary for companies, government departments and universities across Europe to use a wide range of software products from different vendors with PC 'clients' running Windows operating system. Any problems of true interoperability between Microsoft's server software and that of its competitors could have been resolved by using open industry standards, add-ons, and reverse engineering. Thus customers were not hindered in using heterogeneous computer networks by the absence of what might be called perfect identity as between different kinds of server.

What Sun requested, by contrast, was information that would effectively allow it to replicate particular features of Microsoft's software.[34] A server not running Active Directory could perform the Active Directory function in a Windows network only if, for directory purposes, it behaved identically to the Windows servers. Thus, it was technically necessary to replicate Microsoft's software functionality to achieve this end. The Commission rejected these arguments. The Commission asserted that although rival server operating systems could compete, they could not compete 'viably' unless given access to Microsoft's technology. According to the Commission, information on Microsoft's technology was indispensable because no other alternative was available for Microsoft's competitors to achieve the same level of interoperability

[34] Notably, Sun's request called for a wider divulgation than the Commission eventually compelled.

as that existing between Microsoft's own software.[35] Though it acknowledged that interoperability was a matter of degree and recognized that a lower level of interoperability existed,[36] it required a near-perfect, 'native' level of interoperability.

The Commission published the *Microsoft* Decision four weeks before the Court of Justice delivered its *IMS* ruling. It used a significantly different approach to the assessment of a refusal to license than that adopted by the Court of Justice. Unsurprisingly, in the course of the proceedings before the Court of First Instance (now the General Court), the Commission disputed Microsoft's assertions that interoperability protocols were covered by IP and that its decision involved compulsory licensing.[37] It argued that the 'automatic' application of the *IMS* criteria would not be appropriate in the *Microsoft* case and that other 'particular circumstances' had to be considered in the assessment of a refusal to license under Article 102 TFEU.[38] The Court did not rule on these matters.

Neither did the Court address the more controversial legal theories advocated by the Commission or rule on the question of how other circumstances may influence the assessment of a refusal to deal under Article 102 TFEU. The Court took the same approach as the Commission in the Decision, which assumed that IP rights were at stake and that they would have to be overridden. It held that a compulsory licence is warranted 'only in exceptional circumstances',[39] which occur in particular when the *IMS* conditions are met, namely that: (1) the refusal relates to a product that is indispensable to the exercise of a particular activity in a neighbouring market; (2) the refusal 'is of such kind as to exclude any *effective* competition on that neighbouring market'; and (3) the refusal prevents the appearance of a new product for which there is potential consumer demand.[40] Once it is established that such circumstances are

---

35 Microsoft Decision (n 3 above) para 669,

> As regards the use of open industry standards implemented in Windows, interoperability within a Windows work group network largely depends on specifications that are proprietary or are extended versions of standard protocols. Therefore, open industry standards fall short of enabling competitors to achieve the same degree of interoperability with the Windows domain architecture as Windows work group server operating system do. Since all major work group server operating system vendors already support most of the open industry standards in Windows, it can be concluded that this degree of interoperability proves to be insufficient for them to viably compete on the market. Therefore, reliance on open industry standards cannot be considered to be at present a realistic substitute to disclosures by Microsoft.

36 See Microsoft Judgment (n 3 above) para 139.

37 *Ibid* paras 267–82. The Commission conceded that the interoperability information could be protected under trade secret laws, but argued that trade secrets did not deserve the same level of protection as other forms of IP and that the EU case law on compulsory licensing of IP rights did not apply to trade secrets.

38 *Ibid* para 316.

39 *Ibid* para 331.

40 *Ibid* para 332 (emphasis added).

present, a refusal to license may violate Article 102 TFEU unless the dominant company proves that its refusal was objectively justified.[41]

The *Microsoft* Court invoked the *IMS* criteria, but its interpretation and application of these criteria were novel.[42] Under *IMS*, the indispensability criterion was quite clear—there could be no actual or potential substitutes for the requested product or service.[43] The *Bronner* ruling, relied upon by the *IMS* Court, specifically states that the requirement of indispensability is not satisfied even if such alternative products or services are 'less advantageous' than those controlled by the dominant company.[44] The *Microsoft* Court inserted an important qualification, which gave the notion of indispensability a new meaning. It held that the indispensability criterion is satisfied when the IP in question is necessary to *compete viably*.

The Court reasoned that Microsoft's protocols were indispensable for Microsoft's competitors, because their products had to work with Windows domain architecture on an equal footing with Microsoft's systems in order to *compete viably* on the market.[45] The Court also found that the requirement of eliminating competition was satisfied, rejecting Microsoft's argument that a refusal is abusive only if it is 'likely to eliminate all competition'[46] and that the prospect of eliminating competition must be 'immediate and strong.'[47] The Court agreed with the Commission that it is sufficient that the refusal creates a *risk* of elimination of all *effective* competition, dismissing the evidence that Microsoft's competitors had a collective market share in the range of 30–40 per cent.[48]

The *Microsoft* case makes it clear that the essential facilities doctrine is not satisfactory as a limiting principle in the application of Article 102 TFEU to refusals to license. It is easier to assess indispensability (yes or no) when it relates to a physical input, in particular infrastructure. The input is indispensable simply when it is impossible to do business without it. In cases involving IP rights, the focus often seems to be whether it is possible without the IP in question to do business *in a manner desired by the company requesting the licence*.

---

41 *Ibid* para 333.

42 As one commentator noted, the *Microsoft* Court interpreted the *IMS* criteria 'surprisingly elastically.' J Vickers, 'Competition Policy and Property Rights' (May 2009) University of Oxford, Department of Economics, Discussion Paper Series No. 436, 22.

43 Case C-418/01 *IMS* (n 25 above) para 28; see also *Bronner* Judgment (n 24 above) paras 43–6. The *Bronner* Court clarified that this would typically be the case if there are 'technical, legal or even economic obstacles capable of making it impossible, or . . . unreasonably difficult' to replicate the product or service (para 44).

44 *Bronner* Judgment (n 24 above) para 41.

45 Microsoft Judgment (n 3 above) paras 230, 248–50 and 369–422.

46 Microsoft Judgment (n 3 above) para 439.

47 *Ibid* para 440.

48 *Ibid* paras 563–4.

The danger of abusing the essential facilities doctrine in the IP context is apparent. Essential is readily transmuted into 'the convenient facilities doctrine',[49] as competition law enforcers attempt to fit a more controversial case into the essential facilities framework.[50]

C. New Product and Dynamic Competition Considerations

The fact that a refusal to license prevented the appearance of a new product was discussed in the three seminal cases—*Magill, IMS*, and *Microsoft*. It has been suggested that the new product criterion could serve as a significant limitation on unmeritorious complaints about refusals to license.[51]

The Court of First Instance reasoned in *Magill* that although the exercise of the exclusive right '. . . *is not in itself an abuse*', it may be abusive if the right is not '. . . exercised in a manner which corresponds to its essential function', '. . . in such a way and circumstances as in fact to pursue an aim manifestly contrary to the objectives of Article [102 TFEU].'[52] This formulation of what is the essence of an abuse was not very convincing as it assumed there are good exercises and bad exercises of legal rights, but neglected the author's right to make first publication. The conclusion was firm: by preventing the emergence of a new product, the TV stations were using their exclusive rights '. . . in order to secure a monopoly in the derivative market of weekly television guides', which was '. . . beyond what is necessary to fulfil the essential function of copyright as permitted by Community law.'[53] On appeal, the Court of Justice more skilfully stated that the refusal to license was abusive because it prevented the appearance of a comprehensive weekly guide to television programmes, which the television companies in question did not offer and for which there was a potential consumer demand.[54] Magill was not attempting to copy the products already offered by the broadcasters. It was offering an improvement. Consumers could purchase a single guide with information on three channels instead of three separate guides.

---

49 See in general D Ridyard, 'Compulsory Access under EC Competition Law—A New Doctrine of "Convenient Facilities" and the Case for Price Regulation' (2004) 11 European Competition L Rev 269.

50 Professor Fox suggested, that although the *IMS* test might be appropriate for judging refusals to license IP that has become a standard, it was '. . . too specific to provide a fitting framework for analysis' of the issues that surfaced in Microsoft. See EM Fox, 'Monopolization, Abuse of Dominance, and the Indeterminacy of Economics: The U.S./E.U. Divide' (2006) Utah L Rev 725, 733.

51 On this point, see IS Forrester, 'Regulating Intellectual Property Via Competition? Or Regulating Competition Via Intellectual Property? Competition and Intellectual Property: ten years on, the debate still flourishes' in CD Ehlermann and I Atanasiu (eds), *European Competition Law Annual 2005: The Interaction Between Competition Law and Intellectual Property Law* (Oxford: Hart, 2007) 83–4.

52 General Court Magill Judgment (n 1 above) para 71.

53 *Ibid* para 73.

54 *Ibid* para 54.

The broadcasters were preventing Magill from delivering a product which the broadcasters themselves were unwilling to supply.

The new product condition also forms part of the *IMS* test,[55] but the rationale changed. The Court of Justice reasoned that the refusal to grant the licence could be abusive if it impairs '. . . the development of the secondary market to the detriment of consumers,'[56] which is when the company requesting the licence '. . . does not intend to limit itself essentially to duplicating the goods or services already offered on the secondary market by the owner of the IP right, but intends to produce new goods or services not offered by the owner of the right and for which there is a potential consumer demand.'[57] Demonstrating that a refusal prevents the creation of a concrete 'new product' is still required, but it is used as a means to accommodate dynamic competition concerns. Though the Court does not comment on the degree to which the new product must be different from the product offered by the IP holder and whether the two products could be substitutable, the judgment reflects a policy choice in favour of a system based on competition by substitution and not on competition by imitation.

In *Microsoft,* the new product condition was again reinterpreted. The Commission reasoned that Microsoft's refusal to share the 'interoperability information' had '. . . the consequence of stifling innovation in the impacted market and of diminishing consumers' choice by locking them into a homogenous Microsoft solution.'[58] The Commission applied a 'balancing' test, weighing Microsoft's incentives to innovate against the incentives of the whole industry to innovate were Microsoft required to make its technology available for licence.[59] In its decision, the Commission did not identify any new products Microsoft's competitors would develop using Microsoft's communication protocols, or show that there was any demand for such a product. Nor did it limit the licence to the making of new products. It merely claimed that Microsoft's competitors could use the interoperability information to develop 'advanced features of their own product.'[60] Indeed, it is not surprising that competitors could develop new features with access to sophisticated technology.

The Commission argued that its decision was not confined to '. . . a mere analysis of the "new product" criterion as defined by IMS.' It analysed this criterion '. . . by

[55] Case C-418/01 *IMS* (n 25 above) paras 37–8.

[56] Case C-418/01 *IMS* (n 25 above) para 48.

[57] *Ibid* para 49.

[58] Microsoft Decision (n 3 above) para 782.

[59] *Ibid* para 783,

> . . . a detailed examination of the scope of the disclosure at stake leads to the conclusion that, on balance, the possible negative impact of an order to supply on Microsoft's incentives to innovate is outweighed by its positive impact on the level of innovation of the whole industry (including Microsoft). As such, the need to protect Microsoft's incentives to innovate cannot constitute an objective justification that would offset the exceptional circumstances identified.

[60] Microsoft Decision (n 3 above) para 695.

reference to the prohibition. . . of abuses of a dominant position that limit technological development to the prejudice of consumers.'[61] On appeal, the Court of First Instance embraced the Commission's interpretation of the new product condition and held that a refusal to license may be abusive not only if it prevents the marketing of a new product, but also when it limits 'technical development.'[62] It was not necessary to identify any particular product that the refusal prevented from coming into being.[63] The Court held that Microsoft's refusal to license limited technical development because it created an 'artificial interoperability advantage', which, together with Microsoft's market position, discouraged the development of competing server operating systems.[64] In fact, the Court noted, consumers considered that non-Microsoft work group server operating systems were better than Windows work group server operating systems with respect to a series of important features. As a result, '. . .the limitation thus placed on consumer choice [was] all the more damaging to consumers.'[65]

The significance of the new product condition has thus changed from case to case. In *Magill* it was used, rather formalistically, to delimit the scope of a copyright; in *Microsoft* it was used to analyse dynamic competition considerations. The meaning of the condition shifted accordingly from a rather clear notion of a new product that the company requesting the licence would develop to a rather abstract notion of preventing technological development. As has been rightly noted, if the prevention of 'technical development' is abusive then every refusal to license by a dominant player will fulfil this criterion.[66] There is always a chance that forcing a dominant company to share its technology will lead to follow-on innovation and eventually advantage for society. But this is a very speculative justification for a huge encroachment on the interests of the right holder.

D. Correcting National IP Legislation?

In *Volvo,* the Court of Justice elected not to pronounce on whether the design of replacement body parts for cars 'deserved' protection. But worthiness continued to feature in subsequent compulsory licensing cases. Both in *Magill* and *IMS*, the IP in question was controversial and the copyrighted work was distributed free of charge to commercial partners or intermediaries. In both cases, there was said to

---

[61] Microsoft Judgment (n 3 above) para 632.
[62] *Ibid*(n 3 above) paras 647–9 (internal citations omitted).
[63] *Ibid* para 647.
[64] *Ibid* para 653.
[65] *Ibid* para 652.
[66] D Howarth and K McMahon, 'Windows has Performed an Illegal Operation: The Court of First Instance's Judgment in Microsoft v. Commission' (2008) 29 European Competition L Rev 117, 123.

be an imbalance between the creative effort of the right holder and the economic advantages flowing from its exercise of the right. In *Magill*, the challenged copyrights were criticized as an anomaly, as they granted TV companies control over information (channel, day, time, and title of programmes) and not just the way in which that information was expressed. Similarly, in *IMS*, the scope of the copyright in question as well as entitlement to the copyright were debated, a debate which is still unresolved today years later. Although the merits of extending copyright protection were not directly considered in *Magill* or in *IMS*, they had significance for the final outcome of each case.[67]

One of the arguments made in the Magill case was that the Commission was using competition law to override or to correct unacceptable or unwelcome features of national copyright law. This was formally incorrect but it may well have been true that the nature of the right had an effect on the decision to intervene and the overall outcome of the case. Correspondingly one might argue (and one of the authors did argue) that the Commission would not be likely to encroach upon a universally recognized IP right.[68] The *Microsoft* case, however, suggests otherwise.

After ascertaining that the circumstances of the refusal justified an imposition of an obligation to license, the *Microsoft* Court turned to the question of whether there was an objective justification for the refusal. The Court did not address the issue whether, as the Commission argued, trade secrets, which protected Microsoft's interoperability information, deserved less respect than other forms of IP. It based its reasoning on the assumption that the technology at issue was innovative and of significant value simply because it was protected by IP rights.[69] Yet, in and of itself, this fact was not a valid reason for a refusal to license.[70] Such refusal would only be justified if compulsory licensing had a negative impact on the dominant company's incentives to innovate *in a particular case*.[71] It did not matter that the patents and trade secrets that protected Microsoft's software were robust from an IP law perspective. Similarly, Microsoft's arguments that it had made a significant investment in the development of its communication protocols and that it did not wish its competitors to use that technology to compete with it were rejected as 'vague' and 'theoretical'.[72] The Court held that Microsoft did not meet the burden of proof because it failed to provide evidence that a compulsory licence would have a negative impact on the development of specific technologies or products. The relevant questions were whether

[67] See, eg, V Korah, 'The Interface between IP and Antitrust: The European Experience' (2002) 69 Antitrust L J 801, 810–13.

[68] See IS Forrester, 'EC Competition Law as a Limitation on the Use of IP Rights in Europe: Is There Reason to Panic', in CD Ehlermann and I Atanasiu (eds), *European Competition Law Annual 2003: What is an Abuse of a Dominant Position?* (Oxford: Hart, 2006) 517.

[69] Microsoft Judgment (n 3 above) paras 693–5.

[70] *Ibid* para 690.

[71] *Ibid* paras 696–711.

[72] *Ibid* para 698.

Microsoft's rights were necessary to protect Microsoft's incentives to innovate and, whether, should a compulsory licence be ordered, it would negatively affect Microsoft's incentives to innovate.

The evidence of negative impact on the incentives to innovate would have to outweigh the negative impact of preserving the dominant company's exclusivity.[73] The Court reasoned that since sharing interoperability information is 'normal practice for operators in industry', it was unlikely that imposing such an obligation on Microsoft would adversely affect its incentives to innovate.[74] In this context, the fact that compulsory licensing would not allow rivals to 'clone' Microsoft's products was important.[75] It was also relevant that Microsoft had agreed to make disclosures of certain communication protocols under the United States settlement, which, as Microsoft acknowledged, did not have a negative effect on its incentives to innovate. This suggests that the evidence of previous dealings with rivals, including dealings forced by regulatory authorities in other jurisdictions, as well as general industry practices could be used against a dominant IP holder.

E. Where Does the Law Now Stand?

Only a few compulsory licensing cases have so far been decided at EU level. They each created new waves of controversy. The more liberal or minimalist approach clashed with the more interventionist or formalistic approach to competition law enforcement.

In practice, EU courts appear to be more cautious when stating the law for the guidance of national courts as to when a refusal to deal is illegal, than in scrutinizing Commission decisions in the same field. The deferential scope of the judicial review of complex economic and technical appraisals made by the Commission is at least a partial explanation of this phenomenon.[76] In *IMS*, a preliminary reference case, the Court of Justice established a test, consisting of four, rather strict, conditions to assess whether the refusal to licence was deemed to be abusive. Using this test to decide if the Commission's decision should stand in *Microsoft*, the Court of First Instance significantly relaxed the *IMS* conditions.

The challenge for competition law is to establish a realistic and predictable framework to determine those rare circumstances where the holder of the legal monopoly must share the IP right with others. So far, attempts to develop predictable limiting

[73] *Ibid* paras 709–11.
[74] *Ibid* para 698.
[75] *Ibid.*
[76] In *Microsoft*, the General Court held that the review of complex economic and technical appraisals made by the Commission is limited to 'whether the facts have been accurately stated and whether there has been any manifest error of assessment or misuse of power.' *Ibid* para 87.

principles have failed. As 'exceptional circumstances' have been redefined, they hardly seem to have any determinable meaning. As invoked in *IMS* and *Microsoft*, the criteria have not been applied consistently enough to provide meaningful guidance to dominant IP right holders.

Analysing EU compulsory licensing case law may lead to the conclusion that all is left to prosecutorial discretion. The lack of certainty inherent in the approach advocated by the Commission means that a dominant IP holder faced with a request to license will find it difficult to ascertain if his refusal to license is legitimate. Unlike in the United States, where there are hundreds of relevant decisions, EU case law is quite thin. 'Negative' case law, showing the circumstances where a request for a licence could be safely denied, is lacking. It is unlikely that we will have many such 'negative' decisions in the future, considering that cases challenging a rejection of complaint by the European Commission are procedurally very difficult to win.

F. Murky Guidance from the Abuse of Dominance Guidance

The Abuse of Dominance Guidance does, sad to say, nothing whatever to remedy the obscure state of the law.[77] The document states the Commission's enforcement priorities and 'is not intended to constitute a statement of the law.' Of course only the court can state the law, but the Commission could have given useful guidance even if this meant limiting its discretionary room for manoeuvre.

Instead of trying to reconcile *Microsoft*'s broad reading of the *IMS* test with a more restrained approach taken in other compulsory licensing cases, the Commission proposed yet a different approach, which appears to be based on the essential facilities theory. According to the Guidance, the Commission will consider a refusal to deal as its 'enforcement priority' if (1) 'the refusal relates to a product or service that is objectively necessary to be able to compete effectively on a downstream market,' (2) 'the refusal is likely to lead to the elimination of effective competition on the downstream market', and (3) 'the refusal is likely to lead to consumer harm.'[78]

The Commission mentions the *IMS* conditions adopted by the *Microsoft* Court and (understandably) leaves for itself a significant degree of discretion in their application. The Guidance provides that indispensability means that there is '. . . no actual or potential substitute on which competitors in the downstream market could rely so as to counter—at least in the long-term—the negative consequences of the refusal.'

[77] For a more detailed comment on the Abuse of Dominance Guidance (n 6 above) see Centre for European Policy Studies 'Treatment of Exclusionary Abuses under Article 82 of the EC Treaty: Comments on the European Commission's Guidance Paper. Final Report of a CEPS Task Force' (2009) <http://www.ceps.eu/book/treatment-exclusionary-abuses-under-article-82-ec-treaty-comments-european-commissions-guidance> accessed 9 September 2010.

[78] Abuse of Dominance Guidance (n 6 above) para 81.

It clarifies that an input would be indispensable if competitors cannot 'effectively duplicate' it 'in the foreseeable future'[79] and that 'an input is likely to be impossible to replicate when it involves a natural monopoly due to scale or scope economies, where there are strong network effects or when it concerns so-called single source information.'[80] The Commission stresses that indispensability does not entail that 'no competitor could ever enter or survive on the downstream market' without the requested input. The test is not whether it is impossible to duplicate the input, but whether it is impossible *effectively* to duplicate the input.

The Guidance's take on the elimination of competition criterion is little better in terms of clarity—the test is whether the refusal to supply is 'generally liable to eliminate, immediately or over time, effective competition in the downstream market.'[81] Yet, aside from the rather obvious statement that the higher the market shares of the dominant company in the downstream market the more likely that competition will be eliminated, the Commission provides little guidance as to when competition in the downstream market is robust enough that a dominant company does not have to be concerned with the legality of its refusal to deal with a competitor.

With respect to the new product criterion, the Guidance departs from the existing case law. It treats prevention of the emergence of a new product or technological development only as indicative examples of 'consumer harm', a new condition making a refusal to deal illegal.[82] The 'consumer harm' criterion resembles the balancing test from the *Microsoft* Decision—the refusal is likely to lead to consumer harm if '. . . the likely negative consequences of the refusal to supply in the relevant market outweigh over time the negative consequences of imposing an obligation to supply.' This test is remarkably vague as the basis for an economic crime. It is hard to predict how it will be applied. The more so since abuse is supposed to be an objective concept. Should the abstract possibility of an increase in innovation and consumer choice be sufficient to establish consumer harm?

The Guidance states that a dominant company may offer evidence that its,

> . . . own innovation will be negatively affected by the obligation to supply, or by the structural changes in the market conditions that imposing such an obligation will bring about, including the development of follow-on innovation by competitors.[83]

In and of itself, the test does not appear to be objectionable. It seems, however, that no level of effort would suffice to prevail on this defence. The *Microsoft* case suggests that the burden of proof placed on the IP right holder would be severe and that, for

[79] *Ibid* para 83.
[80] *Ibid* para 58.
[81] *Ibid* para 85.
[82] *Ibid* paras 85–6.
[83] *Ibid.*

a company with high market shares, prevailing on this defence would be next to impossible. No level of intellectual or technical effort could assure immunity.

The Abuse of Dominance Guidance confirms that the merits of the IP in question will be taken into account when assessing whether the refusal to license was justified, but does not clarify the questions that the *Microsoft* Court left open. The Guidance states that the Commission will consider whether the refusal is '. . . *necessary* to allow the dominant undertaking to realise an *adequate* return on the investments required to develop its input business, thus generating incentives to continue to invest in the future, taking the risk of failed projects into account.'[84] This would indicate that evidence of significant expenditure with respect to the development of the technology may justify a refusal to license. But the Commission offers no guidance as to how it would assess whether IP protection is necessary to provide an *adequate* return on the right holder's investment. The test leaves significant scope for discretionary judgments. It is hard to predict how it would be applied in a specific case.

Sorry to say, the Guidance is almost worthless, save in that it confirms the Commission will retain the right to find an abuse when all the circumstances so suggest; and one cannot deny that the cases justify its taking such a line.

Conclusion and a Few Guiding Principles

The current state of the law is unsatisfactorily vague. The criteria coined by the EU Courts in the *Magill* and *IMS* cases offered a principled basis to determine whether a refusal to license could be challenged under Article 102 TFEU. The coherence of this approach has been made uncertain by the *Microsoft* judgment. The Abuse of Dominance Guidance appears to have been drafted to ensure maximum flexibility for the Commission in the future.

It may be impossible to create an exhaustive list of 'exceptional circumstances' making a refusal to licence abusive. But a more principled approach to assessing refusals to license is certainly possible. We take the liberty of suggesting some principles and rules of thumb which may be useful until the law has become more satisfactory.

First principle—verify the facts

What it is that the requesting company wants? What would the dominant company be expected to do? What rights are at stake? These questions must be decisive. One should avoid the temptation to concentrate on minor details matching an attractive legal theory and leave inconvenient truths out of the analysis. In all compulsory licensing cases decided at the EU level, there have been significant errors by several

[84] *Ibid* para 89.

commentators in recounting the facts, and significant misappreciations in deciding what the key factual points were. Facts should never be stretched to help launch new cases. Rigorous sober enquiry must be the point of departure.

Second principle—presumption of legality

The right holder should, in principle, be allowed to exclude others from using its IP and should be able to choose freely whom to license. Only truly egregious circumstances should create a need for antitrust intervention.

Rules of thumb

We suggest that the following list of considerations, some rooted in case law, all based on what appears to the authors to be common sense, should be helpful in assessing whether intervention is warranted:

Is the requested input indispensable? Is the requested input objectively essential for entering the market, or would it just be very convenient? Is the input necessary to do business in the downstream market, or is it only *necessary for doing the business in a particular way* favoured by the complainant? Does the lack of the IP right offer the would-be licensee absolutely no alternative? There may be cases where a licence is given despite less than utter necessity, but we suggest that it must be right to assess where the facts lie on the spectrum from desirability to indispensability.

How big has been the investment in time, skill, effort, and in resources to create the protected subject-matter? The most oft-cited rationale for IP rights refers to the need to ensure an appropriate reward for IP creation.[85] Thus proper consideration should be given to the circumstances preceding the creation of the protected subject-matter. How much investment went into developing it? Is it a fruit of skilled or laborious efforts? Is its real value as a gatekeeper or is it intrinsically valuable? We presume that a patented cure for cancer that is the fruit of years of research and huge financial expenditure would not be treated in the same way as TV listings, but the law at this stage does not clearly say so.[86]

Is the requested information secret? Is the input secret or is it widely dispersed? If it is already widely accessible, perhaps free of charge, it seems less justified to restrict access to it, less shocking to compel its wider reproduction.

[85] See, eg, K Arrow, 'Economic Welfare and the Allocation of Resources for Inventions' in R Nelson (ed), *The Rate and Direction of Economic Activity: Economic and Social Factors* (Princeton: Princeton University Press, 1962) 619; W Nordhaus, *Invention, Growth, and Welfare: A Theoretical Treatment of Technological Change* (Cambridge, MA: MIT Press, 1969); WM Landes and RA Posner, 'An Economic Analysis of Copyright Law' (1989) 18 J Legal Studies 325.

[86] Such an approach was argued by the Commission in *Magill*, by Microsoft in *Microsoft*, and (obliquely) by the complainant in *IMS*.

How intrusive is the proposed intervention? In most competition cases, the remedies imposed by the Commission consist of demanding that the dominant company terminate the infringing conduct. In compulsory licensing cases, however, the intervention in the commercial freedom goes further, as it involves the creation of a positive duty to license. In some cases, this duty may be less onerous than in others. Notably, both in *Magill* and in *IMS*, the copyrighted subject-matter was widely reproduced and, at least under certain constraints, available free of charge. In *Microsoft*, the dominant company had to draft 'specifications', which involved effectively writing a detailed description of how to make a competing product incorporating Microsoft's technology. In our view, the presumption of legality should be rebutted more easily in cases where the encroachment on the IP rights and the commercial freedom of the right holder is relatively minor.

Are there immediate, direct, and identifiable benefits for consumers? Is the compulsory licence likely to create specific benefits for consumers? We think that this condition is most likely to be satisfied when as a result of the compulsory licence, the consumers would get access to a completely new product, which the dominant company perhaps may, but does not, offer.

The questions listed above could be cumulative, though in most cases answering only one or two will reveal the answer.

We would hope that such mushy, adjectival concepts as 'exceptional circumstances' or 'objective justification' may become obsolete, if the foregoing factors were prudently analysed. Economic crimes demand articulated and foreseeable criteria. We submit that the approach we propose would allow to identify, with mild confidence even if not with legal certainty, meritless requests. Even the large and the prosperous deserve a modicum of legal predictability in running their affairs. As more cases are decided, we will be able to predict with more confidence with the benefit of a richer jurisprudence.

8

REFUSAL TO GRANT ACCESS TO TRADE SECRETS AS AN ABUSE OF MARKET DOMINANCE

Josef Drexl

Introduction

The topic of this chapter is inspired by a debate which took place in Brussels in June 2006 at the public hearing on the European Commission's Discussion Paper on Article 82 EC (now Article 102 TFEU).[1] At that occasion my presentation of the comments of the Max Planck Institute for Intellectual Property, Competition and Tax Law[2] triggered somebody from the audience to claim that the standard of competition law intervention in a case of a refusal to disclose trade secrets, as compared to a refusal to license other intellectual property, should be highest since trade secrets nowadays are among the most valuable assets of undertakings.

Given that, at the time of the hearing, the *Microsoft* case was pending before the Court of First Instance (CFI, now the General Court),[3] this intervention attracted great attention. The Commission, in its decision of 2004,[4] had ordered Microsoft to

---

[1] 'DG Competition Discussion paper on the application of Article 82 of the Treaty to exclusionary abuses, December 2005', <http://ec.europa.eu/comm/competition/antitrust/art82/discpaper2005.pdf> accessed 10 September 2010 (hereinafter 'Discussion Paper').

[2] J Drexl, B Conde Gallego, S Enchelmaier, M Leistner and M-O Mackenrodt, 'Comments of the Max Planck Institute for Intellectual Property, Competition and Tax Law on the Directorate-General Discussion Paper on the Application of Art. 82 of the EC Treaty to Exclusionary Abuses' (2006) 37 Intl Rev Intellectual Property and Competition L 558.

[3] Case T-201/04 *Microsoft v Commission* [2007] ECR II-3601 (hereinafter 'CFI judgment').

[4] *Microsoft* (COMP/C-3/37.792) [2007] OJ L32/23 (summary of the Commission Decision of 24 March 2004); note that all subsequent references to the Commission Decision are to be construed as references to the full text version of the Commission Decision, C(2004)900 final, <http://ec.europa.eu/competition/antitrust/cases/decisions/37792/en.pdf> accessed 10 September 2010 (hereinafter 'Commission Decision').

provide competitors with the interoperability information contained in Microsoft's Windows operating system[5] to enable competitors to produce compatible work group server operating systems.[6] It also seems that the Commission had formulated its Discussion Paper on Article 82 EC (now Article 102 TFEU) in the light of the *Microsoft* case.[7] There, in contrast to the above-cited opinion expressed at the hearing, the Commission argues that the standard for intervention in case of a mere refusal to grant access to trade secrets could be lower than in the case of a refusal to grant a licence for an intellectual property right (IPR). As regards interoperability information contained in a computer program, the Commission states as follows,

> Even if such information may be considered a trade secret it may not be appropriate to apply to such refusals to supply information the same high standards for intervention as those described in the previous subsection.[8]

In its judgment of September 2007, the CFI confirmed the Commission's decision also in regard to the order to provide competitors with interoperability information.[9] However, the CFI did not clarify whether the standard of intervention can indeed be lowered for trade secrets. Already the Commission had based its decision on the assumption that the interoperability information, as argued by Microsoft, was protected as intellectual property[10] and assessed the case under the principles governing the application of Article 102 TFEU (ex-Article 82 EC) on a refusal to license an intellectual property right.[11] Since the CFI found that the Commission had applied these principles correctly, the Commission decision was upheld. Thereby, the CFI, on the one hand, confirmed that the standard for intervention in case of trade secrets is certainly not higher than in a case involving IPRs.[12] On the other hand, the CFI left the two questions unanswered: (1) whether the interoperability information was protected by IPRs in the first place, and (2) whether the standard of intervention in trade secrets could be lower than for intellectual property. It is the latter question in particular which will be discussed in this chapter.

---

[5] 'Operating systems' are '... system software products that control the basic functions of a computer and enable the user to make use of such a computer and to run application software on it.' See Commission Decision (n 4 above) para 37.

[6] Work group operating systems are '... operating systems designed and marketed to deliver collectively file, print and group and user administration services to relatively small numbers of client PCs linked together in a small to medium-sized network'. See Commission Decision (n 4 above) para 345.

[7] Discussion Paper (n 1 above) paras 241 *et seq.*

[8] *Ibid* para 242.

[9] CFI judgment (n 3 above).

[10] Commission Decision (n 4 above) para 546. The Commission states as follows: '... [I]t cannot be excluded that ordering Microsoft to disclose such specifications and allow such use of them by third parties restricts the exercise of Microsoft's intellectual property rights.'

[11] CFI judgment (n 3 above) para 242.

[12] The CFI judgment (n 3 above) para 284 stated that the Commission, under the assumption that the interoperability information was protected by intellectual property rights applied, '... chose the strictest legal test and therefore the one most favourable to Microsoft.'

In the following sections, I will first consider whether Microsoft had convincing arguments to claim IP protection before the CFI (in section A below). Then, I will discuss the standard for intervention as it appears after the two decisions of the ECJ in *IMS Health*[13] and the CFI in *Microsoft* (in section B below). In the most important section, I will clarify that there is no reason for applying a substantially different, or even lower, standard of intervention to a case of a refusal to disclose trade secrets, but that the peculiarities of trade secrecy protection, as compared to intellectual property, need to be taken into account in applying the general standard (in section C below). The analysis will be concluded with a short analysis of the Commission's Guidance Paper of December 2008[14] with the objective of considering how the Commission would deal with the interoperability issue in the future (in section D below).

A. Intellectual Property Protection for Interoperability Information

How to define the standard for intervention in case of a refusal to disclose trade secrets is most relevant for establishing interoperability in the software industry. Due to network effects, Microsoft's Windows operating system has emerged as the *de facto* standard in the market for such systems.[15] Whoever seeks access to the adjacent market for compatible application programs depends on access to the interoperability specification (also 'interoperability information') implemented in the Windows program.[16]

Whether and under which standard the Commission was authorized to order Microsoft to disclose the interoperability specification under Article 102 TFEU may depend on the availability of IP protection and the kind of IP right. If that interoperability information, namely the description of what an application program must

[13] Case C-418/01 *IMS Health v NDC Health* [2004] ECR I-5039 (hereinafter '*IMS Health*').

[14] 'Communication from the Commission—Guidance on the Commission's enforcement priorities in applying Article 82 of the EC Treaty to abusive exclusionary conduct by dominant undertakings' [2009] OJ C45/3 (hereinafter 'Commission Guidance Paper'). For a concise overview and evaluation, see 'The new guidance on Article 82—does it do what it says on the tin' (2009) Oxera Agenda. <http://www.oxera.com/cmsDocuments/Agenda_January09/New guidance on Article 82.pdf> accessed 10 September 2010. For another, more general analysis, see L Finell, 'The Article 82 Reform: Increasing the Incentives to Comply?' (2009) 32 World Competition 283.

[15] See Commission Decision (n 4 above) paras 472, 479; on the network effects *ibid* para 438. On network effects in IT markets see, for instance, M-O Mackenrodt, 'Assessing the effect of intellectual property rights in network standards' in J Drexl (ed), *Research Handbook on Intellectual Property and Competition Law* (Cheltenham: Edward Elgar Publishing Limited, 2008) 80.

[16] On the understanding of interoperability information see Commission Decision (n 4 above) paras 37–9.

achieve for being compatible,[17] was protected by IPRs, Microsoft's refusal to disclose that information would amount to a refusal to grant a licence. Before the Commission and the CFI, Microsoft claimed protection of the relevant interoperability specification under copyright, trade secret, and patent laws.[18]

Patent protection

EU competition law, including the law controlling abuse of market dominance, only protects trade between the Member States and, hence, competition in the internal market, as now explicitly formulated in Article 102 TFEU. It does not apply so far as effects in third states are concerned. Therefore, EU competition law can only collide with patent protection if Microsoft is able to claim patent protection for the territory of the EU or at least for the territory of individual Member State.[19] This is why Microsoft was required to name respective patents.

Indeed, Microsoft claimed that competitors who wanted to integrate the Windows file-sharing protocols into their own programs would have to ask for a licence for Microsoft's European Patent No 0669020.[20] This patent, however, does not provide the information that would be required for producing compatible programs; it only describes a '... computer method and system for passing a pointer to an interface from a server process to a client process.' This is why the Commission argued that, without access to the protocols, it was not able to verify whether or to which extent Microsoft's reliance on patent protection was justified.[21] It seems that Microsoft, for quite some time during the investigation, only generally claimed patent protection, and it was only due to the Commission's insistence to substantiate such protection that Microsoft finally named the above-mentioned European patent.[22]

Under a more thorough assessment of the patent situation, the question would not only be whether the interface specification is protected by Microsoft's patent but also

[17] 'A specification is a description of what a software program must achieve, whereas the implementation relates to the actual code that will run on the computer.' See Commission Decision (n 4 above) para 24.

[18] Compare with Commission Decision (n 4 above) para 190 fn 249. The Commission quotes Microsoft as arguing that 'The interoperability information requested by Sun constitutes valuable intellectual property protected by copyright, trade secret laws and patents.'

[19] However, it has to be observed that the Commission's order also produces extraterritorial effects on markets outside the EU. Once Microsoft has provided the interoperability information, competitors are able to use that information in the global market.

[20] Compare Commission Decision (n 4 above) para 249. On this patent see 'A Method and System for Marshalling Interface Pointers for Remote Procedure calls' <http://v3.espacenet.com/publicationDetails/biblio?CC=EP&NR=0669020&KC=&FT=E> accessed 10 September 2010. Based on a PCT application, the patent was first granted in the US (on 23 April 1996; US5511197) and then on 2 April 1997 as a European Patent for Germany, France, and the UK. Meanwhile, Microsoft tries to rely on additional patents; see P Mühlbauer, 'Danaergeschenk aus Luxemburg' (23 September 2007) <http://www.heise.de/tp/r4/artikel/26/26252/1.html> accessed 10 September 2010.

[21] Commission Decision (n 4 above) para 249.

[22] See CFI judgment (n 3 above) para 278.

whether the integration of such specification in the programs of competitors would amount to a use of the patent. The answer to this latter question depends on the patent claims that define the scope of the patent.[23] Assessing the patent situation for the Microsoft case more concretely would go beyond what can be achieved in this chapter. It seems more important to note that Microsoft's own conduct clearly demonstrates that Microsoft's primary aim was to guarantee continued secrecy of the interface specification rather than to prevent a patent infringement. Patent law is not a law guaranteeing the secrecy of specific information. Quite on the contrary, it is meant to provide society with the 'new' knowledge contained in the invention. Under the 'disclosure' theory of patent law, the patentee is granted exclusivity for a limited period of time for having disclosed the invention to society and for having thereby contributed to the common stock of knowledge of mankind.

Copyright protection

At first glance, Microsoft's allegations regarding copyright protection seem more convincing. On the one hand, there can be no doubt that the Windows operating system as a computer program is protected under European copyright law.[24] Copyright protection is granted automatically with the creation of the program. On the other hand, it cannot be excluded from the beginning that a competitor would reproduce copyrightable parts of the Windows program in the process of creating a new, compatible program and, thereby, infringe Microsoft's copyright.

However, copyright law does not protect all parts of a computer program; it only gives rise to a right to prohibit the reproduction of creative elements of the program.[25] Article 2(1) of the Computer Programs Directive[26] explicitly states that only the 'expression' of a computer program is protected. In contrast, underlying ideas and principles, including the ideas underlying the interfaces, fall into the public domain and may therefore be used without the consent of the copyright holder. Hereby, the European legislature follows the generally accepted copyright concept according to which copyright law should enhance spreading information by only protecting the expression of ideas rather than leading to the monopolization of such ideas. Beyond unprotected ideas and principles, according to Article 2(3) of the Directive, a computer program is only protected to the extent that it is 'original', that is, the author's 'own intellectual creation'. Although the Directive does not allow Member States to maintain high levels of creativity requirements, it is equally clear that an infringement

[23] See W Cornish and D Llewelyn, *Intellectual Property: Patents, Copyright, Trade Marks and Allied Rights* (6th edn, London: Sweet & Maxwell, 2007) Ch 4 para 37.

[24] Article 1 of Directive 91/250/EEC of the Council of 14 May 1991 on the legal protection of computer programs [1991] OJ L 122/41 (hereinafter 'Computer Programs Directive' or 'Directive').

[25] For a more complete analysis see A van Rooijen, *The Software Interface between Copyright and Competition Law* (Kluwer Law International, 2010) 50–58 (hereinafter '*Software Interface*').

[26] Directive (n 24 above).

presupposes a reproduction of original elements of the program. Since the interoperability specification of a program is purely of a functional nature,[27] it often turns out that this specification is not original in the sense of copyright law.[28]

Whether a copying of only limited parts of a computer program constitutes a copyright infringement, therefore, requires a highly complex assessment of the individual case.[29] In the *Microsoft* case, two issues would need to be clarified—first, the question is whether the Windows interface specification can be considered creative and is therefore copyrightable. Second, the mere fact that competitors depend on access to such specification for producing compatible programs does not automatically prove copyright infringement. Copyright only protects against unauthorized copying and does not guarantee the secrecy of the creative elements of the work. Therefore, the second question is whether the production of compatible programs would necessarily involve a copying of creative elements of the Windows program.[30] With regard to both issues, the Commission and the ECJ expressed serious doubts as to whether Microsoft had made a substantiated copyright claim.

During the proceedings, Microsoft had argued that the complaint of Sun Microsystems against Microsoft was aimed at the disclosure of the Windows source code and its integration into Sun's programs.[31] This would indeed have led to using Microsoft's copyright. However, the Commission argued that establishing interoperability would only require disclosure of the interface specifications and not of the source code as such. The specifications only fix *what* has to be achieved in establishing interoperability, but not *how*, ie through a concrete combination of algorithms, this has to be achieved.[32] Also, Microsoft's arguments did not appear very consistent, since in the US the company itself only offered licences limited to the interface specifications under its US Communications Protocols Licensing Program.[33] The Commission's conclusion is straightforward—'The present Decision does not contemplate ordering Microsoft to allow copying of Windows by third parties.'[34]

The Commission even goes a step further. It explains that competitors need to know the interface specifications only for designing the interfaces of their own programs.

[27] See n 17 above.

[28] See van Rooijen, *Software Interface* (n 25 above) 56 *et seq.*

[29] So far, such cases mostly appeared before US courts in the late 1980s and 1990s; see J Drexl, *What is protected in a Computer Program? Copyright Protection in the United States and Europe* (New York: VCH Publishers, 1994).

[30] It is not sufficient that the Commission's order to disclose the interoperability information would facilitate copying. If production of compatible programs without using the copyright were an option, the Commission's order could not read as an order to grant a copyright licence to competitors. If the competitor, nonetheless, copies the program, it would be for Microsoft to sue the competitor for copyright infringement.

[31] Commission Decision (n 4 above) para 568.

[32] *Ibid* para 570.

[33] *Ibid* para 571.

[34] *Ibid* para 572.

Therefore, in the process of creating a compatible program, Microsoft's specifications should not be '... reproduced, adapted, arranged or altered'; competitors should only be enabled 'to write their own specification-compliant interfaces.'[35] Although the Commission does not explicitly draw such a conclusion, it is therefore not possible to identify any act of use of Microsoft's copyright.

The Commission's arguments on the copyright situation were obviously shared by the CFI.[36] The fact that the Commission and the CFI nevertheless based their analysis on the assumption that the order to disclose the relevant information had an impact on Microsoft's intellectual property rights, hence, appears as a particular favour granted to Microsoft, but may also be explained by potential protection of the interoperability information as a trade secret.

Protection as a trade secret

Microsoft relied on the protection of the interoperability information as a trade secret on the same level with patent and copyright protection as part of its intellectual property rights. In the proceedings, Microsoft stated that 'The interoperability information requested by Sun constitutes valuable intellectual property protected by copyright, trade secret laws and patents.'[37]

It is not clear at all from its decision whether the Commission was in support of this equation of trade secret protection with intellectual property. In one part of the decision, the Commission first mentions trade secret protection in addition to intellectual property; later on, however, the Commission introduces a concept of intellectual property protection against disclosure. The Commission states as follows,

> It is implicit in Sun's request that Sun intends to be provided with specifications that it will then be able to implement in its products. It is possible that such a use could be prevented by Microsoft relying on *intellectual property rights. Furthermore*, the specifications could constitute innovations that are currently not disclosed and are *protected by trade secrecy*. More generally, it is possible that ordering Microsoft to disclose such specifications and allow their use for the purposes of implementation by third parties in their products would constitute an interference with Microsoft's intellectual property rights. Microsoft in fact *invokes such intellectual property rights over the relevant information to resist disclosure and the subsequent use that could be made of it.*[38]

Despite an obvious lack of conceptual precision, with the last sentence of the quote, the Commission highlights the core of the intellectual property dimension of the *Microsoft* case. Microsoft did not defend itself against any copyright or patent use,

---

[35] *Ibid* para 1004.
[36] CFI judgment (n 3 above) para 286 *et seq*.
[37] Quoted in the Commission Decision (n 4 above) fn 249.
[38] Commission Decision (n 4 above) para 190 (emphasis added).

but tried to maintain the secrecy of the interoperability information which typically falls into the ambit of trade secrecy protection.

In comparison to the Commission, the CFI adds more conceptual clarity. Whereas the CFI first uncritically repeats the statement of the Commission as quoted above,[39] it finally concludes as follows,

> It follows from the foregoing considerations that the appraisal of the merits of the first part of the plea must precede on the presumption that the protocols in question, or the specifications of those protocols, are covered by intellectual property rights *or constitute trade secrets* and that those secrets *must be treated as equivalent to intellectual property rights*.[40]

In this quote, the CFI does not define trade secrecy protection as intellectual property, but still 'equates' it with intellectual property. This 'equation' is not the result of a thorough analysis of the nature of intellectual property. Rather, the CFI follows the Commission approach to apply the principles developed for a refusal to license an intellectual property right under Article 102 TFEU also to trade secrets. For the future, this does not exclude the possibility of applying a comparably lower standard for intervention in cases of a refusal to disclose trade secrets.

In the following sections, I will therefore aim to identify the criteria that should be applied to a duty to disclose trade secrets under Article 102 TFEU.

B. Requirements for a Duty to License under Article 102 TFEU

Whether the requirements for a duty to disclose trade secrets have to be the same as, or possibly less stringent ones than for a duty to license an intellectual property right depends on the justification of the latter standard and whether that justification is also convincing for trade secrets. Before I get into the details of this analysis, I will summarize the case law on the standard for controlling a refusal to license an IPR.

IMS Health and the cumulative theory

It was in the *IMS Health* case that the ECJ,[41] for the last time, defined the requirements for a duty to license under Article 102 TFEU. According to this judgment, a refusal to license an intellectual property right can only be considered an abuse 'in exceptional circumstances.'[42] Such exceptional circumstances are held to exist if four *cumulative* requirements are fulfilled:[43]

---

39 CFI judgment (n 3 above) para 285.
40 *Ibid* para 289 (emphasis added).
41 *IMS Health* (n 13 above).
42 *Ibid* para 35.
43 *Ibid* para 38.

(1) Use of the right must be *indispensable* for the petitioner to carry on a particular business.
(2) The refusal to license *prevents the emergence of a new product* for which there is potential consumer demand.
(3) The refusal to license is *unjustified.*
(4) The refusal to license would *exclude any competition in a secondary market.*

Microsoft and its flexible reading of the *IMS Health* requirements

In contrast to its earlier decision in *Magill*,[44] where the Court for the first time confirmed a duty to license under Article 102 TFEU (erstwhile Article 86 TEC), the ECJ clarified in *IMS Health* that the prevention of the emergence of a new product has to be understood as an indispensable requirement for a duty to license (so-called 'cumulative theory'). This holding, which arguably made it more difficult to justify a duty to license, was very important for the pending *Microsoft* case. Only shortly before the ECJ delivered its judgment in *IMS Health*, the Commission had issued its decision in *Microsoft*, without being aware of the ECJ's upcoming preference for the cumulative theory. It was then for the CFI to respect the ECJ's holding in *IMS Health* and to deliver a decision on whether the Commission decision in *Microsoft* was in line with the cumulative theory. Indeed, it was anything but clear whether the requirement of the prevention of the emergence of a new product (the so-called 'new product rule') would be fulfilled in the *Microsoft* case. The question here was whether the work group server operating systems of competitors could be viewed as 'new products', although Microsoft also supplied such systems to customers in the relevant market.

In fact, the Commission had not applied the cumulative theory in its decision. In citing the *Magill* requirement of exceptional circumstances, the Commission did not mention at all the new product rule.[45] Rather, the Commission relied upon the wording of Article 102(b) TFEU according to which any '... limitation of technical development to the prejudice of consumers' has to be considered an abuse.[46] In *Magill*, it was through reliance on the wording of a 'limitation of production', that the ECJ required the prevention of the emergence of a new product to the prejudice of consumers for an abuse. In application of this test, the Court then held that the comprehensive TV guide that would be preferred by consumers to the single-programme guides distributed by TV stations would constitute such a new product.[47] Whereas in *IMS Health* the ECJ did not even mention Article 102(b) TFEU, in *Microsoft*, the Commission relied again on the wording of this provision and argued that Microsoft had '... limited technical development to the prejudice

[44] Joined Cases C-241/91 P and 242/91 P *RTE et al v Commission* ('*Magill*') [1995] ECR I-743.
[45] Commission Decision (n 4 above) para 551.
[46] *Ibid* para 551. See also A Bartosch, 'EuG-Urteil in der Rechtssache Microsoft' (2007) Recht der internationalen Wirtschaft 908, 915 *et seq.*
[47] *Magill* (n 44 above) para 54.

of consumers'.[48] Based on evidence that competitors would have offered computer programs with technologically superior features, the Commission subsequently held that Microsoft had committed an abuse. The Commission explained that interoperability with Windows would not be the only feature that matters for consumers. Rather, consumers would also be interested in the reliability and availability of the program and security features. These features were not defined by the interface specifications but through the source code which is produced by competitors independently once they get to know the interface specifications. In addition, a technical expert confirmed that Microsoft's programs were technologically inferior with regard to such features.[49] What was obviously most important for the Commission was to maintain incentives for innovation *within* the *de facto* standard controlled by Microsoft. If Microsoft were allowed to keep the interoperability information secret, competitors would be excluded from the market for work group server operating systems and Microsoft would be the only market player left to invest in innovation within its standard.[50] Microsoft did not convince the Commission with its argument that a duty to disclose the interoperability information would deter competitors from competing with Microsoft for the provision of better interoperability specifications. The Commission rejected this argument, since the emergence of Windows as a *de facto* standard prevented other companies from competing in the first place.[51]

The Commission was willing to accept only Microsoft's pro-competitive theory, according to which Microsoft would be deterred from future investment if it had to share the interoperability information with competitors, in the framework of a potential justification.[52] Thereby, the Commission asked whether the incentives for innovation that underlie the very standard (dynamic competition *for* the standard) would outweigh the limitation of technological progress of Microsoft's competitors regarding innovation within the standard (dynamic competition *within* the standard).[53] However, by taking account of Microsoft's arguments only in the framework of a justification, the CFI put the burden of proof on Microsoft for its pro-competitive allegation. The Commission finally held '. . . that it is dubious

[48] Commission Decision (n 4 above) paras 693–701.

[49] *Ibid* para 699.

[50] *Ibid* para 700.

[51] *Ibid* para 697.

[52] *Ibid* paras 709–29. On the argument that innovation-intensive firms should not be forced to grant licences under competition law because of risk that these firms will not be able to recoup their investment (so-called 'recoupment argument'), see A van Rooijen, 'The Role of Investments in Refusals to Deal', (2008) 31 World Competition 63. The author concludes that the recoupment argument is not always a valid one. Often, the investment would have been made even if a duty to share were foreseeable.

[53] Commission Decision (n 4 above) para 712. On the distinction of competition for the market and competition in the market in interoperability cases, see also van Rooijen, *Software Interface* (n 25 above) 128–34.

whether an order to supply in this case would have any negative impact on Microsoft's incentives to innovate.'[54] By assessing the case in the light of the incentives to innovate of both Microsoft and its competitors, the Commission applied a kind of the so-called 'incentive balancing approach' suggested also by economists.[55] The one critical point of the Commission's approach is that it did not grant equal weight to the incentives for both parties by putting the burden of proof on Microsoft for the deterrent effect of competition law intervention.[56]

It did not come as a surprise that Microsoft, after the ECJ had handed down its seemingly favourable judgment in *IMS Health* with the recognition of the cumulative theory, did everything to convince the CFI that the interoperability specifications were protected by intellectual property rights and that it had not prevented the emergence of a new product. In doing so, Microsoft argued that the work group server operating systems distributed by its competitors would only 'mimic' the functions of Microsoft's work group server programs.[57] Hence, Microsoft criticized that the Commission was not able to name any 'new product' competitors would offer consumers if they had access to the interoperability specifications of the Windows program.[58]

For the CFI the challenge indeed consisted in justifying the Commission's holdings in the light of the cumulative theory. In contrast to the ECJ in *IMS Health*, the CFI followed the example of the Commission and predominantly relied upon the wording of Article 102(b) TFEU. The CFI held as follows,

> The circumstance relating to the appearance of a new product, as envisaged in *Magill* and *IMS Health* . . . cannot be the only parameter which determines whether a refusal to license an intellectual property right is capable of causing prejudice to consumers within the meaning of Article 82(b) EC [now Article 102 TFEU]. As that provision states, such prejudice may arise where there is a limitation not only of production or markets, but also of technical development.[59]

In short, the CFI reintroduced more flexibility into applying the new product rule by relying on the wording of the European competition law provisions. In fact, it was time to remind the ECJ of Article 102(b) TFEU as the source of the new

---

[54] Commission Decision (n 4 above) para 729.

[55] See F Lévêque, 'Innovation, leveraging and essential facilities: Interoperability licensing in the EU Microsoft case' in F Lévêque and HA Shelanski (eds), *Antitrust, Patents and Copyright: EU and US Perspectives* (Cheltenham: Edward Elgar Publishing Limited, 2005) 103, 106–10 (recommending this approach as an alternative to the new product rule).

[56] On the problem of dealing with uncertainty in this case, see in more detail J Drexl, 'Real Knowledge is to Know the Extent of One's Ignorance: On the Consumer Harm Approach in Innovation-Oriented Competition Cases' (2010) 78 Antitrust Law Journal 677.

[57] CFI judgment(n 3 above) para 623.

[58] *Ibid* para 624.

[59] *Ibid* para 643.

product rule.[60] It seems more important to note, however, that the CFI accepted the approach of the Commission according to which the interest in promoting follow-on innovation within a standard merits particular consideration. The CFI may even have had the intention to support the Commission's position expressed in its Discussion Paper on Article 82 EC as far as follow-on innovation is concerned. In any instance, the paragraph of the Discussion Paper dealing with follow-on innovation reads like the essence of the later CFI decision,

> A refusal to license an IPR protected technology which is indispensable as a basis for follow-on innovation by competitors may be abusive even if the licence is not sought to directly incorporate the technology in clearly identifiable new goods and services. The refusal of licensing an IPR protected technology should not impair consumers' ability to benefit from innovation brought about by the dominant undertaking's competitors.[61]

This test can hardly be seen to be in compliance with the cumulative theory applied in *IMS Health*. Nevertheless, it may be explained with the wording of Article 102(b) TFEU. It remains to be seen whether the ECJ will accept the test developed by the CFI. Until the ECJ has to express itself on that test, the requirements as developed in *IMS Health* should be applied as interpreted by the CFI in *Microsoft*.

C. Should the Standard of Intervention be Even Lower in Cases of a Refusal to Disclose Trade Secrets?

Since, in *Microsoft*, the Commission applied the above-mentioned principles based on the assumption that the interoperability specifications are protected by intellectual property rights, the CFI did not have to clarify the issue raised by the Commission in its Discussion Paper on Article 82 EC of whether the standard of intervention should be lower for refusals to disclose trade secrets than for refusals to grant a licence for an intellectual property right as developed in *Magill* and *IMS Health*.[62] For discussing this so far unanswered question in this section, I will, first, analyse differences regarding the nature of intellectual property rights on the one hand and of trade secrecy protection on the other hand and, second, aim to identify the appropriate standard of intervention in the light of innovation economics as expressed by the Commission and the CFI in the *Microsoft* case.

---

[60] See also T Körber, 'Wettbewerb in dynamischen Märkten zwischen Innovationsschutz und Machtmissbrauch - Zum Microsoft-Urteil des EuG vom 17. 9. 2007' (2007) Wirtschaft und Wettbewerb 1209, 1213 (supporting the CFI); D Howarth and K McMahon, '"Windows has performed an illegal operation": The Court of First Instance's Judgment in Microsoft v Commission' (2008) 29 ECLR 117, 123 *et seq* (criticizing the court with the argument that now the slightest technical variation may justify a new product).

[61] Discussion Paper (n 1 above) para 240.

[62] See text to n 8 above.

The legal nature of the protection of trade secrets

The above-mentioned requirements for a duty to license have been developed by the European courts specifically for intellectual property rights. Hence, it has to be asked whether the protection of trade secrets has to be attributed to the field of intellectual property in this regard.

The question is a complex one. For the application of European competition law the problem is already that—contrary to the situation of many intellectual property rights—there is no secondary European legislation on trade secrets and, hence, no generally accepted concept of trade secrecy protection in the European Union. This, however, is not to say that EU law does not offer any indication on the legal nature of this protection. Specific regulation of trade secrets can be found in the TRIPS Agreement[63] which is also binding on the European Union as a Member of the World Trade Organization (WTO).[64] Article 39 of the TRIPS Agreement contains a provision on the protection of 'undisclosed information'. As a first step, Article 39(1) TRIPS obliges the WTO Members to protect undisclosed information according to the rules on unfair competition in the sense of Article 10 *bis* of the Paris Convention on the Protection of Industrial Property.[65] More importantly, the concept of the protection of undisclosed information is further explained in Article 39(2) TRIPS. The provision reads as follows,

> Natural and legal persons shall have the possibility of preventing information lawfully within their control from being disclosed to, acquired by, or used by others without their consent in a manner contrary to honest commercial practices so long as such information:
>
> (a) is secret in the sense that it is not, as a body or in the precise configuration and assembly of its components, generally known among or readily accessible to persons within the circles that normally deal with the kind of information in question;
> (b) has commercial value because it is secret; and
> (c) has been subject to reasonable steps under the circumstances, by the person lawfully in control of the information, to keep it secret.

From a purely formal standpoint, the reference contained in Article 39(1) TRIPS to Article 10 *bis* Paris Convention would argue in favour of conceiving this form of trade secrecy protection as a form of 'industrial property'. According to Article 1(2)

---

[63] Agreement on Trade-Related Aspects of Intellectual Property Rights, Annex 1C to the Agreement Establishing the World Trade Organization of 1994 <http://www.wto.org/english/docs_e/legal_e/legal_e.htm#finalact> accessed 10 September 2010.

[64] On the legal status and the impact of the TRIPS Agreement within the EU legal order see Josef Drexl, 'Internationales Immaterialgüterrecht' in *Münchener Kommentar Bürgerliches Gesetzbuch*, Vol 11 (5th edn 2010) paras 82–105.

[65] Originally concluded on 20 March 1883. The most recent version, which is also applicable for the EU due to the so-called Paris-plus obligation of Article 2(1) TRIPS to respect the core provisions of the Paris Convention, as of 14 July 1967 (Stockholm version), with minor changes dating from 28 September 1979.

Paris Convention, the 'repression of unfair competition' is listed as a form of 'industrial property'.

However, this result may not be maintained from a substantive perspective. Notwithstanding certain limitations, intellectual property rights, like patents, copyrights, trade marks or design rights are characterized by their exclusive nature. In this regard, general protection against unfair competition and, more specifically, protection of undisclosed information or trade secrets is quite different. Whereas exclusive intellectual property rights grant protection against any person who uses the subject-matter of protection without authorization by the right-holder in a way specified by statutory provisions, trade secrets are only protected against unfair conduct. Hence, trade secrecy protection does not belong to the field of exclusive protection of intangible goods; trade secrecy laws, by focusing on the conduct of the competitor instead of defining an area of exclusive use of the subject-matter of protection in favour of the right-holder, belongs rather to the field of torts in the common law tradition and delicts in the civil law tradition. One expression of this different nature, as also to be found in Article 39(2)(a) TRIPS, is the loss of protection once the trade secret becomes generally known. In such situation, claims for violation of trade secrecy protection may only be directed against the person who disclosed the trade secret, but not against any other person who, later on, may use the information. The lack of exclusivity is also highlighted by another wording used in Article 39(2) TRIPS, according to which the protected person '. . . shall have the possibility of preventing' certain conduct, whereas IP laws usually formulate that nobody may use the subject-matter of protection without the authorization of the right-holder. Under trade secrecy laws there is usually no right to act against unauthorized specific use of information, but only to act against such acts that have to be considered unfair.[66]

The description provided above corresponds with the legal situation in most EU Member States. In Germany, for instance, trade secrets are only protected under sections 17–19 of the Act Against Unfair Competition[67] against betrayal and spying. A person who uses information without having violated these rules herself does not act illegally. This is especially so if the relevant information, after violation of said rules, has been disclosed to the public and thereby loses its character as a trade secret.[68]

However, legal protection of trade secrets may come close to a form of exclusive intellectual property protection. In Italy, for instance, the most recent Code on Industrial Property—the *Codice sulla proprietà industriale* (CPI) of 2005[69]—has considerably

[66] In this sense also UNCTAD-ICTSD, *Resource Book on TRIPS and Development* (New York: Cambridge University Press, 2005) 528.

[67] *Gesetz gegen unlauteren Wettbewerb* of 3 July 2004, [2004] BGBl I 1414.

[68] See, for instance, H Harte-Bavendamm, in W Gloy and M Loschelder (eds), *Handbuch des Wettbewerbsrechts* (3rd ed 2004) § 48 para 10.

[69] *Decreto legge* No 30 of 10 February 2005; see G Ghidini and V Falce, 'Recent Developments in Italian Regulation of Trade and Industrial Secrets: A Patent Contradiction of the Patent Regime?', in

strengthened the protection of trade secrets. While Article 98 CPI adopts the terminology of undisclosed information (*informazioni segrete*) as contained in Article 39(2) TRIPS, Article 99 CPI stipulates an absolute prohibition against everybody to disclose, to acquire, and to exploit the information without requiring any element of unfairness.[70] The Italian model also goes beyond trade secrecy protection in the US. There, the Uniform Trade Secrets Act requires the application of 'improper means'.[71] However, also for the legal situation in Italy it has to be noted that protection only lasts as long as the trade secret is not disclosed to the public. Hence, also in Italy, there is no protection against third parties who acquire and exploit the information after this information has become generally known to the public. It may be concluded that trade secrecy protection in Italy comes very close to a form of exclusive IP protection without, however, being completely established as a clear-cut property right.

Against the backdrop of this short comparative analysis, the question is whether the difference between protection of trade secrets as part of unfair competition law on the one hand and protection of exclusive intellectual property rights on the other hand should lead to different standards of intervention under Article 102 TFEU. An analysis of the origins of the European case law on the duty to license may well suggest such a differentiation.

In this regard, it is most important to take account of the two early judgments of the ECJ in *Volvo*[72] and *CICRA*[73] concerning the question of whether car manufacturers may be forced to license their design rights in spare parts to independent spare part manufacturers.

In these two judgments, the ECJ established the principle that a refusal to license cannot 'in itself' constitute an abuse of market dominance.[74] The ECJ explains this principle by arguing that a duty to license would deprive the right-holder of the 'substance of the exclusive right'.[75] In referring to this argument, the ECJ, in the later

G Ghidini and LM Genovesi (eds), *Intellectual Property and Market Power*, ATRIP Papers 2006–2007 (Ciudad de Buenos Aires: Universidad de Buenos Aires, 2008) 585, 586 (criticizing this kind of protection which goes beyond what is required by Article 39 TRIPS).

[70] Article 99 CPI may be translated as follows: 'For the protection against unfair competition it is prohibited to disclose information in the sense of Article 98 to any unauthorized person or to acquire or to exploit such information.'

[71] Available at <http://www.ndasforfree.com/UTSA.html> accessed 10 September 2010. The Uniform Trade Secrets Act has been adopted as the law of 46 States in the US.

[72] Case 238/87 *Volvo v Veng* [1988] ECR 6211 (hereinafter '*Volvo*').

[73] Case 53/87 *Consorzio Italiano della Componentistica di Ricambio per Autoveicoli and Maxicar v Renault* [1988] ECR 6039 ('*CICRA*').

[74] *Volvo* (n 72 above) para 8.

[75] *Ibid* para 8.

judgments of *Magill*[76] and *IMS Health*,[77] developed the other principle that a duty to license can only be affirmed in 'exceptional circumstances'.

From this case law, one may be tempted to draw the conclusion that the *IMS Health* requirements only apply to cases in which an exclusive right is concerned. Although the holder of a trade secret may decide to disclose the information to certain individuals—and not to others—and may enter into contractual arrangements on such secrets, above all in form of know-how licensing agreements, such capacities are not protected by law in any absolute form as is explained by the fact that protection expires with the disclosure of the information to the general public. This again demonstrates that trade secrecy protection is a form of protection against tortious conduct rather than a form of exclusive protection of the relevant information in the sense of a property right. The information is not owned by the holder; it is only protected against conduct that affects its secrecy. This significantly distinguishes trade secrecy protection from patent law in particular, but also from data base protection, which gives rise to exclusive rights on the protected information for the statutory term of protection.

However, this analysis does not provide any concrete standard that would most adequately have to control a refusal to disclose trade secrets under the prohibition of abuse of market dominance under Article 102 TFEU. Identifying such a standard also turns out to be very difficult, since the ECJ, so far, has not had any opportunity to define the standard of intervention for general refusals to deal that are unrelated to IP. In *Bronner*,[78] the ECJ seemed to rely on the test that it had developed in *Magill* some time earlier. Yet, in *Bronner*, the ECJ was not required to make a decision on transferring the requirement of the prevention of a new product to general refusal-to-deal cases, since the Court was able to reject the indispensability of the competitor's home delivery system to which the applicant sought access to this competitor's supply opportunities in the relevant market for daily newspapers.

Defining the standard of intervention in the light of innovation economics

As will be seen in the following paragraphs, the question concerning the most appropriate test for general refusals to deal—and for refusals to disclose trade secrets—may not have to be answered at all. Rather, it will be shown that the decision of the Commission and the judgment of the CFI in *Microsoft* indeed contain the basis for most appropriately assessing cases that only involve a refusal to disclose trade secrets. The essential criterion for intervention cannot depend on the 'form' of protection, ie whether the legislature grants an exclusive right or relies on concepts of

[76] *Magill* (n 44 above) para 50.
[77] *IMS Health* (n 13 above) para 35.
[78] Case C-7/97 *Oscar Bronner v Mediaprint* [1998] ECR I-7791.

unfair competition. Rather, the challenge is how to reach, in the light of the goals of trade secrecy law and competition law, the most appropriate result in the individual case.

The basis of this approach has to be found in the paradigm of complementarity of the two fields of intellectual property and competition law. According to this paradigm, intellectual property and competition are not in a fundamental conflict, but pursue, although with different techniques, complementary goals. This paradigm is also acknowledged by the Commission in the Guidelines on Transfer of Technology Agreements. These Guidelines read as follows,

> Indeed, both bodies of law share the same basic objective of promoting consumer welfare and an efficient allocation of resources. Innovation constitutes an essential and dynamic component of an open and competitive market economy. Intellectual property rights promote dynamic competition by encouraging undertakings to invest in developing new or improved products and processes. So does competition by putting pressure on undertakings to innovate. Therefore, both intellectual property rights and competition are necessary to promote innovation and ensure a competitive exploitation thereof.[79]

However, the paradigm of complementarity does not guarantee that intellectual property always promotes dynamic competition and that, therefore, reliance on competition law is not needed. IPRs may also be used as a means to restrain competition. They do *not always* promote innovation.[80] This is why competition law is needed in individual cases to protect dynamic competition.

As to trade secrets, it has to be pointed out that this form of protection is granted without any regard to whether the information as such can be considered 'innovation'. In contrast to the protection of inventions through patent law or of creative works through copyright law, trade secret protection does not depend on any qualitative standards concerning the subject-matter of protection. The law only requires that the specific information is kept secret. Accordingly, in *Microsoft*, the Commission points out that the relevant specifications, that were kept secret at that time and at least enjoyed protection as trade secrets, '. . . can constitute innovation'.[81] Similarly, the CFI quotes the Commission in the sense that 'it is possible' that the interoperability specifications contain innovation and constitute trade secrets.[82] The reasons for this

[79] Commission Notice, 'Guidelines on the application of Article 81 of the EC Treaty to technology transfer agreements' [2004] OJ C101/2.

[80] The most recent example of growing awareness of the use of IPRs which lowers incentives for innovation and is harmful to competition is the allegation that even the application for blocking patents, especially in the pharmaceutical industry, may violate Article 102 TFEU. On this topic, see L Kjølby, 'Article 82 EC as Remedy to Patent System Imperfections: Fighting Fire with Fire?' (2009) 32 World Competition 163 (arguing in favour of solving such problems within the ambit of patent law). For a recent, more general and very valuable analysis of the relationship of competition and intellectual property, see H Ullrich, 'Propriété intellectuelle, concurrence et regulation—Limites de protection et limites de contrôle' (2009) Revue Internationale de Droit Economique 400.

[81] Commission Decision (n 4 above) para 190.

[82] CFI judgment (n 3 above) para 285.

reservation are obvious. In the case of trade secrets, the law does not guarantee that the protected information contains innovation.

In the light of the definition of 'undisclosed information' as protected by Article 39(2) TRIPS, the question remains whether the second requirement, according to which protection is only granted if the undisclosed information has commercial value because it is secret, has to be considered in the framework of the competition law assessment. This requirement reminds us of the opinion expressed at the public hearing on the Commission's Discussion Paper on Article 82 EC, as reported at the very beginning of this chapter, according to which trade secrets merit particular protection, given their economic value that may even be superior to those of intellectual property rights, including patents.

Yet, against the background of the paradigm of complementarity, this argument has to be rejected. In the market economy, values depict preferences and prices. Prices are formed in markets. Or to put it differently: the value of a trade secret depends on the market power it confers on the holder. In this regard, trade secrets are similar to the subject-matter of protection of intellectual property laws.[83] The reasons for which trade secrets confer market power and economic value can vary considerably. In one case, they may indeed contain important innovation, which is particularly valued by consumers and, therefore, confer considerable advantages in competition with other firms. In other cases, however, the reasons for market power can be completely unrelated to the innovative character of the underlying information. In *Microsoft*, it does not matter for the economic value of the interoperability information whether this information contains any innovation. Indeed, in the *Microsoft* case, the interoperability information does not necessarily play a major role in achieving software innovation and may even appear as a necessary by-product of the programming activity. Nonetheless, keeping that information secret produces high economic value. This value is purely related to the function of that information, and is not at all influenced by the degree of its innovative character. Its value is essentially explained by the establishment of Windows as the *de facto* standard. Its economic value only emerged after several years during which Microsoft provided that information to the producers of application software for free. This was part of Microsoft's strategy to tip the market and to promote the establishment of Windows as the *de facto* standard. Once most application programs were produced for the Windows standard, it was rational for Microsoft to change its strategy and to try to exclude producers of application software, particularly those who offered work group server operating systems, who are most likely to challenge the Windows standard.[84] In short, the economic value of

[83] On the relationship of intellectual property law and the degree of market power they confer see J Drexl, 'The relationship between the legal exclusivity and economic market power: Links and limits' in I Govaere and H Ullrich (eds), *Intellectual Property, Market Power and the Public Interest* (Brussels: PIE. Peter Lang, 2008) 13.

[84] On the strategy of Microsoft, see Lévêque (n 54 above) 113 *et seq*.

secret information is not necessarily the result of technological superiority, but may simply indicate a monopoly situation which is secured by keeping the information secret.

To sum up, the standard of intervention for a refusal to disclose trade secrets under Article 102 TFEU should not differ from the one for a refusal to license intellectual property rights. However, in applying this standard, competition law enforcers should take account of the peculiarities of trade secrecy protection. First, it is not at all guaranteed that the underlying information contains innovation. This should lead to a closer analysis of the nature of that information and may facilitate intervention under competition law. Second, it has to be noted that trade secrecy protection involves a high risk of monopolization of markets. Wherever the law protects against access to information, there is a risk that markets are automatically monopolized if competitors depend on access to that information for entering the relevant market.

The *Magill* case shows that similar situations may not even be excluded in the field of copyright law, although this law is expected to promote the spread of information. In *Magill*, copyright protection on the mere listings of TV programmes attracted criticism from a copyright perspective, since these listings were only the by-product of the broadcasting activity without requiring any additional creativity. Simultaneously, as was also pointed out by the ECJ,[85] copyright protection for the listings leads to a monopolization of the underlying information for the TV stations, which enabled them to fence off independent publishers of TV guides from the market. The *Magill* example also demonstrates that the competition problems that emerge in situations of a refusal to disclose trade secrets are of a more general nature and may also be present in a case of a refusal to license an IP right that controls access to information. Hence, it is of utmost importance that the legislature applies high caution in granting protection to information. This is the more so where, like in the case of trade secrets, the law does not define any qualitative conditions for protection. Conversely, application of Article 102 TFEU may be more easily justified in such cases. By taking account of the particular interest in promoting dynamic competition within the standard controlled by Microsoft, the Commission and the CFI managed to achieve an appropriate balance.

D. The Commission's Guidance Paper on Article 82 EC

It took the Commission quite some time to publish the final document on the process launched by the Discussion Paper on Article 82 EC (now Article 102 TFEU). In part, this can certainly be explained by the pending *Microsoft* case. The result of the Commission's work, however, is rather disappointing. According to the Guidance

[85] *Magill* (n 44 above) para 47.

Paper of 3 December 2008,[86] the Commission only aims at defining its enforcement priorities and, thereby, to enhance the predictability of the Commission's future practice.[87] The Commission explicitly declares that the Guidance is '. . . not intended to constitute a statement of the law' and that they do not prejudice the interpretation of Article 102 TFEU by the European Courts.[88] In line with the Discussion Paper of 2005, the Commission Guidance Paper deals with various forms of abuses. However, in the chapter on refusals to deal,[89] only one paragraph is devoted to refusals to license and to the problems associated with access to interface information. With regard to these two cases, the Commission only points out that these cases constitute subcategories of a refusal to deal.[90]

After the landslide victory in *Microsoft* before the CFI, such reservation on the part of the Commission comes quite as a surprise. It is hoped that the Commission Guidance Paper does not constitute a departure from the fundamental assumptions and considerations in the Discussion Paper of December 2005. However, such departure does not seem unlikely. For identifying its priorities, the Commission now intends to focus on such cases in which the refusal is likely to lead to consumer harm.[91] In contrast, the CFI in *Microsoft* still refused to require evidence of consumer harm for a violation of Article 102 TFEU and simultaneously reminded that, according to the case law of the ECJ, Article 102 TFEU prohibits also those acts '. . . which indirectly prejudice [consumers] by impairing an effective competitive structure.'[92]

Conclusion

This chapter recommends caution vis-à-vis generalized reliance of firms on intellectual property rights and the economic value of trade secrets in order to evade enforcement of Article 102 TFEU. Instead of insulating intellectual property law against the application of competition law, it is crucial to assess the effects on incentives for innovations in all directions and to balance them appropriately (the so-called incentive balancing test). According to this effects-based approach, it does not matter in principle whether the case under consideration is one on a refusal to license an

[86] Commission Guidance Paper (n 14 above).
[87] *Ibid* para 2.
[88] *Ibid* para 3.
[89] *Ibid* paras 74–89.
[90] *Ibid* para 77.
[91] *Ibid* para 81.
[92] CFI judgment (n 3 above) para 664. Thereby, the Court follows the recent holding of the ECJ in Case C-95/04 *British Airways v Commission* [2007] ECR I-2331 para 106. This reference to the protection of the effective competition structure belongs to one of the most criticized parts of the *Microsoft* decision; see, for instance, P Marsden, 'Microsoft versus the Commission: with great power comes great responsibility' (October 2007) <http://www.oxera.com/cmsDocuments/Agenda_Oct 07/Microsoft versus the Commission.pdf> accessed 10 September 2010, 2 *et seq* (criticizing the position of the court as 'ordoliberalism').

exclusive intellectual property right or merely one on a refusal to disclose trade secrets. However, trade secrets may more easily pose competition law problems, since protection—unlike in the field of patent law—does not depend on the fulfilment of any qualitative requirements concerning the subject-matter of protection and since trade secrecy protection may result in the control of information on which competitors depend for being able to enter the relevant market.

9

COMPETITION LAW AND INNOVATION: TECHNOLOGICAL INTEGRATION

Hedvig K S Schmidt

Introduction

Technological integration has only recently been granted greater attention by competition authorities due to fast-growing software industries, rapid developments, and innovation in high-technology markets. The *Microsoft* cases on both sides of the Atlantic are the first cases to deal intimately with the concept of technological integration.

In contrast to contractual tying, technological integration can have different effects upon the market. First, it may create a new product or at least offer an add-on feature of a known product. Second, although courts may rule that the technological integration is unlawful, splitting the product into two may not necessarily be to the benefit of the consumer.[1] Technological integration therefore generates a new set of concerns, which need to be given careful consideration before a legal doctrine is developed.

In relation to intellectual property rights (IP rights), under contractual tying, an excessively narrow view has been applied to define when a component of a product constitutes a product on its own, based on the scope of the IP right instead of an assessment of the actual product. Applying a similar approach to technological integrated products containing an IP right will undoubtedly ensure that the IP right is not misused as a leverage to gain market power in another market. However, the

[1] For instance, *Microsoft* (Case COMP/C-3/37.792) [2005] 4 CMLR 965 para 1011, the remedy applied to the case is somewhat a testament to that argument, see also paras 1012–4 (hereinafter '*EU Microsoft I* Commission Decision'); and US Department of Justice 'Competition and Monopoly: Single-Firm Conduct Under section 2 of the Sherman Act (2008)' <http://www.usdoj.gov/atr/public/reports/236681.htm> accessed 10 September 2010, 88 (hereinafter '*Section 2 Report*').

consequences to innovation and product development are more severe, because sufficient attention will not be given to the understanding of the actual product. Instead the presence of an IP right will be the decisive factor as to whether the products are integrated.

Nevertheless, *EU Microsoft I*[2] and *II*[3] show that the EU Courts and the Commission have without greater concerns merely applied the same approach to technological integration as to contractual tying. The competition authorities are therefore risking causing great harm to innovation and product development by not making allowance for the special issues that arise in relation to technological integration, despite the enhancement of innovation and product development playing an important role in EU competition regulation to ensure the maintenance of competitive markets. In comparison, the US Courts have applied a more cautious approach to technological integration, which has at times created a safe haven for technological integrated products in the software industry.[4]

The problem arises because of the harsh treatment tying itself has received under both EU competition law and US antitrust law, despite being a well-established business strategy. In the US an adapted version of the per se illegality approach is applied to tying arrangements under the Sherman Act, section 1 and the Clayton Act, section 3, and in the EU tying is banned under both Article 101 of the Treaty on the Functioning of the European Union (TFEU) (ex Article 81 EC) if applied as part of an agreement between undertakings, and Article 102 TFEU (ex Article 82 EC) if used by a dominant company.

This chapter considers the different approaches to technological integration in the EU and US. The chapter will, by distinguishing between contractual tying and technological integration, demonstrate that to promote and protect innovation it is necessary to adopt a more flexible approach to technological integration.

The current assessment of tying and technological integration consists of a five-step test:[5]

(1) Identification of two separate products
(2) The establishment of dominance in the tying product
(3) Coercion—the limitation of consumer choice

---

[2] Case COMP/C-3/37.792 *Microsoft* [2005] 4 CMLR 965 (hereinafter '*EU Microsoft I* Commission Decision') confirmed in Case T-201/04 *Microsoft Corp v Commission* [2007 ECR II-3601 (hereinafter 'Case T-201/04 *Microsoft*') (hereinafter '*EU Microsoft I*' when referred to collectively).

[3] *Microsoft* (COMP/C-3/39.530) [2009] OJ C242/20 (the decision on tying) (hereinafter '*EU Microsoft II*').

[4] *US v Microsoft Corp* 253 F 3d 34 (DC Cir 2001) 85 (hereinafter '*US Microsoft III*'); and *United States v Microsoft Corp* 147 F 3d 935 (DC Cir 1998) (hereinafter '*US Microsoft II*').

[5] See cases such as Case T-201/04 *Microsoft* (n 2 above) paras 839–71, and *EU Microsoft I* Commission Decision (n 2 above) para 794; *Jefferson Parish Hospital District No. 2 v Hyde*, 466 US 2, 104 S Ct 1551, 80 L Ed 2d 2 (1984), 8, 18, and *US Microsoft III* (n 4 above) 85.

(4) Anti-competitive or foreclosure effects of the tying arrangement
(5) The rejection of any suggested objective justifications

This chapter will commence by outlining the current approaches to tying and technological integration on both sides of the Atlantic, in sections A and B. The focus will then turn to the two separate products-test and the justifications for technological integration in section C, as these two issues are crucial to whether a technological integration will be permitted under EU competition law or US antitrust law.

A. EU Case Law

In principle only three cases have played a major role in shaping the current approach to tying and technological integration under EU competition law—*Hilti*[6]and *Tetra Pak II*,[7] which dealt with contractual tying; and *EU Microsoft I*,[8] which revolved around technological integration. Interestingly, all three involved some form of IP right and this has undoubtedly influenced the outcome of these cases and to some degree the approach taken to tying arrangements. Other cases, which have had some limited influence, are *IBM Undertaking*,[9] *British Sugar*,[10] and *EU Microsoft II*.[11]

Hilti and *Tetra Pak II* were contractual tying cases, meaning that there was an actual restriction upon the customer choice of where to purchase the tied product.[12] The cases applied a five-step test to the tying conduct in order to assess whether it was abusive. *EU Microsoft I* confirmed the use of the five-step test to tying, but also broadened it, as the General Court (GC), earlier the Court of First Instance, took little note of the fact that *EU Microsoft I* was about technological integration and the test previously only had been applied to contractual tying cases. The Commission took advantage of this in *EU Microsoft II* when finding Microsoft had tied its Internet

[6] Case C-53/92P *Hilti AG v Commission* [1994] ECR I-666; [1994] 1 CMLR 590 (hereinafter 'Case C-53/92P *Hilti*'); Case T-30/89 *Hilti AG v Commission* [1991] ECR II-439, [1992] 4 CMLR 16 (hereinafter 'Case T-30/89 *Hilti*'); *Eurofix-Bauco/Hilti* (Case IV/30.787 and 31.488) [1988] OJ L65/19 (hereinafter '*Hilti* Commission Decision') (hereinafter '*Hilti*' when referred to the decisions collectively).

[7] Case C-333/94 *Tetra Pak International SA v Commission* [1996] ECR I-5951, [1997] 4 CMLR 662 (hereinafter 'Case C-333/94 *Tetra Pak II*'); Case T-83/91, *Tetra Pak Rausing SA v Commission* [1990] ECR II-309, [1991] 4 CMLR 334 (hereinafter 'Case T-83/91 *Tetra Pak II*'); and *Elopak Italia/Tetra Pak* (Case IV/31043) [1991] OJ L72/1, [1992] 4 CMLR 551 (hereinafter '*Tetra Pak II* Commission Decision') (hereinafter '*Tetra Pak II*' when referred to the decisions collectively).

[8] Case T-201/04 *Microsoft* (n 2 above) and *EU Microsoft I* Commission Decision (n 2 above).

[9] *IBM Undertaking* (Case IV/30.849) [1984] OJ L 118/24.

[10] *Napier Brown/British Sugar* (Case No IV/30.178) [1988] OJ L284/41.

[11] *EU Microsoft II* Commission Decision (n 2 above).

[12] R O'Donoghue and AJ Padilla, *The Law and Economics of Article 82 EC* (Oxford: Hart Publishing, 2006) 492.

Explorer to Windows operating system and thus avoided the technological integration discussion.[13]

An interesting feature of these cases is the narrow market definitions applied. Thus in *Hilti,* where the defendant saw one market, the Commission identified three and in *Tetra Pak II* four markets were established in comparison with the two markets Tetra Pak had argued. Importantly, the IP rights held by Hilti and Tetra Pak played a significant role in the market definitions concluded by the Commission.[14] In the Commission Decision *IBM Undertaking*, which in principle is also a technological integration case, a narrow market definition again permitted the Commission to find an illegal tie.[15]

Closely related to the definition of the relevant market in tying cases is the assessment of whether there are in fact two products. The Commission and EU Courts have mainly applied the presence of independent manufacturers as a benchmark for whether there are two products, rather than assessing the tying arrangement and product(s) itself (themselves).

Importantly, neither the assessment of dominance or abuse has been given exceptional attention in tying cases. However, it is worth noting two points in respect of dominance—first, all companies found guilty of tying have so far held market shares above 70 per cent in the tying product market.[16] Second, once two products have been identified and the defending company has been found dominant in one of these product markets, it becomes almost impossible for the company to justify the tying or technological integration, resulting in a foregone conclusion that the tie is anticompetitive. The establishment of dominance thus appears to weigh more heavily than the abuse assessment. This is of course in sharp contrast to recent case law in the US where it was clarified that the market power requirement, albeit an essential requirement, was *no more* than one of the conditions necessary to finding the tying conduct illegal.[17]

In relation to IP rights, such a finding is crucial in particular, when the relevant markets are often defined along the scope of the IP right by the competition authorities thereby making the IP owner dominant in that market. It essentially means that the presence of an IP right can play an influential (often negatively from the IP owner's

[13] *EU Microsoft II* Commission Decision (n 2 above) paras 33–7.

[14] *Hilti* Commission Decision (n 6 above) paras 55, 66; Case T-30/89 *Hilti* (n 6 above) para 93, and Case T-83/91 *Tetra Pak II* (n 7 above) para 10.

[15] *IBM Undertaking* (n 9 above).

[16] Case T-30/89 *Hilti* (n 6 above) para 85; *Hilti* Commission Decision (n 6 above) para 70; *Tetra Pak II* Commission Decision (n 7 above) para 104; *EU Microsoft I* Commission Decision (n 2 above) para 432; and *EU Microsoft II* Commission Decision (n 3 above) para 24.

[17] See *US Microsoft III* (n 4 above) 89–95.

perspective) role on the outcome of whether a tying arrangement or technological integration is illegal.

The question of abuse has centred on the key issue of coercion and whether consumers were realistically restricted from purchasing the tied product separately and thus hindered from making such a choice.[18] This was not discussed at great length in either *Hilti* or *Tetra Pak II.* Nor was the question of whether the dominant companies' behaviour created any anti-competitive effects. In fact these were more or less assumed.[19] The cases thus followed the standard of proof for anti-competitive effects set in *British Sugar.*[20] In that case, British Sugar was found guilty of having tied the sale of sugar to its delivery. The Commission did so by simply assuming that anti-competitive effects were present because British Sugar by its pricing policy effectively had reserved the sugar delivery market to itself.[21] *EU Microsoft I* Commission Decision, on the other hand, contains a thorough assessment of the current and potential anti-competitive effects of the tie-in. Nevertheless, the Commission still found that Microsoft's tying of its Windows Media Player (WMP) with Windows operating system had the *potential* to risk foreclosure in the future.[22] The Commission itself admitted that the case was unique owing to the fact that despite the tie, consumers were able to obtain and use a third party media player through downloads on the internet with Microsoft's operating system and therefore foreclosure could not be assumed.[23]

The reason for the extensive analysis was therefore not the technological integration but the market conditions; the fact that the tie did not fully exclude the use of other independent media players (the tied product) with the operating system (the tying product).

Yet the GC rejected the notion that a downloaded media player was compatible with a pre-installed media player. It did so, because it found it was less efficient to download and less attractive for consumers to use because the downloaded media player would always play second fiddle to WMP.[24] Instead, it highlighted that in the absence of tying, consumers would be encouraged to choose from all available media players.[25] It therefore agreed with the Commission that Microsoft's tie was likely to foreclose the media player market. Interestingly, the GC further indicated that the Commission

[18] *Hilti* Commission Decision (n 6 above) para 75; and Case T-83/91 *Tetra Pak II* (n 7 above) para 137.

[19] See *Hilti* Commission Decision (n 6 above) para 75; and C Ahlborn, DS Evans and AJ Padilla 'The Antitrust Economics of Tying: A Farewell to Per Se Illegality' (2004) 49 Antitrust Bulletin 287, 313–14.

[20] *Napier Brown/British Sugar* (n 10 above).

[21] *Ibid* para 46.

[22] *EU Microsoft I* Commission Decision (n 2 above) paras 842–954.

[23] *Ibid* para 841.

[24] Case T-201/04 *Microsoft* (n 2 above) paras 1045, 1049–50.

[25] *Ibid* para 1041.

had gone beyond what was necessary to establish anti-competitive effects in its ruling.[26]

Despite such comments, the Commission chose again in *EU Microsoft II* to give a comprehensive examination of the foreclosure effects and importantly, the market conditions in the web browser market.[27] Unsurprisingly, however, it still found that Microsoft's stronghold in the operating systems market impeded competition in the web browser market and in particular,

> . . . the tying of Internet Explorer, in addition to reinforcing Microsoft's position on the market for client PC operating systems, created artificial incentives for web developers and software designers to optimise their products primarily for Internet Explorer.[28]

One point that has gone relatively unnoticed in respect of tying is the element of intent. EU case law is ambiguous as to how much of a role intent plays under Article 102 TFEU. When it comes to technologically integrated products, it is nevertheless a critical factor in determining whether the tie should be permitted, because integration for pure strategic reasons that does not add any technical value to the products is likely to restrict consumer choice just as much as a contractual tying arrangement. In *EU Microsoft I*, an email from a Microsoft executive to Bill Gates revealed Microsoft's real intentions behind the tie, namely to force competitors to compete directly against Windows instead of only the WMP.[29] From this email together with Microsoft's lack of defence, it was clear that the Commission and the GC were correct in finding Microsoft's integration an illegal tying arrangement under Article 102 TFEU.

In *Hilti* and *Tetra Pak II* some consideration was however given to the defence of the tying arrangement, although the outcome for both cases was negative. The cases made it clear that although an objective justification can in principle save the tying arrangement; the threshold for this is extremely high.[30] In T-83/91 *Tetra Pak II*, the GC held that when there were less restrictive measures available that would not infringe Tetra Pak's IP rights, Tetra Pak was not allowed to impose a tying arrangement to protect its reputation and product quality.[31] A similar statement had been made in *Hilti*,[32] and this has now also found its way to the *Article 82 Guidance*,

[26] *Ibid* paras 857–9.
[27] *EU Microsoft II* Commission Decision (n 3 above) para 34.
[28] *Ibid* para 37.
[29] Case T-201/04 *Microsoft* (n 2 above) para 911.
[30] Case C-333/94 *Tetra Pak II* (n 7 above) para 37.
[31] *Ibid* paras 83–4, 138–9.
[32] *Hilti* Commission Decision (n 6 above) paras. 89.4–90, 92–93, see also Case T-30/89 *Hilti* (n 6 above) paras 118–19.

> It is not the task of a dominant undertaking to take steps on its own initiative to exclude products which it regards, rightly or wrongly, as dangerous or inferior to its own product.[33]

The high threshold was later confirmed in T-201/04 *Microsoft* where the GC held that the dominant company must demonstrate that the integration offers '. . . superior technical product performance' in comparison to the use of the products together when purchased separately.[34]

The GC's statement is the only indication in EU case law of some, albeit very limited, leeway towards technological integrated products. The Commission has in its *Article 82 Guidance* indicated that it will also consider potential benefits to consumers from the integration of products.[35] Nevertheless, it maintains a rigid approach to tying and states that the risk of foreclosure is expected to be greater for technological integration.[36] There is therefore some discrepancy between the case law approach to technological integration and the Commission's policy on the matter.

B. US Case Law

US case law on tying has a much longer history in comparison to EU case law and one that is closely intertwined with IP case law.[37] However, as in the EU, a handful of cases, *IBM Corp*,[38] *International Salt*,[39] *Jefferson Parish*,[40] *Eastman Kodak*,[41] and *US Microsoft III*, have been the most influential in shaping the approach the US Courts have adopted to tying.

Each of these cases added to the test applied today. *International Salt* established the need for market power. This was confirmed in *Jefferson Parish*, where the Supreme Court held that without market power it was not possible to coerce customers into the tying arrangement.[42]

---

[33] 'Communication from the Commission—Guidance on the Commission's enforcement priorities in applying Article 82 of the EC Treaty [now Article 102 TFEU] to abusive exclusionary conduct by dominant undertakings' [2009] OJ C45/3 para 29 (hereinafter '*Article 82 Guidance*').

[34] Case T-201/04 *Microsoft* (n 3 above) para 1161.

[35] *Article 82 Guidance* (n 33 above) para 62.

[36] *Ibid* para 53.

[37] H Hovenkamp 'IP ties and Microsoft's rule of reason' (2002) 47 Antitrust Bulletin 369, 372.

[38] *International Business Machines Corp v United States* 298 US 131, 56 S Ct 701, 80 L Ed 1085 (1936).

[39] *International Salt Co v United States*, 332 US 392, 68 S Ct 12 (1947).

[40] *Jefferson Parish* (n 5 above).

[41] *Eastman Kodak Co v Image Tech Serv*, 504 US 451, 112 S Ct 2072 (1992).

[42] *Jefferson Parish* (n 5 above) 27 and *Illinois Tool Work Inc v Independent Ink Inc* 547 US 1, 126 S Ct 1281 (2006) 16.

In *International Salt,* the market power was in fact more or less assumed due to the patent International Salt held on its salt processing machine.[43] The Supreme Court held that it was more a question of the 'uniqueness' of the tying product which determined the finding of market power, than any form of serious assessment of market shares as has been applied under section 2 of the Sherman Act.[44] In subsequent case law, US courts continued to assume market power when IP rights, and in particular patents, were present. *Illinois Tool* was the first case to break with this tradition when the Supreme Court held that irrespective of the patent the plaintiff must demonstrate that the defendant holds market power in the tying product market.[45]

In *International Salt,* the Supreme Court also assessed whether a substantial amount of commerce had been affected in the tied product market. It clarified that the effect in the tied product market could be minimal and the tie would still be considered illegal.[46] Moreover, it was also '. . . immaterial that the tendency is a creeping one rather than one that proceeds in full gallop'.[47] In *Northern Pacific Railway*[48] the Supreme Court explained what the 'substantial amount of commerce affected in the tied product market' actually meant when it outlined the two most feared effects of tying. The first effect is the leveraging of market power from the tying product market to the tied product market, which can then hinder free access to the tied product market. Second, the tying arrangement can limit the consumer choice of where to purchase the tied product.[49]

However, it failed to apply these effects as a standard for the requirement. Instead, the Supreme Court moved to look upon the affected dollar-volume in the market to assess whether the market had been foreclosed. This rather imprecise assessment of anti-competitive effects was abandoned in *Jefferson Parish.* Instead, the Court reverted to the two effects it had listed in *Northern Pacific Railway* as a standard for whether a not insubstantial amount of commerce had been affected,[50] and maintained that both the actual and potential risk of foreclosure of the tied product market would be considered. In other words, the threshold for finding foreclosure is relatively low, similar to the approach taken under EU law.

IBM Corp acknowledged that there could be an objective justification for the tying arrangement; however, the threshold for such justification was set at a high level in the case. IBM had claimed that it had engaged in a tying arrangement to ensure that

---

43 *International Salt* (n 39 above) 395.

44 *Ibid* 395; see also *Times-Picayune Pub Co v United States* 345 US 594, 73 S Ct 872, 97 L Ed 1277 (1953) 611; and *United States v Loew's Inc* 371 US 38, 83 S Ct 97, 9 L Ed 2d 11 (1962) 46.

45 *Illinois Tool* (n 42 above) 16.

46 *International Salt* (n 39 above) 396.

47 *Ibid.*

48 *Northern Pacific Railway Co et al v United States,* 356 US 1, 78 S Ct 514 (1958).

49 *Ibid* 6.

50 *Jefferson Parish* (n 5 above) 14.

the quality and functioning of its machines was maintained, when it required users to use only its cards. The Supreme Court rejected the argument stating that first, section 3 of the Clayton Act does not allow for any exceptions and second, there were less restrictive measures available to IBM to maintain the quality of its machine.[51] The US Courts have upheld this high threshold in subsequent case law.[52] Only in a few cases has a tying arrangement been permitted where the Courts felt that the threshold had been met. This was the case in *Jerrold Electronics*[53] in which Jerrold Electronics was allowed to apply a tying arrangement to enter into a new industry in order to ensure quality control.[54]

Jefferson Parish is the pivotal case for tying. Here, the Supreme Court for the first time outlined the test that it had instinctively applied in previous cases to tying arrangements, but never clearly worded.[55] However, the Supreme Court also acknowledged that tying may not always be harmful.[56] As a result, the Supreme Court disagreed upon whether tying should be dealt with from a per se illegality perspective or by adopting a *rule of reason* approach.[57] Although the majority confirmed that the per se illegality approach should remain, the judgment opened up the possibility of a more lenient approach to future tying arrangements—an attitude adopted by the Court of Appeals in the subsequent case of *US Microsoft III*.

In *Jefferson Parish*, the Supreme Court also included a new element in the objective justification assessment, when it analysed the market conditions and in particular whether the conduct was part of commercial usage, ie whether other non-dominant companies in the market were also tying.[58]

In *US Microsoft III*, the Court of Appeals recognized the problem with the current five-step test applied to tying when employing it to technological integration.[59] In particular, it found the separate product test inadequate to assess whether a product is technologically integrated.[60] In fact, the case highlights not only that the current separate product test and the strict illegal per se approach is inappropriate for certain tying arrangement, it also clarifies that there is a need for a different approach to technological integration to ensure that innovation and product development are not stifled by competition rules which are too strict.

---

[51] *IBM Corp* (n 38 above) 134, 138–40.

[52] *International Salt* (n 39 above) 396–8.

[53] *United States v Jerrold Electronics Corp* 187 F Supp 545, 560 (ED Pa 1960), aff'd per curiam, 365 US 567 (1961).

[54] *Ibid* 557.

[55] *Jefferson Parish* (n 5 above) 18.

[56] *Ibid* 27.

[57] *Ibid* 9–11.

[58] *Ibid* 22.

[59] *US Microsoft III* (n 4 above) 85–9.

[60] *Ibid* 89.

Interestingly, the Department of Justice (DoJ) agreed with this view in a very extensive report on section 2 of the Sherman Act. It noted here that,

> . . . the Department agrees with courts and panellists urging restraint in the area of product design and believes that great caution should be exercised before condemning a technological tie under antitrust laws. . .remedying anticompetitive technological ties appropriately can often be difficult, requiring courts to make judgments about unusually complicated, forward-looking business issues and thereby heightening the risk that a remedy will hurt, rather than help, consumers.[61]

However, the views of the report were short-lived—the report was withdrawn by the DoJ just eight months after it was published as no longer being the DoJ's policy.[62] The recommendations made in the Report were found to be too cautious and requiring the competition authorities to rely too heavily upon self-adjustment by the market. In light of the recent economic developments, the DoJ found this no longer a sufficient tool to protect competition and consumers.[63]

The future approach to tying and technological integration is therefore a little uncertain. However, case law indicates that the US Courts are willing to distinguish between contractual tying and technological integration and that in their eyes the latter deserves a more lenient approach where the specific market conditions are taken into account to ensure that innovation and product development is encouraged.

C. The Two Separate Product Test and Possible Defences

In both the US and the EU, consumer demand has been applied to measure whether the tying and tied products are in fact distinct products. In the EU, the GC has outlined that the two separate products assessment is a two part test—that of the product's nature, physical appearance, and commercial usage and second, a consumer demand of the tied product.[64] However, in case law the focus has mainly been on the second part and evidence of independent manufacturers of the tied component have played a significant, if not determining role.[65]

The first part, on the other hand, is established when the relevant product market is identified. Article 102 TFEU case law demonstrates the use of a narrow market

---

61 *Section 2 Report* (n 1 above) 88.

62 DoJ, Press Release 09-459 'Justice Department Withdraws Report on Antitrust Monopoly Law' (11 May 2009) <http://www.justice.gov/opa/pr/2009/May/09-at-459.html> accessed 10 September 2010.

63 *Ibid.*

64 Case T-201/04 *Microsoft* (n 2 above) paras 925, 927.

65 Case C-333/94 *Tetra Pak II* (n 7 above) para 36; Case T-30/89 *Hilti* (n 6 above) para 67; Case T-201/04 *Microsoft* (n 2 above) para 927; *Article 82 Guidance* (n 33 above) 51.

definition in general. Where IP rights have been involved, the scope of the IP right has been followed to define the market rather than the product itself, as demonstrated in *Hilti* and *Tetra Pak II*. Consequently, only when components of a product are protected under the same IP right will they be viewed by the competition authorities as one product; otherwise the tying of them will be an infringement of Article 102 TFEU.[66] As the GC noted, unless it infringes the IP right, independent manufacturers will remain 'quite free' to manufacture consumables for use in other manufacturers' products.[67] In fact, case law shows the EU Courts will actively seek to protect this possibility.

Although such a policy rightly will restrict the abuse of an IP right beyond its exclusive right, it can also cause harm in an industry where innovation and product development is key to progress and may in fact exaggerate the market power conferred by the IP right, which can be detrimental to technologically integrated products. Moreover, it also implies that two components of a product remain separate from a legal perspective.

The use of a narrow market definition affects tying arrangements in two ways—first, it makes it easier to establish dominance; and second, products are more likely to be seen as tied rather than components of a product, because they appear to be in separate markets.

In *Hilti*, the Commission and the EU Courts agreed that Hilti had engaged in illegal tying by amongst other things forcing its customers to purchase its nails when purchasing its cartridge strips to be used with the Hilti nail gun. Independent manufacturers producing Hilti-compatible nails, complained to the Commission claiming that Hilti's behaviour was pushing them out of the nails market. Hilti argued that the nails were part of a powder-actuated fastening system.[68]

The Commission, however, saw the nails as separate to the nail gun and as a part of an aftermarket to the gun as both nails and cartridge strips were purchased independently of the nail gun.[69] Importantly, the presence of independent manufacturers provided 'sound evidence' that Hilti-compatible nails were a separate product.[70]

Similarly, in C-333/94 *Tetra Pak II*, the Court of Justice confirmed that the evidence of independent manufacturers in the tied market played a significant role for the finding of a separate product.[71] The Commission had identified four separate

[66] S Anderman and H Schmidt, *EU Competition Law and Intellectual Property Rights, The Regulation of Innovation* (Oxford: Oxford University Press, 2010) Chapter 4.

[67] Case T-83/91 *Tetra Pak II* (n 7 above) paras 83–4; and see also Case T-30/89 *Hilti* (n 6 above) para 68; Anderman and Schmidt (n 66 above) Chapter 4.

[68] Case T-30/89 *Hilti* (n 6 above) para 93.

[69] *Hilti* Commission Decision (n 6 above) para 57.

[70] *Ibid* paras 57, 66–7; and Case T-30/89 *Hilti* (n 6 above) para 67.

[71] Case C-333/94 *Tetra Pak II* (n 7 above) para 36.

markets—the markets for aseptic cartons and machines, and non-aseptic cartons and machines, instead of an integrated packaging system for packaging liquid food as Tetra Pak argued.[72] The reason for these divisions was mainly the patents that Tetra Pak held on the technology of the machines, cartons, and processes.[73]

In its *Article 82 Guidance,* the Commission has retained the consumer demand as the standard to assess whether two products are distinct.[74] Thus it sees two products as distinct,

> . . . if, in the absence of tying or bundling, a substantial number of customers would purchase or would have purchased the tying product without also buying the tied product from the same supplier, thereby allowing stand-alone production for both the tying and the tied product.[75]

In contrast, the GC's standard in T-201/04 *Microsoft* appears higher as it held that only in the absence of independent demand for the tied product could the products be seen as integrated.[76]

In comparison to *Hilti* and *Tetra Pak II,* the market definition was not a major issue in the case. The GC accepted the Commission's finding of two markets—the operating systems market and the media player market. The Commission had found a separate consumer demand for the media player, because there were independent media player suppliers and that stand-alone media players could be installed on a PC.[77] Microsoft, on the other hand, argued that the two were technologically integrated and that Windows without WMP would weaken Windows' performance.[78]

Although the GC toyed with the idea of the Windows operating system with WMP being one product, it found them to be separate products for three particular reasons. First, one product was a software system and the other a software application.[79] Second, there were independent manufacturers of media players and third, a '. . . not insignificant number of customers' continued to obtain media players from Microsoft's competitors.[80]

The case demonstrates that the Commission and the GC were reluctant to look beyond the market definition when assessing the product. Just because there may be two markets involved, does not mean that the products cannot be technologically integrated to create a new product or one, as in this case with add-on features.

---

[72] Case T-83/91 *Tetra Pak II* (n 7 above) para 34.
[73] *Ibid* para 10.
[74] *Article 82 Guidance* (n 33 above) para 51.
[75] *Ibid.*
[76] Case T-201/04 *Microsoft* (n 2 above) para 918.
[77] *EU Microsoft I* Commission Decision (n 2 above) paras 405, 804, 825.
[78] Case T-201/04 *Microsoft* (n 2 above) paras 886–8, 1165.
[79] *Ibid* para 926.
[80] *Ibid* paras 927, 932.

Moreover, the fact that an operating system can take more than one media player was not discussed before concluding that the products were separate. This important feature of software products has a crucial effect upon the assessment of foreclosure of the tied product market, because even by integrating the two products as Microsoft did, the tied market cannot be fully foreclosed by the integration. Importantly, the Commission acknowledged this in *EU Microsoft II* and conducted a consumer and business survey to assess the restrictions on switching web browsers through downloading and the likely foreclosure effects on the web browser market.[81] This is a move away from the more rigid approach taken by the GC in T-201/04 *Microsoft*, where although it acknowledged that market conditions could play a role in the assessment of whether there were two separate products, it did not provide a sufficiently detailed analysis of it. The GC held that even complementary products, where no separate demand for the tying product exists without the tied product, cannot in themselves justify tying, because customers may want to obtain the products from different suppliers.[82]

This means that the case law provides very little scope for product development and consumers' changing view of a product. Consumers' perception of products is likely to and will change in response to innovation and product development, but not always necessarily at speed. Therefore, although there may be an independent demand for a product, this may diminish in future, because another more advanced product is developed or it is combined with another to enhance its abilities. Examples are those of the DVDs and Digital TV Recorders, which allow you to pause, record, and rewind live TV replacing the VHS videos, and fax machines being integrated with photocopying machines. Applying a too strict approach to technological integration will risk stopping such developments or the very least delaying them.

The *Article 82 Guidance* appears to have taken this into account and thus permits an efficiency defence of the tying arrangement. It particularly notes that '. . . it may also examine whether combining two independent products into a new, single product might enhance the ability to bring such a product to the market to the benefit of consumers.'[83]

The statement indicates that the Commission will be willing to accept and allow a defence of technologically integrated products; however, it remains ambiguous as to the precise standard of proof.

Currently, the defence of a tying arrangement and technological integration takes place after the assessment of abuse. This means that once the Commission or the EU Courts have concluded that there are two separate markets, it is almost impossible to

[81] *EU Microsoft II* Commission Decision (n 3 above) paras 48–54.

[82] Case T-201/04 *Microsoft* (n 2 above) paras 921–2; see also *Article 82 Guidance* (n 33 above) 58.

[83] *Article 82 Guidance* (n 33 above) para 62.

defend a tying arrangement or technological integration.[84] In particular, the defence of the technological integration will always be coloured by the previous finding of two product markets. Consequently, an objective assessment of the technological integration as a product in its own right does not always take place.[85]

In comparison to the US Sherman Act, section 1, there is a built-in defence in the wording of Article 102(d) TFEU as it states that only where contracts are subject to supplementary obligation which '. . . by nature or according to commercial usage' has no connection with the subject of the contract, then the conduct is illegal. This form of defence if taken literally requires an assessment of the market conditions and the product(s) in question. However, so far no company has successfully argued that the tying was part of commercial usage. In C-333/94 *Tetra Pak II,* the Court of Justice noted that other companies in the market (a mere 12 per cent) operated in a different manner to that of Tetra Pak and even if a dominant company's behaviour was consistent with commercial usage, ie that all firms in the market tied, the tying could still be abusive when performed by a dominant company.[86]

The Commission applied a similar line of arguments when it rejected Microsoft's plea of 'commercial usage' in *EU Microsoft I* Commission Decision.[87] In other words, although Article 102(d) TFEU provides an exemption, the Commission and the EU Courts have applied a strict approach, which in practice eliminates any form of exemption that relies upon the notion of commercial usage.

Instead, when it comes to technological integration, the Commission stated in *EU Microsoft I* Commission Decision that Microsoft should have demonstrated that the tie was 'indispensable' to achieve superiority.[88] Such a standard would be very high, if not unreachable and in practice mean that no technological integration would ever be approved.[89]

The GC, on the other hand, merely noted that Microsoft had not proven that the integration of the WMP into Windows created 'superior technical product performance'.[90] This standard for technological integrated products is comparatively lower to the Commission's suggestion above.

If the GC's standard of proof prevails, it is very much in line with the approach taken by the District Court of Appeals in *US Microsoft II* discussed below.[91]

[84] See *EU Microsoft II* Commission Decision (n 3 above) para 34.
[85] H Schmidt, *Competition Law, Innovation and Antitrust, an Analysis of Tying and Technological Integration* (Cheltenham: Edward Elgar Publishing Ltd, 2009) 97.
[86] Case C-333/94 *Tetra Pak II* (n 7 above) paras 36–37.
[87] *EU Microsoft I* Commission Decision (n 2 above) paras 822–4.
[88] *Ibid* para 963.
[89] J Apon, 'Cases against Microsoft' [2007] ECLR 327, 331.
[90] Case T-201/04 *Microsoft* (n 2 above) 1146–7.
[91] *US Microsoft II* (n 4 above).

The Commission's statement in its *Article 82 Guidance* and the GC's 'superior technical product performance' standard both reflect a positive attitude towards technological integration. Yet they should be contrasted with another comment by the Commission in the *Article 82 Guidance* stating that a lasting tie, such as technological integration is likely to be more harmful to competition than contractual tying and the Commission's no nonsense approach in *EU Microsoft II*.[92]

EU case law paints a mixed picture of the approach to technological integration. Although, there is a clear reluctance in case law to treat technological integration differently from tying, *EU Microsoft I* can, after closer scrutiny, be interpreted as a diminutive step in the right direction.

Narrow market definitions shaped around IP rights, overreliance on the presence of independent manufacturers in the tied market, a high standard of proof for when products cease to be separate, and a high standard of proof in respect of the defence means that the strict attitude applied to tying in general will remain the prevailing policy for technological integration in years to come as *EU Microsoft II* demonstrates.

In comparison, the US approach has been more flexible and its threshold lower. In early case law, the approach was 'purely linguistic' under the assumption that no tie existed without two products.[93] In fact, if there are not two products involved, the anti-competitive behaviour can only be assessed as an exclusive dealing, vertical non-price restraint or boycott, all of which falls under a *rule of reason* approach.[94]

In *Jefferson Parish,* the Supreme Court focused upon whether there was a specific demand for the tied product.[95] In the case, the Supreme Court looked at the products, the hospital service and the anaesthesiology service, and found that the anaesthesiology service could be offered separately and was indeed billed separately. However, in this case it found nothing wrong or anti-competitive with the tie as the hospital held no market power.[96] Justice O'Connor, in a concurring judgment, had a slightly different view on the two separate product issue as she noted that it was difficult in reality to see the anaesthesiology service as a separate product as few people would choose to have an operation without it, and thus it must instead be regarded as a component of a product.[97]

---

[92] *Article 82 Guidance* (n 33 above) para 53; and *EU Microsoft II* Commission Decision (n 3 above) para 34.

[93] *US Microsoft III* (n 4 above) 85; and *Jefferson Parish* (n 5 above) 21.

[94] Hovenkamp (n 37 above) 400.

[95] *Jefferson Parish* (n 5 above) 21–2.

[96] *Ibid* 25.

[97] *Ibid* 43.

The Supreme Court rejected this view in *Eastman Kodak*, where Kodak had tied the sale of service for its photocopying machines to the sale of parts.[98] It stated that even where the products are functionally linked and there is no use of the one without the other, a tying arrangement can still be found illegal.[99] This is similar to GC's opinion in T-201/04 *Microsoft* discussed above.

The Supreme Court confirmed the application of the consumer demand test applied in *Jefferson Parish* but noted that when the demand for one of the products is so low that it is no longer economically viable to offer it separately, the two products can be sold as one and the tying of them is permitted.[100] This assessment acknowledges the fact that some demand will always remain for the tied product, even if there has been a product development. This test therefore adheres to both the consumer, her perception of the market and products, and the practice of the defendant and its competitors in the market.[101]

However, in respect of technologically integrated products, the Court of Appeals in *US Microsoft III* recognized that the test has its flaws. Previous case law had in fact created a loophole for technologically integrated products, treating these more leniently than contractual tying cases by finding that the technological integration was beneficial and common market behaviour.[102] For instance, *US Microsoft II*[103] demonstrates that the Courts were not keen on entering into a detailed technical assessment of the product to identify whether the product was in fact two.[104] The case revolved around the interpretation of a consent decree that Microsoft had previously entered into, which stated that Microsoft was prohibited from entering into a licensing agreement, which was conditional upon the licensing of any of Microsoft products covered by the consent decree, although it excluded from this, Microsoft's ability to develop integrated products.[105] The Court of Appeals established a two-part test for assessing whether the products are integrated. The first step was to identify whether the integration took place on the design or installation stage. If at the latter stage, then consumers were unlikely to gain any benefits from the integration as they

98 *Eastman Kodak* (n 41 above) 451.

99 *Ibid* 463.

100 *Ibid* 462.

101 *Jefferson Parish* (n 5 above) 22–4 and *Eastman Kodak* (n 41 above) 462.

102 For instance *Telex Corp v International Business Machines Corp* 510 F 2d 894 (10th Cir, 1975), *Response of Carolina, Inc v Leasco Response, Inc* 537 F 2d 1307 (5th Cir, 1976), *ILC Peripherals Leasing Corp v International Business Machines Corp* 448 F Supp 228, (ND Cal, 1978) (hereinafter '*ILC Peripherals Leasing I*'), *ILC Peripherals Leasing Corp v International Business Machines Corp* 458 F Supp 423 (ND Cal, 1978) (hereinafter '*ILC Peripherals Leasing II*').

103 *US Microsoft II* (n 4 above), the case distinguishes itself from other antitrust cases by focusing upon the violation and interpretation of a consent decree in comparison to a normal antitrust case, 946.

104 *US Microsoft II* (n 4 above) 950.

105 *US v Microsoft Corp.* Civil Action 94-1564, 15 July 1994, Final Judgment, Section IV(E)(i), <http://www.usdoj.gov/atr/cases/f0000/0047.htm> accessed 10 September 2010.

would be able to achieve the same results from two stand-alone products combined. However, if the integration was at the design stage, the integration was likely to create benefits. The second part of the test was to establish that the integration brought some 'technological value' or some advantages rather than the mere bolting of the products.[106] The Court of Appeals found that Microsoft had met that standard of proof.[107]

Although, the Court itself believed that this approach to technological integration was in line with tying case law, the Court does in fact move away from the consumer demand test applied in *Jefferson Parish* and *Eastman Kodak* and has instead applied the 'New Product Rationale test' suggested by Areeda et al.[108]

The case carves out a safe haven for technological integration without actually addressing the real issue, the standard of proof for whether a product is integrated. The Court neatly avoided a detailed assessment of this question as it held it was beyond the courts' 'institutional competence' to embark upon a technological assessment of the integrated products.[109]

The judgment has been greatly criticized.[110] One such criticism came from the dissenting Judge Wald, who rightly stated that creating a safe haven for software based on the mere reluctance to engage in a technological assessment was wrong. Instead the fact that software products were predisposed to separation and integration was a call for closer scrutiny of such products.[111]

In *US Microsoft III*, the Court of Appeals also rejected the use of the *Jefferson Parish* consumer demand test to technological integration, as it noted that the test was '. . . backward-looking and therefore systematically poor proxies for overall efficiency in the presence of new and innovative integration'.[112] Instead, it held that greater examination of the impact on competition created by technological integration was needed before it could be classified as per se illegal and thus remanded the question for a *rule of reason* assessment by the lower court.[113]

---

106 *US Microsoft II* (n 4 above) 949–50.

107 *Ibid* 950.

108 *Ibid* 950 and P Areeda, E Elhauge and H Hovenkamp, *Antitrust Law, An Analysis of Antitrust Principles and their Application* (Vol X, 2nd edn, New York: Aspen Publishers 2004) 224–9.

109 *US Microsoft II* (n 4 above) 949.

110 WH Page, JE Lopatka, 'The Dubious Search for "Integration" in the *Microsoft* Trial' (1999) 31 Connecticut L Rev 1251; MW De Vries, 'United States v Microsoft' (1999) 14 Berkeley Technology L J 303; and Brief of Prof Lawrence Lessig as *Amicus Curiae*, in *United States v Microsoft Corp.* DC of Columbia, Civil Action No. 98-1233 (TPJ), 1 February 2000.

111 *US Microsoft II* (n 4 above) 958.

112 *US Microsoft III* (n 4 above) 89.

113 *Ibid* 94.

The Court of Appeals clarified,

> ... [there] may also be a number of efficiencies [generated from the technological integration of Windows operating system with Windows Internet Explorer] that, although very real, have been ignored in the calculations underlying the adoption of a per se rule for tying. We fear that these efficiencies are common in technologically dynamic markets where product development is especially unlikely to follow an easily foreseen linear pattern.[114]

In other words, commercial usage and market conditions need to be taken into consideration when discussing the pro-and anti-competitive effects of technological integration. Interestingly, the Court of Appeals merely stated that it is 'common' practice within the software market to integrate. It did not as the EU Courts require that *all* firms in the market must be engaged in tying before the commercial usage condition is fulfilled.[115]

In *US Microsoft III*, the Court of Appeals highlighted the core problem with the consumer demand test, when it stated that the first company to integrate two previously separate products will always be caught under the per se rule because the tied products will inevitably have a distinct demand for it until consumer perception changes.[116]

This stands in stark contrast to the view of the GC, who requires that the independent demand for the tied product has ceased before it accepts the two products as integrated.[117]

The Court of Appeals noted that it is inadequate to look only at the demand for the products, the supply of the products also play a part in the assessment of whether the products can be seen as integrated.[118] This is because companies without market power will engage in tying when '... the cost savings from joint sale outweigh the value the consumers place on the separate choice'.[119] Thus market conditions and the behaviour of the competitors are important elements in judging whether integration is beneficial and welfare enhancing or anti-competitive and restrictive to consumer choice.

In the case, the Court of Appeals found that not only was technological integration common in the software industry, but common among non-dominant companies.[120] However, it did not find Microsoft's integration 'welfare enhancing',[121] and therefore

[114] *Ibid* 94.
[115] Case C-333/94 *Tetra Pak II* (n 7 above) paras 36–7.
[116] *Ibid* para 92.
[117] Case T-201/04 *Microsoft* (n 2 above) para 918.
[118] *US Microsoft III* (n 4 above) 88.
[119] *Ibid* 88.
[120] *Ibid* 93.
[121] *Ibid* 89.

the technological integration would properly not have reached the required standard of proof.

That said, the Court of Appeals has in *US Microsoft III* taken a crucial step towards the protection of innovation, when it rejected the use of the consumer demand test for complex tying cases such as technological integration, as well as requiring an assessment under the *rule of reason*. The *rule of reason* assessment would not only mean a more thorough assessment of the technological integration, it would also allow the defendant to demonstrate the efficiencies gained from the integration and that these were likely to outweigh any '... distortions to consumer choice'.[122] Although the latter assessment never took place as the DoJ dropped the tying element of the case to save time,[123] the Court of Appeals has set an important precedent for technological integration cases. This is clearly reflected in the DoJ's comments in its, albeit withdrawn, *Section 2 Report*. The DoJ stated that although it was not willing to see technological integration as legal per se, it did acknowledge that it should not be per se illegal either. In fact '... great caution should be exercised before condemning a technological tie'.[124]

US Microsoft II and *US Microsoft III* demonstrate in principle two different approaches to the same problem of how technological integration should be dealt with, but then they did approach the subject from different angles. In *US Microsoft II*, the Court of Appeals suggests an assessment of the actual integration, which requires the demonstration that the technological integration produces efficiencies and benefits which are beyond the mere bolting of two products. However, the case also highlights the concern of the Court to be involved in a technical assessment of a product. In truth, it raises an important question—namely, whether a court is best placed to determine whether one product is more superior to another.

In *US Microsoft III*, the Court of Appeals left this question unanswered, although it clarified that given the form of tying and the specific market conditions, a more thorough assessment was necessary. The main reason being that the current consumer demand test and the strict per se illegality approach risked restricting innovation and product development in technologically dynamic markets, such as the platform software market.[125]

---

122 *Ibid* 92.

123 DoJ Press release, 'Justice Department informs Microsoft of Plans for Further Proceedings in the District Court' (6 September 2001) <http://www.justice.gov/atr/public/press_releases/2001/8981.pdf> accessed 10 September 2010.

124 *Section 2 Report* (n 1 above) 88–9.

125 *US Microsoft III* (n 4 above) 89–90, 94.

Conclusion

The above analysis highlights four important issues, which should be considered in order for innovation to be sufficiently protected and promoted through product development and technological integration.

First, it is clear that technological integration is distinguishable from contractual tying and therefore should also be treated differently. However, shielding technological integrations in the software industry within a safe haven as has been the practice in US case law is a far too radical approach.[126] In contrast, the EU Courts and the Commission have so far made limited efforts to treat technological integration differently from contractual tying;[127] allowing little consideration to the benefits and efficiencies that the former can generate.

Second, case law on both sides of the Atlantic also clarifies that the hurdle for technological integration lies with the question of whether there are two separate products and the defence of this. Consumers perceive technologically integrated products as just one product, whereas contractual tying still permits the awareness of two products. As the Supreme Court noted in *Jefferson Parish* for tying to exist there must be two products.[128] Therefore when a case involves products that are technologically integrated, the discussion must move away from the traditional assessment of tying and instead focus upon whether the products are in fact integrated. Such an assessment requires more sophistication and detailed analysis of the integration and market conditions than is provided by the current consumer demand test.

Moreover, in relation to IP rights, it would require the competition authorities to look beyond the scope of the IP right to the product itself, an assessment currently obstructed by the narrow market definition employed in particular by the Commission and the EU Courts.

In the *Microsoft* cases on both sides of the Atlantic, the Courts suggested that an argument of two products being technologically integrated would require the defendant to demonstrate that the integration offers 'something more' than the mere bolting of two products.[129] In T-201/04 *Microsoft*, the test was that the integration must lead to '... superior technical product performance'.[130] In *US Microsoft II*, the Court of Appeals held that the combination of the products must be '... different from

[126] *Response of Carolina, Inc v Leasco* (n 102 above) 1330; *ILC Peripherals Leasing I* (n 102 above); *ILC Peripherals Leasing II* (n 102 above) 439; *Memorex Corp v IBM* 636 F 2d 1188 (9th Cir 1980); and *US Microsoft II* (n 4 above) 950.

[127] Case T-201/04 *Microsoft* (n 2 above) paras 850–9.

[128] *Jefferson Parish* (n 5 above) 21.

[129] *US Microsoft II* (n 4 above) 948, also fn 12; and Case T-201/04 *Microsoft* (n 2 above) paras 1146–7.

[130] Case T-201/04 *Microsoft* (n 2 above) para 1159.

what the purchaser could create from the separate products on his own' and the integration should be of some 'technological value'.[131] Although there are some clear similarities here, both Courts avoided greater clarification of the actual assessment of the integration and thus interpretation of these tests, and importantly, the discussion of the standard of proof required.

That said, these suggested tests in the *Microsoft* cases are important building blocks on which a sensible legal doctrine on technological integration can be established, because they do indeed focus upon the imperative issues, not only of what the technological integration brings to the consumer, but also and underpinning this, whether the integration is purely strategic or innovative. It goes without saying that if the integration is purely strategic, it is unlikely to offer anything of technological value to the consumer and thus the consumer's choice is significantly restricted from the integration. In such a case, integration is harmful and should be illegal.

On the other hand, where the company can indeed demonstrate that the integration is innovative, consumer choice would be enhanced by permitting the integration.

However, it does not suffice to merely look at the product; market conditions must also be considered. The starting point is naturally consumer demand, but it must be stressed that the presence of independent manufacturers cannot be the only evidence, and the demand for both the tied and tying product needs to be assessed. If there is no demand for the latter without the tied component, then it is clear that an integrated version can be beneficial for the consumer.[132]

Additionally, the supply of the products must also be assessed, in particular, whether it is commonplace within the market or neighbouring markets to integrate. Requiring all companies to do so, is undoubtedly too strict; instead the trends within the industry should be assessed as the Commission did in *EU Microsoft II.* This would allow for innovation and product development, but would not ignore the effects the integration would have upon the market.[133]

Third, case law flags up an important difference in the approach to tying under EU competition law compared to US antitrust law in relation to the defence of the tying arrangement. In the EU, the possibility of defending a tying arrangement and indeed a technological integration only occurs after the finding of abuse. The standard of proof is extremely high for any justification and it has to be balanced against the finding of anti-competitive effects where the standard of proof is comparably low.

---

131 *US Microsoft II* (n 4 above) 949.

132 Albeit, this does not rule out a requirement of offering both an integrated and an unbundled version. This was the remedy resorted to in *EU Microsoft I* Commission Decision (n 2 above) para 1025; and Case T-201/04 *Microsoft* (n 2 above) para 1229.

133 See DS Evans and M Salinger, 'Why Do Firms Bundle and Tie? Evidence from Competitive Markets and Implications for Tying Law' (2005) 22 Yale J Regulation 37.

The result is in practice that once two separate products are found and dominance established, the technological integration is found abusive.

In comparison, in US case law, a defence for the technological integration is made at the stage of assessing whether there are two separate products. Although, this appears to be a simple technical difference of the two legal systems, the result is that the US approach has shown to be more sensitive to the benefits that technological integration can bring such as product developments and innovations. This critical detail needs to be taken into consideration when developing a legal doctrine for technological integration.

Fourth, although it is clear from the first three points that a more lenient approach to technological integration is called for to ensure and protect innovation and product development, it is at the same time crucial to ensure that greater leeway does not give way to abuse and strategic integration. Courts therefore cannot shy away from difficult technical assessments of integrated products. They must assess whether the consumers ultimately benefit from the integration or whether it is a purely strategic exercise by the company to exclude competitors, foreclose a market, or even avoid being found guilty of tying. Such a requirement raises the vital question of whether the courts are in truth the best placed to assess technological integrated products.

A discussion of this issue is beyond the scope of this chapter, nevertheless, without present alternatives, it is essential that a legal doctrine for technological integration is developed, which distinguishes itself from that of tying by providing a more flexible and sophisticated understanding of both products and markets in which integration takes place, to ensure that innovation and product development can continue to thrive in the 21st century.

10

MARGIN SQUEEZES IN TELECOMMUNICATIONS MARKETS

*Robbie Downing and Alison Jones**

Introduction

An issue which has recently come to the forefront in both the EU and the US is the question of whether, and if so when, a 'margin squeeze' or 'price squeeze'[1] violates Article 102 of the Treaty on the Functioning of the European Union (the 'TFEU')[2] or section 2 of the Sherman Act 1890 respectively. A margin, or price squeeze occurs when a vertically integrated firm, which is dominant in an upstream market, supplies the upstream input to its rivals at such a high price that, when compared to the price the dominant firm charges on the downstream market, it does not allow an as efficient, or perhaps a reasonably efficient, competitor sufficient margin to trade profitably. The vertically integrated firm can, therefore, simultaneously raise the upstream price of inputs and/or cut the retail price of the finished good with '. . . the effect of "squeezing" the profit margins of any competitors in the retail market.'[3]

---

* The authors would like to thank Peter Alexiadis and Keith Jones for their helpful comments on a previous draft of this chapter.

[1] The two phrases are used interchangeably in this chapter as the US literature tends to refer to price squeezes and the EU literature tends to refer to margin squeezes.

[2] Art 102 TFEU (formerly, Article 82 of the Treaty establishing the European Community) prohibits any abuse by one or more undertakings of a dominant position within the common market or a substantial part of it in so far as it may affect trade between Member States.

[3] Section 2 of the Sherman Act of 1890 prohibits monopolization, attempts to monopolize, and conspiracies to monopolize. The key offence of monopolization has two elements: 'the possession of monopoly power in the relevant market, and the wilful acquisition or maintenance of that power as distinguished from growth or development as a consequence of a superior product, business acumen, or historical accident.' *United States v Grinnell Corp* 384 US 563, 571–2 (1966), see below accompanying text to n 38.

Although in both the EU and US there are some older cases dealing with price squeezes, these types of claims have proliferated in regulated utility markets in the last twenty years. Indeed, in the EU, the increasing number of cases arising (both at the EU and national level) has resulted from the fact that, despite liberalization of the telecommunications and energy markets and privatization of many of the state-owned monopolies operating in these sectors, the previous incumbents have retained large market shares and, frequently, the control of networks or infrastructure necessary to provide retail services. This has enabled dominant firms in the EU to engage in pricing practices on the wholesale and retail markets which make it extremely difficult, or impossible, for rivals to operate and compete on downstream retail markets in the long-run. Many of the price squeeze cases in the US have, since the 1990s, also arisen in energy and telecommunications markets.

This chapter focuses on cases that have involved price squeezes in telecommunications markets. Most of these have involved practices affecting broadband markets. Although the chapter does not, therefore, examine the interface between intellectual property and competition laws, and extreme care must be exercised before applying tests and principles developed in one sphere to another, the issues raised in these cases do have relevance to cases arising at the intellectual property and competition law borderline, in particular those involving refusal to license and refusals to license on fair, reasonable, and non-discriminatory terms. Underpinning both groups of cases is the difficulty of determining whether, when, and/or on what terms, owners of bottlenecks should be required to provide access to that bottleneck (whether an intellectual property right or a network) in order to allow proactive competition to develop downstream. In both sets of cases a carefully balanced approach needs to be adopted to ensure that development of competition downstream is not encouraged at the expense of innovation, incentives to invest upstream, or pro-competitive low cost pricing downstream.

The burgeoning numbers of margin squeeze cases that have arisen in the telecommunications sector have raised a number of difficult problems for resolution. An initial matter to be decided is how a price squeeze claim should be analysed under section 2 of the Sherman Act or Article 102 TFEU and how it relates to other rules governing potentially anti-competitive exclusionary conduct by dominant firms. One view is that an independent, standalone price squeeze action can be constructed simply on the basis of the 'margin' between the two prices charged by the dominant firm. Others, however, are fearful that a price squeeze doctrine too broadly drawn does not allow sufficient account to be taken of the potential pro- and anti-competitive effects of the conduct and may, consequently, potentially deter pro-competitive behaviour. These types of anxieties have led to a growing belief that a price squeeze allegation should be analysed *not* as a separate offence but in the same way as other potentially exclusionary behaviour and, in particular, refusals to deal or predatory pricing.

A second question arising is whether it is appropriate to attempt to control price squeezes through *ex post* application of the competition rules. As with refusal to deal and excessive pricing cases, it may not be sufficient to deal with an unlawful price squeeze simply by a prohibition of the behaviour and issuing a cease and desist order. Rather, an effective remedy may require both careful consideration of the appropriate terms of dealing (especially pricing) and monitoring of that behaviour in the future. Arguably, these may not be tasks to which courts or general competition agencies are well-suited. If so, a question which follows is whether price squeezes might be better dealt with through regulation.[4] This latter point is, of course, particularly relevant in the area of telecommunications where, in both the EU and the US, sophisticated regulatory frameworks exist. This point also has resonance for cases involving intellectual property. In both situations, the application of the competition laws may prohibit conduct which has been authorized, or permitted, by a separate more specialist legislative regime.

A third issue in telecommunications margin squeeze cases is, therefore, whether competition laws should be applied *at all* to situations where regulators are empowered to deal with access issues and pricing conduct (should the regulatory rules cover the field or should they exist side by side with, and be supplemented by, the application of the competition law rules). Regulators' functions and powers may be more extensive than those of a competition agency. Not only may they have broader powers to develop competition on a market by, for example, mandating access to facilities, interconnection between networks, and by imposing stringent obligations on firms with market power, but they may be empowered to control the profitability of a dominant firm, to adopt other measures designed to promote competition in the market, and to consider broadly not only the welfare of consumers and the competitive process but other objectives, such as social goals. Being closer to the market, regulators may have a better understanding of the wider issues affecting the market, the market's complexities, and the players on it. Clearly, however, any regulator's ability to prevent, or remedy, a price squeeze will depend upon the precise regulatory framework that exists and the powers conferred upon them. That structure may not always be ideal. In the EU, for example, the legislative process can be tortuous and is frequently held up by political wrangling and/or institutional disagreements. In the context of telecommunications, this has meant that the legislative framework has not always adjusted to developments taking place in the market as easily and quickly as it should have done and that regulators implementing the scheme have not always

[4] For the view that the control of price squeezes is essentially a regulatory matter which makes best sense as an aspect of price regulation in an industry already subject to duties to deal and to control by institutionally competent regulators, see eg JG Sidak, 'Abolishing the Price Squeeze as a Theory of Antitrust Liability' (2008) 4(2) J Competition L and Economics 279.

been able to act nimbly to adjust to market changes.[5] In addition, where regulators are empowered to examine pricing issues *ex ante*, they may not be able to predict precisely how events will pan out in the future. It may therefore be of utmost importance that competition authorities (or regulators[6]) retain the ability to control conduct *ex post* on the market. Another difficulty in a regulated sector may result from the closeness of the regulator's relationship with the players on the market or with the government (which in some EU Member States have retained a financial stake in the fixed incumbent telecommunication operators[7]). The possibility that regulators might be 'captured' impacts on how objective and effective they can remain.

In order to discuss these issues, this chapter is divided into four further parts. Section A analyses the concept of a price squeeze more closely and briefly outlines the regulatory context of cases that have arisen in the telecommunications sector. Section B looks at US and EU cases which support the existence of an independent price squeeze offence whilst section C goes on to examine how the concept of a price squeeze has been rethought in more recent years and how the regulatory context of the more recent cases has affected the assessment. Section C also considers whether either regulatory or traditional competition law remedies can adequately deal with a bottleneck problem, or whether a strategy of 'separation' might be a feasible and preferable means of dealing with the problem presented. In the last section, some conclusions are drawn.

A. The Nature of a Price Squeeze Offence

Price squeezes, predatory pricing, and refusals to deal

The question of whether, and if so when, price squeezes may violate the competition rules governing the unilateral conduct of dominant firms, has caused considerable controversy in both the US and the EU. It has already been seen that a price squeeze occurs when a vertically integrated firm, dominant in an upstream market, either raises the upstream price of inputs and/or cuts the retail price of the finished good, so squeezing the profit margins of rivals on the downstream market and making it impossible for an equally, or reasonably, efficient competitor to compete—for example, where a dominant supplier of a raw material sells to competitors downstream at prices which do not enable the competitor to compete with the dominant firm's

[5] See eg, Francesco Rizzuto 'Reforming the "Constitutional Fundamentals" of the EU Telecommunications Regulatory Framework' [2010] Computer and Telecommunications L Rev 44.

[6] Sometimes regulators have the power to apply both regulatory and competition law rules. This is the case in the UK, for example.

[7] For example, the German State holds 30.92 per cent of shares directly, and 12.13 per cent indirectly, in the capital of Deutsche Telekom AG, the incumbent telecommunications operator in Germany, see Case T-271/03 *Deutsche Telekom v Commission* [2008] ECR II-477 para 1.

prices in the market for a finished product (made from the raw material)[8] or where a firm which has control of a network or infrastructure necessary to provide retail services, charges a wholesale price so close to the retail price that it makes it impossible for rivals to operate and compete on the downstream retail markets in the long-run.

It can be seen from this description that price squeezes have similarities to both predatory pricing and refusal to deal claims. Predatory pricing occurs when a firm with market power (or hope of acquiring market power) prices below cost to eliminate or discipline a rival. Although the lower prices are good in the short-run, consumers are harmed in the long-run if rivals, even equally efficient ones,[9] cannot remain in the market so that the dominant firm is, after the predatory siege, able to raise prices above competitive levels. Like predatory pricing, a margin squeeze case involves an allegation that a dominant firm is pricing at a level on the downstream market which an efficient competitor cannot match.

There are some core differences with traditional predatory pricing cases, however. First, in a margin squeeze case, the dominant firm may not need to price below cost to be able to exclude an equally efficient competitor on the downstream market (so there may be no short-run benefit to consumers).[10] Take, for example, the following hypothetical example in which broadband is made available using new equipment plus associated services over a traditional network. Suppose a sole firm provides the traditional network but that a number of firms (including the monopolist) are engaged in providing the equipment and services enabling broadband. Assume the cost of providing the network is €12 per month. Assume the cost of the broadband equipment plus other services is €8 per month. Assume also that the dominant firm sells the broadband service (involving both traditional network and additional broadband costs) at €25 per month but sells the network input to other rival broadband suppliers at €18 per month. Clearly this pricing structure will 'squeeze' a rival broadband provider with the same broadband costs as the dominant firm (an equally efficient competitor). The dominant firm is, however, making a €5 per month profit on its broadband service. Its pricing can only be considered to be 'below' cost, if the downstream arm of the vertically integrated firm is hypothetically *assumed* to pay the same cost for the network input as its rivals (ie it is also paying the 50 per cent mark up on network costs from €12 to €18 per month).

---

[8] See discussion of *United States v Aluminum Co of America (Alcoa)* 148 F2d 416 (2d Cir 1945); and *Napier Brown-British Sugar* [1998] OJ L284/41 below, accompanying text to nn 37–53.

[9] Although exclusion of a less efficient competitor may harm consumer welfare in certain circumstances, it may '. . . be beyond the practical ability of a judicial tribunal to control without courting intolerable risks of chilling legitimate price cutting', *Brooke Group Ltd v Brown & Williamson Tobacco Co* 509 US 209, 223 (1993). See also DW Carlton, 'Should "Price Squeeze" be a Recognized form of Anticompetitive Conduct?' (2008) 4(2) J Competition L and Economics 271, 273–4.

[10] See Case COMP/38.784 *Wanadoo España v Telefónica* (Commission Decision of 4 July 2007) para 558 ('*Telefónica*'), Case T-337/07 (judgment pending).

This leads on to a further important difference between predatory pricing and margin squeeze. In most predatory pricing cases the below cost pricing occurs in the market in which the predator is dominant and is designed to protect and preserve its market power in that market.[11] In contrast, in margin squeeze cases, the allegation is that the pricing practice is designed to exclude competition in a downstream market. A margin squeeze, like a refusal to deal, involves a form of 'leveraging', in which a dominant firm is able to use its market power in a dominated market to affect competition or gain a competitive advantage in a neighbouring or downstream market. Although a price squeeze is, of course, distinct from the situation where the dominant firm simply refuses outright to supply an essential input to a downstream competitor, forcing them to exit the downstream market, the squeeze will eventually operate in the same way as a refusal to deal as, in the long-run, rivals will not be able to remain on the downstream market and will have to exit from it. A growing view is, therefore, that as the conduct operates 'effectively' as a refusal to deal it should be analysed as such—the question of whether the pricing is squeezing rivals simply constitutes one step in the process which requires proof *both* that the margin does not allow an equally efficient competitor to operate profitably at the downstream level *and* that the 'constructive' refusal to deal is unlawful.[12]

The regulatory background in telecommunications cases

Another difference is that a number of the more recent margin squeeze cases have arisen in the regulated telecommunications sector.[13] An issue which has had to be decided is whether, in this context, there is room for the application of the competition rules or whether the regulatory rules 'cover the field'. Although a close examination of the EU and US regulatory rules is outside the scope of this chapter,[14] this section provides an extremely brief outline of the regulatory regimes and their core differences in order to provide the backdrop and context to the competition law cases that have arisen.

The important features of the EU communications regulatory framework (introduced in 2002) for the purposes of the EU cases in this area have been the

[11] But see eg, Case C-333/94 P *Tetra Pak International SA v Commission (Tetra Pak II)* [1996] ECR I-5951.

[12] See below section C.

[13] But see eg Case COMP/38.233 *Wanadoo Interactive* (Commission Decision dated 16 July 2003) (Commission finding that Wanadoo had engaged in predatory pricing on the French market for high-speed internet access), upheld by the General Court and the ECJ in Case T-340/03 *France Télécom v Commission* [2007] ECR II-107 and Case C-202/07 P *France Télécom v Commission* [2009] ECR I-2369 respectively.

[14] But see eg, I Walden, *Telecommunications Law and Regulation* (3rd edn, Oxford: OUP, 2009), P Roth and V Rose (eds) *Bellamy & Child, EC Competition Law*, (6th edn, Oxford: OUP, 2008) ch 12 and ABA, *Antitrust Law Developments* (6th edn, Chicago: ABA, 2007) ch 15, s B.2.

following—first, it comprises of a package of Directives,[15] which includes an Access Directive on access to, and interconnection of, electronic communications networks and associated facilities.[16] Second the regime is administered at the national level by *independent*[17] National Regulatory Authorities ('NRA')[18] which have extensive obligations to promote and introduce competition into the sector,[19] to identify operators on markets with significant market power and to impose appropriate regulatory obligations upon those operators,[20] such as an obligation to meet reasonable requests for access to, and use of, specific network elements and associated facilities (including unbundled access to the local loop).[21] Importantly, as the NRAs work in a 'forward looking' context and are limited (under the regulatory framework[22]) to remedying *ex ante* market failures, more frequently at the wholesale level,[23] some NRAs have proved to be powerless to control price squeezes.[24] Third the current framework will be replaced by a new telecoms reform package in 2011 which is designed to address technological and market developments, to achieve a single market for EU telecoms operators,[25] to improve consumer rights (recognizing consumers' fundamental right to internet access), to increase security on communications networks, to accelerate broadband access for EU citizens, to encourage competition in the sector by spurring investment in new infrastructure (next generation access networks) and to reinforce the independence of NRAs and eliminate political interference in their day-to-day duties. Importantly, the package will increase the

[15] See especially Directive 2002/21 on a common regulatory framework for electronic communications networks and services [2002] OJ L108/33, Directive 2002/20 on the authorization of electronic communications networks and services [2002] OJ L108/21, Directive 2002/22 on universal service and users' rights relating to electronic communications networks and services OJ 2002 L108/51, Directive 2002/58 concerning the processing of personal data and the protection of privacy in the electronic communications sector OJ 2002 L201/37.

[16] Directive 2002/19 [2002] OJ L108/7.

[17] Independent of entities offering goods and/or services in the telecommunications sector *and* of *government* that might have an interest in the performance of a part-owned or recently privatized entity.

[18] Telecommunications Equipment Directive, art 6.

[19] Framework Directive, art 8. They also have to act impartially and transparently, Telecommunications Equipment Directive art 3.

[20] In certain circumstances they may also impose obligations on undertakings without significant market power, eg where necessary to ensure access and interconnection, see Access Directive, art 5.

[21] See Access Directive, arts 8–13.

[22] Some NRAs may, under national rules, also have power to apply competition laws, see n 6 above.

[23] See eg, P Alexiadis and A Shortall, 'Diverging but Increasingly Converging: The US Supreme Court in *linkLine*—A European Perspective' (April 2009) Competition Policy International. See also n 6 above.

[24] In contrast, the Commission has, when applying Article 102, been able to rely on *ex post* conduct data, *ibid* and see *Telefónica* (n 10 above).

[25] See generally, eg, Press release dated 20 November 2009, 'EU Telecoms Reform: 12 reforms to pave way for stronger consumer rights, an open internet, a single European telecoms market and high-speed internet connections for all citizens' MEMO/09/513.

Commission's role in the regulation of dominant telecommunications operators[26] and will give NRAs power, as a last-resort remedy, to oblige telecoms operators to separate communications networks from their service business (functional separation).

In the US, competition was not introduced into local telecommunications markets (each controlled by regional Bell operating companies ('RBOCs')[27]) until the enactment of the Telecommunications Act of 1996, which sets out a pro-competitive, deregulatory framework for telecommunications markets based on a goal of the promotion of competition.[28] Important features of the US regime are that, in order to introduce competition into local markets and to break monopolies in the local access for narrowband services, section 251 of the Act imposes a general duty on telecommunications carriers to interconnect[29] and an obligation on incumbent local exchange carriers ('ILECs') to provide non-discriminatory access to network elements on an unbundled basis to requesting telecommunication carriers on just, reasonable and non-discriminatory terms.[30] The Act specifies retail and wholesale rates for interconnection and access to network elements;[31] and the access provisions are implemented by the Federal Communications Commission (the 'FCC')[32] which, for a number of years required ILECs to provide access to network elements (such as local loops) to allow the development of competition in narrowband and broadband services.[33]

It can be seen from this short description that both the EU and US regulatory systems seek to deal with the problem of access to network elements essential to develop downstream markets. A critical difference, however, is that the US regime derives, like the antitrust laws themselves, from federal legislation adopted by Congress which is principally administered by a federal regulatory agency, the FCC. The policy underpinning the Telecommunications Act was thus developed by Congress and the courts have declined to let that independent policy influence the interpretation of the antitrust laws. Even though, therefore, the Act contains an 'express antitrust saving clauses' (providing that the Act does *not* exclude the application of the antitrust

[26] In particular, it will allow the Commission to oversee NRAs' regulatory remedies and, ultimately, to veto them where a draft remedy would create a barrier to the single market. It will also be able to adopt harmonization measures to prevent divergences in regulatory approaches.

[27] In 1982 AT&T agreed, in a consent decree concluded following protracted antitrust proceedings against it, to divest itself of its RBOCs—structural separation to prevent it operating as a vertically integrated player, see also below accompanying text to nn 133–143.

[28] See ABA *Antitrust Law Developments* (n 14 above) 1321.

[29] Telecommunications Act 1996, s 251(a).

[30] Telecommunications Act 1996 s 251(c).

[31] Telecommunications Act 1996 ss 252(2)(d)(A)(B) and 251(c)(4)(d)(3).

[32] Telecommunications Act 1996 s 251(d)(2). The FCC was created by the Communications Act of 1934.

[33] But see n 111 below.

laws),[34] the US courts have been reluctant to impose antitrust liability in circumstances where a specialist regime has been devised to deal specifically with the conduct under scrutiny and has made it clear that the Telecommunications Act does not create claims that go beyond existing antitrust standards. In the EU, in contrast, it is the Commission, the body that enforces the competition laws, that has instigated and driven the process of the liberalization of the telecommunications markets and which puts the necessary legislative proposals for the telecoms packages before the Council and/or European Parliament. Despite the fact that the EU regulatory regime depends for implementation upon NRAs (operating under national laws derived from and compatible with EU telecoms rules), the Commission has, therefore, taken, and continues to take, a keen interest in the development of competition in telecommunications markets. Indeed, it has sought to ensure that network operators and service providers in these markets are regulated sufficiently both by ensuring that NRAs correctly apply the regulatory framework and also by applying the competition laws to anti-competitive agreements and conduct which nonetheless occurs. Further, the new telecoms reform package will, when in force, increase the Commission's powers, allowing it to become more closely involved in regulatory remedies chosen by NRAs (which will include the possibility of an order of functional separation).

B. A Standalone Price Squeeze Offence? *Alcoa* and *Deutsche Telekom*

Introduction

A critical starting point in margin squeeze cases has been to determine what legal test should be used to determine whether they violate the competition laws and, in particular, whether the conduct is '. . . subject to condemnation under the antitrust laws because it leads to some harm that is not actionable under other antitrust causes of action and, if so, under what circumstances?'[35] The sections below establish that, in the EU, case law supports the view that a 'margin squeeze' in itself contravenes Article 102 TFEU. Further, that prior to the Supreme Court's opinion in *Pacific Bell Telephone Co v linkLine Communications, Inc*,[36] case law in the US accepted the existence of a distinct price squeeze offence. In section C of this chapter, however, it is seen that the debate about the proper role and scope of both Article 102 TFEU and section 2 of the Sherman Act has led to a call for a less formalistic approach to the analysis of margin squeezes (as well as other single-firm conduct) and rather for one which focuses more closely on its effects. Further, it seems that the development of

[34] See s 60(1)(b). In *Trinko* the Supreme Court held that the Act also does not provide implied immunity from the antitrust laws, see n 106 below.

[35] Carlton (n 9 above) 271.

[36] 129 S Ct 1109 (2009).

the law has been affected by the regulatory context of more recent cases. These factors have led to the analyses adopted in both jurisdictions being refined.

The *Alcoa* case

Until *linkLine* the most important US judgment on price squeeze was Chief Judge Learned Hand's 1945 opinion in *United States v Aluminum Co of America (Alcoa)*.[37] This case, which dealt with Alcoa's conduct on the market for aluminium ingot, was an extremely influential judgment on monopolization generally, which shaped much of the developing law. Building on this ruling, subsequent cases have clarified that the offence of monopolization requires both the possession of monopoly power in the relevant market and the wilful acquisition or maintenance of that power as distinguished from growth or development as a consequence of a superior product, business acumen, or historical accident.[38]

One monopolization allegation which had to be considered in *Alcoa* was a claim that Alcoa had used its monopoly position in the supply of virgin aluminium ingot to put competing manufacturers of rolled aluminium sheet out of business (Alcoa being the largest maker of aluminium sheet). The plaintiffs theory was that '"Alcoa" consistently sold ingot at so high a price that the "sheet rollers", who were forced to buy from it, could not pay the expenses of "rolling" the "sheet" and make a living profit out of the price at which "Alcoa" itself sold "sheet".'[39] Although the trial judge found that Alcoa had not attempted to monopolize the sheet market, Judge Hand found that the figures established that in some instances the price of ingot together with Alcoa's cost of rolling was greater than the price at which Alcoa was selling sheet and in other cases the sheet price only barely covered the cost. Further, he concluded that, once Alcoa had been put on notice that there was no adequate spread for sheet rollers, it was unquestionably unlawful to set the price of sheet so low and at the same time the price of ingot higher than a fair price. The margin between the price of the ingot and its own resale price for the rolled sheets had to be sufficient to allow it to recover its costs incurred in the rolling process.[40] *Alcoa*, thus come to stand for the proposition that,

> a price squeeze violates Sherman Act Sec. 2 . . . when (1) the firm conducting the squeeze has monopoly power at the first industry level, (2) its price at this level is

---

[37] *United States v Aluminum Co of America (Alcoa)* 148 F2d 416 (2d Cir 1945). In this case the Supreme Court lacked, because of recusals, the necessary quorum to hear the case and so referred the appeal to the Court of Appeals for the Second Circuit where Learned Hand was the Chief Judge. Many of the key passages of the judgment were subsequently endorsed by the Supreme Court in *American Tobacco Co v United States* 328 US 781 (1946) (but see the discussion of the Supreme Court decision in *linkLine*, text accompanying (n 118 below)).

[38] *United States v Grinnell Corp* 384 US 563, 571–2 (1966).

[39] *Alcoa* (n 37 above) 437.

[40] 'In other words Judge Hand applied a somewhat primitive version of an "equally efficient rival" test' HJ Hovenkamp and EN Hovenkamp, 'The Viability of Antitrust Price Squeeze Claims' University of Iowa Legal Studies Research Paper, Number 08-33 (2009) <http://ssrn.com/abstract=1156974> accessed 15 September 2010.

'higher than a "fair price,"' and (3) its price at the second level is so low that its competitors cannot match the price and still make a 'living profit'.[41]

Napier Brown and *Deutsche Telekom*

The European Commission also developed a similar theory of when a margin squeeze may constitute an 'abuse' of a dominant position. In *Napier Brown-British Sugar*,[42] for example, the Commission found that a dominant supplier of raw and granulated sugar had sold its sugar at prices which left 'insufficient margin for a packager and seller of retail sugar, as efficient as [British Sugar] itself in its packaging and selling operations, to survive in the long-term.'[43] In *Industrie des poudres sphériques SA v Commission*[44] the General Court confirmed that although a price squeeze might occur in one of two ways—by low pricing of its own products in the downstream market,[45] or by charging high upstream prices to competitors[46]—a separate abuse did exist, based on the margin between the upstream and downstream prices.

In *Deutsche Telekom*,[47] the Commission for the first time used a price squeeze theory to condemn the behaviour of a fixed incumbent telecommunications operator which held a dominant position on the market for access to its fixed network.[48] In this case the Commission found that Deutsche Telekom ('DT') had engaged in an illegal margin squeeze by pricing wholesale access to its local loop (necessary for rivals to provide a variety of services to end users) higher than its retail access prices and rejected DT's view that there could not be abusive pricing unless below-cost selling or predation was established.

On appeal, both the General Court and the ECJ[49] affirmed the Commission's decision and its finding that an abusive margin squeeze exists,

> . . . if the difference between the retail prices charged by a dominant undertaking and the wholesale access prices it charges its competitors for comparable service is negative, or insufficient to cover the product-specific costs to the dominant operator of providing its own retail services on the downstream market.[50]

The General Court confirmed that the abusive nature of the dominant firm's pricing conduct should be determined '. . .on the basis of its own charges and costs, rather

---

41 *Town of Concord v Boston Edison Co* 955 F2d 17, 18 (1 Cir 1990).

42 [1988] OJ L284/41. See also *National Carbonising* [1976] OJ L35/6.

43 *Napier Brown-British Sugar* (n 42 above) para 65.

44 Case T-5/97 [2000] ECR II-3755 para 178.

45 See discussion of the *Deutsche Telekom* case below.

46 See eg, *Telefónica* (n 10 above).

47 [2003] OJ L263/9.

48 But see now also *Telefónica* (n 10 above). The Commission is also investigating possible abusive behaviour (including margin squeezes) by the Polish and Slovak telecommunications operators.

49 Case T-271/03 *Deutsche Telekom v Commission* [2008] ECR II-477 para 193 and Case C-280/08P *Deutsche Telekom v Commission* 14 October 2010.

50 *Deutsche Telekom AG* (2003) OJ L263/9 para 107.

than on the basis of the situation of actual or potential competitors'[51] and that the abuse resulted from the 'spread' between the wholesale and retail prices and did *not* require proof that the retail prices were predatory.[52] It also affirmed the methodology that the Commission had adopted to find the margin squeeze and, importantly, rejected the applicant's argument that the Commission had not demonstrated that the margin squeeze had an *effect* on the market. Rather, the General Court considered it implicit, that, as the applicant's wholesale services were indispensable to enabling a competitor to enter competition with it downstream, a margin squeeze would in principle hinder the growth of competition in the downstream markets as '. . . a potential competitor which is just as efficient as the applicant would not be able to enter the retail access services market without suffering loss.'[53] The ECJ dismissed the appeal from the General Court's judgment in its entirety.

C. Rethinking the Price Squeeze Offence

Background: the general need for a more effects-based approach under section 2 and Article 102?

Problems with the Alcoa and Deutsche Telekom tests

The judgment in *Alcoa* was delivered at the start of an era of aggressive enforcement of section 2 in the US[54] and a period of 'structural consensus',[55] during which relatively broad limits were imposed on the conduct of dominant firms.[56] Similarly, it is well known that the EU courts'[57] judgments, and older Commission decisions, support a broad interpretation of Article 102, which has resulted from a teleological

[51] Case T-271/03 *Deutsche Telekom* (n 49 above) para 188, relying on Case C-62/82 *AKZO v Commission* [1991] ECR I-3359. 'It must be added that any other approach could be contrary to the general principle of legal certainty' para 192, aff'd Case C-280/08P (n 49 above), especially paras 167–197.

[52] Case T-271/03 *Deutsche Telekom* (n 49 above) para 167.

[53] Case T-271/03 *Deutsche Telekom* (n 49 above) para 237. The court also assumed harm from the fact that, despite liberalization, DT's competitors in the retail market had acquired only small market shares in the retail access market, aff'd Case C-280/08P (n 49 above), paras 250–260.

[54] At this time there was a view that the provision should be used expansively against monopolies and to tackle and break up concentrated industries, see eg E Rostow, 'The New Sherman Act: A Positive Instrument of Progress' (1947) 14 U Chicago L Rev 567 and E Levi, 'The Antitrust Laws and Monopoly' (1947) 14 U Chicago L Rev 153, C Kaysen and D Turner *Antitrust Policy* (Cambridge: Harvard University Press, 1959).

[55] The 'structuralists' placed emphasis on market structure, believing that it strongly affected conduct and performance, see eg, Kaysen and Turner (n 54 above), J Bain, *Barriers to New Competition: Their Character and Consequences in Manufacturing Industries* (Cambridge: Harvard University Press, 1956) and E Mason, *Economic Concentration and the Monopoly Problem* (Cambridge: Harvard University Press, 1957).

[56] See eg *United States v Griffith* 334 US 100, 107 (1948).

[57] The courts of the EU comprise the Court of Justice (the ECJ), the General Court (formerly the Court of First Instance) and specialized courts. 'It shall ensure that in the interpretation and application of the Treaties the law is observed.' Treaty on European Union, art 19

construction of the provision in the light of the Treaty's aims and objectives, particularly the goal of single market integration and open market access. The jurisprudence thus establishes that the conduct of a dominant firm may violate Article 102 not only where it has anti-competitive effects to the prejudice of consumers, but also where it manipulates or prevents the flow of goods between the Member States or precludes access to market and sourcing of products from other suppliers.[58] A core criticism of the approach, which is apparent in the judgments of *Alcoa* and *Deutsche Telekom*, is that it is too formalistic. It does not allow for a consideration of the efficiencies which might flow from the conduct and does not require an assessment of whether, and if so how, the conduct harms competition and consumers—ie it does not require the demonstration of a coherent theory of harm. In consequence, concern has been expressed that the rules set out in unilateral conduct cases, and specifically in these price squeeze cases, are liable to be over-inclusive and, consequently, liable to deter pro-competitive behaviour which generates efficiencies. This has led to a call on both sides of the Atlantic for a more refined and principled approach, based on a consumer welfare objective, to be adopted toward single firm conduct and price squeezes.

Call for a more effects-based approach

In the US, the change in thinking commenced much sooner. By the end of the 1970s, criticism of the structuralist assumptions of the old Harvard School was being amassed, principally by the Chicago School. The arguments put forward led to a growing belief that big firms drive competition and innovation, the dream of monopoly and the ability to earn monopoly profits provides the incentive for competition and attracts good businesses into the market, even dominant firms should be encouraged to compete vigorously and that over-enforcement of section 2 was deterring pro-competitive behaviour of dominant firms. As a consensus emerged that the goal of the Sherman Act is consumer welfare, courts became mindful of the need to achieve an efficient outcome, to preserve incentives to compete, and to enunciate clear rules which allow firms to plan their behaviour in a way which complies with the law.[59] It has thus been clarified that to be condemned as exclusionary, '. . . a monopolist's act must have an "anticompetitive effect." That is, it must harm the competitive process and thereby harm consumers. In contrast, harm to one or more competitors will not suffice.'[60] This acceptance has led the Supreme Court to reconsider and dramatically redraw the boundaries of liability under section 2.

[58] For the influence of the 'ordoliberal' school, see eg, D Gerber 'Constitutionalizing the Economy: German Neo-liberalism, Competition Law and the "New Europe"' (1994) 42 American J Comparative L 25. See also eg Case 6/72 *Europemballage Corporation and Continental Can Co Inc v Commission* [1973] ECR 215 para 26, Case C-95/04P *British Airways plc v Commission* [2007] ECR I-2331 and Joined Cases C-501, 513, 515 and 519/06 *GlaxoSmithKline Services Unlimited v Commission* (Judgment of the ECJ dated 6 October 2009) paras 62–3.

[59] *Reiter v Sonotone Corp*, 442 US 330, 343 (1979). But see eg, H Hovenkamp, 'The Harvard and Chicago Schools and the Dominant Firm' University of Iowa Legal Studies Research Paper, Number 07-19 (2007) <http//ssrn.com/abstract=1014153> accessed 15 September 2010.

[60] *United States v Microsoft Corp* 253 F3d 34, 58 (DC Cir 2001).

In the EU, the jurisprudence does not reflect such a dramatic evolution as that seen in the US. Rather, the EU courts have consistently attached (and continue to attach) importance to the need to preserve the process of competition and to protect the structure of the market and competition as such.[61] Conduct of a dominant firm that distorts competition and inhibits the fragile competition that already exists on a market as a result of its dominance may therefore constitute an abuse[62] even where anti-competitive effects in an economic sense have not been proved. No concrete prejudice to consumers need be demonstrated—it is sufficient that the conduct tends to restrict competition (by making it more difficult for competitors to enter the relevant market) or is capable of having, or likely to have, such an effect.[63] In spite of the case law, however, the Commission has listened to criticism and to calls for it to abandon its formalistic methodology and, instead, to evaluate the effect of single-firm conduct on consumer welfare. Indeed, following its review of the working and application of Article 101 TFEU and the European Merger Regulation, the Commission opened a debate on the reform of Article 102 TFEU[64] and subsequently published Guidance on the Commission's Enforcement Priorities in Applying Article [102 TFEU] to Abusive Exclusionary Conduct by Dominant Undertakings (the 'Guidance Paper')[65] indicating that its priority in Article 102 cases is to identify conduct of a firm which impairs effective competition by '. . . foreclosing their competitors in an anticompetitive way, thus having an adverse impact on consumer welfare, whether in the form of higher price levels than would have otherwise prevailed or in some other form such as limiting quality or reducing consumer choice.'[66]

Distinguishing exclusionary behaviour from legitimate competition on the merits

Even if a consumer welfare goal is accepted, it nonetheless remains extremely difficult to identify unlawful 'exclusionary' conduct and to distinguish it from competitive conduct which, frequently, appears similar[67]—how is an unlawful margin squeeze to be distinguished from legitimate low cost pricing in the retail market or conduct designed to protect innovation and investment upstream?

---

[61] See discussion in text accompanying n 58 above.

[62] See Case C-85/76 *Hoffmann-La Roche v Commission* [1979] ECR 461 especially paras 9 and 91.

[63] See eg, Case C-95/04P *British Airways plc v Commission* [2007] ECR I-2331 paras 106–7 and Case T-203/1 *Michelin v Commission* [2003] ECR II-4071 paras 241–4.

[64] See Report by the Commission's economic advisory group, 'An economic approach to Article 82'(July 2005) 8–9 and DG Comp's Staff Discussion Paper on the Application of Article [102 TFEU] to Exclusionary Abuses (December 2005).

[65] Guidance on the Commission's Enforcement Priorities in Applying Article [102 TFEU] to Abusive Exclusionary Conduct by Dominant Undertakings C(2009) 864 final, [2009] OJ C45/7 (the 'Guidance Paper').

[66] Guidance Paper (n 65 above) para 19.

[67] F Easterbrook, 'When Is It Worthwhile to Use Courts to Search for Exclusionary Conduct?' (2003) Columbia Business L Rev 345, 346.

In both the US and EU a public debate and formal discourse about this matter has been conducted—in the US within the forums of the Antitrust Modernization Commission and workshops on unilateral conduct held jointly by the DOJ and FTC[68] and, in the EU, following the publication of the Staff Discussion Paper. This debate has revealed passionately different views on the question of how to put in place accurate,[69] coherent, predictable, and workable mechanisms for determining when conduct will lead to the exclusion of a competitor in a way which will adversely affect consumer welfare. Although it seems to be accepted that it is not possible to utilize a 'single' test to identify *all* unlawful exclusionary behaviour[70] and that conduct-specific tests should be put in place to deal with particular types of conduct, an important point of disagreement is whether it is preferable to adopt tests which require a more exhaustive analysis of the market and the conduct at issue in each case (a rule of reason style analysis) or whether simpler bright-line standards should be applied. A more comprehensive rule of reason style analysis allows courts to take account of all relevant factual, economic, theoretical, and empirical evidence but is complex, time-consuming, and extremely costly to carry out. Further, because of the complexity, the analysis does, in fact, risk error as competition agencies, judges, and juries struggle to apply the myriad of potentially relevant factors. In contrast, bright-line standards are less costly to apply but may, being more rigid, be more likely to result in error. Further, exactly how the rule is constructed will plainly determine whether a greater number of Type I errors (or false positives) or Type II errors (or false negatives) is more likely.[71] The choice between a rule which may be more likely to under-deter or one which may over-deter may depend on an analysis of the particular type of exclusionary conduct at issue, and an assessment of how frequently the relevant conduct is likely to be harmful or beneficial. Where, for example, the chance of

---

[68] The joint hearings culminated, in 2008, in the DOJ publishing a report on single-firm conduct: *Competition and Monopoly: Single-Firm Conduct Under Section 2 of the Sherman Act (2008)*, (the 'Single-Firm Conduct Report') which set out an analytical framework for assessing the legality of single-firm conduct. The report drew severe criticism in particular for downplaying the risk of under-enforcement, see eg T Rosch, 'Thoughts on the Withdrawal of the DOJ Section 2 Report' speech of 25 June 2009, <http://www.ftc.gov/speeches/rosch/090625roschibareport.pdf> accessed 15 September 2010. The Obama administration withdrew the Report on 11 May 2009, see <http://www.justice.gov/atr/public/press_releases/2009/245710.htm> accessed 15 September 2010.

[69] Any rule against exclusionary behaviour should aim to minimize both Type I and Type II errors so it does not either over or under-deter, but see text accompanying n 71 below.

[70] The Commission states in its Guidance Paper (n 65 above) para 23, that in relation to price-related exclusionary conduct, it will normally only intervene to prevent anti-competitive behaviour where the conduct concerned has already been, or is capable of, hampering competition from competitors which are considered to be 'as efficient' as the dominant undertaking.

[71] Those taking the view that exclusionary conduct is rare, may be more concerned that Type I errors will chill aggressive behaviour, keep prices high, discourage innovation and encourage less efficient rivals to misuse section 2 when losing out in the competitive battle, see eg RH Bork, *The Antitrust Paradox: A Policy at War with Itself* (New York: The Free Press, 1993). Others may be concerned that under-inclusive rules will allow dominant firms to place rivals at a disadvantage, raise their costs, so as to preserve their position of dominance, see eg TG Krattenmaker and SC Salop, 'Anticompetitive Exclusion: Raising Rivals' Costs to Achieve Power over Price' (1986) 96 Yale LJ 209.

the conduct harming consumers is very slight, it may be preferable to adopt a rule which under-deters so as not to subject a whole broad class of conduct to antitrust scrutiny on the grounds that it may sometimes be anti-competitive. An over-inclusive test in this situation may, in contrast, cause a substantial social loss by prohibiting or discouraging pro-competitive conduct.[72]

In the US, a number of cases, such as *Aspen Skiing Co v Aspen Highlands Skiing Corp*,[73] *Lorain Journal Co v United States*,[74] and *United States v Microsoft Corp*,[75] suggest that a more balanced analysis should be adopted in specific cases, tailored to the type of conduct but involving a weighing of perceived pro-competitive and anti-competitive aspects of the dominant firm's conduct.[76] In later cases, however, the Supreme Court has appeared to move away from this approach and to become more fearful of wrongly condemning legitimate competition and so chilling aggressive competition, the very conduct the antitrust laws are designed to protect.[77] It has thus avoided '... constructions of § 2 that might chill competition, rather than foster it'[78] and held that '... the cost of false positives counsels against an undue expansion of section 2 liability'.[79] This viewpoint has profoundly affected the development of the case law on predatory pricing, refusal to deal, and price squeezes.

In *Brooke Group Ltd v Brown & Williamson Tobacco Co*,[80] for example, the Supreme Court replaced the rather expansive and unclear basis for liability for predatory pricing set out in its previous opinion in *Utah Pie Co v Continental Baking Co*[81] (that predatory intent combined with unreasonably low pricing was sufficient to prove predatory pricing) with a clearer test which considerably raised the bar for those seeking to prove a predatory pricing violation. Believing that predatory pricing schemes are rarely tried and, even more rarely successful, the court stressed that antitrust rules should not be drawn so as to put in place obstacles to price cutting.[82] It thus held that two prerequisites to recovery had to be established by the plaintiff—(1) that 'the prices complained of are below an appropriate measure of rival's costs'; and (2) 'that the competitor had a reasonable prospect, or, under § 2 of the Sherman Act, a dangerous probability, of recouping its investment in below-cost prices ... "For the

[72] Carlton (n 9 above) 273.

[73] 472 US 585 (1985).

[74] 342 US 243 (1951).

[75] 253 F3d 34 (DC Cir) *cert denied* 534 US 952 (2001).

[76] See C Varney, 'Vigorous Antitrust Enforcement in this Challenging Era' Remarks Prepared for the United States Chamber of Commerce, 12 May 2009.

[77] *Matsushita Electric Industry Co v Zenith Radio* 475 US 574, 594 (1986).

[78] *Spectrum Sports, Inc v McQuillan*, 506 US 447, 458–9 (1993).

[79] *Verizon Communications v Law Offices of Curtis V Trinko LLP* 540 US 398, 414 (2004).

[80] 509 US 209 (1993) ('*Brooke Group*'). See also *Weyerhaeuser Co v Ross-Simmons Hardwood Lumber Co Inc*, (No 05-381) (2007).

[81] 386 US 685 (1967).

[82] *Brooke Group* (n 80 above) 224. It would be ironic if standards for predatory pricing liability were so low that antitrust suits themselves became a tool for keeping prices high.

investment to be rational, the [predator] must have a reasonable expectation of recovering, in the form of later monopoly profits, more than the losses suffered."'[83] In so ruling the court made it clear that the antitrust laws were not concerned about 'unfair' behaviour or losses imposed on a target of predatory pricing if competition was not injured[84] and that it favoured a rule which would be more likely to be under-inclusive, would not lead to false positives or make firms fearful of low cost pricing.

Similarly, in *Trinko*, the Supreme Court advocated caution in imposing section 2 liability, when determining whether a refusal to deal could constitute unlawful monopolization. The court stressed the right of traders to decide independently who they should deal with[85] and the tension that results from compelling firms to share an advantage with the antitrust laws—by lessening incentives for both the monopolist and rival to invest,[86] by requiring the courts to act as central planners, identifying the proper price, quantity, and other terms of dealing[87] and by potentially facilitating collusion. Although the court stated that the right to refuse to deal with other firms was *not* unqualified, it held that the court had been 'cautious' in recognizing exceptions, '... because of the uncertain virtue of forced sharing and the difficulty of identifying and remedying anticompetitive conduct by a single firm.'[88] It now seems that it is difficult for a plaintiff to establish liability in a refusal to deal case unless, for example, it can demonstrate that the dominant firm has willingly forsaken profits with the expectation of recouping them after competition has been eliminated.[89]

In the EU, the courts have not adopted such high bars to establishing liability under Article 102. With regard to predatory pricing, the ECJ in *AKZO*[90] held that prices below average variable costs must be regarded as abusive and that prices between average variable and average total cost may be abusive if part of a plan for eliminating a competitor. Subsequent cases have clarified that a claimant does not need to prove recoupment of losses as a precondition to a finding of predation and that a dominant firm does not have the right to meet competition of new rivals by aligning prices with

[83] *Brooke Group* (n 80 above) 224, relying on *Matsushita Electric Industry Co v Zenith Radio Corp* (n 77 above) 588–9.

[84] 'It is axiomatic that the antitrust laws were passed for "the protection of competition, not competitors."' *Brooke Group* (n 80 above) 224, relying on *Brown Shoe Co v United States*, 370 US 294, 320 (1962).

[85] An obligation to deal interferes with a firms' freedom of contract and right to dispose freely of its property, see *United States v Colgate & Co* 250 US 300, 307 (1919).

[86] Discouraging innovation both by (a) the dominant firm (if it thinks it will be forced to share its innovations) and (b) its competitors (by removing the incentive to develop their own inputs).

[87] A task to which they were ill-suited, *Trinko* (n 79 above) 408.

[88] *Trinko* (n 79 above) 408.

[89] Perhaps where cutting off a prior (profitable) course of dealing without apparent explanation supports an inference of profit sacrifice with anti-competitive motivation (see *Eastman Kodak Co v Image Technical Services Inc* 504 US 451 (1992), *Aspen Skiing Co v Aspen Highlands Co* 472 US 585 (1985), *Otter Tail Power Co v US* 410 US 366 (1973), *Lorain Journal Co v US* 342 US 143 (1951)).

[90] Case C-62/82 *AKZO Chemie BV v Commission* [1991] ECR I-3359.

them, if that would mean selling below cost.[91] The EU courts thus *assume* that prices are designed to exclude rivals and to allow the predator to raise prices and hurt consumers at the end of the predatory siege where (i) the dominant firm is not covering its variable costs or (ii) it is covering variable but not total costs and that pricing was clearly adopted as part of a predatory strategy. Actual proof that the predatory strategy is likely to be successful and, consequently, that consumers will be harmed is not required. Similarly, in the context of refusal to supply, the EU courts have held that a vertical integrated dominant firm should supply a rival downstream where,[92]

(1) access to its product, service or facility is 'indispensable' for the rival to compete (there are no actual or potential substitutes for it);
(2) the refusal is likely to eliminate all competition on the secondary market (even if not imminent);
(3) in the context of refusal to license an intellectual property right at least, the party seeking the licence wishes to offer a new product, not duplicating that offered by the dominant firm, for which there is a consumer demand; and
(4) there is no objective justification for the refusal.

Again the EU courts have been willing to assume that such conduct will harm the competitive structure and, possibly, consumers through the elimination of rivals on the downstream market.[93]

Town of Concord and the Commission's Guidance Paper

The policy which underpinned and influenced the US Supreme Court opinions in *Brooke Group* and *Trinko* evidently sat at odds with the price squeeze test set out in *Alcoa* and it became increasingly apparent that the test might not meet the more rigorous standards favoured by later Supreme Court opinions. The change in approach manifested led lower courts, even prior to *Trinko*, to express concern that the test for liability enunciated in *Alcoa* might lead to Type I errors and that '. . . a general theory of price squeeze to create antitrust liability is likely to chill competition and harm consumers'.[94]

[91] See eg, Case C-202/07 *France Télécom v Commission* (Judgment of the ECJ dated 2 April 2009) paras 37 and 45–58.

[92] See especially Joined Cases C-241and C-242/91P *RTE & ITP v Commission* [1995] ECR 1-743, Case 7/97 *Oscar Bronner v Mediaprint Zeitungs- und Zeitschriftenverlag GmbH & Co KG* [1998] ECR I-7791.

[93] See also n 53 above and accompanying text.

[94] Carlton (n 9 above) 278,

> . . . because the gains from deterring an anticompetitive price squeeze seem quite modest but the costs are likely to be high, a sensible weighing of the gains and losses from adopting a price squeeze theory could lead one to reject that theory.

In *Town of Concord v Boston Edison Co*,[95] for example, (then) Chief Judge Breyer in the First Circuit emphasized the importance in price squeeze cases of both comparing the likely anti-competitive effects of alleged price squeeze conduct (from, for example, protecting the position of dominance in the upstream market or removing the incentive for non-price competition in the downstream market) with its potentially legitimate business justifications and taking account of the fact that antitrust rules are court administered rules. The court was thus concerned that the *Alcoa* test might be over-inclusive since it did not require consideration of whether the conduct would cause harm through the creation or extension of monopoly power (which did '. . . not necessarily, nor in an obvious way, give a firm added power to raise prices'[96]) or whether efficiencies would result (for example, by allowing the primary level monopolist to realize efficiencies from carrying out its second-level activities and/or by allowing the monopolist to squeeze out a second-level firm which is itself a monopolist and eliminate double marginalization). Consequently, the doctrine might affect innovation and incentives to invest,[97] deter the dominant firm from engaging in pro-competitive low cost pricing at the downstream level, encourage it to withdraw completely from the downstream market or discourage upstream firms from dealing at all.[98] Further, Judge Breyer was concerned that Judge Hand's price squeeze test would not be sufficiently clear for businesses to comply with or administrable by courts who would not be able to determine a 'fair price' for the primary product without acting like a rate-setting regulatory authority or determining the proper size of the price 'gap'.[99] The Court thus concluded that before determining that a price squeeze was exclusionary, a court should be clear that its anti-competitive risks outweighed both the possible benefits and the adverse administrative considerations.

In the EU, the Commission has also recognized the need for it to adopt an 'economic and effects-based approach to exclusionary conduct'.[100] Indeed, in its Guidance Paper it has accepted the importance in margin squeeze cases of assessing not only the state of competition downstream, but the impact of a reduction of downstream competition on consumers and the impact of any competition law intervention on investment incentives upstream.[101]

95 915 F2d 17 (1st Cir 1990).

96 The 'one monopoly profit theory', *Town of Concord* (n 95 above) 23.

97 The focus on the downstream costs of production ignored the upstream level and investment incentives on it.

98 AM Panner, 'Are Price Squeezes Anticompetitive?' ABA Antitrust Law Section Spring Meeting 2009.

99 *Town of Concord* (n 95 above) 25.

100 See Article 82 Review <http://ec.europa.eu/competition/antitrust/art82/index.html> accessed 15 September 2010.

101 See further below text accompanying n 123 ff.

The 'new approach'—price squeeze cases in regulated markets

Town of Concord, Trinko, and linkLine

In *Town of Concord v Boston Edison Co*[102] the First Circuit, in addition to its concerns set out above about endorsing a standalone price squeeze offence, expressed uncertainty about the sense of developing a price squeeze offence which would require courts to get involved in direct price administration and to act like a rate setting agency. The complexity of such a task was reinforced in this case by the fact that a regulatory scheme (the case arose in the electric power sector[103]) had been created to perform exactly this job—Boston Edison's wholesale and retail prices were regulated by federal and state regulators respectively. The alleged price squeeze under consideration was not therefore simply an 'ordinary' price squeeze case.[104] Rather, the court considered that the existence of full price regulation (at wholesale and retail) dramatically altered the calculus of antitrust harms and benefits and led it to conclude that 'a price squeeze in a fully regulated industry such as electricity will not normally constitute "exclusionary conduct"'.[105] The court noted, however, that the result might be different where a monopolist, which was regulated only at one level, sought to dominate a second, unregulated level, in order to earn at that second level the very profits that regulation forbade at the first.

The Supreme Court's opinion in *Trinko* lent further support to courts reluctant to apply the antitrust laws to price squeeze claims occurring in a regulated sector. Not only did the court express general concern about the scope of section 2 but its conclusion was, in spite of a finding that the Telecommunications Act did not exclude the operation of the antitrust rules,[106] heavily influenced by the presence of the regulatory regime specifically designed to remedy the anti-competitive conduct alleged (the rules imposed both an obligation to deal and an obligation to interconnect at a cost-based rate of compensation[107]). In declining to add further exceptions to the general rule that there is no duty to aid competitors, the Supreme Court emphasized that the existence of this regulatory structure diminished the likelihood of major antitrust harm and indicated that the benefits to competition provided by antitrust would be small. The slight benefits of intervention were, therefore, outweighed by its

---

[102] *Town of Concord* (n 95 above).

[103] See eg ABA *Antitrust Law Developments* (n 14 above) 1359. See also *City of Anaheim v Southern California Edison Co* 955 F2d 1373 (9th Cir 1992) and *City of Mishawaka v American Electric Power Co* 616 F2d 976 (7th Cir 1980).

[104] The court, however, first discussed,

> the ordinary price squeeze case ... Only if the reader understands the major antitrust harms, benefits, and administrative considerations in that ordinary case, and sees how they are rather evenly balanced, will he understand why price regulation tips that balance so significantly towards legality. *Town of Concord* (n 95 above) 23.

[105] *Town of Concord* (n 95 above) 29.

[106] Telecommunications Act, s 60(1)(b).

[107] Telecommunications Act s 251(c)(3) and *Trinko* (n 79 above) 409.

costs and risks. In a number of subsequent cases, defendants charged with engaging in an unlawful price squeeze in the telecommunications sector argued therefore that, where they were compelled to deal under the federal communications laws, the plaintiffs' claims were foreclosed by *Trinko*. If they had no antitrust duty to deal with their competitors, how could they be compelled to deal with them on specified terms?[108]

linkLine involved price squeeze allegations by plaintiff Internet Service Providers (ISPs) against a defendant ILEC.[109] The defendant owned and operated the infrastructure and facilities needed to provide Digital Subscriber Line (DSL) service[110] (the 'last mile' lines that connect homes and businesses to the telephone network) and also participated in the DSL retail market. The defendant was required by the FCC to provide the wholesale service at a price no greater than its retail price for DSL service.[111] Chief Justice Roberts, giving the majority opinion, analysed the components of a price squeeze case: raising the wholesale price of inputs while cutting its own retail prices. With regard to the wholesale prices, the court concluded simply that '... [a] straightforward application of our recent decision in *Trinko* forecloses any challenge to AT&T's wholesale prices.'[112] In *Trinko*, a defendant required by statute to lease its network to competing firms at wholesale rates did not violate section 2 when it failed to provide a customer with sufficient service.

> Given that Verizon had no antitrust duty to deal with its rivals at all, we concluded that 'Verizon's alleged insufficient assistance in the provision of services to rivals' did not violate the Sherman Act ... *Trinko* thus make clear that if a firm has no antitrust duty to deal with its competitors at wholesale, it certainly has no duty to deal under terms and conditions that the rivals find commercially advantageous.[113]

---

[108] This defence met with differing success in the circuits, contrast *Covad Communications Co v Bell Atlantic Corp* 398 F3d 666, 673 (DC Cir 2005) with *Covad Communications Co v BellSouth Corp* 374 F3d 1044 (11th Cir 2004) and *Pacific Bell Telephone Co v linkLine Communications, Inc* 503 F3d 876 (2007).

[109] The District Court denied a motion to dismiss the price squeeze claim and the Court of Appeals affirmed, concluding that as regulation existed only at the upstream level antitrust law could deal with the interplay between wholesale and retail prices, *linkLine* (n 108 above) 883.

[110] A method of connecting to the Internet at high speeds over telephone lines.

[111] Until recently, the [FCC] required incumbent phone companies such as AT&T to sell transmission service to independent DSL providers, under the theory that this would spur competition. See *In re Appropriate Framework for Broadband Access to Internet over Wireline Facilities*... In 2005, the Commission largely abandoned this forced-sharing requirement in light of the emergence of a competitive market beyond DSL for high-speed Internet service; DSL now faces robust competition from cable companies and wireless and satellite services.

Pacific Bell Telephone Co v linkLine Communications, Inc. 129 S Ct 1109 (2009). AT&T, in this case, however, remained bound, as a condition of a merger, to provide wholesale DSL transport service to independent firms at a price no greater than the retail price of its own DSL service, *In re AT&T Inc and Bell South Corp* 22 FCC Rcd 5662, 5714 (2007).

[112] 129 S Ct 1109, 1119 (2009).

[113] 129 S Ct 1109, 1119.

As there was also no antitrust duty to deal in *linkLine* (sharing was compelled by FCC Regulations),

> . . . [t]he nub of the complaint in both . . . is identical—the plaintiffs alleged that the defendants (upstream monopolists) abused their power in the wholesale market to prevent rival firms from competing effectively in the retail market. *Trinko* holds that such claims are not cognizable under the Sherman Act in the absence of an antitrust duty to deal.[114]

The Supreme Court also held that the second component of a price squeeze claim, excessively low retail prices, required the plaintiff to establish that the prices were below cost and satisfied the other predatory pricing conditions set out in *Brooke Group*. The Supreme Court cited numerous statements of the court stressing the benefit that ensued to competition from price cuts and low pricing and the consequent need carefully to circumscribe the circumstances in which a plaintiff can state a Sherman Act violation by alleging that prices are too low (so avoiding chilling aggressive price competition)—

> Recognizing a price-squeeze claim where the defendant's retail price remains above cost would invite the precise harm we sought to avoid in *Brooke Group*; firms might raise their retail prices or refrain from aggressive price competition to avoid potential antitrust liability.[115]

The court then starkly concluded that the plaintiff's,

> . . . price squeeze claim, looking to the relation between retail and wholesale prices, is thus nothing more than an amalgamation of a meritless claim at the retail level and a meritless claim at the wholesale level. If there is no duty to deal at the wholesale level and no predatory pricing at the retail level, then a firm is certainly not required to price *both* of these services in a manner that preserves its rivals' profit margins.[116]

The court considered its conclusion to be fortified by institutional concerns—the need for clear rules in antitrust law; the fact that the courts were ill suited to act as central planners, identifying the proper price and terms of dealing; and the need for firms to have safe harbour for their pricing practices and clear standards to work to (how could a fair or adequate margin be determined?). Although the court did not deny that a price squeeze might harm competition in certain circumstances (for example by raising entry barriers to, and fortifying, the upstream monopolist's position or by impairing non-price competition and innovation in the downstream market), it held that no independent competition harm had been identified beyond that which would result from a duty to deal or predatory pricing violation. The court did

114 129 S Ct 1109, 1119. 'For if AT&T can bankrupt the plaintiffs by refusing to deal altogether, the plaintiffs must demonstrate why the law prevents AT&T from putting them out of business by pricing them out of the market.' (1123).

115 129 S Ct 1109, 1120.

116 129 S Ct 1109, 1120.

not therefore '... endorse a new theory of liability to prevent such harm'[117] '... never before recognized by this Court.'[118]

Deutsche Telekom, the Guidance Paper, and Telefónica

The latest court ruling in the EU contrasts starkly with the Supreme Court's Opinion in *linkLine*. Not only did the General Court and the ECJ in *Deutsche Telekom* endorse the existence of a margin squeeze offence, based on the spread between the wholesale and retail prices (the ECJ did confirm, however, that although the Commission had been wrong to decide that the very existence of the margin squeeze constituted an abuse, the General Court had been correct to find both that the pricing practice would exclude equally efficient competitors (wholesale local loop access was indispensable to penetration of the retail market) and that it had in fact done so), but the EU courts were not deterred from applying the competition laws because German regulatory rules existed which required DT to offer fully unbundled access to its local loop and to have wholesale and retail prices approved by the German regulatory telecoms authority ('RegTP').[119] DT had argued that it had insufficient scope to avoid a margin squeeze (it did not have scope to fix either its charges for wholesale access (this point was accepted by the Commission) or its retail prices) and that the regulatory system transferred responsibility for maintaining the structure of the market from the regulated undertakings to the regulatory authority (which had checked DT's charges). The General Court accepted that the competition rules only applied to the anti-competitive conduct of firms engaged in on their own initiative, but held that the Commission had been correct to find that DT had sufficient discretion in setting its retail charges (despite RegTP's intervention) to have avoided the margin squeeze. The national legislation thus left open the possibility of competition which could be, and had been, prevented by the conduct of DT.[120] The fact that the applicant's

[117] 129 S Ct 1109, 1122.

[118] 129 S Ct 1109, 1124. In footnote 3 the Supreme Court noted that,

> ... [g]iven developments in economic theory and antitrust jurisprudence since *Alcoa*, we find our recent decisions in *Trinko* and *Brooke Group* more relevant to the question before us

(but see n 37 above and accompanying text). It seems likely that the Obama administration will seek to inject new life into section 2 by bringing and winning some section 2 cases. It is possible, therefore, that the DOJ will seek to test the extent of the doctrine set out in *linkLine* by bringing a price squeeze case, perhaps in an unregulated sector.

[119] Decision No 223a of the Federal Ministry of Post and Telecommunications, 28 May 1996.

[120] The rules would not apply if anti-competitive conduct originates solely in the national law, see Joined Cases T-191, 212–214/98 *Atlantic Container Line v Commission* [2003] ECR II-3275 para 1130. If national law merely encourages or makes it easier for undertakings to engage in autonomous anti-competitive conduct, those undertakings remain subject to Articles 101 and 102 TFEU, see eg, Case C-198/01 *Consorzio Industrie Fiammiferi (CIF) v Autorità Garante della Concorrenza e del Mercato* [2003] ECR I-8055 para 56.

charges had to be approved by RegTP did not absolve DT from responsibility under Article 102 TFEU.[121]

The Commission's decision in *Deutsche Telekom* predated the publication of both its Discussion and Guidance Papers. Since that time the Commission has indicated that it will adopt a more nuanced approach to margin squeeze cases.[122] The Discussion Paper provides that the Commission will analyse a margin squeeze[123] in the same way as it analyses refusals to supply to existing or new customers, refusal to license intellectual property rights or refusal to grant access to an essential facility or network. Further, that it will *ordinarily* be wary of intervening in these cases, in particular, as intervention might undermine firms' incentives to invest and innovate and harm consumers. Consequently, the Commission indicates in paragraph 81 of the Guidance Paper that it will generally only consider a margin squeeze to be an enforcement priority if,

- it relates to an upstream product or service that is objectively necessary to be able to compete effectively downstream ie there are no actual or potential substitute on which downstream competitors can rely on and they cannot effectively duplicate the input;
- the margin squeeze is likely to lead to the elimination of effective competition on the downstream market (more likely the higher the dominant firm's market share in the downstream market); *and*
- the margin squeeze is likely to lead to consumer harm, for example, by preventing foreclosed rivals from bringing innovative goods or services to market, by stifling follow on innovation or by extracting monopoly profits in an unregulated downstream market where the upstream market is regulated.[124]

In determining whether to intervene, the Commission will also consider claims that the conduct allows the dominant firm to achieve efficiencies, for example by realizing an adequate return on its investment upstream.

---

[121] Indeed, DT would not have been absolved from responsibility even if RegTP had checked the charges for compatibility with Art 102, see Case C-344/98 *Masterfoods and HB* [2000] ECR I-11369 para 48.

[122] The Commission has stated that this new approach is manifested in its margin squeeze decision in *Telefónica* (n 10 above). See also the opinion of Mazák AG in Case C-52/08 *Konkurrensverket v TeliaSonera AB* 2 September 2010 (margin squeezes constitute a form of refusal to deal).

[123] Defined as a situation where a dominant firm charges a price for a product on the upstream market which, compared to the price it charges on the downstream market, does not allow even an equally efficient competitor (based on the long-run average incremental costs of the downstream division of the integrated dominant firm) to trade profitably in the downstream market on a lasting basis, Guidance Paper (n 65 above) para 80.

[124] The margin squeeze might also be designed to exclude rivals downstream so as to protect market power in an upstream market. See also nn 129–130.

The Commission states on its website that it applied the new, effects-based approach in its *Telefónica* decision,[125] relating to a margin squeeze in the Spanish broadband market. In that decision, therefore, having determined that there was a margin squeeze,[126] the Commission, went on to establish that Telefónica's conduct had caused prejudice to consumers.[127] Rivals could not compete, consumers were harmed (broadband retail prices were the highest in 15 EU Member States (on average 20 per cent higher) and less widely available) and insufficient competitive constraint was exercised by eg cable or satellite operators.[128] Further, the Commission found no objective justification for the conduct (in particular, it was not necessary for infrastructure investments and the network had been funded by monopoly rents prior to liberalization) or efficiencies to offset the long lasting consumer harm identified. Again, the Commission was not deterred from acting by reason of the fact that the Spanish regulator had looked at pricing and determined that there was no margin squeeze taking place—Telefónica still had sufficient freedom to avoid the margin squeeze.

Although the Commission's willingness to modernize its approach to Article 102 in this way may be welcomed, two caveats about the approach advocated for margin squeezes should be mentioned. First, the Commission's Guidance Paper envisages a *less cautionary* approach in a sector where regulation already imposes an obligation to supply. Although in the US, therefore, the courts consider that the existence of a regulatory regime makes the likelihood of antitrust harm lower (and less likely that antitrust intervention is necessary), the Commission believes that the existence of a regulatory regime makes the case for *ex post* application of the competition law rules *more* compelling. Where a regulatory regime compatible with EU law already imposes an obligation to supply on a dominant firm, the Commission assumes that the necessary balancing between incentives has been made and that imposing an obligation to supply, or prohibiting a margin squeeze, is not capable of having a negative effect on the input owner's incentive to invest and innovate upstream.[129] In such cases, therefore, the Commission simply looks for '. . . likely anticompetitive foreclosure, without considering whether the three circumstances referred to in paragraph 81 are present.'[130] This suggests that in regulated markets, the approach in *Deutsche Telekom* will remain intact—the margin squeeze will be presumed to harm

---

[125] *Telefónica* (n 10 above).

[126] Applying an equally efficient competitor test, based on Telefónica's long-run average incremental costs, see *Telefónica* (n 10 above) paras 310–96.

[127] See *Telefónica* (n 10 above) section E 'Impact on competition' and paras 543–4 and see also eg, *Intel* (Commission Decision dated 13 May 2009). See also Case C-280/08P *Deutsche Telekom* (n 49 above), paras 250–260.

[128] Contrast the position in *linkLine*, text accompanying n 111 above.

[129] This could also be the case where the upstream market position of the dominant undertaking has been developed under the protection of special or exclusive rights or has been financed by state resources, Guidance Paper (n 65 above) para 82. See also opinion of Mazák AG (n 122 above).

[130] Guidance Paper (n 65 above) para 82.

consumers (as the purpose of the obligation to grant access was designed to increase competition downstream without creating disincentives to innovate upstream) and the practice will be prohibited unless the dominant firm can objectively justify the conduct or demonstrate that it produces efficiencies that outweigh the consumer harm.

Secondly, and more generally, extreme care needs to be taken before too much reliance is placed on the approach articulated in the Guidance Paper in so far as it is at variance with the law set out by the EU Courts. It is, of course, the Court of Justice which is charged with responsibility for interpreting the provisions of the Treaty. There is a danger, therefore, that, insofar as inconsistencies in approach exist, the Guidance may create expectations[131] that conduct does not violate Article 102 when it does (or could), and that this inconsistency will create both uncertainty for business (which may still feel obliged to comply with the law) and difficulties for the national courts of the Member States and national competition authorities (and some NRAs) which are obliged to interpret Article 102 consistently with EU jurisprudence, not to take decisions counter to those adopted by the Commission or to '... prejudice the full and uniform application of Community law or the effects of measures taken or to be taken to implement it'.[132]

Remedies: structural or functional separation

It has been noted that one of the Supreme Court's reasons for caution in this area is that price squeezes, like refusal to deal, are especially hard for courts to remedy: the courts have felt it inappropriate for them to act as 'central planners' involved in the construction of 'fair' access terms or the setting of 'fair' prices or spreads between prices.[133] A question which might be raised, however, is whether a more drastic way to side-step these difficulties could be adopted. If separation (whether structural, accounting, or functional)[134] of the vertically integrated firm were ordered, the incentive for the upstream player to foreclose other downstream players would be

[131] Guidelines, may not be regarded as,

> rules of law . . . , they nevertheless form rules of practice from which the administration may not depart in an individual case without giving reasons that are compatible with the principle of equal treatment.

Cases C-189/02 P, C-202/02 P, C-205/02 P to C-208/02 P and C-213/02 P *Dansk Rørindustri A/S v Commission* [2005] ECR I-5425 para 209. By naming the Guidance Paper in terms of enforcement priorities, the Commission has tried to avoid the creation of such legitimate expectation, see LN Lovedahl Gormsen, 'Why the European Commission's enforcement priorities on Article 82 EC should be withdrawn' [2010] European Competition L Rev 45.

[132] Case 14/68 *Walt Wilhelm v Bundeskartellamt* [1969] ECR 1 para 9.

[133] See n 87 and the accompanying text above.

[134] Where a vertically integrated firm is divided into smaller separate firms (structural), where separation between the upstream and downstream divisions is 'simulated' through accounting practices (accounting) or where the upstream and downstream activities are placed in separate 'ring fenced' divisions of an entity with separate managers (functional), see eg I Walden (ed) (n 14 above) 63–4.

taken away. Unlike conduct remedies, such orders may have the advantage of being relatively easy to implement and may be more immediately effective without need of monitoring.

Although when bringing section 2 enforcement proceedings,[135] the DOJ has, in exceptional circumstances in the past, been prepared to seek a structural remedy from the court[136] (in *Standard Oil Co of New Jersey v United States*,[137] for example, the Supreme Court ordered the dissolution of the mighty Standard Oil trust and in *United States v AT&T*[138] AT&T consented, following antitrust proceedings, to sell off its local exchange business), there is today little appetite for the use of such remedies and their use in a margin squeeze would seem a most unlikely outcome.[139] Indeed although in *United States v Microsoft Corp*,[140] the District Court initially ordered that Microsoft be broken into two parts (one for the operating system and one for the applications), the Court of Appeals, despite upholding the finding that Microsoft had violated section 2, reversed and remanded the finding that Microsoft should be split into two. The lower court (1) had not held a remedies-specific hearing; and (2) had not provided adequate reasons for the decreed remedies. Further, the Court of Appeals held that as it had revised the scope of Microsoft's substantive liability, it was impossible to determine what impact this would have had on the remedies provisions and stressed that structural relief should only be granted if the plaintiff could establish a sufficient causal connection between the anti-competitive conduct and the creation or maintenance of its monopoly power in the operating systems market.

At first sight, a structural solution would also seem unlikely in the EU as the Commission has never used its express power to impose '... structural remedies which are proportionate to the infringement committed and necessary to bring the

[135] Sherman Act section 2 is enforced in civil proceedings, brought either by the DOJ, the FTC (the FTC cannot enforce the Sherman Act directly but enforces it indirectly through use of section 5 of the Federal Trade Commission Act of 1914), State Attorneys General and/or private litigants.

[136] In the 1950s and 1960s however there was strong support for the view that section 2 would provide an effective and simple mechanism for deconcentrating both monopolistic *and* oligopolistic markets and rapidly introducing new competition into a market, see eg WE Kovacic, 'Failed Expectations: The Troubled Past and Uncertain Future of the Sherman Act as a Tool of Deconcentration' (1989) 74 Iowa L Rev 1105, RA Posner, *Antitrust Law* (2nd edn, Chicago: The University of Chicago Press, 2001) 106 (Table 5).

[137] 221 US 1 (1911).

[138] *United States v AT&T* 552 F Supp 131 (DDC 1982). See n 27 and accompanying text above.

[139] One concern is that use of structural remedies has not been particularly successful in the past (although evaluating their efficacy is difficult) partly because of the length of time that antitrust proceedings take. See Single-Firm Conduct Report (n 68 above) 157 and RA Posner (n 136 above) 106. In addition, a divestiture remedy risks loss of efficiencies gained from vertical integration and may be extremely hard to structure see *United States v Microsoft Corp* 253 F3d 34 (DC Cir), *cert denied*, 534 US 952 (2001) and *United States v United Shoe Machinery Corp*, 110 F Supp 295, 348 (D. Mass 1953).

[140] *United States v Microsoft Corp* 253 F3d 34 (DC Cir), *cert denied*, 534 US 952 (2001).

infringement to an end'.[141] Nonetheless, when negotiating commitment decisions, the Commission has extracted remedies of a structural nature from dominant companies in the energy sector designed to open the market to competition and to ensure that competitors have sufficient access to infrastructure (eg gas pipelines).[142] There is therefore clearly a possibility that the Commission might, in the future, seek to impose, or negotiate, a structural solution in the telecommunications sector if it considers that separation is the only effective way of preventing unlawful pricing practices by an incumbent telecommunications operator. In addition, the new telecommunications regulatory package will give NRAs power to order functional separation of a telecommunications firm if it considers that the market would be better served by a services-based industry model.[143]

Conclusions

The cases discussed in this chapter indicate that, although at one time the approach under US and EU antitrust law to margin squeezes were similar, the law on the different sides of the Atlantic has evolved in diverse directions. In the US the Supreme Court, being fearful that antitrust rules will be used as a tool to discourage pro-competitive conduct by dominant firms, such as innovation upstream and/or low pricing downstream, has significantly raised the bar for those alleging that a defendant's pricing behaviour has squeezed it from a downstream market. Further, the existence of a specific regulatory regime designed to deal with the conduct complained of has made the courts even more reluctant to intervene. In the EU, in contrast, the authorities are concerned that the regulatory framework for telecommunications will not always allow NRAs to identify and prohibit margin squeezes ex ante and take the view that competition law should fill the gaps that remain. In addition, although the Commission accepts that in principle caution should be exercised before intervening in margin squeeze cases, so as to ensure that innovation and investment incentives and pro-competitive pricing downstream is not undermined, it considers that these risks are not relevant in a regulated market

[141] Regulation 1/2003 [2003] OJ L1/1, art 7(1) (structural remedies can only be imposed either where there is no equally effective behavioural remedy or where any equally effective behavioural remedy would be more burdensome for the undertaking concerned than the structural remedy).

[142] In *RWE* (IP/09/41), for example, the Commission adopted a decision rendering legally binding commitments offered by RWE to divest its gas transmission network following investigations into possible abuse of RWE's dominant position on the gas transmission network in Germany through refusal to supply gas transmission services and margin squeeze. See also eg, *GDF* (MEMO/09/536) (agreement to release a large share of its long-term reservations of gas import capacity into France).

[143] In the UK, the UK communications regulator, OFCOM negotiated, in 2005, the functional separation of the UK incumbent, BT, to create a functionally separate service division, Openreach, with the objective of ensuring that access products are provided on an open and non-discriminatory basis to all wholesale customers.

where a decision has already been taken to open the market to, and promote, competition downstream.

It is difficult to extrapolate clear lessons from the cases discussed in this chapter for those arising at the interface of intellectual property and competition. The divergent US and EU approaches to refusal to deal and margin squeezes, however, appear to confirm that the two jurisdictions are likely to continue to pursue distinct approaches both to cases involving refusals to license and those involving refusal to license on reasonable terms. The continuation in *linkLine* of the cautious approach to section 2 promoted in *Trinko*, undoubtedly provides support for US case law which suggests that in refusal to license cases, consumer welfare will be best promoted by encouraging the investment the intellectual property rules are designed to protect and to courts unwilling to get involved in central planning, compulsory sharing and/or the proper price and terms of dealing. In contrast, the broader approach to refusals to deal and margin squeezes in the EU, reflects the greater willingness of the EU authorities to assess the anti-competitive effects flowing from a refusal to deal or license on downstream markets and to monitor and enforce compliance with obligations to deal.

Part III

PHARMACEUTICAL SECTOR

11

THE PHARMACEUTICAL SECTOR, INTELLECTUAL PROPERTY RIGHTS, AND COMPETITION LAW IN EUROPE

Simon Priddis and Simon Constantine

Introduction

The pharmaceutical sector—perhaps more than any other industry—sits on the fault lines dividing a number of policy objectives—competition policy, intellectual property policy, state regulation, and social welfare. The sometimes unclear interaction between these objectives—and the relevant legal regimes—has a material impact on how the sector works.

Intellectual property laws play a fundamental role in incentivizing firms to develop new products that meet the changing clinical needs of ageing populations (eg, dementia, obesity, oncology). Governments have sought to encourage the application of market disciplines to incentivize firms to engage in product development and also price competition on the resulting marketed products.[1] However, market forces are not allowed free rein. Although governments promote a strong and vibrant pharmaceutical industry to develop new products, at the same time, they also need to cap (or even reduce) public health expenditure, including through making difficult choices as to which product to buy.[2] Many governments seek to balance these objectives through direct price-setting intervention in pharmaceutical markets.

[1] See for example 'Final Government Response to recommendations aimed at Government contained in the OFT Report "The Pharmaceutical Price Regulation Scheme",' Department for Business, Innovation & Skills, June 2009, available at <http://www.bis.gov.uk/files/file51657.pdf>, accessed on 12 September, 2010.

[2] The UK Pharmaceutical Price Regulation Scheme ('PPRS') is a good example of the way in which a national government has sought to balance these competing objectives. The PPRS seeks to cap industry profits made on pharmaceutical drug sales in the United Kingdom, while at the same time ensuring that the industry makes a sufficient return on investment to incentivize it to engage in R&D.

The role of competition policy in pharmaceutical markets is itself complex. Competition policy is often at the sharp end of tensions between competing policy objectives (incentivizing firms to develop new products by granting them time-limited monopoly rights to exploit products versus promoting price competition to reduce health expenditure). The debateable balance between vigorous enforcement of antitrust rules and proper regard for intellectual property rights comes into sharp focus in the pharmaceutical industry.

Two factors have given this long-running debate added impetus—one economic and one legal. First, the changing economics of healthcare-related markets have increased the financial pressures on pharmaceutical companies, and thus increased the incentives to test the fault-lines. As governments have sought to increase their 'value for money' from healthcare expenditure (for example, by seeking increased price competition from generic pharmaceutical companies), innovator companies have inevitably sought to mitigate the impact of this trend by relying on patent rights to maximize the scope, extent, and duration of revenue streams. This has coincided with a large number of industry 'blockbuster' products losing patent protection (coming off-patent) with the consequent rapid fall in revenues from those products as they become exposed to price competition from generic products. Second, the European Commission's 2008–9 sector inquiry into the pharmaceutical industry (hereinafter, the 'Pharma Inquiry') has obscured the fault-lines by raising doubts about the compatibility with antitrust laws of a substantial number of patent-related practices.[3] There are concerns that the potentially wide-ranging conclusions and recommendations in the Commission's resulting final report (hereinafter, the 'Pharma Report'[4]) could have the effect of dulling innovator incentives to engage in R&D.

In this chapter, we review how principles of EU antitrust law impact on the pharmaceutical sector and influence the economic decisions of pharmaceutical companies. We also consider how the apparent policy shifts in European Commission antitrust enforcement policy laid out in the Pharma Report might impact on the same

See joint publication from the Association of the British Pharmaceutical Industry and the Department of Health. 'The Pharmaceutical Price Regulation Scheme, 2009', available at <http://www.dh.gov.uk/prod_consum_dh/groups/dh_digitalassets/documents/digitalasset/dh_098498.pdf>, accessed on 12 September 2010.

[3] See Pharmaceuticals Sector Inquiry <http://ec.europa.eu/competition/sectors/pharmaceuticals/inquiry/index.html> accessed 12 September 2010, for more details on the Pharma Inquiry.

[4] The Pharma Report comprises a Communication from the European Commission, Executive Summary of the Pharmaceutical Sector Inquiry Report (hereinafter, the 'Communication') and a Commission Staff Working Document (Technical annex to the Commission Communication), Pharmaceutical Sector Inquiry Final Report (8 July 2009) <http://ec.europa.eu/competition/sectors/pharmaceuticals/inquiry/index.html> accessed 12 September 2010. References in this chapter to the Pharma Report are to the Technical Annex.

economic decisions in the future. We explain why, in respect of the pharmaceutical sector, the contours of the EU antitrust concept of 'exceptional circumstances' need particular definition, and why expansion of that concept to remedy perceived failings of other legal concepts (such as the criteria for patentability of an invention) would be unwise for both legal and economic reasons.

A. Market Structure and Competition in the EU Pharmaceuticals Sector

An understanding of the way in which the pharmaceutical industry functions is central to any assessment of the intersection of competition and IP laws. This requires consideration of the central role played by product R&D; the source and timing of revenue streams; the commercial relationships among the spectrum of market participants; and the ambiguous role of the state.

The costs, risks, and returns of drug development and commercialization

The development of new medicines by innovative pharmaceutical companies (hereafter 'originators') has brought substantial public health benefits to patient populations. However, the economics of researching and developing new pharma products are not simple.

Originators are incentivized to innovate to develop new products (whether for an entirely new clinical therapy or as an enhancement of existing therapy) by the promise of monopoly rights to exploit their marketed innovation.[5] This is vital given the scale of the risks facing originators, which must contend with very low success rates on innovation projects and long timelines from therapy discovery to marketed product. The economic context facing any pharmaceutical innovator can be broadly described as follows:

- The average cost of developing a new drug and bringing it to market has been estimated at over $1.3 billion—a tenfold increase since 1975.[6] Some of this increase has been driven by a shift from medicines based on 'traditional' chemical compounds to biotechnologies, which has necessitated expensive re-tooling for originators. R&D accounts for a significant proportion of development costs—the Pharma Report found that, in the period 2000–2007, originators spent an

[5] See Communication (n 4 above), 2.

[6] European Federation of Pharmaceutical Industries and Associations (EFPIA), 'Submission to the European Commission in relation to the pharmaceutical sector inquiry' (13 June 2008) <http://www.efpia.org/content/default.asp?PageID=559&DocID=4892> accessed 12 September 2010 para 50, citing JA DiMasi and HG Grabowski, 'The Cost of Biopharmaceutical R&D: Is Biotech Different?', (2007) 28 Managerial and Decision Economics 469–79.

average of 17 per cent of their global turnover from prescription medicines on R&D.[7] Indeed, total pharma industry R&D spend exceeds that of any other sector.[8]

- Balanced against such costs, successful products can more than repay their development costs. So-called 'blockbuster' drugs generate annual global revenues in excess of US $1 billion.
- Blockbusters are few and far between—many products fail at the development stage. Only one in 5,000–10,000 compounds tested reaches market.[9] Even if a product ultimately reaches market, the costs involved in its commercialization (manufacturing scale-up, new product promotion, regulatory pricing, and reimbursement adjudications, etc) mean that 80 per cent of marketed products fail to recoup the average capitalized R&D expenditure made on them.[10] As the English patent law judge Robin Jacob put it—'Most research leads nowhere. The few winners must pay for all the losers.'[11]
- Products reaching market do so only after a long gestation. Typically, it will not be until 10–12 years after initial discovery and patenting of a compound that a product will be launched on the market—in many cases, it will be significantly longer. There are two consequences of this:
  - because so much of the patent-protected time period can expire before a product reaches market, the period of time within which a pharma innovator can recoup its development costs is substantially less than the 20-year exclusivity period envisaged by the European patent protection system. The system of Supplementary Protection Certificates ('SPCs') recognizes this problem, and provides an additional period of exclusivity protection for certain

[7] Pharma Report (n 4 above) para 72. The Commission received evidence from one stakeholder that for biopharmaceutical companies, this proportion increased to approximately 40 per cent (Pharma Report, para 56).

[8] European Commission, *Monitoring Industrial Research: the 2007 EU Industrial R&D Investment Scoreboard* (October 2007) <http://iri.jrc.ec.europa.eu/research/docs/2007/eur22928en.pdf> accessed 12 September 2010, 13.

[9] PhRMA, 'Pharmaceutical Industry Profile' (2008), 3, figures 2, 4, and 6, also citing JA DiMasi, RW Hansen and HG Grabowrski, 'The Price of Innovation: New Estimates of Drug Development Costs', Journal of Health Economics (2003) Vol 122, no 2, 151 at 162.

[10] EFPIA 'Submission to the European Commission in relation to the pharmaceutical sector inquiry' (n 6 above) para 53.

[11] Speech by Sir Robin Jacob on 'Patents and Pharmaceuticals' at the Presentation of the Directorate-General of Competition's Preliminary Report of the Pharma sector inquiry (28 November 2008) <http://ec.europa.eu/competition/sectors/pharmaceuticals/inquiry/jacob.pdf> accessed 12 September 2010, para 4.

medical innovations.[12] For the same reason, originators also benefit from extended regulatory data exclusivity;[13] and

- because of the delay bringing a new product to market, patent protection (or more generally exclusivity protection) may expire while a particular therapy retains its market attractiveness (or has not been superseded by further innovation). This allows for entry by generics firms marketing legitimate 'me-too' copies of the formerly patented product. Such entry often carries low commercial risk (because demand for the patented product is already well-established) and can be undertaken at relatively low cost, certainly when compared with the initial cost of entry borne by the originator company.[14]

- Even products ultimately reaching the market are not guaranteed financial success. Rather than grant reimbursement status to every new authorized product, governments are increasingly seeking to limit healthcare expenditure by only granting reimbursement status to products that demonstrate an appropriate level of 'value-add'. In other words, there should be an appropriate cost-benefit trade-off in favour of the new pharma product before physicians are allowed to prescribe it to patients at national health system cost.

Pharmaceutical companies therefore look to the promise of monopoly profits earned on that small proportion of marketed products to finance the costs of bringing successful products to market, as well as funding the ultimately unsuccessful projects (which comprise the vast majority of pharma R&D). IP rights are critical therefore in ensuring that the risks and rewards of innovation are appropriately balanced.

The corollary is that the race to develop the next treatment for a currently unmet clinical need (whether involving a wholly new compound or a significant enhancement of an existing therapy) is a significant driver of competition in the pharmaceutical industry. It is here that pharma innovators stand the best chance of securing the innovative 'leap forward' that delivers sufficient 'value-add' in order to be granted

---

[12] Council Regulation (EEC) No 1768/92 concerning the creation of a supplementary protection certificate for medicinal products (as amended) [1992] OJ L182/1. If granted, an SPC provides the patent holder with an additional, post-patent period of exclusivity which is the shorter of: (i) five years; or (ii) the time between patent application and marketing authorization, less five years.

[13] Under the EU marketing authorization regime a company manufacturing a bioequivalent, generic version of a drug need not repeat the pre-clinical and clinical trials if that drug has been authorized in the EU for at least eight years. However, the generic producer must wait until ten years have elapsed from the granting of the initial marketing authorization before it can market the product. This is known as the '8 + 2' rule. Such 'data exclusivity' is independent of, and runs parallel to, any period of patent protection.

[14] Estimates of the cost of bringing a generic product to market vary significantly depending on the market concerned and the manufacturing processes and products involved. However, even higher estimates show those costs to be very low compared with the originator's original development costs (eg Bernstein Research, *Biogenerics: Long Rainbow, Small Pot of Gold; Limited Risk for Biotechs*, Research Weekly Notes (27 October 2006) (estimating the cost of generic market entry in the US as being US$8m–30m.))

reimbursement status. One obvious difficulty lies in identifying the new 'breakthrough'—many diseases are now well-researched and understood, greatly reducing the chances of achieving the medical breakthrough that leads to the development of a new blockbuster treatment. Attention of pharma innovators inevitably becomes focused on the same, smaller number of diseases where that major breakthrough has yet to occur (eg, diabetes, Alzheimer's disease). These areas of research are, by definition, among the most difficult and challenging to pursue, hence the R&D expenditure incurred is also very significant.

In short, the success of the research-based pharmaceutical industry is founded on the exclusivity for innovators provided by the patent system.[15] The time-limited monopoly profits afforded to successful innovators by the IP laws are vital to ensuring continued investment in R&D.[16] Those rights need to be sufficiently durable to enable an originator to recover the costs of both successful and failed R&D, and also to earn a reasonable return on that investment. Absent such profits, why would a firm engage in R&D, particularly where the R&D is in respect of new compounds or areas of unmet clinical need where the risks of failure are higher? Indeed, it was estimated that around 65 per cent of the drugs on the market in 1999 would not have been developed at all absent patent protection.[17] The preamble to the European SPC legislation neatly puts the point,

> . . . medicinal products, especially those that are the result of long, costly research will not continue to be developed in the Community and in Europe unless they are covered by favourable rules that provide for sufficient protection to encourage such research.[18]

[15] See I Haracoglou, *Competition Law and Patents: A Follow-on Innovation Perspective in the Biopharmaceutical Industry* (Cheltenham: Edward Elgar, 2008) 29, citing DOJ/FTC Proceedings on the pharmaceutical industry, *Business perspectives on the use and the role of patents in the biotech industry*, 26 February 2002. See also Remarks of J Thomas Rosch, Commissioner, Federal Trade Commission, 'The Role of Static and Dynamic Analysis in Pharmaceutical Antitrust' at the Fifth Annual In-House Counsel Forum on Pharmaceutical Antitrust (18 February 2010) <http://www.ftc.gov/speeches/rosch/100218pharmaantitrust.pdf> accessed 12 September 2010,

> Thus, it appears that patent protection may stimulate innovation in the pharmaceutical industry to a far greater degree than in other industries. This is not entirely surprising, given the large upfront costs and degree of risk developing a new product and the relative ease of copycat products.

[16] European patent laws are not fully harmonized across the EU Member States. The interpretation of such laws by national courts has also caused further divergence. Nonetheless, the effect of international agreements such as the TRIPS agreement and the European Patent Convention 2000 and the adoption of many of the principles in the draft Community Patent Convention have resulted in significant *de facto* harmonization. Thus, unless otherwise noted, the principles discussed in this chapter can be assumed to apply equally across the EU.

[17] OECD Committee on competition law and policy, *Competition and Regulation Issues in the Pharmaceutical Industry* (February 2001), DAFFE/CLP(2000)29, <http://www.oecd.org/dataoecd/35/35/1920540.pdf> accessed 12 September 2010, 34, citing R Levy, *FTC Bureau of Economics Staff Report—The Pharmaceutical Industry: A Discussion of Competitive and Antitrust Issues in an Environment of Change* (March 1999).

[18] Recital 2, Council Regulation (EEC) 1768/92 n 12 above.

This is not to say of course that the scope of IP rights should be unconstrained. Unconfined IP rights might block other firms from engaging in welfare-enhancing (socially beneficial) pharmaceutical innovation and undermine key competitive dynamics in the sector.

Competitive dynamics in the pharmaceutical sector

To a certain extent, the difficulty of analysing the interaction of competition laws and intellectual property laws in the pharmaceutical sector arises because competition in the sector takes a wide variety of different forms. In broad terms, we can think about that competitive interaction among firms occurring in four different ways.

First, originators compete through innovation to bring a patented product (be it an entirely new compound or an enhancement of an existing product) to the market.[19] Originators seek to develop products that will have limited substitutability—thus, they seek to compete 'for the market' because the developed product is the only marketed product available to treat a particular clinical condition. In response, other originators seek to compete in treating the same clinical condition by innovating around the patent(s) of originators already on the market or developing their own alternative products for that condition (the need to innovate around products in development of course being important because of the long period of time between patent registration and a product actually becoming available to patients).

Second, there can be direct 'in market' competition among suppliers of the same patented pharmaceutical product. There is a thriving parallel trade in ethical on-patent pharmaceutical products in the EU. This often results in distributors of products in higher-priced countries facing competition from parallel traders that have sourced products from distributors in lower-priced countries, and are arbitraging those price differences. This results effectively in intra-brand competition, but only in respect of certain products.

Third, there can also be direct 'in market' competition among originators' respective patented products for a particular treatment, where some degree of therapeutic substitutability exists between those products. In this circumstance, originators compete on factors such as efficacy, absence of side-effects, and patient convenience, as well as on price (to the extent permitted by national health service price regulation).

Fourth, as the market exclusivity of an originator's product expires, originators also begin to face competition from manufacturers of generic equivalents of their

[19] We do not discuss the originator's decision to compete via 'make' or 'buy'. Should the originator itself engage in the R&D, or wait for a smaller pharma company or other laboratory to undertake the research, make the discovery, and then 'buy' the product to bring to market (or perhaps even through the later stages of development)?

patented product.[20] Generics firms typically, and logically, seek to market versions of those products that generate the greatest revenues for originators. Given that they have not had to invest in the development of the product, generics are able to offer that product at significantly lower prices. Generics firms argue that, as well as resulting in short-term price benefits to customers, their involvement in the pharmaceutical marketplace itself stimulates originator innovation. As generic competition limits the period over which an originator can charge a premium for its product, originators are incentivized to search for new, patentable medicines in order to maintain their revenue flows.

State intervention

Pharmaceutical markets are subject to wide-ranging state intervention, which exists not only to ensure that safe and efficacious products are brought to market, but also to regulate the ways in which pharmaceutical products are sold, the prices at which they are sold, and by whom. Such intervention impacts on both the supply and demand sides of pharmaceutical markets—it would not therefore be accurate to describe pharmaceutical markets as functioning 'normally' since competition takes place in a highly regulated and state-influenced environment.

On the supply-side, governments can determine which firms are able to supply into a market. At one level, there is a public safety aspect to intervention. For example, national agencies control the marketing of pharmaceutical products by requiring products to be approved and registered with either national agencies or, more likely within the EU, with the European Medicines Agency. However, national governments can also seek to encourage competition among pharmaceutical producers once a product has come off patent by directly encouraging generic market entry, thereby generating direct price competition in the branded product. In the EU, this process is facilitated by entitling generics producers to copy an innovators' clinical data in marketing authorization applications.[21]

On the demand-side, although the precise scope of intervention varies among Member States, it is typical for governments to limit or even set directly the price for prescription drugs (often after negotiations with stakeholders) and also to set demand for those products by determining which products will be reimbursed in part, or in full, by national health schemes. The impact of state interventions on the demand-side of the market is even more important than those on the supply-side. Since the

[20] Typically, generics will enter the market only after the originator's product has lost market exclusivity. Sometimes, a generic will seek to enter the market before such time, if it believes that the originator's patent(s) are invalid or would not be infringed by its version of the medicine in question.

[21] See n 13 above.

state is usually the largest purchaser of pharmaceuticals, decisions made by the state have a direct impact on the economic incentives of pharmaceutical companies.

- First, the differing ways in which governments across the 27 EU Member States control prices—described by the WHO as an 'unpredictable lottery'[22]—causes originators significant uncertainty as to the price at which they will be able to market their product and when they will be able to do so. States use pricing mechanisms to incentivize firms to produce products which, based on commercial considerations alone, would not be produced (by paying higher prices for certain innovative medicines). For example, firms can be supported in developing treatments for diseases that are rare, but which are typically fatal or highly injurious, as well as (potentially more lucrative) treatments for more widespread, but milder, conditions.
- Second, state pricing intervention is most commonly used to cut down public health expenditure. In the economic crisis, a number of EU Member States (Greece particularly) imposed unilateral price reductions on pharmaceutical companies in an effort to contain public expenditure on pharmaceutical products. EU Member States also seek to manage costs of health provision by encouraging the use of cheaper generic products rather than more expensive branded products. In some cases (eg, Germany), this takes the form of mandating that physicians always prescribe an available generic product when it is cheaper than the equivalent branded product.
- Third, as described above, many European Member States will no longer grant reimbursement status to every pharmaceutical product brought to market. Reimbursement status is determined on the basis of cost-effectiveness (ie, dependent on a drug's therapeutic benefit, innovativeness, and/or overall economic benefit).[23] These sorts of mechanisms are said by originators to increase substantially the commercial risks of product development and thus undermine incentives to innovate—which company would invest potentially hundreds of millions of pounds developing a new product, only for it to be refused reimbursement status at the end of the process?[24]

Incentivizing innovation is therefore critical for all industry stakeholders, be they originators, generics, governments, or patients. Equally clear is that it is difficult to balance originator incentives to innovate against competing pressures to (a) manage healthcare budgets and (b) ensure that innovative effort is directed towards areas of

[22] World Health Organisation Report, *Priority Medicines for Europe and the World* (November 2004) <http://archives.who.int/prioritymeds/report/final18october.pdf> accessed 12 September 2010.

[23] In the UK, the National Institute of Clinical Excellence (NICE) performs this function.

[24] See for example, EFPIA 'Submission to the European Commission in relation to the pharmaceutical sector inquiry' (n 6 above) para 117.

unmet clinical need. In this complex matrix of incentives, it is understandable that much debate has centred on the role and scope of patent protection; perceived by the originator segment of the industry as the primary mechanism that protects its ability to earn a return on investment, and perceived by generic firms as the primary barrier to competition and lower costs for patients and national health services.

B. The Proper Extent of Antitrust Intervention in the Pharmaceutical Sector

Laws designed to protect competition and intellectual property are not inherently antagonistic. Both fields of law share a similar policy goal—to promote innovation that leads to better products being developed.[25] Tensions arise between competition and intellectual property laws because each seeks to meet this goal in contrasting ways.

- Intellectual property law (and patent law specifically) seeks to balance competing interests by granting the 'prize' of (time-limited) exclusivity as a *quid pro quo* for the inventor disclosing the details of its innovation. Such disclosure encourages further innovation based on the patented invention[26] and permits imitation (and thus competition) upon expiry of the patent. In other words, we accept that competition to the patented product is muted for a period of time in order to provide an incentive to innovate.
- Competition law seeks to promote innovation by proscribing agreements or conduct that restrict incentives to innovate, seeing development of new products as a way in which firms compete to win business. In other words, we accept that the patent protection must be circumscribed otherwise no firm would ever have an incentive to compete away the advantage obtained by the innovator.

Thus, in a virtuous circle, a firm would develop a new product, for which it would obtain patent protection. Since the new product would mark a material step in competition in the relevant marketplace, the innovator would win a substantial volume of new business—the profits from which would be protected by the patent, enabling the innovator to reap the reward for its research effort. Following expiry of the patent, the innovator would be exposed to competition on the same product by imitators, who would compete on price to win business from the original innovator ('static competition'), thus competing away any remaining monopoly profits and restoring competitive equilibrium in the market. Firms in the marketplace would then compete again ('dynamic competition') to find the next substantial innovation to capture

[25] See Commission Notice, Guidelines on the application of Article 81 of the EC Treaty [now Article 101 TFEU] to technology transfer agreements [2004] OJ C 101/2 para 7.

[26] Many patent laws contain a form of 'research exemption', allowing third parties to use the information disclosed in a patent immediately for the purpose of further research.

all business in the market. This of course assumes that, during the course of the competitive cycle described above, another market player has not introduced its own new innovation and sought to 'leapfrog' the original innovator to capture the patent-protected monopoly profits.

To achieve this virtuous circle, the equilibrium between competition and intellectual property law enforcement needs to be finely balanced. IP practitioners challenge competition law encroachment on the scope of patent protection as undermining incentives to innovate.[27] Conversely, competition law practitioners see overly-broad IP protection as undermining both short-run price competition and long-run dynamic competition.

- The US Federal Trade Commission, arguing for a restrictive understanding of the scope of patents, put the point crisply, '. . . properly interpreted, [the patentability criteria] tend to award patents only when necessary to provide incentives for inventions, their commercial development, or their disclosure.'[28] In other words, unduly broad patent protection would disproportionately mute competition beyond the level needed to incentivize innovation.
- However, the difficulty of striking the correct balance between competition and innovation has been recognized by the European Commission. When considering the possible consequences of imposing on an IP rights holder an obligation to license the IP right, the Commission noted that, '. . . intervention on competition law grounds requires careful consideration where the application of Article [102 TFEU] would lead to the imposition of an obligation to supply on the dominant undertaking. The existence of such an obligation—even for a fair remuneration—may undermine undertakings' incentives to invest and innovate and, thereby, possibly harm consumers.'[29]
- The European courts, in recognition of the same difficulty of balancing competing incentives, have therefore concluded that the exercise of an intellectual property right should only be considered to be contrary to competition law in 'exceptional circumstances'.[30]

[27] The view that antitrust enforcement acts as a disincentive to innovation is not universally held, however. See I Haracoglou, 'Competition Law as a Patent "Safety Net" in the Biopharmaceutical Industry' (2004) 1 (2) Competition L Rev 65, 76 *et seq*.

[28] FTC Report, *To Promote Innovation: The Proper Balance of Competition and Patent Law and Policy* (October 2003) <http://www.ftc.gov/os/2003/10/innovationrpt.pdf> accessed 12 September 2010, 4.

[29] Communication from the Commission, Guidance on the Commission's Enforcement Priorities in Applying Article 82 of the EC Treaty to Abusive Exclusionary Conduct by Dominant Undertakings [2009] OJ C45/07 (hereinafter, 'Article 82 Guidance') para 75.

[30] See Case T-201/04 *Microsoft v Commission* [2007] ECR II-3601 ('*Microsoft*') para 331 (citing, at paras 329–30, Case C-418/01 *IMS Health GmbH v NDC Health GmbH* [2004] ECR I-5039 ('*IMS Health*'); at paras 325–28, Case C-7/97 *Oscar Bronner GmbH & Co. KG v Mediaprint Zeitungs- und Zeitschriftenverlag GmbH & Co. KG* [1998] ECR I-7791; and, at paras 322–24, Joined Cases C-241/91 P and C-242/91 P *RTE and ITP v Commission* [1995] ECR I-743 ('*Magill*')). In *Microsoft*, the General Court considered, but ultimately rejected, Microsoft's arguments that the imposition of a compulsory

- Whatever the correct position, the OECD noted that 'Investment in innovation requires a predictable legal system and, as a result, antitrust policy should be formulated to ensure that incentives to innovate are not unnecessarily weakened or destroyed.'[31] If inventors do not have certainty at the outset that their research efforts and discoveries will be adequately protected from copying by others, there will be little incentive for them to undertake that research.

The interaction between competition law and IP law in the pharmaceutical industry is perhaps even more complex (and the risks of over- and under-intervention correspondingly more substantial) than is suggested by the quotations above. Partly, this is a function of the direct correlation discussed above between patent protection and innovation incentives in the pharmaceutical industry. It is also a function of the fact that the antitrust armoury, and in particular the Article 102 TFEU prohibition on abuse of dominance, struggles to accommodate the complexity of the pharmaceutical sector. In addition, remedies for abuse of dominance are often blunt (eg, licensing remedies).

Evidence of these difficulties emerges from the Pharma Inquiry. In brief, the Commission concluded in its Pharma Report that originators use a 'toolbox' of patent-related practices that contribute to delays in generic entry and a decline in the number of innovative new drugs coming to market.[32] The implication of the Pharma Report was that these practices might infringe EU competition law (under either or both of Articles 101 and 102 TFEU). The Pharma Report did not seek to explain specifically the circumstances in which any practice would in fact infringe the EU competition laws. Nor did it seek to explain the circumstances in which a European competition authority could begin to mount a competition law case against an originator company, particularly under Article 102 TFEU.

Although the Commission's Pharma Report made its most forthright criticisms of 'reverse' or 'pay for delay' patent settlements (ie, anti-competitive agreements reviewed under Article 101 TFEU), the Commission's initial focus in the Pharma Inquiry was on *unilateral* conduct by originators, and thus on potential violations of Article 102 TFEU. Competition law enforcement under Article 102 TFEU is

licence would reduce its incentives to innovate (para 697–701). See also Commission decision in Case COMP/C-3/37.792 *Microsoft*, at paras 713–29 and fn 874.

[31] OECD, *Competition, Patents and Innovation 2006* (2008) DAF/COMP(2007)40 <http://www.oecd.org/dataoecd/26/10/39888509.pdf> accessed 12 September 2010, 8.

[32] A detailed discussion of the progress and evolution of that sector inquiry is beyond the scope of this paper. Further critical analysis can be found in David Rosenberg, 'A Critique of the European Commission's Sector Inquiry into the Pharmaceutical Sector', (2010) 24 (2) Antitrust 35 (Spring 2010). In particular, the Commission's provisional conclusions were subject to sustained criticism for excluding the impact of the regulatory regime from consideration. While this defect was to some extent addressed in the Pharma Report, the Commission did not seek to quantify the extent to which the adverse effects identified might in fact have been properly attributable to any of the patenting or competitive practices identified.

notably difficult—there are numerous reasons for this, but the necessity of showing dominance or market power is fundamental to many of these reasons.[33]

Establishing dominance

Establishing that a firm holds a dominant position on the relevant market is a core step in showing an infringement of Article 102 TFEU. When addressing dominance in the context of an IP-driven business, there is no direct correlation between the possession of a patent and the possession of market power amounting to dominance. While the existence of a patent in a pharmaceutical market might well result in the patent holder accounting for a significant share of sales,[34] intellectual property rights do not, of themselves, give rise to a dominant position.[35] The burden will still be on the Commission to prove, on the facts of an individual case, that market power amounting to dominance does in fact exist.

The starting point for any assessment of dominance is the definition of the market relevant to the pharmaceutical product in question. Previous Commission cases involving pharmaceutical products (principally merger investigations) have used 'ATC Level 3',[36] which groups medicines according to their therapeutic/pharmacological indication, as the appropriate starting point for a market definition analysis. ATC Level 3 classes are typically broad, and can encompass a wide range of possible treatments (including some that are indicated for a number of therapeutic indications) yet appear only in a single ATC class. As a result, market definition based only on ATC Level 3 class can be either too broad or too narrow.

A more sophisticated analysis of market definition, focused principally on demand-side considerations, would need to take into account not only the existence of other

[33] It may be in part for this reason that the two main competition investigations commenced by the Commission since the conclusion of the Pharma Inquiry both involve allegations of Article 101 TFEU, as well as Article 102 TFEU infringements—Case COMP/39.612 *Perindopril (Servier)* and Case COMP/39.226 *Lundbeck*. The dawn raids in both cases involved inspections at the premises of generics, as well as originators. The same is true of raids carried out by the Commission in October 2009. Those raids have yet to result in formal proceedings, but press reports suggest that these were again aimed at investigating settlement agreements (see C Jacob, 'Sanofi says doors "wide open" to EU drugs probe', Reuters (8 October 2009) <http://www.reuters.com/article/idUKL813828120091008>, accessed on 12 September 2010).

[34] EU case law has long been clear that possession of a substantial market share (above 50 per cent) gives rise to a rebuttable presumption of dominance. See Case C-62/86 *Akzo Chemie BV v Commission* [1991] ECR I-3359.

[35] DG Competition, Discussion paper on the application of Article 82 of the Treaty to exclusionary abuses (December 2005) <http://ec.europa.eu/competition/antitrust/art82/discpaper2005.pdf> accessed 12 September 2010 para 40. See also, for example, the decision of the US Supreme Court in *Illinois Tool Works Inc v Independent Ink, Inc*, 126 S Ct 1281 (2006), in which the Court rejected the notion that existence of a patent presumptively conferred market power on the patent holder. The Court adopted instead an approach which reviews the economic effect of the patent and its impact on competition.

[36] Anatomical Therapeutic Chemical (ATC) classification is an international standard for classifying medicines. There are fourteen main groups (Level 1), each with one pharmacological/therapeutic subgroup (Level 2). Levels 3 and 4 are chemical/pharmacological/therapeutic subgroups and Level 5 represents the specific chemical substance.

patent-protected products that could compete for the same therapeutic indication as the product under investigation, but also factors such as patient usage, dosage levels, competition from off-label prescribing, and possibly also threats from generic products being used in the marketplace.[37] The Commission followed a similar analytical approach in its *AstraZeneca* decision, now upheld by the General Court. There, the Commission found that the relevant market definition was at the ATC4 level (proton pump inhibitors) since the ATC3 class contained products (H2 blockers) that were not entirely substitutable for the same therapeutic indication—a conclusion supported by the General Court.[38]

In an economically-focused market definition analysis, the existence of a patent is not therefore decisive as to market definition, and not alone decisive as to the existence of dominance. The existence of strong patent rights is of course a relevant factor in assessing dominance, but has to be considered in the round along with other evidence on competitive dynamics in the market.

Even if a patent-holder were found to be either the sole or majority provider of product for a particular therapeutic indication, the presence of monopsonist buyers in the form of Member State governments, health services, and insurance funds raises a further question as to the ability of even a patent-protected monopoly seller to exercise market power. European competition authorities have reached different conclusions on this issue when assessing the countervailing buyer power exercised by different Member State health services.[39] Most recently, in *AstraZeneca*, the General

[37] See for example, the analysis of market definition in Case COMP/M.5476 *Pfizer/Wyeth*, Commission Decision of 17 July, 2009, at paragraphs 21–38. For a contrasting view, note that one Commission official has suggested that market definition should be reconsidered upon a particular product losing exclusivity, and implies that it should, from that point on, be based on ATC level 5, equating to the specific (previously patented) molecule alone. See D Schnichels, 'The Application of European Competition Law to the Pharmaceutical Sector—Some Personal Thoughts', Paper produced for the Fordham Annual Conference on International Antitrust Law and Policy 2009, 13–14. Recent merger cases involving generic companies, in which the Commission has defined the market as limited to the molecule in question, are cited to support such an approach. However, this also misses the point that competition between molecules can continue (and may even increase) post-patent expiry depending on relative price movements. It is submitted that a rigid analytical approach of the type proposed would step away from a true economic analysis of competitive constraints.

[38] Case T-321/05 *AstraZeneca v Commission* (Judgment of 1 July 2010, not yet reported) available at <http://curia.europa.eu>, paras 73, 107, and 148–55 (hereinafter '*AstraZeneca*'), now an appeal to the European Court of Justice. The latter paragraphs deal in particular with evidence on functional substitution that narrowed the market definition below the ATC3 classification, and also with the argument that the Commission's prior precedents had all initially focused on ATC3 as a starting point for market definition.

[39] In the decision under appeal in *AstraZeneca* (n 38 above), the Commission had rejected the defendant's buyer power arguments and found that the existence of health services was not sufficient to prevent the defendant's exclusionary practices or effects—a position upheld by the General Court. The UK competition authorities have reached a similar conclusion with respect to the NHS (OFT decision of 27 March 2003 *Genzyme Ltd*). By contrast, the Spanish authorities have previously cited the buying power of the domestic health service in finding that no market power existed, despite significant market shares (Case 628/07 *Preparados Farmacéuticos* and Case 56/2004 *Spain Pharma/Smithkline Beecham* (10 May 2007)).

Court upheld the Commission's conclusion that AstraZeneca's purported dominance was not defeated by the presence of very large, national health service purchasers.[40]

Patent-related practices as abuses under Article 102 TFEU

In the pharmaceutical industry, one of the most complex interactions between the competition and IP laws concerns the circumstances in which the exercise of lawful IP rights (normally patents) can constitute an abuse of dominance. This debate has historically focused only on the exercise rather than the existence of IP rights, since Community law has long been clear that the very existence of an IP right does not infringe EC competition law. More recently, the Commission's enforcement agenda has suggested that it is keen to revisit this approach. The Pharma Report for example appears to challenge the very existence of certain patent rights as well as the way in which they are exercised. Similarly, the Commission has pursued Article 102 enforcement action against both AstraZeneca and Boehringer Ingelheim in relation to the existence of particular patent rights.[41] The upshot of this apparent expansion of the reach of Community competition law has been to challenge views on the boundaries of IP and competition law and, more practically, to impugn a range of *prima facie* lawful patent-related behaviours, from initial acquisition of a patent through to its enforcement in court and on, ultimately, to development of patented improvements on a marketed product.

Competition and patent laws interact at the following points: acquisition of a patent (both for an initial discovery and for subsequent 'follow-on' innovations); enforcement of a patent; licensing of the patent discovery/technology; and settlement of patent litigation claims. Each of these is addressed in turn below.

Patent acquisition

The extent to which an invention is awarded a patent is an objective question determined on the basis of established patentability criteria: novelty, inventiveness, and industrial application. If each of those criteria is met, then an invention is granted a patent.[42] It is possible to think about pharma patent grants in two broad (and admittedly simplistic) categories.

---

[40] See *AstraZeneca* (n38 above) paras 265–6.

[41] *Ibid* para 362,

> The Court rejects the applicant's argument that a finding of an abuse of a dominant position requires that an exclusive right obtained as a result of misleading representations has been enforced . . . The mere possession by an undertaking of an exclusive right normally results in keeping competitors away, since public regulations require them to respect that exclusive right.

In other words, the very existence of an improperly obtained IP right is sufficient to affect competition adversely and thus could expose firms to Article 102 enforcement action.

[42] For some, the fact that an invention meets the patentability criteria precludes any role for competition law, insofar as the requisite 'public interest' that competition law seeks to protect has already been taken into account in the very foundation of the patent system,

First, a patent is obtained for an initial discovery (ie the 'base' or 'primary' patent). The competitive impact of the patent at this stage is on originator-originator competition. Logically, if one originator obtains a patent over the substance of a particular compound or molecule, then another originator is precluded from copying that same formula. Where competition problems are said to arise, however, is either where the claimed patent is overly broad in scope (so competition is precluded in areas outside the true scope of the patent) or where defensive patenting strategies are pursued, which the Commission says '... mainly focus on excluding competitors without pursuing innovative efforts'. Such conduct may serve to block efforts by other originators to develop products that compete with the patented drug, particularly if coupled with a refusal to license unused patents. In its Pharma Report, the Commission claimed to have found evidence of this problem, together with the related creation of patent 'clusters' or 'thickets', ie, the creation of a group of interrelated patents which make it difficult for a competing originator (or indeed a generics manufacturer) to be sure that their developed product does not in some way infringe one of the patents covering a specific product.

Second, originators may also obtain further 'secondary' patents that relate to the same molecule or compound subject to the primary patent—for example, a patent relating to the manufacturing process used to produce the molecule (a 'process patent') or a patent that relates to a new formulation of the same product (eg, for a capsule rather than a tablet version, or an extended release version of the product). The Commission has identified this practice as a barrier to generic entry, and said in the Pharma Report that documents gathered from originators in the course of that investigation '... confirm that an important objective of this approach is to delay or block the market entry of generic medicines',[43] including by '... extend[ing] the breadth and duration of their patent protection'.[44]

> If the invention is new, involves an inventive step, and is susceptible of industrial application, it is patentable. Patent law has determined that, in such circumstances, the grant of the exclusive right is in the public interest and competition law should not second guess this choice. It is not in the consumer interest to turn against successful innovators once the invention has proven its value and use competition rules to chip away patent protection.

See L Kjølbye, 'Article 82 EC as Remedy to Patent System Imperfections: Fighting Fire with Fire?' (2009) 32(2) World Competition L Economic Rev 163 181.

[43] Communication (n 4 above), para 10. There has been much criticism of the Commission's apparent misconception in its preliminary findings that the filing of additional patent applications extends the length or breadth of the patent covering the original molecule (the base patent), whereas, in fact, such applications relate to further inventions, building upon that base patent. Once the base patent has expired, generics are free to copy it, irrespective of how many other patents have been filed. The Pharma Report seeks to acknowledge this, and notes that its concerns instead relate to the extension of a *marketed product's* effective period of market exclusivity by the acquisition of additional, later patents, which the Commission alleges are then invoked after the original molecule patent has expired. Nonetheless, its use of overly broad language, such as that quoted here, may serve to perpetuate misconceptions as to how the patent system operates.

[44] Communication (n 4 above) 10. The Commission also identifies the filing of voluntary 'divisional patent' applications, potentially extending the examination period of the patent office, and, '... under certain conditions,... add[ing] to legal uncertainty for generics.' Communication (n 4 above) 11.

Underpinning future competition policy approaches to both of these patent categories are two central questions:(a) To what extent is the intention of the patent applicant relevant to assessment of whether the obtaining of a particular patent is abusive? (b) To what extent is the value of the patent in question relevant to the characterization of that patent as abusive?

(a) The intention of the patent applicant When an applicant applies for a patent, they are by definition intending to bring into existence a legal right that will to some extent exclude competition. That is the whole object of a patent. Thus, for many IP practitioners it makes no sense to use the concept of exclusionary object or intent as the distinguishing factor between abusive and non-abusive conduct. The only purpose of a patent is to provide an exclusionary right and no-one, applying for patent protection with their eyes open, could have any intent other than an exclusionary one. This concern is shared by the European Patent Office which has expressed its '. . . discomfort in [the Commission] classifying behaviours on the basis of the applicant's intent in applying for patent rights.' It further emphasized that,

> . . . it is a cardinal principle of European patent law that issues going to the intent of the applicant are irrelevant in terms of obtaining patent rights. The [European Patent Convention] requirements are strictly objective, and therefore predictable and easy to apply, thus enhancing legal certainty.[45]

While the Commission does acknowledge this fact, it nonetheless refuses to accept that its ability to employ competition laws in the patent context is consequently limited,

> . . . It goes however, without saying that the description of the underlying intentions [of a patent applicant] is relevant to understand how companies use existing legislative framework for their purposes. *The intention can also be taken into account in competition law assessments.*[46]

Were this to be an accurate statement of the law, it would expand the parameters of competition law review considerably. To understand this, we need to review—(i) the circumstances in which the existence of a pharma patent could be subject to scrutiny; and (ii) how intent evidence might be used to understand the implications of that conduct. It is easiest to consider these two issues together and to consider them first in the context of an application for a base or primary patent, and then in the context of a follow-on or secondary patent.

Obtaining a base/primary patent for a newly discovered molecule or compound As a general proposition, obtaining such a patent should not infringe the competition

---

[45] European Patent Office, 'European Commission Pharmaceutical Sector Inquiry Preliminary Report—Comments from the EPO' (2009) <http://ec.europa.eu/competition/consultations/2009_pharma/european_patent_office.pdf> accessed 12 September 2010, 6.

[46] Pharma Report (n 4 above) fn 375 (emphasis added).

laws—even if the applicant's stated intention (recorded in internal documents) is to exclude competition on that discovery. This is because the potential theories of harm that might ground a competition law challenge to a primary patent application all struggle to articulate why a particular patent application should be treated as an abuse of dominance, and then to distinguish accurately between problematic and non-problematic applications. Indeed, there is a range of considerations here which all suggest that competition law should approach questions about primary patent scope with the greatest of caution.

- The Commission has identified concerns arising from claims that an initial application is overly broad (or imprecise) and thus might hinder competition from rival originators. Such patents are described by the Commission as 'defensive' patents over '. . . inventions which the applying company considers to have little or no prospect of being developed and/or commercialised and/or which, once granted, the company holds primarily to protect itself against actual or potential competition.'[47] However, at the time of seeking a patent, the precise definition and future use of the patented product may still be unclear—indeed, further therapeutic uses for a molecule may emerge later. Penalizing pharmaceutical firms in competition law (or even potentially threatening to do so) for an unavoidable step runs the risk of undermining incentives to engage in R&D.
- Similarly in relation to divisional patents, the Commission said in the Pharma Report that '. . . divisionals seemingly serve to prevent or delay generic entry'. Divisional patents can serve a useful purpose in allowing pharmaceutical firms to obtain appropriate patent protection for a molecule that is found to have distinct applications during the patent application procedure. If a firm could not divide its initial patent application and obtain protection for additionally discovered uses of a molecule, then that would undermine the incentives of pharmaceutical firms to continue researching particular molecules once they are discovered. This could well deprive patients of valuable new products or improved versions of existing products.
- A competition law challenge on the basis that a patent has not been used could also harm innovation incentives. The Commission's approach in the Pharma Report appears to contemplate antitrust liability based on whether the originator considered the relevant invention to have '. . . little or no prospect of being developed and/or commercialised' at the time that the relevant patent application was filed. However, at the point that a firm needs to file for patent protection for a new discovery, it will simply not know whether that particular discovery is ultimately going to reach market. In fact, the proportion of new molecules and products that never reach market vastly exceeds the proportion of successfully developed and marketed products. The fact that a particular patent is not exploited by a firm does

[47] *Ibid* 1118.

not, on its own, indicate any intention to use the patent as a blocking measure. If that were to be the risk (which may well be unquantifiable at the time of filing the relevant patent application), firms might well be reluctant to file for patent protection. This, in itself, may deter them from investing in R&D.

- Competition concerns in relation to patent clusters centre on the complaint that clusters (or thickets) make it harder for potential competitors to understand exactly the scope of the patent(s) relating to a specific product, compound or molecule, and thereby hinder rival innovation and entry—'While this, during the period of exclusivity, is generally in line with the underlying objectives of patent systems, it may in certain cases only be aimed at excluding competition and not at safeguarding a viable commercial development of own innovation covered by the clusters.'[48] The primary intention of an originator in filing patents to protect any aspect of a particular molecule is directly linked to the need to protect an uncertain investment—at such an early stage of product development, the originator may not know which variant on a patent is likely to be most valuable, or whether the process innovation discovered to manufacture the product is in fact likely to be the most important and valuable innovation in the future. Therefore, critiquing patent clusters on the grounds that the intention of an applicant must have been to exclude competition simply misunderstands the way in which pharmaceutical R&D operates.

Given the difficulties and the potential adverse consequences of allowing too easy an interference with initial patent applications, in what sort of exceptional circumstances could one imagine interfering with a granted patent to find that the very fact of obtaining the primary patent was abusive conduct?

Is evidence of an intent to exclude competition via the primary patent enough to identify a set of exceptional circumstances? Does this question distinguish effectively between legitimate pro-competitive conduct and abusive competition? The Commission has implied so. In the Pharma Report, the Commission relied on selective quotes from originators' internal documents collected during the course of the pharma inquiry to evidence purportedly exclusionary 'intentions' on the part of patent applicants.[49] As the EPO has noted, the question of,

> . . . whether a patent is applied for in any given case to block the potential development of another invention by third parties rather than to protect the exploitation of

---

[48] *Ibid* 523.

[49] *Ibid* 493–5; 503–5; 1122–6. In general terms, the use of internal documents to assess the competitive impact of a particular course of conduct is now commonplace. The Commission's guidance on its Article 102 enforcement priorities states that its assessment may take into account—among various other factors—'direct evidence of any exclusionary strategy' (eg, in companies' internal documents) insofar as this 'may be helpful in interpreting the dominant undertaking's conduct.' See Article 82 Guidance (n 29 above).

> the invention as contained in that application by the applicant is a matter which may be difficult to elucidate prior to grant.

Seeking to make such an assessment after filing remains problematic and would cause considerable uncertainty for all patent holders and innovators. The very fact of a patent application is per se exclusionary, so the exclusionary nature of a statement does not in itself differentiate between legitimate use of the patent system and inappropriate anti-competitive use of the system.[50]

One may for example be able to justify interference where the primary patent at issue was obtained fraudulently—in other words, where the monopoly patent right was obtained as a result of a misrepresentation to the patent office. In that circumstance, the patent would not have been obtained as a result of any conduct that might be described as a 'legitimate' use of the patent system; hence, competition law might legitimately intervene in the patent system and sanction the applicant for the adverse competitive effects arising from the deception. For example, in *AstraZeneca*,[51] the Court concluded that,

> (355)...[T]he submission to the public authorities of misleading information liable to lead them into error and therefore to make possible the grant of an exclusive right to which an undertaking is not entitled, or to which it is entitled for a shorter period, constitutes a practice falling outside the scope of competition on the merits which may be particularly restrictive of competition ... (356) [T]he misleading nature of representations made to public authorities must be assessed on the basis of objective factors and that proof of the deliberate nature of the conduct and of the bad faith of the undertaking in a dominant position is not required for the purposes of identifying an abuse of a dominant position.

It is submitted that with evidence of misleading conduct, fraud, or deception, competition law can sanction improper use of the patent system if the effect of that conduct is to exclude rivals.[52] However, it does not follow that conduct which excludes

[50] Similar concerns arise from the Commission's focus on the 'primary' motivation behind originators' interventions before marketing authorization bodies regarding generic equivalents of their product. It can legitimately be argued that if the originator has bona fide concerns about a product's safety or efficacy, it should be entitled to raise such concerns even if that might delay market entry of that product. That is particularly so because the originator is often best placed to identify those concerns, given its access to the relevant scientific research.

[51] For example, AstraZeneca was found to have made 'misleading representations' to national patent authorities in order to gain extended SPC protection. As the General Court has now ruled in *AstraZeneca* (n 38 above), the fact that the same outcome might be obtained under both competition and IP laws is not a good objection to use of competition law sanctions in these circumstances.

[52] The UK submission to the OECD's 2009 Policy Roundtable, *Competition, Patents and Innovation II*, DAF/COMP(2009)22, drew an analogy from EU case-law on vexatious litigation (discussed later in this chapter) to limit intervention to cases of 'bad faith',

> There may well be an analogous argument that competition authorities ought only consider intervening in cases of strategic patent applications where there is evidence of a clear anti-competitive purpose behind patent filings, such as evidence of misleading the patent office ... or if there is other evidence of bad faith with regard to the patent application. (138).

(or hinders) rivals is by definition abusive, and competition law should not readily characterize as abusive conduct which can be explained by the particular way in which pharma R&D and product development works. In this situation, antitrust over-enforcement could have a material disincentivizing effect on originators by disturbing the uneasy balance between competition and IP laws.

Obtaining a secondary patent in relation to the same molecule A further important feature of innovation in the pharmaceutical industry is the incremental nature of product development—many products are only developed by building on existing patent rights.[53] This sort of innovation can take the form either of expansion of the existing therapeutic indications of a particular molecule (eg, the development of Viagra from treating heart conditions to treating male erectile dysfunction), or the discovery of more effective ways of formulating the same product or making it easier to take (eg, the development of extended release versions of products that allow patients to take one tablet a day, rather than several, thus increasing compliance with medication regimes and hence drug efficacy). Incremental innovations can, by their nature, often deliver greater 'social value' than the initial discovery of a particular molecule.[54]

At first sight, an originator seeking to protect follow-on discoveries is simply obtaining a patent—the existence of which should not be in question under EU (or indeed, national) competition laws. However, the Commission's approach in the Pharma Report was somewhat different from this, distinguishing between patentable primary discoveries on the one hand, and patentable secondary innovations on the other, and suggesting that the latter give rise to greater antitrust concerns than the former.

Competitive concerns arising from secondary patents centre on the complaint that such patents can be used to facilitate so-called 'second-generation' switching, whereby an originator, as it develops patentable improvements to its original 'first generation' product, focuses its marketing efforts on the improved product. As a result, on expiry of the initial primary or base patent, a generic will be free to imitate the first generation product, but not the second, which may at that point have usurped the first generation product in the market.[55] In other words, the practical effect of the incremental innovation may have been to extend the duration (and possibly also scope) of

[53] EFPIA, 'Submission to the European Commission in relation to the pharmaceutical sector inquiry' (n 6 above) para 49: 'Innovation in the pharmaceutical sector takes many forms. There is rarely a "Eureka" moment. Rather, pharmaceutical development is a process of gradual discoveries, improvements and incremental changes.'

[54] See, US National Research Council, *Prospectus for National Knowledge Assessment* (1995) 10:

> The cumulative effect of numerous minor incremental innovations can sometimes be more transformative and have more economic impact than a few radical innovations or 'technological breakthroughs'.

[55] This might be particularly the case if the launch of the second product has been timed to ensure switch-over to the second generation product sufficiently before patent expiry on the first product to guarantee that the second product can launch unchallenged into the market.

the originator's patent protection until long after the expiry of the base patent filed for the original chemical compound. For this reason, the Pharma Report argued that secondary patents '... may be used as one of the means to extend the exclusivity period of a product to the detriment of health insurance schemes and ultimately consumers'.[56]

Competition challenge to the *existence* of secondary patents raises a number of issues. First, European patent law draws no distinction between primary and secondary patents. Any innovation is assessed individually against the objective criteria for grant of a patent, consistent with the proposition that IP law rewards and protects innovative endeavour.[57] In line with the analysis above, that should also suggest that provided an innovation satisfies the conditions for grant of a patent, competition law should not intervene unless very exceptional circumstances (such as fraud) are in evidence.

Second, the fact that patent protection is available for secondary innovations itself promotes competitive interaction among firms. If it is right that firms innovate in an effort to escape competition from others marketing similar products, incremental innovation is the natural consequence of this. Such incremental innovation stimulates further pharmaceutical research by competitors, creating a 'virtuous' circle of research, product discovery and development, and ultimately competition on the market.[58]

Against this background, the Commission's identification of intent evidence as the differentiator between legitimate innovative incremental competition and illegitimate anti-competitive use of secondary patents is understandable. But that approach faces similar analytical flaws to those discussed above in relation to base or primary patents.

- Challenging the intention of an originator in filing for a patent on a secondary innovation does not of itself differentiate between pro- and anti-competitive activity since, by definition, a secondary patent aims to exclude competition on that innovation.

[56] Pharma Report (n 4 above) 1514.

[57] In its response to the Commission's preliminary report in the pharma inquiry, the EPO commented that 'Provided the invention falls within the definition of patentable subject-matter, neither the nature of the invention, its importance or usefulness for mankind, its commercial potential or the purpose for which the application has been filed, are relevant to the decision-making process within the European patent granting system' European Patent Office, 'European Commission Pharmaceutical Sector Inquiry Preliminary Report—Comments from the EPO' (n 45 above).

[58] Under Indian law, no patent protection is granted to incremental pharmaceutical innovations that do not improve a drug's 'efficacy'. Many commentators argue that this approach has stunted innovation. (White & Case LLP and Dua Consulting, *The Value Of Incremental Pharmaceutical Innovation: Benefits For Indian Patients and Indian Business*, Report Prepared for the US-India Business Council (June 2009)) <http://dev.ahealthyindia.org/wp-content/uploads/2010/03/USIBCIncremental InnovationReportFinal.pdf>, accessed on 12 September 2010).

- A secondary innovation—even if implemented as part of an intended exclusionary strategy—will not work unless the innovation in question delivers clinical benefits over and above those of the original product. In essence, the marketplace will not pay the higher price of a new patent protected product unless justified by clinical advantages over the (now lower-priced) generic version of the original product.

The discussion above in relation to both base (primary) patents and secondary patents demonstrates the theoretical and practical difficulties of seeking to rely on 'intent evidence' to differentiate between legitimate and anti-competitive use of the patent system: a difficult step when the patent right in question by definition has exclusionary effects. The fact that this question arises in relation to steps taken by originators to create patent rights makes this question especially complex since it appears to undermine a formerly cardinal principle of EC competition law—namely that the competition laws bite only on the exercise of an IP right and not on its creation.

(b) The 'strength' of the patent When scrutinized carefully, competition concerns about the creation of patent rights relate not only to the intention of the originator when applying for the patent, but principally to the underlying strength of a patent.

If an originator intends to exclude competition on the basis of a legitimately gained, valid patent, that conduct (and indeed intent) should not be impugned by the competition laws. However, if the creation of a patent right could be challenged by the competition laws in circumstances where the underlying patent is not legitimately gained (eg, because it is known to be an ill-founded patent), we should also question whether the set of 'exceptional circumstances' in which a patent could be challenged can be expanded to include other instances where the underlying patent is poorly founded.

The Commission raised this issue squarely in the Pharma Report by identifying internal documentary evidence from originators conceding that underlying patents (in this case, in relation to patent clusters) might not in fact be strong.[59] Such concessions might arise in various ways—whether related to the primary or base patent, or possibly more likely in relation to secondary patents (which the Pharma Report treats as being inherently weaker than primary patents). There are a number of problems with the Commission's focus on such 'concessions'.

It is not immediately clear that simply because one patent might be thought to be weaker than another then it is not deserving of patent protection. Patents are granted

---

59 See Pharma Report (n 4 above) 503–5. Such an implication stems from the juxtaposition, in both the Communication (n 4 above) at 10 and the Pharma Report (n 4 above) at 201, of, on the one hand, the potential effects of patent clusters on market entry and, on the other, originators' views of the 'strength' of their patents:

> When the number of patents and in particular of pending patent applications is high (patent clusters), this can lead to uncertainty for generic competitors—affecting their ability to enter the market. Statements in internal documents collected in the context of the sector enquiry point at the awareness by patent holders that some of their patents might not be strong.

on the basis of a set of objective criteria; application of which results in a binary 'granted' or 'not granted' decision. There are well-developed appeal processes for both applicants and objectors to challenge decisions of the EPO.

Indeed, it is difficult to see how, other than in the most exceptional circumstances, competition law could on an *ex post* basis reach a judgment that a granted and valid patent was not sufficiently valid to be a legitimate use of the patent system. On a practical level, how is the strength of a patent's validity to be judged, and by whom? The subjective evaluation of a patent application's 'strength' under competition laws risks substituting the competition authority's view for that of the applicant and of the relevant expert patent office.[60]

One might infer from the above that competition challenges are really a reflection of a concern that either: (a) patent offices grant patents too easily, and that a 'raising' of the bar is required;[61] or (b) established patentability criteria (even if properly applied) are inadequate to ensure that an invention is 'worthy' of a patent monopoly. In theoretical terms, there is some logic to this approach—it would mean that a discovery could be patented only when it could be demonstrated that it delivered some incremental therapeutic or clinical gain. It is submitted that this would be inappropriate for a number of reasons.

First, it would mean that patenting criteria for pharmaceutical-related discoveries would be different from those applicable to other industries. In no other industry is the availability of patent protection linked to the ultimate utility of the patented invention. If indeed that were the case, many fewer patents would be granted. Incentives for firms to engage in R&D would be greatly reduced since it would be far less clear that any R&D effort could obtain any return at all. The complexity of patent assessments also means that some patents may be granted which are less robust than others. This is true of all sectors, and not just the pharmaceutical market. In many respects, it is inherent in any system that seeks to apply universal criteria. Even if a 'fix' for this problem exists, however—and it is not apparent that one does[62]—that solution likely lies within the patent system itself and not with the competition laws.

---

[60] The head of the Commission Task Force running the pharma inquiry, speaking in a personal capacity, has himself acknowledged that:

> . . . competition authorities are well advised not to believe that they could or should—outside specific circumstances—try to correct failures in the patent system and even in these exceptional cases they do not take over the role of the patent office. D Schnichels (n 37 above) 8

[61] In the Pharma Report (n 4 above), the Commission expresses explicit support for recent efforts by the EPO to '. . . raise the patentability bar'.

[62] As Sir Robin Jacob, the English patent court judge commented:

> In truth a Patent Office is a kind of coarse filter—rejecting clearly bad cases but having to allow those which may be good. One should not feel too depressed by this major truth—particularly in the pharma industry where the legal costs of both innovating companies and generic companies are only a minor proportion of what is stake.

R Jacob (n 11 above).

Second, applying an additional 'efficacy' criterion in the pharmaceutical industry would be extremely difficult since, as noted earlier, an originator will not know the likely efficacy of a product at the point of seeking a patent. As explained at the beginning of this chapter, in order to protect the results of R&D effort, a pharmaceutical company needs to seek patent protection for a discovery at the earliest possible moment in time—any delay could render all of the R&D investment nugatory. At that point in time, the precise therapeutic indication for that chemical compound may still be unclear, as will certainly the compound's actual efficacy in treating that therapeutic indication, and its safety record in doing so.

Third, it is also unclear how the EPO (or indeed any other patent office) would adjudicate on claims of efficacy. Specialist agencies exist at both European and national level to review efficacy claims for drugs. This task requires a set of skills quite distinct from answering the key scientific questions on which a patent claim is based—novelty and inventive step. Nor is it apparent how or when any such test of strength/efficacy would be applied, or the evidence on which any evaluation would be made.

Fourth, and perhaps most importantly, it is far from clear that the outcome of an efficacy-related patenting criterion would be increased competition (which is of course the ultimate policy goal). In fact, what may result is a reduction in levels of innovation among pharmaceutical firms that are unwilling to risk the vast amount of money needed to develop and test a product before a patent becomes available.[63] If the initial discovery is not protected by the patent, then there is little incentive to continue developing the product for up to ten years in order to ascertain its efficacy and be granted a patent.

These four main difficulties demonstrate that any re-evaluation of patent strength (validity) raises material problems for pharmaceutical firms, patent offices and courts in seeking to understand when the existence of a patent might fall foul of the competition laws. While this may be appropriate in rare cases of wilfully deceitful behaviour (similar to those circumstances discussed above), outside of such extremes it is difficult to see how such 'evidence' can usefully be analysed. Indeed, when translated into a practical 'test' that firms can apply in order to determine the risk of application for a particular patent later being held to be abusive conduct, it is difficult to identify the degree of confidence that a firm must have in the strength of its application for that application, or any subsequent attempt to enforce the granted patent, to be guaranteed not to be abusive.

It is therefore submitted that, absent the misleading representations or fraudulent behaviour referred to above and found in, for example, *AstraZeneca*, any use of Article 102 TFEU in relation to the acquisition of patents should be

[63] See n 58 above.

carefully circumscribed.[64] The discussion above suggests that the original EU law principle (that competition laws bite only on the exercise of a right and not on its existence) has much to commend it. Efforts to look behind that principle and consider the circumstances in which a patent has been gained or the underlying validity of the patent raise very substantial legal and practical problems. Those problems are exacerbated when one considers the potential remedies for abusively bringing a patent into existence—potentially, the relevant remedy would invalidate the patent, thus depriving the company of *any* patent protection. This outcome is even more harmful to that company's interests than a compulsory licence (as discussed below, itself justified only in exceptional circumstances) which at least permits the company to obtain a reasonable royalty.[65]

Enforcement of patents

The value of a patent lies in the fact that it can be enforced against infringers, thus maintaining monopoly rights over the relevant product or compound. The European Council, when legislating for IP rights, put the position as follows,

> . . . without effective means of enforcing intellectual property rights, innovation and creativity are discouraged and investment diminished . . . the means of enforcing intellectual property rights are of paramount importance.[66]

Accordingly, regulatory authorities have been historically reluctant to intervene in litigation brought by or against patent holders to enforce or defeat, respectively, exclusive IP rights. The reluctance to intervene stems also from the sheer complexity of much (if not all) patent litigation and from the recognition that firms enjoy a fundamental right of access to the courts.[67] Antitrust authorities historically have been reluctant to find the commencement of litigation abusive except in very limited circumstances. Those circumstances arise when litigation is deemed 'vexatious', namely where the litigation,

(1) cannot reasonably be considered to be an attempt to assert what the dominant firm reasonably believes to be its rights, and can therefore be intended solely to harass or hinder another firm; and
(2) is part of a plan whose aim is to eliminate competition.[68]

---

[64] The OECD has warned against any involvement of competition authorities in the patent grant process:

> . . . [f]or several reasons, such as a lack of relevant technical expertise and limited resources, it does not appear to be prudent for competition authorities to assume responsibilities related to the initial review of IP applications.

OECD, *Competition, Patents and Innovation 2006* (n 31 above) 36.

[65] L Kjølbye, (n 42 above) 179.

[66] Council Directive (EC) 2004/48 on the enforcement of intellectual property rights [2004] OJ L 195/16, Recital 3.

[67] Under the European Convention on Human Rights.

[68] See Case T-111/96 *ITT Promedia v Commission* (1998) ECR II-2937, paras 55 and 57. Non-competition specific legal provisions also exist at national level to sanction litigants for unmeritorious

The difficulty, however, is that neither limb of this two-part test can be easily applied to complex patent infringement litigation, such as that involving pharmaceuticals. In the first place, the complexity of the patents typically involved in pharma patent litigation means that there will almost invariably be a genuine dispute (or at least a reasonable issue) about infringement. These disputes sometimes arise because the issue concerns secondary patents filed by an originator some years after a primary (or base) patent. In these circumstances, the litigation nearly always raises material issues as to the scope of the patent and the ability of a rival originator or generics firms to innovate around the claimed patent. Second, by its very nature, any enforcement action by an originator is intended to protect the monopoly created by the patent, and thus to eliminate competition within the scope of the patent. A further problem is that a proportion of patent litigation between generics and originators does not fit the mould envisaged by the above criteria—this is because the litigation is often commenced by generics firms seeking either declarations of non-infringement or declarations that the relevant patent is invalid.

The practical difficulty of implementing the above criteria as a test for vexatious (and therefore abusive) pharma litigation is amply demonstrated in the Pharma Report. Here, the Commission did not assess whether the patent litigations identified were prompted by an originator's genuine attempt to assert what it could reasonably consider to be its rights. Rather it focused principally on, for example, whether it was the originator that initiated the litigation (as opposed to the generic), whether the action related to primary or secondary patents, and which party ultimately prevailed in the litigation. The Commission's analysis did not draw an explicit link between issues of litigation initiation, litigation success and the lawfulness of pursuing a patent action; it did however assert that originators '. . . may consider litigation not so much on its merits, but rather as a signal to deter generic entrants.'[69]

It is respectfully submitted that this Commission statement does not represent an accurate statement of the law, nor would this be an appropriate 'direction of travel' for Commission enforcement. This is because the fundamental issue underpinning the Commission statement is that the strength of the patent is directly relevant to the question whether litigation in respect of that patent is legitimate or abusive. Put differently, this approach would place the emphasis on the dominant firm's view of the prospects of litigation success. It would mean that, in pharmaceutical and other complex patent litigation cases, to avoid a vexatious litigation challenge a dominant firm could be required to show not only that it initiated the litigation in a genuine attempt to assert what it reasonably considered to be its rights, but also that it believed

claims, eg, summary judgments that there is no arguable case, the striking out of a claim for disclosing no cause of action, and adverse costs awards.

69 See Communication (n 4 above) 11 and Pharma Report (n 4 above), para 549.

it had good prospects of success in that litigation. There are a number of straightforward objections to this approach:

- This analysis requires a firm to disclose privileged advice it has received from its counsel as to the prospects of success in the litigation.[70]
- It again places emphasis on the 'value' of the underlying patent. As discussed earlier in this chapter, decisions about the 'value' to be attributed to the patent(s) at issue do not fall easily within the competence of a competition authority—however sophisticated. It is unclear how a competition authority would decide that one litigation process was more (or less) legitimate than another based on the patent at issue and the relevant issue in the litigation.

It is submitted therefore that competition authorities should intervene only in exceptional circumstances to find that a firm's litigation enforcement of its patent constitutes an abuse of dominance. One can for example envisage limited situations where a patent on which originator sued a generic was subsequently found to be entirely without merit and the originator's stance in the litigation due solely to its knowing desire to maintain the unmeritorious patent and thus exclude competition on the patent for a longer period, consequently excluding a competitor from the marketplace. However, identification of patent litigation brought entirely without merit is a difficult analysis to undertake. One straightforward approach to this question would be to confine such a finding to situations where a patent defence or challenge has been dismissed by a court in terms suggesting that the defence/challenge should never have been mounted.

The difficulties of pursuing Article 102 TFEU investigations in this context are amply illustrated by the ongoing *Servier* saga. In 2009, Les Laboratoires Servier lost a patent infringement claim against Apotex Inc, which it was alleged had infringed Servier's patent on its heart drug, Coversyl (generic name perindopril). When upholding the High Court's decision invalidating the patent, the Court of Appeal criticized Servier in trenchant terms. Jacob LJ commented that Servier should have known from the outset that its claimed patents were invalid and that this was precisely the sort of situation where competition law ought to intervene.[71] The Commission opened an Article 102 investigation into Servier's conduct.[72]

---

[70] A discussion of the Commission's approach to legal professional privilege is outside the scope of this chapter.

[71] *Les Laboratoires Servier v Apotex* [2008] Cir 445,

> It is the sort of patent which can give the patent system a bad name... the only solution to this type of undesirable patent is a rapid and efficient method for obtaining its revocation. Then it can be got rid of before it does to much harm to the public interest.... It is the court's job to see that try ons such as the present patent get nowhere.

[72] See Commission press release of 8 July 2009, 'Commission opens formal proceedings against Les Laboratoires Servier and a number of generic pharmaceutical companies' MEMO/09/322 <http://europa.eu/rapid/pressReleasesAction.do?reference=MEMO/09/322&format=HTML&aged=0&language> accessed 12 September 2010. This investigation also relates to agreements between Servier and a number of generics companies including Teva and Mylan.

Subsequently, however, the Canadian Federal Court upheld the validity of a related Servier patent which covers perindopril—a critical step since Apotex is a Canadian firm and manufactures its generic perindopril in Canada. Apotex's manufacture of the product and possibly also export sales are therefore an infringement of the patent. Servier applied to amend its pleadings in the English litigation to take into account this fact. The Court of Appeal (Jacob LJ again) permitted Servier to do so,[73] but refused to order Apotex to repay damages and costs that it had previously been awarded. The judge noted that the Canadian litigation pre-dated the commencement of the English litigation, so Servier's application to amend its pleadings was extremely late. The litigation continues to resolve both the effect of Servier now being permitted to amend its pleadings and to resolve the costs issues.

This case demonstrates the complexity of applying competition law to pharma patents. Not only is there a dispute as to the validity of the underlying compound patent, there is a question as to the legitimacy of Servier commencing the original English litigation (amplified by Jacob LJ's comments). But all of this seems effectively to have been overridden by the Canadian judgment which essentially converts Apotex's winning UK position into an infringing position. While this may seem an extreme example, in this context characterization of conduct as abusive seems similarly extreme and prone to error.

Refusals to license patented technologies

A relatively unexplored frontier of the competition and IP laws—at least insofar as the pharma industry is concerned—relates to compulsory licensing of patents. Indeed, any use of Article 102 TFEU in relation to the licensing of patents would break new ground.[74] Nonetheless, the Commission has taken the view that '... refusal[s] to grant a license on unused patents ... in particular in situations where innovation was effectively blocked' are a matter that will *'remain'* under antitrust scrutiny.[75]

As a practical matter, licensing of patents is extremely common in the pharmaceutical industry. Many of the leading products now on the market in Europe are the product of extensive cross-licensing arrangements among the various originators that contributed to the development of the product. It is not surprising that the Pharma Report could identify only two cases in Europe in the 2000–2007 period in which an originator had abandoned an R&D project purely on the grounds of the patentee's refusal to grant a licence to a blocking patent.[76] Quite simply, if a patent is not being used by an originator, it will often be a better prospect to license the patent

[73] *Les Laboratoires Servier v Apotex* [2010] EWCA Civ 279.

[74] In *Microsoft* (n 30 above), although Microsoft argued that the property to which access was sought was covered by patents and other intellectual property rights, the CFI followed the Commission's position in holding that its finding was not predicated on whether this was or was not the case.

[75] Pharma Report (n 4 above) 1571.

[76] *Ibid* 1163–6.

in expectation of some royalty flow than to rely entirely on one's own product making it through the development process and to become an effective, marketed product.

Analysis of those cases where compulsory licensing might be in issue is complicated by both the relative lack of instances in which the question has arisen, and also by the comparative uncertainty of the law.

EU law on compulsory licensing of IP rights is in a state of flux.[77] While EU competition law has generally taken the view that IP rights should be compulsorily licensed only in exceptional circumstances, the precise definition and parameters of those 'exceptional circumstances' are uncertain. In some cases,[78] the Court focused heavily on the fact that the intellectual property rights (specifically copyrights) in question were '. . . difficult to justify in terms of rewarding or providing an incentive or creative effort';[79] ie, that they were somehow 'undeserved'. There are clear similarities between this case law and the sort of analysis that the Commission has been seeking to develop in respect of certain pharma patents. This already difficult position has been complicated by the judgment of the General Court in *Microsoft*, which appears to expand the concept of 'exceptional circumstances' beyond situations characterized by weak IP rights.[80] There is no explicit recognition in the Court's judgment of how these principles would apply where the underlying IP right is a patent, leading to concerns that the principles could compel dominant firms to license patents to competitors.[81]

It is submitted that questions commonly raised by licensing of pharma patents are different to the issues with which the Court was grappling in *Microsoft*.

- First, in contrast with the *Microsoft* scenario of using IP rights to leverage protection from competition from one market into another neighbouring market, the most common scenario arising in the pharma sector concerns licensing of patents to permit head-to-head competition with the dominant firm in the same market. Compulsory licensing in this context therefore runs the direct risk of undermining the very purpose for which a patent is granted, ie, allowing an innovator a period of exclusive use of the patented discovery as a reward for their investment and effort.
- Second, it is arguable that analysis of a pharma patenting situation could be analogous to a *Microsoft*-type position where the relevant patent is in a sense 'upstream' of a proposed development—for example, a second originator is seeking to develop a compound using a base patent, where the compound would be used for a second (and distinct) therapeutic indication. In these circumstances, it may be possible to

[77] See Part II in this book: Market Power and Abuse
[78] See *IMS Health* and *Magill* (n 30 above).
[79] Opinion of AG Jacobs in Case C-7/97 *Bronner* [1998] ECR I-7794, para 63.
[80] *Microsoft* (n 30 above).
[81] L Kjølbye (n 42 above).

draw some parallels with the Court's *Microsoft* analysis, but where the second therapeutic area is distinct from the first, it is difficult conceptually to argue that a refusal to license could amount to an abuse unless the patent holder was itself seeking to develop a product in the same second area. Even in this situation, it is submitted that the better view is that no abuse can arise since the essence of the abuse of dominance is to reserve to the dominant firm a substantial position on the second market—as noted above, there is no guarantee that the efforts to develop the second product would be successful and, perhaps more importantly, it is also open to the applying firm to seek to innovate around the patent. In fact, it may therefore be arguable that imposing a licensing obligation may serve to *reduce* innovative effort by the firm seeking the licence (and now presumably complaining about the lack thereof). This is different from *Microsoft* where access to Microsoft's interface information was a critical step in being able to compete in the second market.

- Third, there is also an interesting contrast with the rationale behind the Technology Transfer Block Exemption, which aims to promote innovative effort and the availability of technical advances by exempting in certain circumstances licensing of developed technologies among actual and potential competitors. However, one of the circumstances in which such licensing may not be exempted concerns situations where firms account for a substantial proportion of research and development in a particular sector. It would seem odd therefore that Article 102 TFEU might mandate licensing to a competitor in circumstances where the voluntary entry into that same licensing agreement between potential competitors could infringe Article 101.

Given these potential conflicts in the application of different competition laws, competition agencies should be loathe to impose licensing obligations on firms, especially in circumstances where the empirical evidence has not disclosed any material competition concerns.

Settlement agreements

Settlement agreements are addressed in detail elsewhere, so the discussion here is limited to identifying some common strands of analysis that flow from the nature of a pharma patent and the treatment of such patents generally under EU competition law.[82]

Agreements settling pharma patent disputes have attracted extensive antitrust scrutiny on both sides of the Atlantic. The most explicit competition concerns identified in the Commission's Pharma Report related to agreements between originators and generics to settle patent litigation. This particularly related to agreements where the generic agrees to remain off the market in return for any payment or value transfer that goes beyond mere reimbursement of legal expenses (known as 'reverse' or 'pay

[82] For a detailed analysis, see Chapter 12, in this book: Pat Treacy and Sophie Lawrance, 'Intellectual Property Rights and Out of Court Settlements'.

for delay' settlements). The Commission's enforcement activity since the conclusion of the pharma inquiry suggests that settlement agreements will be its primary area of focus going forward.[83] Former EU Competition Commissioner, Neelie Kroes, described such 'pay for delay' settlement agreements as '... not in line with [the Commission's] policy'.[84] Since the precise terms of that policy are unclear and the Pharma Report does not clarify the position, what might that policy on patent settlement agreements be?

The attitude of EU legislation to settlement agreements generally is ambiguous. On one hand, the treatment of IP litigation settlement agreements under Article 101 TFEU is favourable. The Commission's TTBER guidelines recognize the simple advantages of a settlement agreement in ending uncertainty as to the validity or otherwise of a particular IP right and bringing resulting litigation to an end. This philosophy was endorsed by the Court of Justice in the context of 'no challenge' clauses, holding that,

> In its prohibition of certain 'agreements' between undertakings, Article [101(1) TFEU] makes no distinction between agreements whose purpose is to put an end to litigation and those concluded with other aims in mind.

The rationale for diverging from this position appears related to the idea of weak patents being properly challenged and, if necessary, revoked. A settlement agreement could prevent this happening and thus unduly extend the period of patent protection enjoyed by the right holder. The head of the Commission's pharma inquiry task force has suggested that *any* restriction of generic entry accompanied by a significant

---

[83] See, eg, Case COMP/39.612 *Perindopril (Servier)*. Kroes, Press Conference at the European Commission (8 July 2009) (unpublished transcript):

> ... our first case is now open ... against Les Laboratoires Servier and several generics companies. It concerns suspected breaches of the EC Treaty's rules on both restrictive business practices (or Article [101 TFEU]) and abuse of a dominant market position (Article [102 TFEU]). The case will look at the agreements between Servier and a number of generic companies.

and *Perindopril (Servier) Opening of Proceedings* <http://ec.europa.eu/competition/antitrust/cases/decisions/39612/opening_proceedings.pdf> accessed 12 September 2010. In October 2009, the Commission launched further raids on six pharmaceuticals companies. Press reports suggest that these were again aimed at investigating settlement agreements. See also European Commission, '1st Report on the Monitoring of Patent Settlements (period: mid-2008–end 2009),' published 5 July 2010 available at <http://ec.europa.eu/competition/sectors/pharmaceuticals/inquiry/patent_settlements_report1.pdf>, accessed on 12 September 2010.

[84] Neelie Kroes, Press Conference at the European Commission (8 July 2009) (unpublished transcript). The Commission appears to have concerns that patent settlements in the pharmaceutical sector,

> ... are typically concluded without the most effected [sic] stakeholders being present during the settlement negotiations, namely the consumers or the health schemes representing their interests.

D Schnichels (n 37 above) 14. With respect, it is hard to see how private patent litigation could be resolved in the round—were that to be the case, it is unlikely that any patent litigation would be settled with consequent costs for firms, healthcare systems, and ultimately patients.

value transfer '... deserves antitrust scrutiny' and that originators should have the burden of '... demonstrating the appropriateness' of any value transfer.[85]

There are a number of concerns with this approach. If it were unclear to parties at the commencement of litigation that they could, if needs be, settle the litigation, that would act as a disincentive to commencing litigation at all. It is not the role of competition law to deter parties from litigating, nor to prevent them from settling that litigation if they conclude that it is in their best interests to do so. It would be an odd (and inappropriate) outcome if competition law effectively compelled a party to continue litigating an issue.[86]

Much of the Commission's current policy initiative is based on the approach of the United States Federal Trade Commission. However, the US approach is based on the very specific US regulatory requirements under the Hatch-Waxman Act applicable to generic product entry, and is itself highly controversial. A number of US superior courts have disagreed with the analysis advanced by the FTC, which has in fact experienced relatively little success in challenging pay for delay settlement agreements. US judicial resistance to the FTC's complaints rests on the notion that, provided the effect of the settlement agreement is not to confer on the rights holder a broader protection than was offered by the patent in question, then the settlement agreement cannot be impugned since it has no greater competitive effect than the patent itself does. This is a sensible approach since it preserves the integrity of the patent system, does not introduce odd incentives to litigate (or more likely, not to litigate), and leaves it to firms to decide which patents are in fact worth challenging.[87]

It will be apparent from the above brief discussion of pharma patent dispute settlements that the emerging central theme is founded on the strength of the underlying patent. In other words, competition concerns about settlement agreements appear to be raised when the patent concerned—although still valid—is not deemed 'strong'. This approach raises similar concerns to those discussed above in relation to the underlying patent claims themselves.

Conclusion

The balance between competition and intellectual property laws is difficult to maintain, and the resulting trade-offs persist through all the lifecycle stages of a pharma

85 D Schnichels (n37 above) 22.

86 To test this point, consider the absurdity of seeking to identify the stage at which a litigating party could cease to litigate without incurring competition law risk. If the logic of the Commission's analysis is that a weak patent should be challenged and if appropriate invalidated, would a generics firm incur competition risk by deciding not to litigate beyond a certain stage, or deciding not to appeal?

87 These difficulties have led to legislative proposals in the USA to limit the use of 'pay for delay' settlement agreements.

product's development and commercialization. These trade-offs and the differing incentives to which they give rise mean, in combination, that the risk of analytical error when reviewing a competition law issue is high. A failure to identify properly the underlying competition problem could mean that any subsequent application of an antitrust 'solution' risks being inadequate, inappropriate, or contrary to long-term trends in consumer welfare.

What role should antitrust law play in the pharmaceutical sector? Despite the complexities noted above, the pharmaceutical sector should not be subject to antitrust or IP rules different from other industries.[88] Rather, the specificities of the sector need to be taken into account when applying, or deciding whether to apply such rules. As the Pharma Inquiry has shown, there are risks in competition lawyers seeking to assess laws outside of their areas of expertise, particularly where doing so effectively overrides the views of those experienced and qualified in intellectual property matters.[89] Equally, if antitrust law is to have a role, the limits of that role should be as clearly defined as possible.

The single most important criticism though of the Pharma Report is the uncertainty that it has introduced into whether, or when, identified patent practices will be considered unlawful. It is not necessarily the purpose of a sector inquiry to assess the compliance with competition law of a particular practice, but the lack of such guidance in the Pharma Report, together with its promise of greater antitrust scrutiny of certain broad categories of practice and its implied questioning of the lawfulness of several others[90] leaves companies uncertain as to their legal rights. Given the symbiosis between such rights and ongoing R&D investment, this position risks undermining incentives to innovate.

---

[88] Some commentators have suggested that, when intellectual property rights are at issue, a sectoral approach to antitrust enforcement would be beneficial. Thus pharmaceuticals would be distinguished from, for example, telecoms markets on the basis of the differing roles that patents play in the competitive and innovative dynamics of each (see, for example, Dietmar Harhoff et al, *The Strategic Use of Patents and its Implications for Enterprise and Competition Policies*, Report commissioned by the European Commission (8 July 2007), pp. 91, 128–9, and 136–41, available at <http://www.en.inno-tec.bwl.uni-muenchen.de/research/proj/laufendeprojekte/patents/stratpat2007.pdf>, accessed on 12 September 2010). Others argue that there is no basis for such a distinction, or indeed any differentiation between the application of competition laws to intellectual property, as opposed to any other form of property right (see, eg, I Haracoglou, 'Competition Law as a Patent "Safety Net" in the Biopharmaceutical Industry' (n 27 above)).

[89] The EPO, among others, was very critical of the Commission's mischaracterization in its preliminary findings of several fundamental aspects of the patent system (eg the relevance of the intent of the applicant; the methodology for 'counting' the number of patents relating to an individual product). The Pharma Report does address these criticisms and the Commission purports to have engaged extensively with the EPO throughout the inquiry. However, concerns remain that some of the preliminary conclusions and criticisms which the Commission derived from those misconceptions have, implicitly at least, nonetheless been carried over into the Pharma Report.

[90] See further, S Priddis and S Constantine, 'The findings and wider impact of the EU pharmaceutical sector inquiry' (2010) 24(2) Antitrust 29 (Spring 2010).

It may be that such issues can be addressed through legislative[91] or other incentivizing mechanisms.[92] Even the Pharma Report itself appears to acknowledge this. The majority of the Commission's recommendations for 'the way forward' relate to regulatory or patent-related reforms and it saves its strongest plea for the urgent establishment of a unitary Community patent and single patent court. It is apparent throughout the Pharma Report that the Commission perceives that such unitary systems would help remove many of the concerns regarding patent enforcement strategies in particular, obviating the need for competition law involvement other than in extreme cases.

However, across the board reforms will, by their nature, be less likely to deal with the specificities of the pharma sector. Similarly, the ability of patent authorities to further 'raise the bar' in their review of patents, to ensure that fewer 'bad patents' are granted, has its limits. In this regard, might the Commission simply have to accept that market players will inevitably seek to use the patent system—or indeed any legal system—to their greatest advantage. That does not, however, mean the conduct is necessarily abusive, even if in certain circumstance other parties may be competitively harmed as a result. Of course, that is the essence of competition.

While the Pharma Report identifies a number of ways in which EU pharma markets are said to work less well than they should, it is respectfully submitted that those broad conclusions do not justify a recalibration of the balance between competition and IP laws in the sector. Competition and IP laws currently create a delicately balanced ecosystem in which the pharma industry flourishes. As explained above, that balance can be easily displaced with potentially significant consequences for investment in pharma R&D. The authors respectfully submit that caution is required, and competition law should hesitate before disturbing legitimately gained and held intellectual property rights.

[91] See Regulation (EC) 141/2000 on orphan medicinal products (2000) OJ L18/1, enacted to encourage R&D in rare diseases and increase orphan drug output.

[92] For example through the flexible use of pricing mechanisms (such as the Pharmaceutical Price Regulation Scheme in the UK).

12

INTELLECTUAL PROPERTY RIGHTS AND OUT OF COURT SETTLEMENTS

*Pat Treacy and Sophie Lawrance**

Introduction

This chapter reviews the EU competition law approach to settlements. This review starts by considering the incentives for companies to settle IP litigation, together with an overview of the competing policy objectives that are in play. We also provide a brief summary of the key terms that are typically included in settlement agreements.

We then turn to an in-depth consideration of the relevant legislation (in particular the European Commission's Guidelines on technology transfer agreements)[1] and case law. This is followed by a section on the particular issues that arise in relation to settlements in the pharmaceutical sector following the Commission's Sector Inquiry and—for comparison purposes—the relevant US case law. Finally, we discuss how the competition authorities may apply EU law in this area and the difficulties that they may face, and look at the likely future developments.

* Pat Treacy leads the Competition team at Bristows and is widely recognized for her advice on the competition law issues that affect high technology companies. Sophie Lawrance works within the Competition team at Bristows and advises on all aspects of competition law, but has a particular interest in issues involving the interplay between competition law and IP rights.

[1] Commission Notice, 'Guidelines on the application of Article 81 of the EC Treaty [now Art 101 of the Treaty on the Functioning of the European Union] to technology transfer agreements' OJ [2004] C 101/02 (hereinafter '*Technology Transfer Guidelines*').

A. Preliminary Considerations

Why companies settle IP litigation and why there may be competition law concerns

According to statistics issued by the European Patent Office, some 146,561 European patents were applied for in 2008, with nearly 60,000 being granted in that year.[2] Indeed, the last ten years have seen significant increases in overall patent numbers. With more patents come greater possibilities for enforcement activity. It is thus unsurprising that there has also been an upward trend for patent litigation.[3] There is less evidence that enforcement of other IP rights has increased to the same extent. Nevertheless, the overall frequency of IP litigation means that an increasing number of cases have the potential to settle out of court. Indeed, settlement is a popular option, with a significant proportion of cases settling prior to judgment, and in some cases even before formal legal proceedings have started.[4] This applies to all litigation, not just that relating to intellectual property.

There are a number of reasons why parties may wish to settle threatened or actual litigation, rather than pursue it to its conclusion. Both parties to the dispute risk incurring considerable costs. Litigation expenses are often high, in particular in jurisdictions (such as England and Wales) which provide for documentary disclosure and cross-examination. The losing party may be exposed to a costs order, requiring it to reimburse the successful party. There will also be the cost of management time that is allocated to fighting the case. Both parties are subject to considerable commercial uncertainty. The alleged infringer will not know whether its plan to enter, or remain on, a particular market will still be viable after judgment. The right holder does not know if it may face new competition.

Patent litigation gives rise to particular risks for the parties. For patents, and other registered rights, the litigation usually involves both an allegation of infringement and an attack by the alleged infringer on the validity of the asserted right. Rights owners will rarely be entirely confident that they will not lose such a validity challenge. Although granted IP rights are treated as presumptively valid until revoked,[5]

2 European Patent Office, Annual Report 2008 (2009) <http://documents.epo.org/projects/babylon/eponet.nsf/0/7943587024b8e445c12575a00056831b/$file/epo_annual_report_2008.pdf> accessed 13 September 2010, 72–5.

3 See for example, statistics for the pharmaceutical sector, see European Commission 'Pharmaceutical Sector Inquiry Final Report' (8 July 2009) (hereinafter '*Sector Inquiry Report*') <http://ec.europa.eu/competition/sectors/pharmaceuticals/inquiry/staff_working_paper_part1.pdf> accessed 13 September 2010 para 589.

4 See for example, the UK Patents Court, see UK Ministry of Justice, Judicial and Court Statistics 2008 (2009) <http://www.justice.gov.uk/publications/docs/judicial-court-statistics-2008-full.pdf> accessed 13 September 2010, 34.

5 In the UK, see the Patents Act 1977, s74. The presumption of validity is not as explicit as under US law (Patent Act, 35 U.S.C 282). The exact legal status of the presumption of validity varies around the EU.

patents have also been described as 'probabilistic' rights,[6] signifying that the grant of the right by a patent office merely gives the patentee a chance of being able to exclude those who use technology that appears to be covered by the right. The right to assert the patent does not remove the risk that it may be invalidated by a subsequent attack in litigation. For example, a new piece of prior art may be identified that had not been uncovered during the patent office's review.

Both parties may also discover that they have a common interest in discontinuing the litigation. Rather than fighting, they may be able to agree to work together. For example, a would-be entrant with a generic pharmaceutical product who has been sued by the incumbent pharmaceutical company may agree to become a distributor, rather than launching its own product. This could be desirable for both parties, and may also be an efficient outcome in competition terms. Moreover, national courts often promote settlement for efficiency reasons, for example, by requiring the parties to undertake alternative dispute resolution.[7]

It is less clear that the EU competition authorities and courts have fully recognized the benefits of settlements.[8] The Court of Justice of the European Union ('CJEU') has acknowledged that trade mark delimitation agreements (which also often serve to settle litigation) may be 'lawful and useful'.[9] The Commission has also signalled its acceptance of settlement agreements in the report issued in the Pharmaceutical Sector Inquiry,

> As in any other area of commercial disagreement, the parties concerned may have an interest in ending a dispute, opposition or litigation and instead reaching a settlement as a compromise. The parties may prefer to discontinue the dispute or litigation because it proves to be costly and time-consuming, and might also be unpredictable in its outcome. Settlements are thus a generally accepted way of ending disputes, opposition procedures and litigation.[10]

Of course, there may be settlements which have potential competition law implications. The owner of an IP right which has been attacked in litigation may be facing the prospect of losing a valuable asset. This is particularly true of patents, which often number among a company's most valuable assets. Since the loss of such patents often leads to a price crash on the products that they protected (following market entry by competitors) companies may be willing to take radical measures to protect those assets. The alleged infringer may also have an interest in settling that goes beyond putting an end to the cost and uncertainty of litigation. For example, if it can secure

[6] M Lemley and C Shapiro, 'Probabilistic Patents' (2005) 19(2) J Economic Perspectives 75.

[7] Civil Procedure Rules (England and Wales), r 1.4(e).

[8] In the US there is judicial recognition of the value of settlements, despite the risk of harm to competition: *Re Tamoxifen Citrate Antitrust Litigation* 544 F3d 197 (2nd Cir 2006) 202.

[9] Case 35/83 *BAT Cigaretten-Fabriken GmbH v Commission* [1985] ECR 363 para 33 (hereinafter referred to as '*Dorcet/Toltecs*', being the trade marks in issue).

[10] *Sector Inquiry Report* (n 3 above) para 707.

an assurance from the incumbent IP owner that it will receive some form of preferential treatment, this may give it an advantage over other potential entrants. Where the companies have a common interest in settling, and where such settlements seem likely to have effects on competition between the parties or on the ability of third parties to compete, the competition authorities have taken an interest under the competition rules.

The overall policy objectives of IP rights and competition laws affect the analysis in this area, and should not be ignored. IP rights reward innovation and encourage research and development. However, the competition authorities are concerned to ensure that companies do not reach agreements protecting rights which, in some instances, may be said to have been wrongly granted. Thus, the CJEU has said of patents that '... it is in the public interest to eliminate any obstacle to economic activity which may arise where a patent was granted in error'.[11] The Court has also approvingly reported a view of the Commission that '... a reduction in the number of trade marks on the register would promote the opening up of the markets and increase competition'.[12]

There are thus a number of competing policy objectives in play:

- improving efficiency of the courts by encouraging settlement;
- the general presumption that IP rights have been validly granted (applicable for registered rights, such as patents);[13]
- ensuring that valid IP rights can be fully exploited and protected so as to encourage and reward innovation in the long term;
- allowing (or even encouraging) the revocation of invalid IP rights, which are a barrier to competition; and
- preventing infringements of competition law and encouraging competition in the short to medium term.

This chapter reviews the EU competition law approach to settlements in the light of these different policy objectives. Before turning to consider the applicable legislation and case law, it may be useful to consider a short summary of the key terms that are included in most settlement agreements.

[11] Case 193/83 *Windsurfing International Inc v Commission* [1986] ECR 611 para 92 (hereinafter '*Windsurfing*').

[12] *Dorcet/Toltecs* (n 9 above) para 32. In fact, the extent to which an IP right blocks market access varies between the different rights. The Commission's view arguably overstates the extent to which a particular mark is capable of affecting competition. Others on the market are free to develop different trade marks. If a trade name has become a generic reference for a particular class of goods, there are provisions under trade mark law to declare that this is the case.

[13] The presumption of validity of European patents varies on a country-by-country basis, and is not typically as strong as under US law.

Anatomy of a settlement agreement

When a case settles out of court, it is rare for the parties simply to drop hands. Even if they are not planning any future collaboration, they will usually set out the terms on which they agree to end the dispute. Typical terms are as follows:

- arrangements for withdrawing the litigation (which may be in more than one forum, possibly involving both courts and patent offices, and could involve both infringement and validity issues);
- where there has been an attack on validity, the settlement agreement will include a commitment not to challenge the IP right again, including in other jurisdictions/courts; a positive acknowledgement of a right's validity may also be included;
- in trade mark cases, the parties may agree a geographic or subject-matter delimitation to allow two similar trade marks to coexist;
- other terms relating to the IP rights may be included, such as agreement as to which of two rights has priority or, in a standardized industry, whether a particular patent is essential to the standard;
- the alleged infringer will commit either to stop infringing or threatening to infringe the IP right; alternatively, the parties may agree terms allowing entry, for example by granting a licence;
- the parties may reach some agreement about how the costs of the litigation, and any associated costs (eg due to withdrawing from a market) are to be met.

Where the parties wish to have some future licence or supply relationship, they may include such terms in the same agreement as the settlement, or they may decide to deal with them in a separate contract.

B. The EU Competition Law Approach to IP Settlement

This section considers first the overall approach to settlement agreements, including those where a licence or supply agreement is granted, before turning specifically to no-challenge provisions and trade mark delimitation agreements. This section finally considers the potential applicability of the rule prohibiting abuse of a dominant position.

Overall approach to settlement agreements

The EU competition authorities do not expect litigating parties to fight to a standstill. The Commission's final report in the Pharmaceutical Sector Inquiry indicates that settlements do not automatically fall within the scope of the prohibition on anti-competitive agreements in Article 101(1) of the Treaty on the Functioning of the

European Union (TFEU).[14] Similarly, licences granted in conjunction with a litigation settlement are generally accepted, subject to the usual rules,

> Licensing including cross licensing in the context of settlement agreements and non-assertion agreements is not as such restrictive of competition since it allows the parties to exploit their technologies post agreement. However, the individual terms and conditions of such agreements may be caught by Article [101(1) TFEU]. Licensing in the context of settlement agreements is treated like other licence agreements.[15]

This accords with the general principle expressed in the introductory section of the Technology Transfer Guidelines that '... licensing as such is pro-competitive as it leads to dissemination of technology and promotes innovation'.[16]

However, this approach applies only if there is a genuine dispute about a valid right. If the dispute is a sham, or if the right itself is invalid, any 'settlement' agreement involving restrictions upon market players will be caught by Article 101(1) TFEU. This was the case in *Dorcet/Toltecs*, where BAT agreed to withdraw its opposition to the international extension of a trade mark that it claimed was likely to be confused with its own. As its own trade mark was, however, dormant (ie unused and susceptible to being removed from the register), the CJEU held that the obligation on BAT to withdraw its opposition was 'purely fictitious' and that the agreement, in consequence, allowed BAT to interfere with the conditions of competition.[17] The English Court of Appeal took a similar approach (on the advice of the Office of Fair Trading) in another case, refusing to allow the parties to patent litigation to settle their dispute by agreeing to restore a patent to the register that had been declared invalid at first instance.[18]

Settlement agreements that involve the grant of a licence serve to 'unblock'; ie they remove obstacles to market entry, thus opening up competition. However, insofar as such agreements impose restrictions '... that clearly go beyond what is required in order to unblock',[19] the Technology Transfer Guidelines state that they are likely be caught by Article 101(1) TFEU, as will provisions which lack any connection with the use of the right in question.[20]

The Technology Transfer Guidelines do not identify which restrictions in a settlement agreement 'clearly' do more than unblock. This must be analysed on the basis of general principles and the case law relating to settlement agreements. It is clear

---

[14] Before the Treaty of Lisbon came into force on 1 December 2009, Article 101(1) TFEU was Article 81(1) EC Treaty.

[15] *Technology Transfer Guidelines* (n 1 above) para 204.

[16] *Ibid* para 9.

[17] *Dorcet/Toltecs* (n 9 above) para 35.

[18] *Halliburton Energy Services Inc v Smith International (North Sea) Ltd* [2006] EWCA 185 (hereinafter '*Halliburton Energy v Smith*').

[19] *Halliburton Energy v Smith* (n 18 above) para 207.

[20] eg the prohibition on advertising in *Dorcet/Toltecs* (n 9 above) paras 34, 37.

that the Commission intends the word 'required' in this paragraph to be understood objectively, rather than from the subjective perspective of one of the parties. Thus, for example, restrictions that relate to other IP rights (ie not those in dispute) may well be regarded as going further than needed to unblock. Having said that, there may be situations where IP rights are so closely linked that there could be some justification for dealing with them together (for example, applying the ancillary restraints doctrine).

No-challenge clauses

Probably the most important provision in any settlement agreement is the no-challenge clause. The Commission acknowledges in the Technology Transfer Guidelines that,

> It is inherent in such agreements that the parties agree not to challenge *ex post* the intellectual property rights covered by the agreement. Indeed, the very purpose of the agreement is to settle existing disputes and/or to avoid future disputes.[21]

Accordingly, the Guidelines state that such provisions usually fall outside Article 101(1) TFEU altogether.[22] It may be inferred that the Commission would treat a positive acknowledgement of the right's validity in the same way as an actual obligation not to challenge.

However, Commission guidelines are not binding on the European courts.[23] In fact the Commission's view does not entirely accord with the prior case law of the CJEU.

Windsurfing

The first of the two key cases on this point was decided by the CJEU in 1986. In *Windsurfing*,[24] the CJEU had to consider a number of clauses in a licence, including an acknowledgement of trade mark validity and an obligation not to challenge a patent. The Court held that such clauses did not fall within the specific subject matter of the rights in question and thus fell within the purview of the prohibition on anti-competitive agreements insofar as they restricted competition.[25] The CJEU emphasized that '. . . it is in the public interest to eliminate any obstacle to economic activity which may arise where a patent [is] granted in error'.[26]

The CJEU went on to find that both clauses restricted competition. As (following the rules then applicable) the parties to the agreement had not sought an exemption for

[21] *Technology Transfer Guidelines* (n 1 above) para 209.
[22] *Ibid* para 209.
[23] *Technology Transfer Guidelines* (n 1 above) para 4.
[24] *Windsurfing* (n 11 above).
[25] *Ibid* paras 92–3.
[26] *Ibid* para 92.

the agreement, the CJEU did not decide whether the clauses fulfilled the exemption requirements in Article 101(3) TFEU.

This approach appears to be contrary to the Technology Transfer Guidelines, which treats such clauses as (generally) being outside Article 101(1) TFEU altogether. The difference may be due in part to the fact that the agreements under consideration in *Windsurfing* were not directly connected to the settlement of litigation. There were thus no specific justifications for the no-challenge clauses. Had the main purpose of the agreements been to settle litigation, the outcome might have been different (although see discussion of *Bayer v Süllhofer* below).

Judged by today's standards, which are more informed by the underlying economics, the case now appears somewhat outdated.[27] For example, the CJEU agreed with the Commission that a requirement to acknowledge a trade mark was liable to '. . . restrict the licensees' competitiveness', yet the ways in which this might occur were not analysed. Moreover, under the Technology Transfer Block Exemption, no-challenge clauses are covered by the exemption, provided they go no further than allowing for the termination of the licence in the event that the licensee challenges the licensed IP.[28] Similar provisions, such as an acknowledgement of validity, might now fall to be considered under the ancillary restraints doctrine.[29]

Bayer v Süllhofer

A case directly dealing with a settlement agreement came before the CJEU two years after *Windsurfing*.[30] In *Bayer v Süllhofer*, the Court held that settlement agreements do not warrant special treatment, and that the competition rules apply to agreements putting an end to litigation in the same way as other agreements.[31] It therefore held that a no-challenge clause in such an agreement may be caught by Article 101(1) TFEU.

The Court outlined a couple of exceptions to this general position, neither of which are specific to settlements. Thus, a no-challenge clause will not restrict competition if the licence is free.[32] Even if royalties are payable, there will be no restriction of

[27] B Batchelor, 'EC tones down its final report into the pharma sector, but ramps up enforcement activity' (2010) 31 ECLR 16, 18.

[28] Commission Regulation (EC) No 772/2004 on the application of [Article 101(3) TFEU] to categories of technology transfer agreements, OJ [2004] L123/11, Article 5(1)(c). The parties must also meet the market share thresholds.

[29] As in Commission Guidelines on the application of [Article 101(3) TFEU] OJ [2004] C101/97 paras 28–31.

[30] Case 65/86 *Bayer AG v Maschinenenfabrik Hennecke GmbH & Heinz Süllhofer* [1988] ECR 5249 (hereinafter '*Bayer v Süllhofer*').

[31] *Ibid* para 15.

[32] *Ibid* para 17.

competition if '... the licence relates to a technically outdated process which the licensee undertaking did not use'.[33]

The justification for these dicta is not obvious. The judgment refers to the fact that the licensee will not suffer a 'competitive disadvantage' if it does not have to pay royalties, yet the policy basis for criticizing no-challenge agreements in the *Windsurfing* case was not so much the restriction placed on the licensee, but rather the blocking effect on the wider market that may arise from shielding invalid rights. The fact that a free licence has been given to one licensee does not mean that the licensor will be similarly generous to others. In fact, assuming the licensed technology is not in fact so outdated as to lack any objective value, granting a free licence might be seen as a 'value transfer' (to use the words of the Sector Inquiry Report)[34] to the licensee. As will be discussed further below (section C), such transactions may be regarded as giving rise to quite significant competition issues.

The Commission's observations in the case set out a four-part test embodying its approach. It stated that no-challenge clauses may 'in principle' be considered to restrict competition, but that a particular clause will be compatible with Article 101(1) TFEU provided,

- the purpose of the agreement is to put an end to proceedings pending before a court;
- the existence of the IPR which is the subject of the dispute is genuinely in doubt (presumably the Commission means that there should be a genuine dispute as to validity);
- the agreement includes no other clauses which restrict competition; and
- the provision relates to the IP right in issue.[35]

The CJEU's judgment explicitly dismisses the Commission's approach.[36] Nevertheless, the Commission's view is instructive, given the subsequent divergence (as a result of guidelines and enforcement practice) between the Commission's policy and that of the courts. This test has a good deal in common with the approach now taken in the Technology Transfer Guidelines.[37]

Penneys

One further case merits a mention. *Penneys* concerns a trade mark delimitation agreement entered into after litigation between two companies bearing the same name. The agreement included a no-challenge provision together with a promise by one of

[33] *Ibid* para 18.
[34] *Sector Inquiry Report* (n 3 above) para 742.
[35] *Bayer v Süllhofer* (n 30 above) para 14.
[36] *Ibid* para 15.
[37] Although the *Technology Transfer Guidelines* (n 1 above) explicitly refer to the case, the reference does not address the gap between the Court's approach and the Commission's enforcement policy.

the parties not to use the marks.[38] The Commission cleared the agreement following a modification to limit the duration of the no-challenge obligation to five years. Any challenge after this time was required to be supported by a form of escrow payment by the challenger.

Whilst this case was decided before the Commission's re-appraisal of no-challenge clauses in settlements as generally non-restrictive, the different approach may also reflect the difference between trade marks and patents. Whereas the latter have a limited lifetime, expiring after 20 years, trade marks can be maintained indefinitely, thus potentially increasing the restrictive effects. Equally, the Commission has recognized since the *Penneys* decision that the exclusionary effect of trade marks is limited unless the holder has a strong market position.[39] These differing approaches show how the restrictive effects of a no-challenge provision must be assessed in the context both of the type of right and of the market position of the right holder.

Conclusion on no-challenge clauses

Whilst the judgments in *Windsurfing* and *Bayer v Süllhofer* take legal precedence over the Guidelines and Commission decisions, the Commission has control over which cases are investigated at EU level. National competition authorities around the EU are also likely to follow the spirit of the Guidelines. Until the issue comes before the CJEU again, some ambiguity as to exactly how the issue will be treated must remain.

Trade mark delimitation agreements

Trade mark delimitation agreements are entered into between companies which have similar marks applied to similar goods or services. The risk of confusion is such that the companies may agree—often after a legal dispute—the countries in which the goods may be sold, or to allocate certain types of goods to each party. An agreement of this kind thus has much in common with a classic market sharing agreement.[40]

Nevertheless, the CJEU has acknowledged that,

> . . . agreements known as 'delimitation agreements' are lawful and useful if they serve to delimit, in the mutual interest of the parties, the spheres within which their respective trade marks may be used, and are intended to avoid confusion or conflict between them.[41]

[38] *Penneys* [1978] OJ L60/19.

[39] In *Moosehead/Whitbread* (Case IV/32.736) OJ [1990] L100/32, the Commission cleared a no-challenge provision in an exclusive licence on the basis that the trade mark, which was 'comparatively new in the territory' did not represent an appreciable entry barrier; European Commission, '20th Report on Competition Policy' (1990).

[40] A recent example of a trade mark delimitation being held to conceal a market sharing agreement is Case 13 U 227/02 decided by the Oberlandesgericht Celle (applying Article 101 TFEU), discussed in P Roth and V Rose (eds) *Bellamy & Child European Community Law of Competition* (6th edn, London: Sweet & Maxwell, 2008) [5.055].

[41] *Dorcet/Toltecs* (n 9 above) para 33

In *Dorcet/Toltecs*, the CJEU held that a delimitation agreement was not lawful. Not only was one of the underlying trade marks dormant, the scope of the restriction to be placed on the other party was unclear, and the agreement included additional restrictions (such as a limitation on advertising) which the Court said '. . . [do] not bear even the semblance of a connection with the question of use of the trade mark as such'.[42] The obligations were strongly biased in one party's favour and the agreement was held to be void and unenforceable.The Commission again held a trade mark delimitation agreement to be restrictive of competition in *Syntex/Synthelabo*.[43] This case concerned a geographical limitation agreement intended to end litigation between two companies as to the validity of their trade marks. Some guidance was given on situations where the Commission may intervene,

> The Commission is more often likely to intervene in cases where available evidence points to the inability of the party to legally enforce its claim of invalidity of the other party's trade mark. If it is evident that through such actions one party may legally exclude the other from selling in certain Member States, an agreement between the two companies having the same effect would not restrict competition.

The parties are obliged to examine all options available to them before signing an agreement that would effectively partition the common market. In particular, less restrictive solutions should be examined to determine if, for example, an agreement on how a trade mark should be used would succeed in eliminating the risk of confusion.[44] Leaving aside the question of how the competition authorities are to evaluate the likely outcome of proceedings before a specialist IP tribunal,[45] this commentary makes clear that the Commission regards territorial divisions as particularly problematic. Agreements which lead to the creation of other significant imbalances between the parties are also likely to be viewed as restrictive of competition. For example, the agreement in *Dorcet/Toltecs* appeared to give one party the rights to a more profitable part of the market (by confining the other to technologically inferior products).[46]

Conversely, agreements that are genuinely intended to resolve confusion between the marks of different parties are likely to be acceptable.[47] If the parties are making an effort to find less restrictive solutions—for example, by differentiating their

[42] *Ibid* para 34.

[43] Commission case closure *Syntex/Synthelabo*, as discussed in European Commission, '19th Report on Competition Policy' (1989) para 59.

[44] *Ibid.*

[45] This issue is discussed further in the context of 'reverse payments' in Section C below.

[46] *Dorcet/Toltecs* (n 9 above) para 21. Compare Commission decision in *Racal/Decca* [1990] 4 CMLR 627 where a licence agreement (not involving trade mark delimitation) confining the licensee to 'trailing-edge' products was found to breach Article 102 TFEU.

[47] There is a useful discussion of the cases in this area in Van Bael & Bellis, *Competition Law of the European Community* (5th edn, Kluwer Law International, 2010) 629–31.

products' get-up,[48] incorporating distinguishing marks, shapes, colours or wording,[49] or by entering into a licensing agreement without any territorial restriction[50]—no competition law infringement will be found.

Article 102 TFEU

The cases and guidance discussed above have all related to Article 101 TFEU. In principle, Article 102 TFEU (prohibition on abuse of dominance) may also apply if the party imposing the restrictions has market power. The Commission has already adopted this approach in relation to licensing agreements more generally.[51] Indeed, the recent judgment of the General Court in the *AstraZeneca* abuse of dominance case treats AstraZeneca's ability to conclude patent settlement agreements with potential competitors as itself indicative of the existence of a dominant position.[52]

To date, however, there have been no cases applying Article 102 TFEU directly to a settlement agreement (in *AstraZeneca*, the Court made clear that there was no suggestion that the settlement agreements themselves could be considered to be abusive).[53] It remains to be seen whether the investigations announced by the Commission following its Pharmaceutical Sector Investigation will change this.[54]

---

[48] *Persil Washing Powder* in European Commission, '7th Report on Competition Policy' (1978) paras 138–40.

[49] *Penneys* (n 38 above); *Bayer/Tanabe* in European Commission, '8th Report on Competition Policy' (1979) paras 125–27; *Osram/Airam* in European Commission, '11th Report on Competition Policy' (1981) para 97.

[50] *Herschey/Hershi*, in European Commission, '20th Report on Competition Policy' (1990) para 111.

[51] eg *Racal/Decca* (n 46 above); Case T-139/98 *Amministrazione Autonoma dei Monopoli di Stato v Commission* [2001] ECR II-3413; *Re YKK's Complaint* [1978] 3 CMLR 44.

[52] Case T-321/05 *AstraZeneca*, judgment of General Court dated 1 July 2010, <http://curia.europa.eu/jurisp/cgi-bin/form.pl?lang=EN&Submit=Submit&numaff=T-321/05> accessed 13 September 2010 (not yet reported) para 272.

[53] See n 52 above.

[54] Commission press release (IP/10/8), 'Commission opens formal proceedings against pharmaceutical company Lundbeck' (7 January 2010) <http://europa.eu/rapid/pressReleasesAction.do?reference=IP/10/8> accessed 13 September 2010; and Memo/09/322 'Commission opens formal proceedings against Les Laboratoires Servier and a number of generic pharmaceutical companies' (8 July 2009) <http://europa.eu/rapid/pressReleasesAction.do?reference=MEMO/09/322> accessed 13 September 2010. Both releases refer to Article 102 as well as 101 TFEU. Given the concerns expressed in the *Sector Inquiry Report* (n 3 above), reverse payment settlements may well be the focus for the Commission's investigations. See also G Ruddick 'EU's Neelie Kroes turns to Shakespeare to criticise "rotten" drug industry' Telegraph (London 8 July 2009) <http://www.telegraph.co.uk/finance/newsbysector/pharmaceuticalsandchemicals/5780189/EUs-Neelie-Kroes-turns-to-Shakespeare-to-criticise-rotten-drug-industry.html> accessed 13 September 2010.

C. 'Pay for Delay' Settlements: The US Experience, Rights and Wrongs, Modes of Assessment

In recent years, a 'new' type of settlement has been identified by the competition authorities on both sides of the Atlantic. For various reasons, consideration of such agreements has been most prominent in the pharmaceutical sector, although it is possible to foresee that, given certain market conditions, equivalent issues might arise in other sectors.

The agreements in question—sometimes pejoratively referred to as 'pay for delay' arrangements, or, more neutrally, as settlements involving a 'reverse payment'—typically arise out of a dispute between a pharmaceutical company which holds patents covering a particular medicine,[55] and a generic drug manufacturer which wishes to enter the market with a lower-priced equivalent product. The litigation may start with the generic company seeking to 'clear the way' to market entry, by applying to revoke the pharmaceutical company's blocking patents (or seeking a declaration of non-infringement). Alternatively, infringement proceedings may be started by the pharmaceutical company itself, following a threatened generic launch or entry at risk by the generic company. In the latter case, the generic company usually retaliates by arguing that the asserted patent should be declared invalid.

The parties to such a dispute may settle or pursue the proceedings in the ordinary way as discussed above. However, in some cases the parties have concluded an agreement which involves a promise by the generic company not to enter until some time in the future, coupled with some other benefit flowing from the pharmaceutical company to the generic manufacturer.

There are a number of possible permutations on this scenario,

- the agreed entry date may be no earlier than patent expiry, or it may 'split the difference' between the two companies, setting a date part-way between the date of the agreement and patent expiry;
- provision may be made for entry before the expiry date if another generic enters the same market;
- the benefit passing to the generic manufacturer may be a simple cash payment, perhaps stated to represent litigation costs, or may be given in kind (eg grant of a licence);
- other related agreements may be reached, eg appointment of the generic as (exclusive) distributor for the pharmaceutical manufacturer, possibly coupled with an agreement by the latter not to launch an authorized generic (ie in the

[55] In this chapter, the term 'pharmaceutical company' is used to refer to innovative R&D based companies in the pharmaceutical sector, as distinct from companies that focus on launching generic drugs.

manufacturer's own generic get-up), or a co-promotion agreement, under which the generic company agrees to support one of the branded products.[56]

Agreements of this kind cause the competition authorities concerns which go beyond the maintenance of potentially invalid patents as discussed in cases such as *Windsurfing*. When a generic drug enters into competition with a branded pharmaceutical there is typically a very significant price decrease.[57] To delay this price decrease beyond the period objectively justifiable under a valid patent may thus carry a high cost for health authorities and/or consumers.

However, as we discuss further below, it is not always appropriate to consider such collateral arrangements as 'value transfers', to use the rather pejorative term employed by the US Federal Trade Commission ('FTC') and the European Commission. A genuine transfer of value (or a transfer of value attributable to delayed entry) arises only where the cost to the payer appreciably exceeds what they otherwise receive. Such a cost/benefit analysis in a complex commercial arrangement will not be easy.

D. Basing EU Competition Policy on the US Approach to Reverse Payment Settlements

The European Commission has been clear about its intention to pursue agreements which reward generic companies simply for staying off the market. Settlement agreements of this kind were a key focus of the Sector Inquiry Report. The Commission stated that it found evidence of 'reverse payments' to generic companies totalling in excess of €200 million.[58] It concluded further that '. . . the average time to enter after loss of exclusivity was more than seven months on a weighted average basis',[59] and suggested that savings of some €3 billion could have been made if generic entry had occurred immediately after loss of exclusivity.[60] The report concludes that the

[56] In the US the FTC's reports on pharmaceutical settlements (received pursuant to the Medicare Prescription Drug, Improvement, and Modernization Act, 2003) provide information about the sorts of 'side deals' found in the agreements, eg Report by the Bureau of Competition, 'Summary of Agreements Filed in FY 2006' <www.ftc.gov/reports/mmact/MMAreport2006.pdf> accessed 21 September 2010, 4–5, in which the FTC reported a significant increase in the number of agreements involving some transfer of value to the generic company following *Schering-Plough* (n 66 below).

[57] The *Sector Inquiry Report* (n 3 above) puts the average decrease at about 20 per cent after one year and almost 25 per cent after two years (para 212). The FTC suggests a much greater drop, of around 77 per cent; FTC Staff Study, 'Pay-for-Delay: How Drug Company Pay-Offs Cost Consumers Billions' (January 2010) <www.ftc.gov/os/2010/01/100112payfordelayrpt.pdf> accessed 21 September 2010, 8 (hereinafter '*FTC Pay-for-Delay Study*').

[58] *Sector Inquiry Report* (n 3 above) para 1341.

[59] *Ibid* para 1559.

[60] Arguably this figure is inflated. Entry will rarely occur immediately, even in the absence of conduct that potentially breaches competition rules.

policies and conduct of pharmaceutical manufacturers, as well as certain features of the regulatory framework, contribute to this loss.[61]

As yet, there are no decided cases on agreements of this kind in the EU. However, there is likely to be a higher incidence of such agreements in the US due to the regulatory framework for medicines. The Hatch-Waxman Act[62] provides incentives to the first generic company to challenge a pharmaceutical patent. It does this by giving that generic company standing to bring invalidity challenges before entering the market and by granting a 180-day marketing exclusivity for that company's product upon entry. An unintended consequence has been to align the interests of the patentee and the generic challenger to settle on terms allowing the former to maintain its patent and the latter to benefit from the exclusivity period as well as any reverse payment.[63]

The US competition authorities have been directly or indirectly involved in a number of cases. Agreements of this kind have also attracted political opprobrium, with Senator Obama calling (prior to election as President) for reinvigorated antitrust enforcement.[64] Further, in January 2010, the FTC issued a staff report stating that '... agreements with compensation from the brand to the generic on average prohibit generic entry for nearly 17 months longer than agreements without payments', and estimating that agreements currently in force '... protect at least $20 billion in sales of brand-name pharmaceuticals from generic competition'.[65]

However, as at the date of writing, the US antitrust agencies have yet to gain any significant victories. Rather, the courts have held in most cases that the settlement agreement and payments to the generic company did not result in any restriction on competition beyond that (legitimately) created by the disputed patent itself.[66] Accordingly, they have refused to strike down such agreements, regardless both of the

---

[61] *Sector Inquiry Report* (n 3 above) para 1512.

[62] Drug Price Competition and Patent Term Restoration Act, 1984.

[63] For a summary of criticisms of Hatch-Waxman, see AF Abbott and S Michel, 'Exclusion payments in patent settlements: a legal and economic perspective' (2006) 1 J Intellectual Property Land Practice 207, 220–2; see also speech of FTC Chairman J Leibowitz, 'Pay-for-Delay Settlements in the Pharmaceutical Industry' (June 2009) <www.ftc.gov/speeches/leibowitz/090623payfordelayspeech.pdf> accessed 21 September 2010.

[64] Statement of Senator B Obama for the American Antitrust Institute (27 September 2007) <http://www.antitrustinstitute.org/files/aai-%20Presidential%20campaign%20-%20Obama%209-07_092720071759.pdf> accessed 21 September 2010: 'An Obama administration will ensure that the law effectively prevents anticompetitive agreements that artificially retard the entry of generic pharmaceuticals onto the market...'.

[65] *FTC Pay-for-Delay Study* 2 (n 57 above).

[66] Cases include *Re Androgel Antitrust Litigation* (No II) 1:09-MD-2084-TWT (ND Ga, 2010); *Re Ciprofloxacin Antitrust Litigation* 544 F3d 1323 (Fed Cir, 2008) (a related case is pending before the Court of Appeals for the Second Circuit); *Re Tamoxifen Citrate Antitrust Litigation* 466 F3d 187 (2d Cir, 2006); *Schering-Plough Corp v FTC* 402 F3d, 1056 (11th Cir, 2005) (hereinafter '*Schering-Plough*'); *Valley Drug Co v Geneva Pharmaceuticals Inc* 344 F3d 1294 (11th Cir, 2003) (hereinafter '*Valley Drug*'). For more detail, see below in the context of discussion of the EU approach.

level of the payment to the generic company (which has generally exceeded the US$2 million that the FTC considers to be an acceptable maximum for litigation costs),[67] and of whether there have been a succession of similar agreements. The US Courts of Appeal have emphasized the presumption of patent validity, and have applied this in preference to a presumption of illegality for reverse payment settlements urged upon them by the FTC. The US Supreme Court has to date refused to hear appeals on this point.[68]

As the FTC's recent 'Pay-for-Delay' report suggests, this is not an outcome with which it is happy. It has continued to bring cases before the courts.[69] Until recently the Department of Justice ('DOJ') had supported the courts' approach, but—under political pressure following the change in the US administration and the appointment of democrat Christine Varney as head of the DOJ Antitrust Division—it has now aligned its views largely with the FTC's.[70] The possibility of new laws prohibiting reverse settlement payments also remains open.[71]

Application of EU law to reverse payments

Despite the different regulatory framework (and consequently potentially different incentives), the Sector Inquiry Report indicates that the EU is following the lead of the US in reviewing such agreements. Such actions may well necessitate an analysis of the legality of 'reverse' payments.[72]

Object infringement?

The first key question will be whether such agreements are to be regarded as anti-competitive by object. Object agreements are more serious than restrictions by effect, and are typically easier for the regulator to prove.

The fact that settlements can be desirable from a policy perspective does not remove the risk that the competition authorities will seek to characterize those which incorporate reverse payments as object infringements. An agreement may be

[67] *Schering-Plough* (n 66 above) 11.

[68] eg *Schering-Plough* (n 66 above).

[69] eg *Solvay*: see <www.ftc.gov/os/caselist/0710060/index.shtm> accessed 21 September 2010; *Cephalon*: see <www.ftc.gov/os/caselist/0610182/index.shtm> accessed 21 September 2010.

[70] DoJ Brief for the United States in Response to the Court's Invitation, *Re Ciprofloxacin Antitrust Litigation*, no.05-2851-cv (2d Cir, 6 July 2009), <www.usdoj.gov/atr/cases/f247700/247798.pdf> 21 September 2010 (hereinafter '*DoJ Brief for the United States*').

[71] For more on the US cases and policy developments, see A Synnott and W Michael, 'Antitrust and Intellectual property: Recent Developments in the Pharmaceuticals Sector' (Sept 2009) The Antitrust Chronicle (available online at <http://www.paulweiss.com/files/Publication/58781214-1fda-402d-8165-90d07d0472fb/Presentation/PublicationAttachment/919791fb-2382-47c4-82d5-9113d8860f3d/PW_CP_Sep09.pdf> accessed 13 September 2010); and R Leibenluft and E Stock, 'Pharmaceuticals in the USA' (20 Oct 2009) Competition Law Insight 10–11 (available online at <http://www.hoganlovells.com/newsmedia/newspubs/PubDetail.aspx?publication=4882> accessed 13 September 2010).

[72] See n 54 above.

anti-competitive by object even if it also pursues a legitimate aim.[73] However, the fact that patent rights can, unless revoked, block entry to the market, makes the object test difficult to apply to patent settlement agreements. A patent is already an exclusionary right, so there is no reason why a patentee should not have the intention of excluding infringing products. This is so even in the absence of a formal legal presumption of patent validity as in the US. Moreover, any individual settlement does not prevent others from raising invalidity attacks or seeking to launch at risk.[74]

There are two situations where a finding of a restriction by object might be appropriate. First, if the agreed restrictions clearly exceed the scope of the patent, an object infringement may be found. Examples might include an agreement by the generic company not to enter even after patent expiry (including any legitimate extensions to patent term such as Supplementary Protection Certificates), or to accept limitations on sales of products not covered by the disputed patent.[75]

The second case in which an object infringement finding might be appropriate would be if the patentee had some relevant knowledge about the patent's invalidity. This would have to go beyond the normal uncertainty about the outcome of patent litigation, and arguably, beyond a mere 'likelihood' of losing a trial. In the authors' view, the application of the object rules would be appropriate only in exceptional circumstances, such as where the patentee had previously admitted the invalidity of its patent, or where the granting patent office had been misled, for example by the patentee withholding information about known prior art or previous disclosure of the subject matter of the patent.[76] Internal documents clearly acknowledging patent invalidity would significantly increase the likelihood of an infringement decision on this basis, as might a reverse-payment settlement concluded during the appeal process following an invalidity finding at first instance.[77]

This analysis reflects the approach that has been taken by the US courts, which have held that patent settlements containing reverse payments are not 'per se' violations.[78] Even the US agencies, which are outspoken in favour of antitrust sanctions being applied to reverse payment settlements, have conceded that a rule of reason analysis is required.[79]Indeed, the European Commission has itself acknowledged—by

[73] Case C-551/03 *General Motors v Commission* [2006] ECR I-3173, para 64.

[74] Settling with a succession of patent challengers might increase the likelihood of the regulator inferring that the patentee believes the patent is invalid.

[75] For the approach of the US courts see *Re Cardizem CD Antitrust Litigation* 332 F3d 896 (6th Cir, 2003). See also SP Brankin, 'Patent settlements and competition law: where is the European Commission going?' (2010) 5 J Intellectual Property L and Practice 23, 25.

[76] *AstraZeneca* (n 52 above) (relates in part to inconsistent statements made by AstraZeneca to national patent offices).

[77] This was not permitted by the English Court of Appeal (the case did not involve reverse payments): see *Halliburton Energy v Smith* (n 18 above).

[78] See eg *Re Ciprofloxacin Antitrust Litigation* (n 66 above) 10–5.

[79] *DOJ Brief for the United States* (n 70 above) section 2.

implication—that reverse payment agreements will be (potentially) anti-competitive only by effect, stating in its Sector Inquiry Report that,

> . . . any assessment of whether a certain settlement could be deemed compatible or incompatible with EC competition law would require an in-depth analysis of the individual agreement, taking into account the factual, economic and legal background.[80]

The Commission may ultimately have to distance itself from this statement if it wishes to succeed in this area. As we discuss below, proving anti-competitive effects may not be straightforward. A presumption—like that advocated by the FTC—that a significant payment is unlawful may prove to be the only way for the Commission to succeed in attacking settlement agreements.

Effects on competition

The difficulties outlined above in the context of object infringements also represent an obstacle when it comes to proving effects on competition. As with Article 102 TFEU cases on excessive pricing, it will be difficult to define a 'bright line' for the point at which a reasonable settlement payment switches over to an anti-competitive pay-off. To do so will require an analysis of a number of factors including likely costs of the litigation, the generic company's exposure in terms of wasted preparations for coming to market, the expectations for generic price and market uptake in the event that it had come to market, the time that the generic would have taken to get to market, as well as the remaining term of the patent.[81] Moreover, settlements may be pro-competitive if they permit generic entry before the date of patent expiry.

Another potential difficulty for the competition authority will be the argument that a reverse payment does not itself affect competition, but serves simply to increase the likelihood that the recipient will enter into the agreement. In many cases there will be a strong bias in favour of a settlement involving a payment. The loss to the patentee if its patent is invalidated will usually exceed the potential profits of the generic company if it enters the market (due to the significant price drop on generic entry). In settling, the patentee therefore stands to lose considerably less than it would if its patent were revoked, and the generic will receive more. Both parties will also benefit from certainty. It was argued by the FTC in *Schering-Plough* that without the payment by the patentee, the parties would have agreed on an earlier entry date.

[80] *Sector Inquiry Report* (n 3 above) para 1530.

[81] Dominik Schnichels (head of the Commission's Pharma Task Force) mentions some of these considerations: 'The application of European competition law to the pharmaceutical sector: Some personal thoughts', 36th Annual Fordham Conference on International Antitrust Law and Policy (2009), 22.

However, the US Court rejected this approach. There was no evidence that, in the absence of the agreement, the parties would have settled at all.[82]

It must be assumed that, without a settlement agreement, the litigation would have continued unless there is evidence to the contrary. If this is so, it follows that the counterfactual analysis needs to assess the probable outcome of the patent litigation so as to decide what the effect on potential entry of settlement is likely to have been. This means reaching a view on the strength of the disputed patent and on the likelihood of it being held invalid. Despite the obvious difficulties with a competition authority pre-judging the finding of the patent courts, the Commission places considerable weight on the question of patent strength in the Sector Inquiry Report. For example,

> The fundamental factor considered by originator companies when deciding whether to enter into a settlement agreement with generic companies is the strength of their position in the patent litigation (the expected likelihood of winning). When companies assess their position as strong, they do not consider entering into a settlement agreement. However, if their chances of winning are assessed [as] less strong and there is a great deal at risk, they give careful consideration to the possibility of settling with the other party.[83]

The Sector Inquiry Report does not explain how the Commission would address this issue in the context of an infringement case. To date, EU law has not had to deal seriously with the question of who should make an assessment of the strength of an IP right, or how appropriate it is.[84]

Both *Windsurfing* and *Bayer v Süllhofer* are again relevant. In *Windsurfing*, the Commission carried out an assessment of the scope of the patent in order to determine whether certain obligations imposed on licensees went beyond the patented rights, and were thus potentially caught by Article 101(1) TFEU. Windsurfing (the patentee) argued that the Commission was not competent to make such an assessment. The CJEU's analysis of this point makes a rather fine distinction, holding that,

> . . . the Commission is not competent to rule upon the scope of a patent [but that] it is still the case that it may not refrain from all action when the scope of the patent is relevant for the purposes of determining whether there has been an infringement of Article [101 or 102 TFEU].[85]

[82] *Schering-Plough* (n 66 above) fn 15; Brankin (n 75 above) includes an interesting discussion of how this point played out in *Schering-Plough*, 25.

[83] *Sector Inquiry Report* (n 3 above) para 720.

[84] Under the free movement of goods rules, the concept of 'specific subject matter' of the IP right does not entail any evaluation of the strength of the right.

[85] *Windsurfing* (n 11 above) para 26.

The CJEU therefore accepted the Commission's entitlement to assess a patent's scope, but emphasized the non-determinative nature of any such evaluation.[86] The Advocate-General in *Bayer v Süllhofer* took a more cautious approach, saying,

> The criterion relating to the 'strength' of the right appears attractive in theory . . . A serious disadvantage would, however, be presented by the fact that the assessment thereby required would involve subsequent judicial verification at odds with the conciliatory approach which makes it possible to come to an amicable settlement of a dispute.[87]

In *Windsurfing*, the Commission had to determine whether the patent in question covered only the rig for a windsurfing sailboard, or whether it covered the whole sailboard. Although the German patent authorities had not reached a final decision, there had already been a considerable amount of discussion of this point. The patent had moreover been amended to replace references to 'sailboard' with 'rig for a sailboard'.[88] In effect, the exercise that the Commission needed to carry out in this case was a fairly simple interpretation of the wording of the patent, which described a mechanical invention.

Contrast the likely position for a patent covering a molecule or a method of manufacturing a medicine. Such patents are typically complex and highly technical. It is therefore questionable whether in cases involving pharmaceuticals the Commission will be competent to carry out the sort of assessment that it made in *Windsurfing*. If it needs to rely on experts, this will not only prolong the Commission's inquiries, but it will make it very difficult to avoid cutting across national courts' jurisdiction over patents (which are, of course, national rights). Problems with conflicting expert evidence are likely to arise. Even if a final court judgment has been reached (perhaps in a case brought by another party) holding the disputed patent to be invalid, it does not follow that the result would have been the same between the settling parties, or that the competition authorities should be permitted to take the benefit of hindsight if the judgment in question post-dated the settlement.[89] Indeed, the US courts have held that subsequent invalidity should not be taken into account when considering the antitrust analysis of a particular settlement.[90]

---

[86] In *Dorcet/Toltecs* (n 9 above), there was a discussion of whether the Commission was entitled to assess the risk of confusion between two trade marks (paras 26–30). This raises a similar issue to the assessment of patent strength/scope. However, the Court did not decide the point.

[87] *Bayer v Süllhofer* (n 30 above) Opinion of Advocate-General Darmon.

[88] *Windsurfing* (n 11 above) para 32.

[89] Pending the creation of a Community patent, there are regularly inconsistent findings by the different national patent courts, as well as reversals upon appeal and discussion of different prior art in different cases. I Forrester discusses the controversial use of hindsight in the context of Article 102 TFEU in 'A European Paradox: Imposing Market Reform "Voluntarily"', paper presented at 1st Concurrences Conference in Paris (15 February 2010), 8.

[90] *Valley Drug* (n 66 above) 1306–7; recently affirmed in *Re Androgel Antitrust Litigation* (n 66 above).

An alternative to using experts would be for the Commission to seek internal evidence of the company's beliefs about the strength of the patent. Documents proving that the company expects to lose patent litigation would facilitate the Commission's task as it could be construed as evidence of an intention unfairly to exclude competition.[91] However, given companies' awareness of the importance of pharmaceutical patents, such evidence is likely to surface rarely.[92]

A further possibility would be for an adverse inference about patent strength to be drawn if the pharmaceutical company has settled with a number of different litigants.[93] This would again be controversial as companies should be entitled to decide how to deal with each case on its own individual merits.

Even if the Commission can come to a reasonable view of patent strength, establishing a reasonable counterfactual may not be straightforward. The table below shows some of the possible permutations:

Table 1 Permutations for patent litigation outcomes and counterfactuals

Strength of patent	Incentives for patentee to settle	Incentives for generic to settle	Likely outcome	Counterfactual
Case 1: Strong	Weak	Strong	Either litigation continues, or settles with no/ trivial payment	Generic cannot enter; any payment thus irrelevant
Case 2: Weak	Strong (with or without payment: any payment likely to be lower than loss if patent invalidated)	Weak without payment Strong with payment, especially if payment exceeds likely market income for period of delay	Settlement with (non-trivial) payment likely	Generic can (probably) enter, and would have done so without the payment; it may already have entered at risk
Case 3: Around 50–50	Strong (with or without payment: any payment likely to be lower than loss if patent invalidated)	Fairly weak without payment (less to lose than originator) Strong with payment, especially if exceeds likely profits over period of delay	Settlement with (non-trivial) payment likely	Generic may or may not be able to enter; effect of payment therefore difficult to determine

[91] It has been argued that this could be the only basis for a viable case: Batchelor (n 27 above) 18.

[92] During its sector inquiry the Commission uncovered statements indicating that companies considered certain patent rights to be weak (*Sector Inquiry Report* (n 3 above) section 2.1.2). Companies in this sector are likely to be cautious in future. Assessments made in the context of legal advice on the prospects of patent litigation are likely to give rise to arguments about privilege. The Commission will argue that privilege applies only to legal advice addressing the company's position under competition law. Patent agents' advice is unlikely to be viewed as privileged in EU competition proceedings.

[93] It appears that the FTC may run this argument in the *Cephalon* case: see (n 69 above).

In case 1 above (strong patent), it is apparent that the counterfactual is that the patent would have blocked market entry.[94] Any payment to the generic is thus immaterial to the outcome.

In case 2 (weak patent), the competition authority may be justified in inferring that, but for the payment, generic companies would have been able to enter the market. A question mark remains over how the patent's weakness is established, and at what level it is regarded as 'weak'. (Less than 40 per cent chance of prevailing? Less than 25 per cent?) The logical inference to be drawn is moreover that if the patent is weak (and the parties know that, so the generic will not settle without payment), the patentee should have litigated to a conclusion. This is a very onerous requirement, especially if it was the generic that started the proceedings. It is also at odds with the Commission's general position that patent settlements are acceptable.[95] According to the US court, such a requirement could even '... impair the incentives [in the patent system] for disclosure and innovation'.[96]

To complete the counterfactual analysis in this case it would also be necessary for the competition regulator to carry out a further assessment of the likely date of generic entry. This will not be straightforward, depending not only on the likely duration of the patent litigation, but also on the generic company's market readiness. It should not be assumed that the generic would have been prepared for commercial launch on the date of judgment, either in terms of scaling up its production or in meeting all the necessary regulatory requirements.

Even if a company assesses its own prospects of success as rather low, it is unclear at what level it would be sustainable for the Commission to argue that the company should have refrained from settling the litigation on terms involving a reverse payment. Unless the patent is what has been referred to as a 'sham',[97] the competition authorities are unlikely to criticize a settlement that does not involve any form of 'value transfer' to the generic company,[98] yet the effects on competition are arguably identical even if there is a payment.

Case 3 (approximately even chance of the patentee prevailing) is the most difficult of all. It is also likely to be the most common given the imperfect knowledge that each

[94] The Commission accepted this approach in *Syntex/Synthelabo*, see n 43 above and extract quoted in connection with this.

[95] See extract in Section A above (*Sector Inquiry Report* (n 3 above) para 707). Equally, it would be controversial to suggest that companies should never bring proceedings on the basis of patents that may be weak—as LJ Jacob put it, 'To say that [pharmaceutical companies] should not go to court on a weak case would be a serious threat to access to justice' (November 2008) Speech on release of the Commission's preliminary report in the pharmaceutical sector inquiry available online at <http://ec.europa.eu/competition/sectors/pharmaceuticals/inquiry/jacob.pdf> accessed 13 September 2010.

[96] *Valley Drug* (n 66 above) 1308.

[97] Schnichels (n 81 above) 22. A 'sham' patent is presumably one which has one of the defects identified above in the context of object infringements (patentee's knowledge of patent's invalidity).

[98] See discussion of the general approach to settlements earlier in this chapter (Section B).

party has of the other's case and of the likely view of the judge. The Sector Inquiry Report suggests that the Commission will seek to approach case 3 in the same way as case 2 (the report refers generally to 'less strong' patents).[99] However, this does not take a realistic view of the counterfactual. Arguably, the 'but for' outcome is simply unknowable in this case.

The competition authority might seek to fall back on an alternative assumption. One possibility would be to assess the value of the payment to the generic company. The generic, rather than the patentee, is necessarily the focus of this exercise: it is fair to assume that any payment will be less than the patentee stands to lose as it would otherwise make no economic sense for the patentee.[100]

Subject to difficulties in quantifying the scale at which the generic would have been able to enter, it could be argued that there will be an anti-competitive effect if the payment exceeds the generic's likely profits, whereas there will be no such effect if the payment is lower. This approach would give rise to considerable dispute, including as to issues such as the uplift of discount that should be applied to compare a certain payment under an agreement with speculative future profits. Nevertheless, this may be a route that the competition authorities seek to adopt.

Agreements with other aspects

Thus far the discussion has focused on 'bare' reverse payments, where cash flows from patentee to generic. The situation is complicated still further if the patentee receives something in return for the settlement payment (eg distribution services, the sale of an IP right). Such cases will present additional problems for the regulator, which may have to assess the value of the rights transferred in order to assess whether there is a concealed over-payment designed to induce the generic company to settle.

The competition authorities' approach may have to vary yet again if the settlement involves a 'value transfer' that is 'in kind'. The US case law to date has focused on agreements involving significant cash payments, although the FTC's *Cephalon* action refers to the existence of a series of 'side-agreements' with potential generic entrants, including patent licences, supply agreements and co-development deals.[101] The Sector Inquiry Report also indicates that settlements involving the grant of a licence or distribution right may be within the scope of the competition rules,

> Other types of value transfer include distribution agreements in which the generic company becomes a distributor of a product of the originator company or a 'side-deal' in which the originator company grants a commercial benefit to the generic company,

[99] eg *Sector Inquiry Report* (n 3 above) para 720; see text to n 83 above.

[100] See further Brankin (n 75 above) 25. This assumes reasonably accurate forecasts of future sales trends.

[101] (n 69 above); P Treacy and H Hopson, 'Patent settlements: will the European Commission follow in the FTC's footsteps. . .?' (2008) 3 J Intellectual Property L and Practice 622 includes further discussion of the background to this case.

> for example, by allowing it to enter the market before patent expiry in another geographical area or with another product. Furthermore, value transfer could consist in granting a patent licence to the generic company. A patent licence enables the generic company to enter a market with a product but . . . the commercial freedom of the generic company is limited by the terms of the licence agreement which, for instance, can include limitations on the quantity of the types of products that the generic company may sell . . . The terms of the licence agreement determine the level of the value transfer to the generic company.[102]

A competition infringement finding for these types of agreements would be viewed as highly controversial by pharmaceutical and generic companies alike. Companies need commercial freedom to enter into agreements without fear that the 'correct' value of their transactions will be second-guessed by a competition regulator. Given how difficult it will be to assess the true value of the asset or right that has been transferred to the generic company, in particular where future use of the right will be metered (eg by payment of royalties), contracting companies should be accorded a broad margin of appreciation by competition authorities assessing such transactions. It is strongly arguable that regulators should attack transfers of value—if at all—only with great caution, and only: (1) where the generic company stands to gain something that substantially out-values what it is giving up; and (2) where the payment made (or value transferred) by the patentee can be explained largely by the delay achieved, rather than other commercial benefits such as improved distribution or access to a raw material.

Article 102 TFEU

There are evidently significant challenges for the competition authorities to establish that a patent settlement involving a reverse payment contravenes Article 101 TFEU. As an alternative, and assuming dominance can be established,[103] they may therefore turn to Article 102 TFEU.

There is no precedent for this—the Commission would have to establish a new type of abuse. As with Article 101 TFEU, it will again bear the burden of proof. Many of the same difficulties will arise, although the requirement to show only a potential effect on competition may prove ultimately less problematic than the counterfactual analysis required under Article 101 TFEU. The Commission will be particularly likely to favour the abuse of dominance rules where it can point to a pattern of conduct which affects generic companies' ability to enter the market. It might thus deal with settlement agreements in conjunction with other strategies identified in the Sector Inquiry Report.[104] Such cases may—like the *AstraZeneca* case which also

[102] *Sector Inquiry Report* (n 3 above) 269 (text box below para. 742).

[103] The General Court ruling in *AstraZeneca* (n 52 above) upheld the Commission's adoption of a narrow market definition in the first abuse of dominance case concluded against a pharmaceutical sector company.

[104] eg patent clustering, interventions before regulatory authorities, product switching.

concerned the use of IP rights in the pharmaceutical sector—rely in large part on the Commission's ability to impute to the company under investigation an intention to damage competition.[105]

Conclusion

This is a difficult, but fascinating, area. The competing policy objectives have thus far not been resolved into a particular hierarchy, but continue to exert different pressures on decision makers.

Developments, particularly in the pharmaceutical sector cases, are inevitable. Investigations have already been announced and, in January 2010, the Commission acted on its stated intention to carry out further focused monitoring of pharmaceutical settlements. The monitoring took the form of a questionnaire to companies active in the pharmaceutical industry (both 'originators' and generics) asking them to submit copies of settlement agreements entered into in 2008–09. In its report on the outcome of the monitoring exercise, the Commission noted that the number of 'problematic' patent settlements has decreased.[106] It nevertheless remains possible that the Commission will bring further enforcement actions as a result of the information that it obtains. The Commission stated that it would take into account '... the size of the markets concerned, the value transferred from originator to generic companies and the restriction on independent generic entry ...' when considering whether to take action.[107] It has also confirmed that it will repeat the monitoring exercise in 2011.

The situation in the US is also expected to develop rapidly, with the US 2nd Court of Appeal in the Ciprofloxacin Hydrochloride litigation recently recommending that the claimant's petition for en banc review by the Court, so that the '... "exceptional importance" of the antitrust implications of reverse exclusionary payment settlements of patent infringement suits' can be considered further.[108] To date, such a review has been refused.[109] However, such full panel consideration (whether in this or a future case) could lead to precedents such as *Re Tamoxifen*—in which the court held that the reverse payment settlement was lawful so long as it did not exceed the scope of the patent—potentially being overruled. A further development is unfolding at legislative level. In July 2010, the US House of Representatives approved a Preserve Access to Generics Act that would render 'pay-for-delay' settlements presumptively unlawful. If a change of this kind comes about in US law (in particular if

[105] *AstraZeneca* (n 52 above).

[106] Commission press release IP/10/887, 5 July 2010.

[107] *Ibid.*

[108] *Re Ciprofloxacin Antitrust Litigation* 05-2851-cv(L), 0522852-cv(CON) (Fed Cir, 2010).

[109] The application for full panel review in *Re Ciprofloxacin Antitrust Litigation* was denied by the US Court of Appeals for the Second Circuit on 7 September 2010.

other circuits adopt a similar approach), it could further embolden the competition authorities in the EU.

Given the importance of patents for pharmaceutical companies, it can be expected that any infringement cases will be fiercely contested. It may be some years before the legal approach to these cases is clarified.

13

INTELLECTUAL PROPERTY RIGHTS AND MARKET INTEGRATION

Andreas Heinemann

Introduction

Since the *Spaak* Report of 1956,[1] market integration has always been an important goal of European competition law. The abolition of frontiers for the free movement of goods and services should not be contradicted by private restraints. Intellectual Property Rights (IPRs) have played a major role in this respect—parallel imports of IP-protected goods must be possible in the internal market. Therefore, private practices, based on IPRs, tending to prevent cross-border trade have to undergo antitrust scrutiny. General principles have been laid down in European regulation. Others have been developed by the European Courts. Recently, a more cautious approach may be identified when it comes to the application of competition law to market partitioning practices. On the one hand, there is the more lenient perception of vertical restraints based on the effects-oriented analysis of the Chicago School which, to a certain extent, has found its way into European competition law via the more economic approach. On the other hand, there are sector-specific arguments militating for a larger margin in favour of innovative industries like for example the pharmaceutical sector. In addition, there have been issues of a more general or technical nature, for example regarding the concept of 'agreement' in the sense of Article 101 TFEU. The relevant case law is of great importance when it comes to the leeway undertakings enjoy in defining their distribution strategies.

The following analysis starts by sketching the traditional approach of European competition law to market partitioning practices (A). Next, the recent developments are

[1] Report of the Heads of Delegation to the Ministers of Foreign Affairs ('The Brussels Report on the General Common Market') (Brussels 1956) Introduction, Section II B. The French version of the Spaak Report is available at: <http://www.unizar.es/euroconstitucion/library/historic%20 documents/Rome/preparation/Spaak%20report%20fr.pdf> accessed 15 September 2010.

described which put emphasis on the limits of competition law application in this field (B). This directly leads to the question what impact the specific features of innovative industries should have on the application of competition law (C). From this discussion, consequences will be drawn for the status which the aspect of market integration should enjoy in European competition law.

A. The Traditional Approach to Market Partitioning Practices

The triangle of integration, competition law, and IP protection

The establishment of an internal market is at the heart of the European Union, Article 3 (3) TFEU. Even if the Lisbon Treaty has considerably broadened the foundations and the goals of European unification, the setting up of an area without internal frontiers in which the free movement of goods, persons, services, and capital is ensured (Article 26 TFEU) stays a cornerstone of European integration. As regards competition policy, the reference to a system of competition which is not distorted (ex-Article 3(1)(g) EC) has been removed from the Treaties and shifted to a Protocol.[2] However, as Protocols form an integral part of the Treaties (Article 51 TEU), it is expected that this modification will not have substantive effects on the status of European competition policy.[3] Therefore, competition policy will keep its important function in the context of the internal market.[4] Finally, the role of IPRs has been strengthened by the Lisbon Treaty—whereas in the past, on the level of primary law, intellectual property was only mentioned as a justification for restrictions of the free movement of goods, Article 118 TFEU now provides for an explicit legal basis for the creation of uniform European IPRs, and Article 207 TFEU adds IP protection to the catalogue of EU competences in the field of the common commercial policy.[5]

[2] Protocol No 27 on the Internal Market and Competition. The Protocol clarifies that also in future it will be possible to make use of the flexibility clause in Article 352 TFEU in the field of competition law (as it was done, eg, in the case of the EC Merger Regulation).

[3] However, this optimistic view cannot be taken for granted. Ex-Article 3(1)(g) EC has been used in a considerable number of cases in order to strengthen pro-competitive arguments. In *Eco Swiss*, eg the ECJ based the *ordre public* character of the competition rules on this article, Case C-126/97 *Eco Swiss China Time Ltd v Benetton International NV* [1999] ECR I-3055 para 36.

[4] Article 3 (1)(b) TFEU attributes to the Union the exclusive competence to establish the competition rules necessary for the functioning of the internal market. One could also ask the question if the reference to a 'social market economy' in Article 3(3) TEU strengthens the weight of competition policy. This concept was coined in Germany by Alfred Müller-Armack and Ludwig Erhard and is based on the ordoliberalism of the Freiburg School which gave highest priority to competition policy. See DJ Gerber, *Law and Competition in Twentieth Century Europe: Protecting Prometheus* (Oxford: Clarendon Press, 1998) 259–61.

[5] Moreover, according to Article 262 TFEU, jurisdiction may be conferred on the ECJ in order to judge on disputes relating to uniform European IPRs.

Exhaustion as a means to extenuate tensions between IPRs and market integration

IPRs normally confer exclusive rights including an exclusive import and export right. This may cause a conflict with the free movement rules. In Europe, this conflict has been resolved by the concept of EU-wide (respectively EEA-wide) exhaustion. Once a protected product has been put in circulation somewhere in the EU (or in the EEA) by the IP owner or with his consent, this product is allowed to move untroubled in the EU (and the EEA). The IP owner cannot prevent further transactions; her right of distribution is exhausted. The principle of exhaustion is derived from Articles 34, 36 TFEU.[6] It has been extended to the EEA by EEA-Protocol 28 on Intellectual Property. With respect to some IPRs, the principle of exhaustion has been explicitly included into secondary law.[7]

This rather simple starting point has been elaborated in subsequent case law.[8] For our context, the following specifications are of particular importance—according to the ECJ, Member States are no longer free to opt for international exhaustion.[9] There is no consent of the right holder and hence no exhaustion if the right holder was under a legal obligation to market the product in question (including mandatory licensing). On the other hand, mere ethical obligations are not sufficient to trigger exhaustion. Nor is the existence of price control, eg, in the field of pharmaceutical products, relevant to exclude the exhaustion effect.[10] Exhaustion affects the distribution right (and rights directly related hereto, like, eg the right to use a trade mark for

---

[6] Groundbreaking judgments are Case 78/70 *Deutsche Grammophon/Metro* [1971] ECR 487; Case 15/74 *Centrafarm/Sterling Drug* [1974] ECR 1147.

[7] See, eg, Article 4 (2) of the Directive 2009/24/EC of the European Parliament and of the Council on the Legal Protection of Computer Programs [2009] OJ L111/16:

> The first sale in the Community of a copy of a program by the rightholder or with his consent shall exhaust the distribution right within the Community of that copy, with the exception of the right to control further rental of the program or a copy thereof.

[8] For a detailed overview see S Enchelmaier, *Europäisches Wirtschaftsrecht* (2005) paras 72–82; C Stothers, *Parallel Trade in Europe. Intellectual Property, Competition and Regulatory Law* (Oxford: Hart Publishing, 2007) 40 *et seq.*

[9] Case C-355/96 *Silhouette International Schmied v Hartlauer Handelsgesellschaft mbH* [1998] ECR I-4799 paras 22–7. The Court said so with respect to the Trademark Directive (Council Directive 89/1004/EEC to Approximate the Laws of the Member States Relating to Trade Marks [1989] OJ L40/1). In *Laserdisken* (Case C-479/04 *Laserdisken v Kulturministeriet* [2006] I-8089 paras 19–27), the Court came to the same result in the field of copyright relying on the InfoSoc Directive (Directive 2001/29/EC of the European Parliament and the Council on the Harmonisation of Certain Aspects of Copyright and Related Rights in the Information Society [2001] OJ L167/10). It seems most probable that the Court would apply the same principle to all IPRs. For a broader perspective see T Cottier, 'The Exhaustion of Intellectual Property Rights—A Fresh Look' [2008] Intl Rev Intellectual Property and Competition L 755–7.

[10] Joint Cases C-267/95 and C-268/95 *Merck/Primecrown* [1996] ECR I-6285 (hereinafter '*Merck/Primecrown*').

advertising purposes[11]), but not the reproduction right. Exhaustion applies to IP-protected goods, but (normally) not to services.[12] The case law is not restricted to the fundamental principles of exhaustion, but also deals with specific details. A famous example is the repackaging cases in the pharmaceutical sector. The Court has set up checklists which have to be respected if a firm wants to re-box, re-label, or even re-brand pharmaceuticals in order to make them compatible with the legal requirements in the import state.[13]

Even though the doctrine of exhaustion is now in a very advanced stage, the development is not yet finished. The ECJ is continuously confronted with new questions as are the national courts. As firms have a natural tendency to test the limits of the exhaustion doctrine, national courts have to deal with these questions in numerous cases including the consequences in other fields of law, like unfair competition law. They do not always succeed in striking an appropriate balance.[14]

The role of competition law with respect to the principle of exhaustion

In the traditional view, one of the functions of European competition law is to protect free movement of goods and services against private restraints. Therefore, private agreements or unilateral conduct providing for territorial restrictions have to be examined as to their compatibility with the competition rules. Article 101(1)(b) and (c) TFEU expressly mention agreements which limit, control, or share markets. Firms are not allowed to circumvent the principle of exhaustion by means of agreements or abusive behaviour.[15] An example is the *Nintendo* case—prices for consoles and game cartridges were very uneven in the European Union. In the UK, they were situated 65 per cent below the price level in Germany and the Netherlands. *Nintendo* agreed with seven European distributors upon preventing parallel imports. Distributors which were identified as the source of parallel imports were sanctioned by the reduction or even the refusal to supply furthermore. The European Commission fined all parties involved, not only the producer but also the distributors. The Commission did not accept the argument that the producer had exerted considerable pressure on the distributors. They should have come to the Commission in the first place.[16]

---

[11] Case C-337/95 *Parfums Christian Dior/Evora* [1997] ECR I-6013. An exception applies if the use of the trade mark in the reseller's advertising seriously damages the reputation of the trade mark, ibid para 46. For a critique of this exception see H Ullrich, 'Gewerblicher Rechtsschutz und Urheberrecht im Gemeinsamen Markt' in U Immenga and E-J Mestmäcker (eds), *Kommentar zum Europäischen Kartellrecht*, Vol 1(2) (4th edn, 2007) para 91.

[12] Case 62/79 *Coditel/Ciné Vog Films* [1980] ECR 881.

[13] Case C-348/04 *Boehringer Ingelheim* [2007] ECR I-3391.

[14] For Germany see A Heinemann, 'Gebietsschutz im Binnenmarkt—Bewährungsprobe für ein prekäres Gleichgewicht' in A Ohly et al (eds), *Festschrift für Gerhard Schricker* (2005) 53, 56–9.

[15] B Conde Gallego, *Handelsbezogene Aspekte des Lizenzkartellrechts* (2003) 225, considers the maintenance of parallel imports the main task of international IP antitrust law.

[16] *Nintendo Distribution* [2003] OJ L255/33.

The CFI reduced the fines of some parties for reasons of equal treatment, and upheld the decision as for the rest.[17]

The *Nintendo* case confirms the consistent practice of the European courts (beginning with *Consten and Grundig*[18]) according to which one of the tasks of competition law is to guarantee parallel imports within the Community.[19] According to the case law, absolute territorial protection is anti-competitive whereas a certain territorial protection is compatible with Article 101 TFEU.[20] The details were laid down in block exemption regulations. In vertical distribution agreements, a supplier may allocate exclusive territories to his buyers. The block exemption regulation on vertical agreements[21] contains specific rules on the legality of such restrictions. According to the regulation, the supplier may agree with the buyer on a restriction of *active* sales into the exclusive territories of other enterprises (if the market shares of the parties do not exceed 30 per cent). However, passive sales, ie sales in response to unsolicited orders, must not be excluded.[22] These rules apply to simple distribution agreements in which IPRs only play a minor role. To certain licensing contracts, the block exemption regulation on technology transfer agreements[23] is applicable which—in the interest of IP protection—allows larger restrictions of competition. If the parties of a licensing contract are not competitors (and if their market shares do not exceed 30 per cent), even passive sales into territories allocated to other licensees may be excluded for a period of two years. This only applies to direct sales by the licensee. Once the licensed product has been put on the market, parallel imports must be admitted.

[17] Cases T-12/03 *Itochu/Commission* [2009] ECR II-883, T-13/03 *Nintendo and Nintendo Europe/Commission* [2009] ECR II-947, Case T-18/03 *CD-Contact Data/Commission* [2009] ECR II-1021. In the latter case, an appeal to the ECJ is pending, see case number C-260/09.

[18] Joint Cases 56 and 58/64 *Consten and Grundig v Commission* [1966] ECR 321.

[19] Other recent cases are Case T-67/01 *JCB Service/Commission* [2004] ECR II-49, confirmed in the aspects relevant here by Case C-167/04 P *JCB Service/Commission* [2006] ECR I-8935; European Commission, *Souris/Topps*, 26 May 2004, COMP/C-3/37.980, [2006] OJ L353/5. For an overview on the illegality of direct and indirect export bans see R Whish, *Competition Law* (6th edn, Oxford: OUP, 2009) 622–26.

[20] SD Anderman, *EC Competition Law and Intellectual Property Rights: the Regulation of Innovation* (Oxford: Clarendon Press, 1998) 90 *et seq*.

[21] Commission Regulation (EU) 330/2010 on the application of Article 101(3) of the Treaty on the Functioning of the European Union to categories of vertical agreements and concerted practices [2010] OJ L102/1.

[22] For a definition of active and passive sales see Commission Guidelines on Vertical Restraints [2010] OJ C130/1 para 51.

[23] Commission Regulation (EC) No 772/2004 on the application of Article 81 (3) of the Treaty [now Article 101(3) TFEU] to categories of technology transfer agreements [2004] OJ L123/11. For a comprehensive analysis see SD Anderman and J Kallaugher, *Technology Transfer and the New EU Competition Rules: Intellectual Property Licensing after Modernisation* (Oxford: OUP, 2006).

B. Limits of the Traditional Approach

For a long time, European competition law has been interpreted so as to complement the rules on exhaustion and to safeguard parallel imports. Recent developments have cast doubt on whether this tradition will be followed in the future. Relevant cases involve the question under which conditions private behaviour may be qualified as an agreement or a concerted practice within the meaning of Article 101(1) TFEU. Another problem lies in the requirements concerning the right holder's consent. Finally, the most fundamental question has been raised by economic theory—if vertical restrictions have to be dealt with in a more cautious way, do we have to say farewell to the close link of exhaustion and competition law which has been observed so far?

The concept of 'agreement' and 'concerted practices' in the context of market-partitioning strategies

Article 101 TFEU only applies to agreements, decisions of associations of undertakings, and concerted practices. In case of unilateral conduct, the control is effected by Article 102 TFEU which presupposes a dominant position. Therefore, the impact of competition law is heavily influenced by the concept of agreement and concerted practices. If these notions are defined narrowly, many cases will be subject to a mere control of abuse. The case law is ambivalent. On the one hand, the European courts give a broad definition of concerted practices. In *T-Mobile Netherlands*, the ECJ has come to the result that a single meeting between companies may constitute a concerted practice.[24] This finding is based on the consistent definition of a concerted practice as a:

> ... form of coordination between undertakings by which, without it having been taken to the stage where an agreement properly so-called has been concluded, practical cooperation between them is knowingly substituted for the risks of competition.[25]

On the other hand, when applying the abstract definition to specific cases, the European courts are sometimes very restrictive. One example is the *Adalat* case concerning a pharmaceutical product used to treat cardiovascular diseases. The German company *Bayer* limited the supply of its product *Adalat* to the Spanish and French wholesalers in order to prevent parallel imports to the UK where the price for *Adalat* was much higher. The Commission came to the conclusion that the quantitative limitation was not unilateral conduct, but followed from an agreement between supplier and buyer.[26] The CFI and the ECJ did not share this view and annulled the Commission's decision. The mere coexistence of a sales contract and a unilaterally

[24] Case C-08/08 *T-Mobile Netherlands* ECR [2009] I-4529 para 26.

[25] *Ibid.*

[26] *Adalat* [1996] OJ L201/1.

imposed limitation of supply does not, according to the courts, constitute a restrictive agreement in the sense of European competition law.[27]

This finding has to be put into the context of other judgments trying to identify the thin line between, on one side, genuinely unilateral conduct and, on the other side, coordination even if it is disguised as unilateral behaviour.[28] This distinction is essential, not only for competition law in general, but specifically for the topic of parallel imports since a narrow concept of coordination enlarges the leeway for the holders of IPRs to restrict parallel imports. In addition, according to the rules on the burden of proof, the authorities have to prove that there was a concurrence of wills. In our view, the delimitation has not always been convincing. In the *Adalat* case, eg, the European Commission found that *Bayer* limited supplies to a certain buyer when he had been identified as a source of parallel imports. Thus, the decrease in supply was a sanction aimed at eliminating parallel trading. The buyers reacted to this threat by stopping or at least hiding their parallel exports. This very evident connection of stimulus and response leads to a coordination of behaviour for which in European competition law the category of 'concerted practices' exists. Therefore, even if there is not an agreement between *Bayer* and its wholesalers, there might be a concerted practice which is equally caught by Article 101 TFEU.

Unfortunately, the European courts did not examine the question if there is such a concerted practice. In our view, it should be sufficient for the application of the European cartel interdiction if the supplier, explicitly or implicitly, expresses certain behavioural expectations to the buyer, and if the buyer then complies or at least pretends to comply with the declared intentions of the supplier. Otherwise, it would be possible for the supplier to circumvent the cartel interdiction by taking measures which are unilateral only at first sight. It has to be added that a broader concept of the notion of 'agreement' and 'concerted practice' is not supposed to predetermine the outcome of substantive analysis. It simply allows behaviour to be brought into the scope of Article 101 TFEU which aims at the coordination between undertakings. This approach does not touch the question if the agreement or concerted practice has as its object or effect the restraint of competition, and if such a restraint could be justified under Article 101(3) TFEU.

Consent of the right holder as a prerequisite for exhaustion

Exhaustion only takes place if the product has been put on the market with the consent of the right holder. Thus, the notion of 'consent' is pivotal for exhaustion doctrine. Consent must have been expressed with respect to each individual product for which

[27] Case T-41/96 *Bayer/Commission* [2000] ECR II-3383; Joined Cases C-2/01 P and C-3/01 P *Bundesverband der Arzneimittel-Importeure* [2004] ECR I-23.

[28] Case C-74/04 P *Commission/Volkswagen* [2006] ECR I-6585; Case T-450/05 *Automobiles Peugeot SA and Peugeot Nederland/Commission* [2009] ECR II-2533.

exhaustion is claimed. Therefore the fact, that a firm has put a certain product on the European market at all, does not justify parallel imports of identical or similar products which have been put on the market outside the EEA.[29] Moreover, there is no consent simply because a product was marketed under the same trade mark in another Member State if the identical trade marks belong to different and independent entities. If for example the holder of a trade mark in EEA Member State A puts the trade-marked good on the market of state A, the holder of the identical trade mark in state B can still prevent the import of that good into state B even if it was him who assigned the trade mark for state A to the other firm, as long as there is no economic link between the two firms. Hence, the voluntary transfer of IPRs to another firm does not imply the consent with putting protected goods on the market.[30]

Even if this result is convincing, it is obvious that the separate assignment of IPRs with respect to single Member States may be used to partition the internal market. From the very beginning, ie the *Consten and Grundig* case,[31] the ECJ has clarified that assignment of IPRs must not be used in an anti-competitive manner. In *Ideal Standard*, the Court has explicitly referred to the cartel interdiction,

> It should be added that, where undertakings independent of each other make trade-mark assignments following a market-sharing agreement, the prohibition of anti-competitive agreements under Article 85 [now Article 101 TFEU] applies and assignments which give effect to that agreement are consequently void.[32]

Hence, the entire legal and economic context has to be explored in order to assess the conformity of an assignment with Article 101 TFEU. The relevant question will be if the assignment of IPRs made for part only of the European Union aspires to fully exploit the economic potential of the rights in question (then it is perfectly legal), or if it is part of a plan to allocate markets in an anti-competitive manner.[33] If the latter is the case, the assignment is void (Article 101(2) TFEU) and the assignee has no title to prevent imports into the state for which the right in question ineffectively was

[29] Case C-173/98 *Sebago and Maison Dubois* [1999] ECR I-4103 para 19.

[30] Case C-9/93 *IHT Internationale Heiztechnik/Ideal Standard* [1994] ECR I-2789 (hereinafter '*Ideal Standard*'), especially para 43. According to the court,

> ... the owner of the right in the importing State must, directly or indirectly, be able to determine the products to which the trade mark may be affixed in the exporting State and to control their quality. That power is lost if, by assignment, control over the trade mark is surrendered to a third party having no economic link with the assignor.

For the ECJ to get to this result the long way, see WR Cornish, *Intellectual Property: Patents, Copyright, Trade marks and Allied Rights* (3rd edn, London: Sweet & Maxwell, 1996) [18–06] calling this development the 'common origin adventure'.

[31] *Consten and Grundig* (n 18 above).

[32] *Ideal Standard* (n 30 above) para 59.

[33] The *Ideal Standard* case concerns trademarks. The reasoning can be extended to other IPRs, eg, to patents, compare J Straus, 'Gebietsbeschränkte Übertragung von Patenten und Patentportfolios und Erschöpfung des Patentrechts in der Europäischen Gemeinschaft' in RM Hilty, J Drexl and W Nordemann (eds), *Festschrift für Ulrich Loewenheim* (2009) 309–21.

transferred to him.[34] The same is true for strategies where independent firms file the same IPR in various Member States in order to divide up the internal market among themselves.[35]

The form of consent

According to general legal principles, consent may be declared in an express or implied way. In *Davidoff and Levi Strauss* the ECJ held that '. . . in view of its serious effect in extinguishing the exclusive rights of the proprietors. . ., consent must be so expressed that an intention to renounce those rights is unequivocally demonstrated'.[36] Normally such intention will be shown by an express statement. But it is possible—according to the Court—that consent is declared in an implied manner, ie it may be deduced from facts and circumstances in the context of placing the protected goods on the market.[37] However, consent cannot be inferred from the mere silence of the right holder. Therefore, there is no obligation of the IPR owner to attach an EEA import ban to her products when putting them on markets outside the EEA. Likewise, she is not obliged to impose contractual reservations preventing the import of the goods in question into the EEA.[38]

When is a protected good put on the market?

Although the above-mentioned principles seem clear, the goal of market integration depends to a great extent on their substantiation. When is an IP-protected good put on the market? And what is necessary to affirm the consent of the right holder? As regards the first question, a product is put on the market if there is a transaction allowing the right holder to realize the full economic value of his IPR.[39] This will normally be a sale, but exhaustion should not be restricted to this kind of transaction from the outset. If a protected good is transferred for use only, exhaustion will take place if the duration of the contract covers the economic lifespan of the product in question.[40] However, for exhaustion to take place, it is not sufficient to import the product, to stock it or to offer it with a view to selling it. The product must have actually been transferred, and exhaustion only takes place *after* the transfer has

[34] Discussion of this and of similar problems by O-A Rognstad, 'The exhaustion/competition interface in EC law—is there room for a holistic approach?' in J Drexl (ed), *Research Handbook on Intellectual Property and Competition Law* (Cheltenham: Edward Elgar Publishing, 2008) 427, 441–3.

[35] Case 144/81 *Keurkoop/Nancy Kean Gifts* [1982] ECR 2853 para 28.

[36] Case C-414/99 *Zino Davidoff and Levi Strauss* [2001] ECR I-8691 para 45.

[37] No matter if the products in question are put on the market outside or inside the EEA, Case C-324/08 *Makro Zelfbedieningsgroothandel and others* (ECJ judgment dated 15 October 2009) ECR I-10019 paras 28–34.

[38] Case C-414/99 *Zino Davidoff and Levi Strauss* [2001] ECR I-8691 paras 53–60.

[39] Compare the definition in Case C-16/03 *Peak Holding* [2004] ECR I-11313 para 40.

[40] Otherwise it would be easy to circumvent the effect of exhaustion, compare H Ullrich (n 11 above) para 88 fn 615.

been completed.[41] If the product is not transferred to a third party, but only moves within a group of companies which are under common control, IPRs are not consumed.[42]

Hence, in groups of companies, the time of exhaustion is postponed—only when the protected good leaves the group, ie when it is sold by an affiliated company to an actor outside the group, exhaustion takes place.[43] There are proposals in the literature to retard the time of exhaustion even further. Exhaustion shall not take place after a transaction between the members of a distribution system even if they are not part of the same group of companies. In this view, exclusive rights only exhaust when the product leaves the distribution system.[44] The condition is added that the distribution system has to be compatible with competition law.[45] In the same sense, Rognstad opts for a 'holistic approach'—he criticizes the traditional 'cumulation theory' according to which compatibility with the exhaustion principle and the competition rules has to be examined in a separate way. Instead, he pleads for removing or at least reducing inconsistencies between the two sets of rules.[46] He adds the question if right holders should be entitled to assert their IPRs against imports which are in breach of territorial restrictions compatible with competition law.[47]

Qualified consent?

This leads to the general question if there should be a refinement of the exhaustion principle. In this sense, simple consent would not be sufficient for exhaustion to take place. A qualified consent would be necessary requiring that the protected product is put on the market in conformity with all or at least the most important contractual obligations as far as they are compatible with competition law. This is of importance not only for exclusive and selective distribution but also for licensing agreements. A violation of the contract between producer and wholesaler or between licensor and licensee would have the consequence that there is no consent of the right holder to putting the product on the market. Such a violation could refer for example to territorial limitations, quantitative restrictions or quality specifications. Hence, if a

[41] H Ullrich (n 11 above) paras 38–44 with respect to the Trademark Directive.

[42] The concept of putting on the market is also relevant in transit cases. IPRs can only be infringed by goods in transit if they are put on the market of the state of transit which is not the case if the product is placed under the external transit procedure and if it is not subject to the act of a third party, Case C-281/05 *Montex Holdings* [2006] ECR I-10881.

[43] For qualifications to this general statement see A Lessmann, 'Erschöpfung von Patentrechten bei Konzernvertrieb' [2000] Gewerblicher Rechtsschutz und Urheberrecht 741–50; E-J Mestmäcker and H Schweitzer, *Europäisches Wettbewerbsrecht* (2nd edn, 2004) § 27 paras 53–5.

[44] K-H Fezer, 'Wettbewerbsrechtlicher und markenrechtlicher Bestandsschutz funktionsfähiger Distributionssysteme selektiven Vertriebs vor Aussenseiterwettbewerb' [1999] Gewerblicher Rechtsschutz und Urheberrecht 99, 105.

[45] A Metzger, 'Erschöpfung des urheberrechtlichen Verbreitungsrechts bei vertikalen Vertriebsbindungen' [2001] Gewerblicher Rechtsschutz und Urheberrecht 210, 213–14.

[46] O-A Rognstad (n 34 above), 432–43. The proposals are presented in a cautious way; see p 437—it '. . . is at least worth considering abandoning the cumulation theory in favour of a more holistic or harmonized approach.'

[47] O-A Rognstad (n 34 above) 439.

licensee sells the protected product outside the appointed territory, if he produces more than agreed upon, or if he does not meet the quality standards stipulated in the contract, exhaustion would not take place (provided that the contractual commitments do not infringe competition law).[48] The proprietor then would be able—in addition to his contractual and property claims against the contract partner—to assert his property rights against any party in the distribution chain.

The ECJ did not accept this narrow concept of exhaustion. In the *Copad* case, the Court had to answer the question—with respect to trademark law and based on the Trademark Directive—if the disregard of a provision in a licensing agreement prohibiting the sale of the licensed products to discount stores would preclude consent thus preventing exhaustion.[49] It held that a licensing agreement does not constitute the absolute and unconditional consent of the proprietor. But only the licensee's contravention of clauses triggering trademark-based claims (as opposed to mere contractual claims) would hinder exhaustion.[50]

The starting point of the Court deserves approval. There can be no exhaustion if the licensee exceeds the limits of his competences as regards for example, time or scope of his licence. In these cases, the proprietor did not have the possibility to reap the benefits of her IPRs. In trademark law, the restrictions directly based on the exclusive right have been listed by the legislature.[51] Hence, as regards the examples mentioned above, trade mark rights *erga omnes* would exist if restrictions of the scope of the licence or territorial restrictions are violated, but not in the case of disregard of quantitative restrictions (which are not mentioned in the list of Article 8(2) Trademark Directive).[52] For other IPRs, the competences accruing directly from the exclusive right and having *erga omnes* effect have to be determined by way of interpretation based on the protective goal of the IPR in question.[53]

Likewise, it seems convincing not to reject exhaustion in case of disregard of mere contractual obligations. One of the goals of the exhaustion principle is legal certainty. In commerce, purchasers should be allowed to suppose that IP-protected goods which have been sold by the proprietor or an authorized person now can move on without interference of the right holder. Of course, this certainty

[48] In this sense Dior and the French Government in Case C-59/08 *Copad SA v Christian Dior couture SA* [2009] ECR I-3421 para 39 (hereinafter '*Copad*') '. . . that any contravention of the licence agreement by the licensee precludes exhaustion'.

[49] The licensing agreement concerns luxury goods. The interdiction of selling to discount stores was supposed to maintain the reputation and the prestige of the trade mark.

[50] *Copad* (n 48 above) paras 47–51.

[51] Article 8 (2) of the Trademark Directive. This list is exhaustive, *Copad* (n 48 above) para 20. See also Article 22 (2) of Council Regulation (EC) 40/94 on the Community trade mark [1994] OJ L11/1.

[52] H Ullrich (n 11 above) paras 78–9. Compare opinion of AG J Kokott (Case C-59/08 *Copad* [2009] ECR I-3421 para 20) who leaves the question of quantitative restrictions open.

[53] An analysis differentiating between different types of IP is made by H Ullrich (n 11 above) paras 78–9.

is not absolute.[54] It may always be the case that for example a licensee did not respect obligations covered by the exclusive right so that exhaustion did not take place. But this is not an argument to completely undermine legal certainty by attributing claims *ad rem* where mere contractual obligations were violated. For the sake of legal certainty, it is important to reduce uncertainty as much as possible by defining a simple concept of consent.[55] The exhaustion principle tends to reconcile IP protection and the free movement of goods. The proprietor has the right to decide on the first sale of an IP-protected good. This balance should not be upset by a narrow concept of exhaustion giving the right holder also the power to decide on the second and third sale. It is the task of competition law to determine to what extent resale restrictions shall be legal.

This reasoning may be underpinned by more fundamental reflections. It is true that the free movement rules and the competition rules have certain goals in common and complement each other. But the two sets of rules are situated on different levels: The free movement rules—in the context of IP protection—confine the legal power which may be attributed to right holders whereas competition law limits the use which may be made of these rights in the market.[56] One of the major achievements of modern IP competition law is the emancipation of competition law from the scope of IPRs. IPRs are not immune against the application of competition law. The fact, that a certain behaviour is covered by an exclusive right (for example a refusal to license), does not mean that competition law is inapplicable.[57] Neither is it possible to justify the extension of IPRs with the argument that underlying contracts do not violate competition law. It is not the task of IP protection, even less of competition law, to vest distribution systems with an effect going beyond the privity of contract. IPR owners have sufficient possibilities to obtain this result by limiting licences carefully.

In our view, the ECJ has gone too far as regards the *erga omnes* effect of a breach of contract. In the *Copad* case, the Court held that the disregard by a licensee of a clause prohibiting sales of luxury products to discount stores constitutes a violation of the rights conferred by the trade mark if this damages the 'aura of luxury' of the products in question.[58] Starting point is Article 8(2) of the Trademark Directive which allows the proprietor of a trade mark to invoke the rights conferred by that trade mark

---

[54] O-A Rognstad (n 34 above) 436.

[55] The importance of legal certainty in the context of property rights is underlined by S Enchelmaier, 'Intellectual property, the internal market and competition law' in J Drexl (ed), *Research Handbook on Intellectual Property and Competition Law* (Cheltenham: Edward Elgar Publishing, 2008) 405, 420–23.

[56] H Ullrich (n 11 above) paras 105–9.

[57] Therefore, in our view, the concept of 'specific subject matter' which limits the application of the free movement rules to IPRs, cannot be extended to the application of the competition rules to IPRs, see A Heinemann, *Immaterialgüterschutz in der Wettbewerbsordnung* (2002) 327–8.

[58] *Copad* (n 48 above) paras 30–2.

against a licensee who contravenes any provision in his licensing contract with regard *inter alia* to the quality of the goods manufactured by the licensee. Thus, the Community legislature has, to a certain extent, put a value on a quality function of trade marks complementing the traditional origin function. It has to be stressed that therefore violations of quality specifications in licensing agreements not only trigger contractual and property claims against the licensee but also property claims against third parties since exhaustion is denied.

The *Copad* decision goes one step further—even if the properties of a product perfectly comply with the specifications in the licensing agreement, exhaustion does not take place in case of disregard of selective distribution rules since,

> . . . the quality of luxury goods . . . is not just the result of their material characteristics, but also of the allure and prestigious image which bestows on them an aura of luxury.[59]

As a result, the exclusive right is no longer limited to allow its proprietor to control the first marketing in the EEA, but serves to control further transactions and to stabilize a selective distribution system. This happens in an absolute way since outsiders not standing in a contractual relationship with the right holder can be sued for infringement of the trade mark.[60] This does not seem compatible with the functions of trade mark protection even if the quality function is accepted. Distribution systems should be shaped by contracts *inter partes* within the limits of competition law. It is not the goal of IP protection to waterproof distribution systems. The term 'quality' in the meaning of Article 8(2) Trademark Directive should therefore be restricted to material properties.

The challenge of modern economic theory to the assessment of territorial restrictions

The discussion on the prerequisites and the limits of exhaustion is embedded into the general debate on the appropriate treatment of vertical restraints in the light of modern economics.[61] The starting point is well-known—under the influence of the Chicago School, the assessment of vertical restraints has undergone an important development towards a more lenient competition law treatment. Whereas initially scholars of the Chicago School postulated the per se legality of vertical restraints, today a more qualified position prevails. Vertical restraints may—according to the

---

[59] *Copad* (n 48 above) para 24.

[60] The case concerns a licensing agreement allowing the licensee to produce and to sell certain goods under the licensed trade mark. The situation would be different in a simple distribution agreement: The Court held that the sale by the proprietor of the trade mark automatically leads to exhaustion no matter if restrictions on the further marketing of the good have been stipulated between the seller and the purchaser, Case C-16/03 *Peak Holding AB v Axolin-Elinor AB* [2004] ECR I-11313 paras 50–6.

[61] Compare the remarks of Metzger (n 45 above) 213–14, and Rognstad (n 34 above) 434, 438.

circumstances of a particular case—be beneficial or detrimental.[62] In the context of market integration, it has been stated that strict rules against market segmentation disregard the free-rider problem and encourage vertical integration.[63] Already the former block exemption regulation on vertical agreements recognized the efficiency-enhancing effects of vertical restraints and underlined the importance of market power, ie the degree of interbrand competition in this context.[64] The new regulation has integrated even more recent economic thinking, but, at the same time, it has maintained the negative assessment of absolute territorial restrictions and resale price maintenance.[65]

Competition law has of course to be adapted to economic insights corroborated in theory and in practice. However, reforms should be proportional. The fact that in single cases a free-rider problem may occur, is not sufficient to introduce the per se legality of resale price maintenance or of absolute territorial protection. The current system is flexible enough to take into account efficiency gains via an individual assessment based on Article 101(3) TFEU. The new guidelines on vertical restraints explain in detail under which conditions even hardcore restrictions may be justified, for example if they are necessary for the introduction of a new product.[66] In our context, it is important to note that the goal of efficiency pursued by the more economic approach is not contradictory to the goal of market integration. The rules on exhaustion which—in the field of IP protection—are a key component of market integration, make it possible that an internal market of considerable size is constituted which leads to huge efficiency gains. It is important to keep this context in mind when shaping the details of IP competition law.

C. Special Rules in Favour of Innovative Industries?

To what extent do the general rules apply to innovative industries, like for example the IT sector or the pharmaceutical industry? As regards the IP aspect, it has to be repeated that property rights need a precise definition which must not vary subject to market conditions. Moreover, the exhaustion principle has to be kept simple. Therefore, and in conformance with the case law of the ECJ, exhaustion takes place even if there is state influence into pricing as it is the case in many Member States

[62] M Motta, *Competition Policy—Theory and Practice* (Cambridge: Cambridge University Press, 2004) Ch 6.

[63] G Monti, *EC Competition Law* (Cambridge: Cambridge University Press, 2007) 363–6.

[64] Commission Regulation (EC) 2790/1999 on the application of Article 81 (3) of the Treaty to categories of vertical agreements and concerted practices [1999] OJ L336/21 recital 6–7.

[65] Commission Regulation (EU) 330/2010 on the application of Article 101(3) of the Treaty on the Functioning of the European Union to categories of vertical agreements and concerted practices [2010] OJ L102/1.

[66] Commission Guidelines on Vertical Restraints [2010] OJ C130/1 para 47.

with respect to pharmaceutical products.[67] As regards competition law, aspects of innovation are relevant in all sub-fields. The following reflections focus on Article 101 and Article 102 TFEU.[68]

Restrictive agreements

As regards restrictive agreements, the impact on innovation has become relevant in the context of quantitative restrictions and of 'dual pricing'. As regards quantitative restrictions, the control of supply by fixing allotments for purchasers or by imposing output restrictions in licensing contracts may pursue the aim of eliminating parallel trade. By limiting supply, the other party may get just the number of products necessary to satisfy domestic demand. Thus, no quantities are left which could be exported. The same goal may be pursued by 'dual pricing'—in this case, purchasers have to pay an 'export tax', ie a higher price for goods going to be exported than for goods that are sold on the domestic market.

Limitation of parallel trade as the object or effect of an agreement?

Both strategies restrict intrabrand competition and hence fall under Article 101(1) TFEU if the other conditions of this Article are fulfilled.[69] Although this starting point is widely accepted, the CFI has created some confusion in the Spanish *Glaxo* case. The European Commission prohibited a system of dual pricing stating that the agreement entailed restrictions of competition 'by object', and that, according to settled case law, it was not necessary to take account of the actual effect of this agreement.[70] The CFI disagreed. According to the Court, parallel trade is not to be protected 'as such', but only '. . . in so far as it gives final consumers the advantages of effective competition in terms of supply or price'.[71] The Court denied a restriction 'by object' in this sense, but affirmed a restriction 'by effect' because of the cost savings of patients in the countries of destination.[72]

---

[67] *Merck/Primecrown* (n 10 above). It is of course different if the legislature has opted for another solution. For the 'Specific Mechanism' with respect to the new Member States of Middle and Eastern Europe see C Heath, 'Parallel Imports of Patented Pharmaceuticals from the New EU Accession States' [2004] Intl Rev Intellectual Property and Competition L 776–87. See also Swiss patent law; since 2009, Switzerland has (unilaterally) introduced EEA-wide exhaustion into patent law, but provides for an exception in Article 9a (5) of the Federal Patent Law with respect to products whose price is imposed by the state.

[68] For merger control see A Heinemann, 'Intellectual property rights and merger control: How to secure incentives to innovate in the long run' in J Drexl et al (eds), *Technology and Competition—Contributions in Honour of Hanns Ullrich* (Brussels: Larcier, 2009) 601–26 (hereinafter '*Technology and Competition*').

[69] In the case of quantitative restrictions, the existence of an agreement or a concerted practice has to be shown, see part B above. If there is mere unilateral conduct, only Article 102 TFEU is applicable.

[70] *Glaxo Wellcome* [2001] OJ L302/1 (hereinafter '*Glaxo Wellcome*').

[71] Case T-168/01 *GlaxoSmithKline Services v Commission* [2006] ECR II-2969 para 121 (hereinafter 'Case T-168/01 *GlaxoSmithKline Services*').

[72] *Ibid* para 148 *et seq*, especially para 181 *et seq*.

This approach is not convincing.[73] If it were necessary to show the direct consequences for final consumers in order to prove the restrictive object of an agreement, even hardcore cartels would not have an anti-competitive object if the disadvantages are not passed on to consumers. The participation of consumers in the resulting benefit is expressly mentioned in Article 101(3) TFEU and therefore does not belong in the analysis of the anti-competitive object of the agreement. On appeal, the ECJ has held in this sense—according to the Court,

> . . . for a finding that an agreement has an anti-competitive object, it is not necessary that final consumers be deprived of the advantages of effective competition in terms of supply or price.[74]

Thus, the case law holds true according to which agreements aimed at restricting parallel trade have as their object the restraint of competition. Dual pricing strategies belong to this category and therefore have a restriction of competition as their object. Therefore, a justification is necessary.

Block exemption

The justification may be either general in the form of a block exemption regulation or individual. As regards output restrictions in licensing agreements, the legal situation has considerably changed. Under the 1996 block exemption regulation on technology transfer agreements, quantitative restrictions in licensing contracts were blacklisted in order to avoid the same result as export bans.[75] In the 2004 Regulation[76] the treatment of quantitative restrictions in licensing contracts has been relaxed.[77] Output restrictions are only blacklisted in certain situations.[78] According to the European Commission, the pro-competitive aspects (dissemination of the technology in question) outweigh the anti-competitive risks (reduced intra-technology competition).[79] However, in order to avoid the danger of partitioning national markets, the Commission has added some caveats in the Technology Transfer Guidelines—quantity

[73] See the critique of P Behrens, 'Parallelhandelsbeschränkungen und Konsumentenwohlfahrt' [2008] Zeitschrift für Wettbewerbsrecht 20, 37–9; A Heinemann, 'Geistiges Eigentum im internationalen und regionalen Kartellrecht' in A Ohly and D Klippel (eds), *Geistiges Eigentum und Gemeinfreiheit* (2007) 199, 212–4. As regards the general difficulty of determining the effects on consumer welfare see H Schweitzer, 'The role of consumer welfare in EU competition law' in *Technology and Competition* (n 68 above) 511 *et seq*.

[74] Joined Cases C-501/06 P, C-513/06 P, C-515/06 P, C-519/06 P *GlaxoSmithKline Services Unlimited et al/Commission et al* [2009] ECR I-9291 (hereinafter 'Case C-501/06 P *GlaxoSmithKline Services Unlimited*').

[75] Article 3 (5) of Commission Regulation (EC) No 240/96 on the application of Article 85 (3) of the Treaty to certain categories of technology transfer agreements [1996] OJ L31/2.

[76] Regulation No 772/2004 (n 23 above).

[77] However, it has to be repeated that the block exemption only applies to individual market shares up to 30 per cent (in case of competitors up to a combined market share of 20 per cent), see Article 3 Regulation 772/2004.

[78] That is reciprocal output restrictions between competitors, see Article 4(1)(b) of Regulation 772/2004.

[79] European Commission, Technology Transfer Guidelines [2004] OJ C101/2 para 175 *et seq*.

limitations must not be used in order to restrict passive sales. An indication thereof would be the adjustment of quantities over time to cover only local demand.[80]

Individual exemption

Can market-partitioning practices be justified under Article 101(3) TFEU, and if so, under which conditions? The Commission and the European Courts had to deal with these questions in the Spanish *Glaxo* case. The Commission came to the result that the dual pricing scheme did not meet the conditions for an exemption. According to the Commission, *Glaxo* did not prove a causal link between parallel trade and R&D investments, and could therefore not show that consumers will receive a fair share of the resulting benefit.[81] According to the CFI, the Commission did not sufficiently examine the question if dual pricing improves innovation.[82] The ECJ shared this view.[83] As a result, the Commission has to re-examine the application of Article 101(3) TFEU.

The Court decisions bring up at least two questions. As regards procedure, it has to be asked, how detailed the Commission's analysis has to be when assessing the conditions of Article 101(3) TFEU. And as regards substance, can private restraints aiming at hampering parallel trade be justified by effects on innovation? As regards the first question, the Commission has analysed the determinants of R&D expenditure. It came to the conclusion that there is no evidence that parallel trade has led to cuts or has prevented the growth of the R&D budget.[84] The Courts criticized that the Commission did not analyse sufficiently the economic studies submitted, and that the Commission's conclusions are lapidary.[85] This stands in contrast to the generally accepted starting point that the European Courts only have a limited power of review of complex economic assessments made by the Commission.[86] In our view, the requirements of the European courts in this case go beyond a confined control. An obligation to respond to all details of the studies submitted oversteps the mark and opens the way to defence strategies trying to stall proceedings by the quantity of evidence submitted.

This point of view may get support from the answer to the second question. The starting point is quite simple: Of course, restrictive agreements may be justified under Article 101(3) TFEU by effects on innovation since the promotion of technical or economic progress is expressly mentioned in this article. However, all conditions of this paragraph have to be met. The contribution of the restrictive agreement to the promotion of innovation has to be shown, and there has to be a direct link

[80] *Ibid*, para 98.
[81] *Glaxo Wellcome* (n 70 above) para 153 *et seq*.
[82] Case T-168/01 *GlaxoSmithKline Services* (n 71 above).
[83] Case C-501/06 P *GlaxoSmithKline Services Unlimited* (n 74 above).
[84] *Glaxo Wellcome* (n 70 above) paras 155–69.
[85] See for example paras 275–9 in Case T-168/01 *GlaxoSmithKline Services* (n 71 above).
[86] Case C-501/06 P *GlaxoSmithKline Services Unlimited* (n 74 above) para 146.

between both. It is not sufficient to claim a general context between increased profits and stronger R&D activities. Such a general reflection would be applicable to hardcore price cartels as well.[87] Therefore, it is not sufficient to say that without parallel imports the undertaking in question would earn higher profits which would be invested into R&D.

This leads to the fundamental question on which legal level the innovation aspect of parallel trade should be taken into account. In fact, the rules on exhaustion have multiple effects, not only on the pharmaceutical companies but also on distributors, the national or private health insurances, and on consumers/patients. The long-term effects have to be taken into account as well as the opposing interests of consumers in countries with high or low prices.[88] In our view, these questions have to be solved on the level of the exhaustion rules themselves. If a comprehensive economic analysis had the result that—because of the beneficial effects of regional price discrimination and in the interest of innovation—parallel trade should be prohibited, the principle of national exhaustion should be introduced.[89] If on the other hand, the answer is, with good reason, in favour of parallel trade (including the pharmaceutical sector), the rules on regional exhaustion should be upheld. Then it is consequent for competition law to support this fundamental decision by denying an exemption to agreements which try to suppress the viability of parallel imports. The IP holder still has the possibility of minimizing the effects of parallel trade by unilateral measures. In this field, limitations are only set by Article 102 TFEU which presupposes market dominance and which does not prohibit any restraint of competition but only abuse.

The abuse of a dominant position

Every undertaking is free in formulating its distribution system. It can sell its products via branches, subsidiaries or agents, or over independent wholesalers and retailers. Parallel trade occurs in the latter case. The producer may prevent or at least diminish parallel trade by imposing quota based on the demand in the respective country. The question arises if such allocations, practised by an undertaking having a dominant position, constitute an abuse if their aim is to hamper parallel trade. The ECJ has dealt with this question in the Greek *Glaxo* case. The Court weighs the importance of parallel trade against the existence of price and supply regulation in the pharmaceutical sector. The result is a compromise—price regulation does not justify the complete refusal to supply wholesalers. But it is not abusive to refuse

[87] E-J Mestmäcker and H Schweitzer (n 43 above) § 13 para 48. For further development of this argument see J Drexl, 'Healing with bananas—How should Community competition law deal with restraints on parallel trade in pharmaceuticals?' in *Technology and Competition* (n 69 above) 571, 586–91.

[88] For an economic analysis of parallel trade see P Behrens (n 73 above) 20 *et seq*; J Drexl (n 87 above) 575–86.

[89] Compare J Drexl (n 87 above) 575 who raises the question if the outcome of economic competition law analysis should not influence the design of the exhaustion doctrine as well.

orders which go beyond the 'ordinary'. It is for the national court to determine whether orders are ordinary in the light of the size of the market and of the previous business relations with the wholesalers concerned.[90] As regards the effect on innovation, the Court does not see the necessity to examine the argument that the limitation of parallel exports might be necessary in order to avoid the risk of a reduction of R&D expenditure.[91]

The judgment is ambiguous—on the one hand, the importance of parallel trade is underlined.[92] On the other hand, the IP holder has the right to defend his own commercial interests, as long as the measures taken are reasonable and proportionate.[93] The result is a limited obligation to contract—the orders which have to be met are capped at the level of domestic demand including a certain safety margin. In the first place, this favours dealers which have been active on the market and who see their supplies cut. Orders, which have been met in the past, will probably qualify as 'ordinary' and have to be met in the future, too. However, this should not exclude newcomers. If new firms are interested in being supplied, the dominant firm is obliged to reallocate the 'ordinary' quantity for the respective country on all interested parties.[94] Against the backdrop of Article 102(b) TFEU (limitation of production, supplies or technical development), the result of the *Glaxo* case seems convincing—a dominant undertaking has to satisfy the national demand in a given Member State. It would have been enlightening, though, if the Court had stated more precisely its view on the relationship between parallel trade and the incentives to innovate.

Conclusion: The Status of Market Integration in European Competition Law

There are some indications that the importance of market integration is decreasing. Market-partitioning strategies have been qualified as unilateral so that the cartel interdiction is not applicable.[95] The CFI has held that the effect on final consumers has to be shown in order to establish the anti-competitive object of an agreement.[96]

[90] Joined Cases C-468/06 to C-478/06 *Sot. Lélos kai Sia et al v GlaxoSmithKline AEVE Farmakeftikon Proïonton* [2008] ECR I-7139 paras 58–77 (hereinafter 'Joined Cases C-468/06 et al *Sot. Lélos kai Sia*'). See the in-depth analysis of the *GlaxoSmithKline* saga by J Drexl (n 87 above) who criticizes the Court for the lack of economic reasoning.

[91] Joined Cases C-468/06 et al *Sot. Lélos kai Sia* (n 90 above) para 70.

[92] Albeit in a half-hearted manner: 'Indeed, parallel imports enjoy a certain amount of protection in Community law because they encourage trade and help reinforce competition' (Joined Cases C-468/06 et al *Sot. Lélos kai Sia* (n 90 above) para 37).

[93] Joined Cases C-468/06 et al *Sot. Lélos kai Sia* (n 90 above) para 69.

[94] W Wurmnest, 'Einseitige Beschränkungen des Parallelhandels durch marktbeherrschende Arzneimittelhersteller im Binnenmarkt' [2009] Rabel J Comparative and Intl Private L 962, 983. The ECJ did not have to rule on this question.

[95] See text to n 27 above.

[96] Case T-168/01 *GlaxoSmithKline Services* (n 71 above).

Exhaustion has been denied in the case of luxury products sold to discounters in contradiction to the terms of a licensing agreement.[97] Generally, modern economic thinking has underlined the beneficial effects of territorial segmentation.[98] These developments have provoked contradiction, though—regardless of whether a more economic approach is accepted or not, a microeconomic 'absolutization' of the (consumer) welfare criterion loses sight of the institutional framework which alone guarantees the maintenance of a competitive system in the long run.[99] In recent case law, the ECJ has clearly rejected positions according to which the application of competition law systematically requires the effects on consumers. According to the Court, European competition law,

> . . . is designed to protect not only the immediate interests of individual competitors or consumers but also to protect the structure of the market and thus competition as such.[100]

The same holds true for the goal of market integration. The construction of the internal market has offered far-reaching opportunities to undertakings and to consumers. It would not be convincing to grant free market access within the internal market while reserving the right to create private barriers where it is deemed advantageous.[101] In the IP context, the backup function of competition law with respect to market integration is of particular importance since IPRs provide additional opportunities for segmenting markets according to national borders. This article has tried to illustrate how diverse the legal questions are which have to be assessed in this context. Territorial protection may be necessary for the development or the dissemination of innovations. However, the high value of market integration has to be taken into account whenever restrictions of cross-border trade are to be assessed.

---

[97] See text to n 58 above.

[98] Compare G Monti (n 63 above) 365: 'In sum, the importance of market integration wanes as economic analysis shapes competition policy.'

[99] Compare P Behrens (n 73 above) 37–9; J Drexl (n 87 above) 596; D Zimmer, 'Competition law de lege ferenda' in R Zäch et al. (eds), *The Development of Competition Law: Global Perspectives* (Cheltenham: Edward Elgar Publishing, 2010) 322-324; R Zäch and A Kuenzler, 'Reining in the "more economic approach": Some overriding constraints from constitutional law and economics' in *Technology and Competition* (n 69 above) 541, 550–4.

[100] Case C-08/08 *T-Mobile Netherlands* (n 24 above) para 38. Confirmed by Case C-501/06 P *GlaxoSmithKline Services Unlimited* (n 74 above) para 63.

[101] The opinion is widespread that the integration function of EU competition law is or should be of unchanged importance, compare J Drexl (n 87 above) 597–9; A Heinemann (n 14 above) 67; D Hildebrand, *The Role of Economic Analysis in the EC Competition Rules* (3rd edn, Kluwer Law International, 2009) 16–18. According to others, its significance is diminishing, compare B Rodger and A MacCulloch, *Competition Law and Policy in the EC and UK* (4th edn, Routledge-Cavendish, 2009) 19. The market integration function is underlined by the ECJ (Case C-501/06 P *GlaxoSmithKline Services Unlimited* (n 74 above) para 61) and upheld by the European Commission in the new Commission Guidelines on Vertical Restraints (n 66 above) para 7, 100 (d).

Part IV

STANDARD SETTING AND FRAND COMMITMENTS

14

STANDARD SETTING AND HOLD-UPS UNDER SECTION 5 OF THE FEDERAL TRADE COMMISSION ACT

*Alden F Abbott and Nicholas J Kim**

Introduction

Standard setting has long been scrutinized under US antitrust law. This scrutiny reflects the fact that, although standard setting has the potential to confer substantial benefits on producers and consumers, it also may be misused to limit the terms of competition among rival firms or to enable a firm to achieve monopoly power.

This chapter focuses on the application of Section 5 of the FTC Act's[1] prohibition on 'unfair methods of competition' to single firm conduct in standard setting ('Section 5'). After describing judicially-established and FTC-proposed constraints on the application of Section 5 to business behaviour, we briefly highlight the nature of standard setting and patent 'hold-up' in standard setting contexts (opportunistic behaviour by a patent holder whose patents 'read' on a standard). We then focus on two recent standard setting-related matters involving hold-ups that have prompted FTC enforcement actions, the *Rambus* litigation and the *N-Data* settlement. We then propose three 'limiting principles' for Section 5 that could fruitfully be applied to hold-ups associated with standard setting. We close by assessing the institutional advantages Section 5 may enjoy over other means (particularly private litigations) to combat welfare-inimical hold-ups.

* The views expressed in this article are solely attributable to the authors. They do not necessarily represent the views of the Federal Trade Commission or any individual Federal Trade Commissioner.

[1] Federal Trade Commission Act, 38 Stat 717 (1914), as amended, 52 Stat 111 (1938), 15 USC § 45 (2006). As we will show, Section 5 sweeps more broadly than the Sherman Act. The FTC alone, however, enforces Section 5. In contrast, the Justice Department enforces the Sherman Antitrust Act, and private parties and state attorneys general (in their role as *parens patriae*), also may bring actions under the Sherman Act.

A. Section 5 is Broad, but Clearly not Unbounded

For years, the competition community has eagerly anticipated the FTC's pronouncement of its interpretation of 'unfair methods of competition' within the meaning of Section 5. In the meantime, the debate has been informed in part by an October 2008 FTC public workshop held to '... explore the scope of the prohibition of "unfair methods of competition" in Section 5 of the FTC Act.'[2] On one end of the spectrum, some practitioners and academics argue that the FTC should not apply Section 5 to conduct falling outside the scope of the Sherman or Clayton Acts.[3] However, as reflected in the public workshop and recent commission decisions,[4] the Commission clearly sees a role for Section 5 that reaches beyond the antitrust laws.

The Supreme Court explicitly recognizes that Section 5 encompasses behaviour outside the scope of the federal antitrust laws. In *Sperry & Hutchinson*,[5] the Court addressed whether Section 5 empowered the FTC to define and proscribe unfair competitive practices that did not infringe '... either the letter or the spirit of the antitrust laws.'[6] In addition to holding that Section 5 was broader than the antitrust laws, the Court found that Section 5 empowered the FTC to proscribe conduct '... as unfair or deceptive in their effect upon consumers regardless of their nature or quality as competitive practices or their effect on competition.'[7] Looking to the legislative history of the Trade Commission Act, the Court was persuaded by Congress' deliberate choice *not* to reduce the ambiguity of the phrase 'unfair methods of competition.'[8] In a subsequent case, the Court reiterated that Section 5 encompassed '... not only practices that violate the Sherman Act and other antitrust laws,

---

[2] Public Workshop Concerning the Prohibition of Unfair Methods of Competition in Section 5 of the Federal Trade Commission Act, 73 Fed Reg 50, 818 (28 August 2008); see also '2008 FTC Workshop: Section 5' <http://www.ftc.gov/bc/workshops/section5/index.shtml> accessed 15 September 2010.

[3] See AD Melamed, 'Comments Submitted to the Federal Trade Commission, Workshop Concerning Section 5 of the FTC Act' (14 October 2008) <http://www.ftc.gov/os/comments/section5workshop/537633-00004.pdf> accessed 15 September 2010. AD Melamed was counsel for both N-Data and Rambus before the FTC and strongly advocates against applying Section 5 to prohibit conduct that does not violate the antitrust laws.

[4] See JT Rosch, Commissioner, FTC, 'Welcoming Remarks at the FTC Section 5 Workshop' (17 October 2008) <http://www.ftc.gov/bc/workshops/section5/docs/trosch.pdf> accessed 15 September 2010 (hereinafter 'JT Rosch, *Welcoming Remarks*'); J Leibowitz, Commissioner, FTC, '"Tales from the Crypt"—Episodes '08 and '09: The Return of Section 5', FTC Section 5 Workshop (17 October 2008) <http://www.ftc.gov/bc/workshops/section5/docs/jleibowitz.pdf> accessed 15 September 2010 (regarding submissions by workshop attendees, '... [A]ll of us agree that there are circumstances in which the Commission ought to bring "pure" Section 5 cases. But none of us agree on precisely when the Commission should invoke this statute.').

[5] *FTC v Sperry & Hutchinson Co*, 405 US 233, 92 S Ct 898 (1972) (hereinafter '*Sperry & Hutchinson*').

[6] *Sperry & Hutchinson* (n 5 above) 241.

[7] *Ibid* 241.

[8] *Ibid* 241.

but also practices that the Commission determines are against public policy for other reasons.'[9] The Supreme Court having established that Section 5 gave the FTC power to prevent practices that were not antitrust violations, a trio of subsequent federal appeal court decisions, described below, enunciated three principles that limit Section 5's scope.[10]

Judicial decisions limiting the scope of Section 5

The key judicial limitations on Section 5 may be described as follows. First, Section 5 does not, or should not, sanction commercial conduct when there is a lack of evidence that the conduct in question has anti-competitive effects.[11] In *Boise Cascade*,[12] the Ninth Circuit declined to enforce an FTC order finding southern plywood manufacturers violated Section 5 by adopting and maintaining a system of delivered pricing that involved rail freight charges from the Pacific Northwest to determine the price of southern plywood.[13] The FTC typically challenged delivered pricing systems under a theory of conspiracy by adherence to a formula that makes competitors' prices identical.[14] Analysing the case law and the FTC's own policy, the court held that, in the absence of an overt agreement, the FTC must demonstrate that the pricing system had the actual effect of fixing or stabilizing prices.[15] Holding that substantial evidence did not sustain the FTC's findings, the court found that the FTC provided '... little more than a theory of the likely effect of the challenged pricing practices.'[16]

The court also declined the FTC's invitation to sustain the order on the basis of something less than substantial evidence of anti-competitive effect because of the broader nature of Section 5.[17] As '... the law of delivered pricing is well forged, having been developed by the [FTC] and courts over years of litigation', the FTC had to find either collusion or actual effect on competition to make out a Section 5 violation for delivered pricing.[18] In the court's view, allowing the FTC's claim would '... blur the distinction between guilty and innocent commercial behaviour.'[19] Accordingly, *Boise Cascade* forecloses the availability of a standalone Section 5 action as a backdoor to liability when the conduct in question falls squarely within the ambit of the antitrust laws but the facts fall short of establishing a claim under them.

---

[9] *FTC v Indiana Federation of Dentists*, 476 US 447, 454 (1986).

[10] See JT Rosch, *Welcoming Remarks* (n 4 above).

[11] *Ibid* 2.

[12] *Boise Cascade Corp v FTC*, 637 F 2d 573 (9th Cir 1980) (hereinafter '*Boise Cascade*').

[13] Delivered pricing has the potential to allow competitors to match prices and avoid the rigours of price competition. *Boise Cascade* (n 12 above) 575.

[14] *Ibid* 575.

[15] *Ibid* 577.

[16] *Ibid* 578.

[17] *Ibid* 581.

[18] *Ibid* 582.

[19] *Ibid* 582.

Second, a proper Section 5 claim requires at least some evidence of 'oppressiveness' to be caused by the conduct in question. In *Ethyl*,[20] the Second Circuit set aside an FTC order finding that Du Pont, Ethyl, and other anti-knock gasoline additive manufacturers violated Section 5 through independent and unilateral conduct involving sales at a delivered price that included transportation costs, giving advance notice of price increases beyond that required by contract, and use of a 'most favored nation' clause with sellers.[21] Although the FTC conceded that the practices were not collusive and had a legitimate business purpose, the FTC alleged that the practices *collectively* lessened competition in violation of Section 5 by facilitating price parallelism at levels above the competitive levels.[22] The court recognized that Section 5 was broader than the antitrust laws, but stressed that appropriate standards should be adopted to protect against an arbitrary and capricious application of the FTC's power.[23] According to the court, the FTC owed,

> . . . a duty to define the conditions under which [the questioned conduct] would be unfair so that business will have an inkling as to what they can lawfully do rather than be left in a state of complete unpredictability.[24]

The court found that the FTC decision did not provide any such guidelines, so it provided a minimum standard of its own.[25] Specifically, the court held that, absent a tacit agreement, at least some 'indicia of oppressiveness' must exist.[26] Examples of the indicia included evidence of anti-competitive intent or purpose or the absence of an independent legitimate business reason for its conduct.[27] This minimum standard was influenced by the 'plus factor' test for conscious parallelism under the Sherman Act.[28] Accordingly, *Ethyl* suggests a proper Section 5 action must include an 'unfairness plus factor'.

Third, Section 5 cannot be applied to create an affirmative duty to act where antitrust law expressly provides that no duty exists. In *OAG*,[29] the Second Circuit reversed an FTC order finding that a monopolist publisher of flight schedule guides had a duty under Section 5 to refrain from discriminating between certificated air carriers and commuter airlines.[30] Listing connecting flights in the guide was important and valuable to a user.[31] The publisher's guides listed only the connecting flights of

[20] *E.I. Du Pont De Nemours & Co v FTC*, 729 F 2d 128 (2d Cir 1984) (hereinafter '*Ethyl*').

[21] *Ethyl* (n 20 above) 130.

[22] *Ibid* 130, 135.

[23] *Ibid* 136–8.

[24] *Ibid* 139.

[25] *Ibid* 139.

[26] *Ibid* 139.

[27] *Ibid* 139.

[28] See *Ethyl* (n 20 above) 139 fn10. See also *American Tobacco Co v United States*, 328 US 781 (1946).

[29] *Official Airline Guides, Inc v FTC*, 630 F 2d 920 (2d Cir 1980) (hereinafter '*OAG*').

[30] *OAG* (n 29 above) 921.

[31] *Ibid* 922.

certificated carriers, which paid the publisher hundreds of thousands of dollars to have their flights listed.[32] The FTC found that the publisher's arbitrary refusal to list connecting flights of commuter airlines was in violation of Section 5.[33] The FTC argued that the *Colgate* doctrine,[34] generally permitting refusals-to-deal, must give way where a monopolist arbitrarily exercised its business judgment, even when the affected competition was in a different industry than that of the monopolist.[35] Additionally, the FTC justified its finding based on the essential facilities doctrine.[36] The court, however, found that application of Section 5 as urged would give the FTC too much power to substitute its own business judgment for that of the monopolist when the monopolist's decision arguably affects competition in another industry.[37] Instead, to be actionable, the conduct must be coercive.[38] Accordingly, *OAG* restricts the scope of Section 5 so it cannot be used to 'recapture' behaviour exempted from the antitrust laws without evidence of some pernicious or coercive intent.

Exploring the gap between the antitrust laws and Section 5: judicial precedents and the FTC's Section 5 Report

In light of *Sperry & Hutchinson* and the appellate trio of and *Boise Cascade*, *Ethyl*, and *OAG*, there lies a nebulous region of potentially suspect commercial behaviour beyond the reach of the antitrust laws, with respect to which the marketplace requires more explicit guidance. In fact, although courts are inclined to give the FTC's findings deference and review FTC orders under the substantial evidence standard,[39] courts have emphasized the need for definite standards to guide the FTC's application of Section 5.[40]

The *Boise Cascade-Ethyl-OAG* trilogy may prove to be less of a constraint on Section 5 enforcement than might appear at first blush. For example, the prohibition suggested by *Boise Cascade* may be more applicable to allegations of concerted action (conscious parallelism) among multiple parties, as opposed to unfair acts by a single market participant. Additionally, *Boise Cascade* relied on the 'well forged' application of the antitrust statutes against delivered pricing schemes. Arguably, *Boise Cascade*'s

[32] *Ibid* 922.

[33] *Ibid* 923.

[34] *United States v Colgate & Co*, 250 US 300, 307 (1919) ('In the absence of any purpose to create or maintain a monopoly, the [Sherman Act] does not restrict the long recognized right of trader or manufacturer... freely to exercise his own independent discretion as to parties with whom he will deal.').

[35] *OAG* (n 29 above) 925.

[36] *Ibid* 926.

[37] *Ibid* 927.

[38] *Ibid* 927.

[39] See 15 USC § 45(c); *Universal Camera Corp v NLRB*, 340 US 474 (1951).

[40] See, eg, *Ethyl* (n 20 above) 137 ('... appropriate standards must be adopted and applied'). FTC Commissioner Rosch echoed the courts' concerns in the FTC's Section 5 Workshop; JT Rosch, *Welcoming Remarks* (n 4 above) 3 (citing a need for limiting principles on the application of Section 5).

prohibition against a standalone Section 5 claim is weakened if there is less case law and established agency policy applying the Sherman or Clayton Acts to the questionable conduct. *Ethyl*, based on analogy to Sherman 1, might also be similarly limited. Furthermore, *Boise Cascade*, *Ethyl*, and *OAG* all were cases in which the actions under scrutiny had well recognized potential efficiency justifications, involved the routine exercise of business judgment, and lacked convincing evidence of likely consumer harm. They thus by their own terms were a 'hard sell' for the proposition that they correctly identified 'unfair' methods of competition.

We propose that the FTC respond to these case law precedents by adopting a three-part test for Section 5 that is in line with the *Boise Cascade-Ethyl-OAG* trilogy. Specifically, we suggest that the FTC treat business conduct as violating Section 5 if it: (1) is culpable in that it does not involve competition on the merits, (2) should have been seen as questionable before the fact, and (3) actually causes a net harm to competition, or raises a substantial risk of such harm. By applying these three limiting principles, the FTC could help ensure that Section 5 principles of liability are predictable, minimize applications that are inconsistent with any express standards under the Sherman Act, and also minimize applications that serve only to simplify litigation by omitting elements of proof. Taken as a whole, these principles correspond closely to the judicial concern with limiting Section 5 to 'oppressive' conduct involving violation of duties that have harmful effects and are readily identifiable in advance. As we shall see, the standards setting context provides a plausible setting for competitive concern under these principles.

Analysing Section 5 in the specific context of standard setting hold-up can inform the later development of general principles

Patent hold-up in standard setting provides a unique context in which to evaluate the potential scope of standalone Section 5 enforcement. First, collaborative standard setting has been the subject of particularized FTC and DOJ analysis and guidance. Second, patent hold-up, like its contract law analogue, has recognized economic effects on parties' incentives. Thus, instead of seeking limiting principles for Section 5 outright, analysing the application of Section 5 in this narrower context may illuminate limits to Section 5 generally.

Standard setting

Industry standards, including interoperability and performance standards, have been recognized as '... one of the engines driving the modern economy' by making products less costly for firms to produce and more valuable to consumers.[41]

[41] US Department of Justice and FTC, 'Antitrust Enforcement and Intellectual Property Rights: Promoting Innovation and Competition (2007) <http://www.ftc.gov/reports/innovation/P040101PromotingInnovationandCompetitionrpt0704.pdf> accessed 15 September 2010, 33 (hereinafter '*IP 2 REPORT*').

Standards can also increase innovation, efficiency, and consumer choice.[42] For example, standards in telecommunications make such devices more valuable by fostering interoperability among devices and services provided by distinct sources.[43]

Although *de facto* standards may emerge through market competition and consumer choice, standards may also be set through collaboration by multiple businesses in a standard setting organization (SSO) to create a standard that could be used by the entire industry.[44] SSOs often involve collaboration among competitors, whereby the standard setting process displaces consumer choice and competition.[45] Accordingly, antitrust law carefully scrutinizes SSOs and their activities.[46] This substitution may yield substantial efficiencies, including the reduction of wasteful spending or delays associated with an all-out standards war.[47] Accordingly, conduct which undermines the standard setting process or renders it anti-competitive may be actionable under the antitrust laws.[48]

Patent hold-up in standard setting

Hold-up may be defined as a contractual concept in which one party to a contract extracts additional value or concessions from another party based on circumstances creating an exceptional need.[49] Even the possibility of hold-up creates undesirable effects including a reduction of the incentives to invest in the activity or relationship the contract was designed to support.[50] In the standard setting context, patent hold-up can arise because firms may make significant investments in developing and employing a technology standard that is covered by one or more parties' patent rights.[51] If these investments are not applicable or re-deployable using other IP, a single patent holder can engage in patent hold-up to an extent corresponding to the

---

[42] *IP 2 REPORT* (n 41 above) 33.

[43] *Ibid* 33.

[44] *Ibid* 33–4.

[45] *Ibid* 33–4.

[46] *Ibid* 34–5.

[47] *Ibid* 34–5; see also Opinion of the Commission, *In the Matter of Rambus, Inc* (2 August 2006), <http://www.ftc.gov/os/adjpro/d9302/060802commissionopinion.pdf> accessed 15 September 2010, 36 ('. . . [S]tandard setting can function as an efficient substitute for selecting interoperable technologies through direct competition.') (hereinafter 'Opinion of the Commission, *In the Matter of Rambus, Inc*').

[48] See, eg, *Allied Tube & Conduit Corp v Indian Head, Inc*, 486 US 492, 509–11 (1988) (hereinafter '*Allied Tube*') (affirming court of appeals' reinstatement of a jury verdict awarding damages for a Sherman Act violation where producers and sellers of steel conduit had packed a meeting with new members whose sole function was to vote against a proposal to allow the use of equally viable plastic conduit in the building industry); *American Society of Mechanical Engineers v Hydrolevel Corp*, 456 US 556, 574 (1982) (hereinafter '*ASME*') (finding SSO liable for actions of its agents acting with apparent authority to discourage customers from purchasing one competitor's water boiler safety device, stating that it did not comply with the SSO's safety code, even though it did).

[49] See eg S Shavell, 'Contractual Holdup and Legal Intervention' (2007) 36 U Chicago J Legal Studies 325, 326; *IP 2 REPORT* (n 41 above) 35 fn 11.

[50] See Shavell (n 49 above) 326; *IP 2 REPORT* (n 41 above) 35 fn 11.

[51] *IP 2 REPORT* (n 41 above) 35 fn 11.

magnitude of the switching costs.[52] Specifically, after an SSO chooses technologies for a standard, an incorporated technology may lack effective substitutes because of its inclusion in the standard.[53] The owner of a patent covering this incorporated technology can thereby extract higher royalties or stricter terms than possible *ex ante*.[54] If these additional royalties are passed down through higher costs, consumers could be harmed.[55]

Seeking to mitigate the risk of patent hold-up, SSOs generally require participants to abide by a set of SSO policies and obligations including disclosure rules and licensing rules.[56] Disclosure rules require participants to disclose their respective patent interests, while licensing rules restrict the terms that patent holders can demand of standards adopters.[57] Licensing rules range from general obligations to license any adopter on reasonable and non-discriminatory (RAND) terms to specific disclosures of maximum or strictest terms.[58] Outside of these policies, reputational risk to an SSO participant may also discourage a participant from engaging in patent hold-up.[59]

Let us now assess two recent FTC actions involving patent hold-up in standard setting context. In the first case, Rambus, an SSO participant, was accused of anticompetitive conduct. In the second case, N-Data, a non-party to the standard setting process, was the subject of investigation. From these specific applications of Section 5 in the relatively narrow context of standard setting, some limiting principles begin to appear. Based on these limiting principles, this article then analyses whether Section 5 liability should be broadened to encompass patent hold-up outside this special context.

B. SSO Participant Rambus: Section 5 Lessons from a Failed Section 2 Case

The FTC's investigation into questionable conduct by Rambus was neither a standalone Section 5 case nor a victory by the FTC.[60] The Rambus prosecution did not

---

52 *Ibid* 35 fn 11, 37–8.

53 *Ibid* 36.

54 *Ibid* 36.

55 *Ibid* 36.

56 *Ibid* 36, 42.

57 *Ibid* 42.

58 See DOJ, VITA Business Review Letter (30 October 2006) <http://www.justice.gov/atr/public/busreview/219380.htm> accessed 15 September 2010 (declining to investigate an antitrust investigation based on an SSO policy requiring mandatory disclosure of a patent holder's most restrictive licensing terms).

59 *IP 2 REPORT* (n 41 above) 40.

60 See *Rambus Inc v FTC*, 522 F 3d 456, 459 (DC Cir 2008), *cert. denied*, 129 S Ct 1318 (2009). After the Court of Appeals denied the FTC's petition for rehearing *en banc*, the FTC recently

meet the rigorous requirements for a Sherman Act Section 2 violation, according to the reviewing US court of appeals. Nevertheless, the FTC's pursuit of Rambus provides important guidance regarding a possible role for standalone Section 5 actions (actions that do not allege Sherman Act violations) to police unfairly opportunistic behaviour by an SSO participant.

Allegedly anti-competitive conduct by Rambus before JEDEC

Founded in 1990, Rambus develops, patents, and licenses computer memory technologies.[61] Rambus holds patents and pending applications directed to high-performance dynamic random access memory (DRAM) architecture, Rambus DRAM (RDRAM).[62] In particular, Rambus filed its first DRAM-related patent application in 1990.[63] Beginning in 1992, Rambus filed ten divisional patent applications claiming the benefit of the parent.[64] Along with other continuation applications, Rambus amassed over 20 relevant US patents by June 1996.[65]

DRAM is an essential component in electronic devices ranging from personal computers to fax machines and video games.[66] In 2000, DRAM sales in the United States alone exceeded $12 billion.[67] By the late 1980s, processing technology was progressing faster than memory technology to the point where existing DRAM architectures created a 'memory bottleneck'.[68] Synchronous DRAM (SDRAM) was a promising solution to the memory bottleneck, but its success depended on the adoption of SDRAM standards by the industry.[69]

The Joint Electron Device Engineering Council (JEDEC) Solid State Technology Association was an organization for promoting development and standardization in solid state products such as DRAM.[70] JEDEC membership was freely available to any company or individual that manufactured electronic equipment or provided electronics-related services.[71] To become a member, a company only had to submit an application, pay membership fees, and agree to abide by JEDEC's rules.[72] Starting in October 1993, the JEDEC Manual provided that JEDEC committee chairpersons

petitioned for a writ of certiorari to review the decision by the Court of Appeals adverse to the FTC, but that writ was denied by the US Supreme Court.

[61] Complaint, *In re Rambus Inc.* (18 June 2002) Docket No 9302 <http://www.ftc.gov/os/adjpro/d9302/020618admincmp.pdf> accessed 15 September 2010, 2, 8 (hereinafter 'Complaint').

[62] Complaint (n 61 above) 8.

[63] *Ibid* 9.

[64] *Ibid* 9–10.

[65] *Ibid* 22.

[66] *Ibid* 3.

[67] *Ibid* 3.

[68] *Ibid* 3.

[69] *Ibid* 4.

[70] *Ibid* 4.

[71] *Ibid* 4.

[72] *Ibid* 4.

must call to the attention of meeting attendees '... the obligation of all participants to inform the meeting of any knowledge that they may have of any patents, or pending patents, that might be involved in the work they are undertaking.'[73] The JEDEC Manual also provided that no standards were to include patented items without (1) a technical justification for including the patented item; and (2) express written assurance from the patent holder to license the technology either royalty free or with RAND terms.[74] JEDEC's JC-42.3 Subcommittee on RAM Devices (JC-42.3) began work on standards relating to SDRAM in 1990.[75] Members involved with JC-42.3 included leading domestic and foreign memory designers, makers, and users.[76]

Rambus attended its first JEDEC meeting as a guest in December 1991 and officially joined shortly thereafter.[77] While work was under way on the SDRAM standards, Rambus regularly attended JC-42.3 meetings.[78] Between December 1991 and May 1992, Rambus representatives attended JC-42.3 meetings discussing two key features for incorporation into the first-generation SDRAM standard: (1) programmable CAS latency technology; and (2) programmable burst length technology.[79] In March 1993, a Rambus representative also attended the JC-42.3 meeting at which these technologies were approved for inclusion in the first-generation SDRAM standard and forwarded to the JEDEC Council.[80] Despite attending both meetings, Rambus did not disclose that it might have or later pursue patent rights relevant to these technologies.[81]

Starting in September 2004, Rambus also attended JC-42.3 meetings involving proposals to include a third key feature for clock synchronization, on-chip PLL/DLL technology, in the second-generation SDRAM standard.[82] Although Rambus had presented new claims in a pending patent application in June 1993 covering this technology, Rambus did not disclose this to JEDEC.[83] From December 1991 to April 1992, Rambus also attended JC-42.3 meetings discussing a fourth key feature for data acceleration, dual-edge clock technology.[84] In September 1994, Rambus presented new claims covering the dual-edge clock technology in a pending application which issued in April 1996, but Rambus never disclosed this application

[73] *Ibid* 5. According to the FTC, the existence and scope of these disclosure obligations were commonly known within JEDEC '... throughout the entirety of Rambus' involvement in the organization.'

[74] *Ibid* 6.

[75] *Ibid* 7.

[76] *Ibid* 7.

[77] *Ibid* 11.

[78] *Ibid* 11.

[79] *Ibid* 16.

[80] *Ibid* 16.

[81] *Ibid* 16.

[82] *Ibid* 17.

[83] *Ibid* 17.

[84] *Ibid* 18.

to JEDEC.[85] This technology was eventually incorporated into the DDR SDRAM standard after a formal proposal made at a JC-42.3 meeting in March 1996.[86]

Rambus formally withdrew from JEDEC by sending a letter to the Secretary of JEDEC's JC-42 Committee on 17 June 1996.[87] The letter noted that Rambus would continue licensing its technology, but on terms that may not be consistent with the terms set by standard setting organizations like JEDEC.[88] Rambus also enclosed a letter listing 23 domestic and foreign patents issued to Rambus.[89] Although five of the listed patents were related to Rambus' original 1990 application, none of the patents claimed the SDRAM standards set by JEDEC.[90] The letter did not list Rambus' pending applications, nor did it list the Rambus patent issued in April 1996 directed to dual-edge clock technology incorporated into JEDEC's SDRAM standard.[91]

During its involvement with JEDEC, Rambus did not disclose any of its relevant patents or pending applications.[92] Rambus did present a letter to the JC-42.3 on one occasion in 1995 stating that,

> Our presence or silence at committee meetings does not constitute an endorsement of any proposal under the committee's consideration nor does it make any statement regarding potential infringement of Rambus intellectual property.[93]

Rambus also disclosed one issued patent back in September 1993, but this patent was directed to a clock technology significantly different than those considered by JEDEC.[94]

In November 1993, while Rambus was still a member, JEDEC issued the first-generation SDRAM standard.[95] Starting in 1995, JEDEC-compliant SDRAM began to replace the older generation of SDRAM, asynchronous DRAM.[96] In August 1999, after Rambus' withdrawal, JEDEC published the second-generation SDRAM standard, known as double data rate (DDR) SDRAM.[97] Most of the technologies included in the first-generation SDRAM standard were included in the second-generation DDR SDRAM standard.[98] By 1999, JEDEC-compliant SDRAM had

85 *Ibid* 18–19.
86 *Ibid* 18.
87 *Ibid* 22.
88 *Ibid* 22.
89 *Ibid* 22.
90 *Ibid* 22–4.
91 *Ibid* 23–4.
92 *Ibid* 19.
93 *Ibid* 19.
94 *Ibid* 20.
95 *Ibid* 7.
96 *Ibid* 24.
97 *Ibid* 7.
98 *Ibid* 7.

effectively displaced previous SDRAM architectures in the marketplace.[99] In 1999, Rambus began to assert its patents against major DRAM and chipset manufacturers, alleging that their use of JEDEC-compliant SDRAM infringed Rambus' patents.[100]

The FTC filed its Complaint against Rambus on 18 June 2002, alleging three separate violations of Section 5.[101] In the first count, the FTC alleged that Rambus engaged in a pattern of anti-competitive and exclusionary acts and practices whereby it obtained monopoly power in the SDRAM technology market, in violation of Section 2 of the Sherman Act.[102] In the second count, the FTC alleged that Rambus intended to monopolize the SDRAM market resulting in a dangerous probability of monopolization, also in violation of Section 2.[103] The third count alleged that Rambus unreasonably restrained trade in the SDRAM market through acts and practices constituting unfair methods of competition under Section 5.[104] Citing several internal Rambus emails and business presentations, the Complaint alleged that Rambus embarked on a calculated scheme to 'capture' the SDRAM standards.[105]

Rambus at the FTC

Chief Administrative Law Judge McGuire ('ALJ') initially dismissed the FTC's Complaint, finding that the agency had failed to show that Rambus had improperly withheld material information regarding its patents.[106] Additionally, the ALJ concluded that, even if Rambus had disclosed the IP interests at issue, there was insufficient evidence to show that JEDEC would have chosen alternatives to Rambus' technologies. However, on appeal within the agency, the FTC's Opinion of the Commission vacated the Chief ALJ's Initial Decision and focused exclusively on a claim of monopolization under Section 2.[107]

---

[99] *Ibid* 24.

[100] *Ibid* 25.

[101] *Ibid* 31–3.

[102] *Ibid* 32. The alleged violations included unlawful obtainment of monopoly power, intent to monopolize, and unfair methods of competition with respect to four particular technology markets encompassed in the SDRAM technology market: (1) latency; (2) burst length; (3) clock synchronization; and (4) data acceleration. *Ibid* 31–3.

[103] *Ibid* 31–3.

[104] *Ibid* 33.

[105] *Ibid* 13,

> . . . [A] scheme whereby Rambus would actively seek to perfect patent rights covering technologies that were the subject of an ongoing, industry-wide standardization process, in which Rambus itself was a regular participant, without disclosing the existence of such patent rights (or the pertinent patent applications) to other participants, many of whom, by producing products compliant with the standards, would later be charged with infringing Rambus's patents.

[106] Initial Decision, *In re Rambus Inc* (23 February 2004). Docket No. 9302 <http://www.ftc.gov/os/adjpro/d9302/040223initialdecision.pdf> accessed 15 September 2010, 334.

[107] Opinion of the Commission, *In re Rambus Inc* (n 47 above) 27 fn 124,

> Because we find that Rambus unlawfully monopolized the four relevant markets delineated by Complaint Counsel. . . we need not consider the further allegations that. . . Rambus' conduct otherwise constituted an unfair method of competition.

FTC found that the circumstances justified application of Section 2

The FTC's Opinion first established that the interference with the standard setting process had an anti-competitive effect outweighing any pro-competitive benefit.[108] Applying the DC Circuit's analysis from *United States v Microsoft*,[109] the FTC clearly distinguished the cooperative nature of the standard setting process from the 'rough and tumble' competitive environment in which deceptive practices are less likely to create significant anti-competitive effects.[110] Recognizing antitrust's tolerance of cooperative standard setting on the basis of the relative efficiency it provided,[111] the FTC was concerned by deceptive conduct that might distort the selection of technologies and evade the SSO's own protective measures against the exercise of monopoly power.[112] If the anti-competitive harm from deceptive conduct reduced or destroyed the efficiencies provided by cooperative standard setting, the process itself would no longer be beneficial and justifiable under the antitrust laws.[113] Therefore, applying *Allied Tube*[114] and *ASME*,[115] the FTC justified Section 2 scrutiny of potentially deceptive conduct in the standard-setting context.[116] In particular, the FTC found potentially substantial anti-competitive harm would likely result because of the SSO's cooperative process, wherein participants would be less likely to detect deception and counteract it.[117] The FTC further noted that deceptive conduct might be found in the absence of an express obligation to disclose.[118] In such a case, an obligation to disclose might be inferred by examining how the rules were interpreted by SSO members, as evidenced by their behaviour and their understanding of their disclosure obligations.[119]

The FTC then analysed the potential anti-competitive harm of patent hold-up in standard setting, where a patent holder's market power may be materially enhanced after the patented technology is incorporated into the standard.[120] SSO patent disclosure rules allowed participants to make informed choices of which technologies to include in the standard based on an understanding of the royalties or licences that

---

108 Opinion of the Commission, *In re Rambus Inc* (n 47 above) 30.

109 *United States v Microsoft Corp* 253 F 3d 34 (DC Cir 2001).

110 Opinion of the Commission, *In re Rambus Inc* (n 47 above) 32–3

> . . . [W]e cannot stress too strongly the importance we place on the fact that the challenged conduct occurred in the context of a standard-setting process in which members expected each other to act cooperatively.

111 Opinion of the Commission, *In re Rambus Inc* (n 47 above) 33 ('. . . the efficiency benefits of consensus standard setting can easily outweigh the loss of competition').

112 *Ibid* 33.

113 *Ibid* 33.

114 *Allied Tube* (n 48 above) 500.

115 *ASME* (n 48 above) 571.

116 Opinion of the Commission, *In re Rambus Inc* (n 47 above) 34.

117 *Ibid* 34.

118 *Ibid* 34–5.

119 *Ibid* 35.

120 *Ibid* 35–6.

might be required to avoid liability for infringement.[121] Therefore, disclosure obligations allowed participants to avoid hold-up *ex ante*.[122]

Citing *Aspen Skiing*,[123] the FTC found that the circumstances suggested exclusionary conduct under Section 2 because the alleged course of conduct impaired the standard setting process.[124] Specifically, deception that 'hides the price' of a patented technology is not 'competition on the merits.'[125] Further, deception that 'thwarts informed choice' is not competition on the 'basis [of] efficiency.'[126]

FTC found Rambus' conduct to be deceptive and exclusionary

Turning to the facts of Rambus' conduct, the FTC held Rambus' conduct to be both deceptive and exclusionary.[127] According to the FTC, Rambus' actions before JEDEC included two types of potentially deceptive and exclusionary conduct: (1) making potentially deceptive omissions through continued concealment of its patents and patent applications until after the DDR SDRAM standard was in place, and (2) making outright misrepresentations by giving evasive and misleading responses to questions about its conduct.[128]

To find that Rambus' conduct was *actually* deceptive, the FTC analysed the disclosure obligations of JEDEC members.[129] The FTC found that JEDEC's policies and practices gave JEDEC members a reason to believe the standard setting process would be cooperative and free from deception.[130] Thus, Rambus' conduct was likely to 'infect' the members' technology adoption decisions.[131] The FTC conceded that the JEDEC policies were 'not a model of clarity', and instead relied on an underlying duty of good faith.[132] Although the FTC distinguished its Section 2 case from the state law claim of fraud in *Rambus v Infineon*,[133] the FTC applied *Infineon* by looking to the behaviour, understanding, and expectations of JEDEC members to define the duty to disclose under JEDEC policies.[134] This evidence included testimony of other

121 *Ibid* 35–6.
122 *Ibid* 35–6.
123 *Aspen Skiing Co v Aspen Highlands Skiing Corp*, 472 US 585, 605 (1985) (holding that a firm's attempt to exclude rivals on some basis other than efficiency is predatory) (hereinafter '*Aspen Skiing*').
124 Opinion of the Commission, *In re Rambus Inc* (n 47 above) 36.
125 *Ibid* 36, citing *Aspen Skiing* (n 123 above) 605 fn 32.
126 Opinion of the Commission, *In re Rambus Inc* (n 47 above) 36, citing *Aspen Skiing* (n 123 above) 605.
127 Opinion of the Commission, *In re Rambus Inc* (n 47 above) 36–7.
128 *Ibid* 50.
129 *Ibid* 51–68.
130 *Ibid* 51–2.
131 *Ibid* 52.
132 *Ibid* 52.
133 *Rambus v Infineon Techs*, 318 F 3d 1081 (Fed Cir 2003) (holding that Rambus had not breached its duty to disclose patent interests under JEDEC rules) (hereinafter '*Rambus v Infineon*').
134 Opinion of the Commission, *In re Rambus Inc* (n 47 above) 53.

JEDEC members that they understood that disclosure of patents and applications was expected,[135] patent disclosures by JEDEC members and JEDEC reactions to attempts at hold-up,[136] and the insufficiency of Rambus' actions to give JEDEC members notice of Rambus' patent interests.[137]

Therefore, the FTC concluded that JEDEC's policies and practices and the actions of JEDEC members provided the members with an expectation that the standard setting process would be cooperative and free from deception with respect to '... the patents they possessed or expected to possess.'[138] Rambus played on the expectations of JEDEC members by failing to disclose its patent interests while actively participating in the standard setting process and attempting to obtain patent claims encompassing the developing standard.[139] By distorting JEDEC's choice of technologies and undermining the group's protections against patent hold-up, Rambus' deceptive conduct gave it monopoly power and harmed competition.[140]

FTC finds causation in the alternative

In addition to establishing exclusionary conduct and possession of monopoly power,[141] Section 2 required that Rambus' exclusionary conduct was causally linked to its monopoly position.[142] The FTC found that Rambus' deception was causally linked to JEDEC's adoption of Rambus technologies in its standard, which in turn was linked to Rambus' monopoly power.[143] In the first link, the FTC noted that alternative technologies were available when JEDEC chose the Rambus technologies and could have been substituted for the Rambus technologies had Rambus disclosed its patent interests.[144] In the second link, the FTC found that the market was likely to coalesce around JEDEC's standardized technology.[145] In addressing the causation issue, the FTC relied on the inference that, but for Rambus' deception, JEDEC would have either (1) excluded Rambus' patented technologies from the SDRAM standards; *or* (2) demanded RAND assurances with an opportunity for *ex ante* licensing negotiations.[146] As discussed in the following section, this inference in the alternative would prove crucial in the subsequent vacateur by the DC Circuit.

---

135 *Ibid* 55. Several witnesses testified that disclosure of *planned amendments* was also required, *ibid* 56.

136 *Ibid* 57–9.

137 *Ibid* 59–66.

138 *Ibid* 66.

139 *Ibid* 66–7.

140 *Ibid* 68.

141 The FTC found that Rambus had monopoly power in four key technology markets, as its patents covered these four technologies essential to the practice of the JEDEC SDRAM standard. Opinion of the Commission, *In re Rambus Inc* (n 47 above) 73.

142 Opinion of the Commission, *In re Rambus Inc* (n 47 above) 73–4.

143 *Ibid* 74–9.

144 *Ibid* 76.

145 *Ibid* 79.

146 *Ibid* 74 (emphasis added).

Lastly, in a separate remedial opinion and final order, the FTC declined to order Rambus to license its patents royalty-free because there was insufficient evidence that, but for Rambus' deception, JEDEC would have standardized non-proprietary technologies.[147] Specifically, the Remedy Opinion found evidence which supported that JEDEC might have still incorporated Rambus' patented technology even if it had fully known of Rambus' IP interests.[148] Instead, the FTC required Rambus to license its technology at 'reasonable rates' based on a hypothetical negotiation undertaken before JEDEC set the standards.[149]

The FTC's loss on appeal suggests limiting principles for Section 5

Rambus continued its fight against the FTC in the courts. In a highly anticipated decision, the Court of Appeals for the District of Columbia Circuit held that the FTC failed to sustain its allegation of monopolization.[150] Ironically, this is the same court that issued the *Microsoft* decision on which the FTC's Opinion relied.[151] In holding that the FTC's Section 2 case failed, the appellate decision provides guidance on the scope of a standalone Section 5 action.

The DC Circuit finds no violation of Section 2

On appeal, Rambus argued that the FTC had insufficient basis for liability because it found the consequences of Rambus' non-disclosure only in the alternative.[152] The FTC had found that Rambus prevented JEDEC either (1) from adopting a non-proprietary standard; or (2) from extracting a RAND commitment from Rambus when including its technology in the standard.[153] Rambus argued that the second alternative, conduct preventing a RAND commitment, was not an antitrust violation.[154]

---

[147] Opinion of the Commission on Remedy, *In re Rambus Inc* (5 February 2007) Docket No. 9302 <http://www.ftc.gov/os/adjpro/d9302/070205opinion.pdf> accessed 15 September 2010 (hereinafter 'Remedy Opinion, *In re Rambus Inc*').

[148] Remedy Opinion, *In re Rambus Inc* (n 147 above) 12,

> We have examined the record for the proof that the courts have found necessary to impose royalty-free licensing, but do not find it. Our liability opinion identified two realistic possibilities for what would have occurred had Rambus not engaged in deception of JEDEC members: either (i) JEDEC would have chosen alternative technologies, or (ii) JEDEC would have incorporated Rambus's technologies into the standard but would have demanded, as a precondition of adopting Rambus's technology, that Rambus agree to license the technology on RAND terms. There is evidence in the record to support both possibilities.

[149] Remedy Opinion, *In re Rambus Inc* (n 147 above).

[150] *Rambus Inc v FTC*, 522 F 3d 456, 459 (DC Cir 2008) (hereinafter '*Rambus v FTC*'). On the day of the appellate decision, 22 April 2008, Rambus (NASDAQ: RMBS) traded as high as 24.30 and closed at 23.19, 2.9 per cent over the previous day's close of 22.54 (Source: <http://bigcharts.marketwatch.com/historical/> accessed 15 September 2010).

[151] See text to n 127 above.

[152] Opinion of the Commission, *In re Rambus Inc* (n 47 above) 462. Rambus also argued that the FTC erred in finding that Rambus violated any JEDEC patent disclosure rules.

[153] Opinion of the Commission, *In re Rambus Inc* (n 47 above) 462.

[154] *Ibid* 462.

As Rambus did not dispute that its patent rights gave it monopoly power in relevant markets, the core inquiry was whether Rambus engaged in unlawful exclusionary conduct prohibited by Section 2.[155]

The court was persuaded by Rambus' arguments regarding the FTC's alternative reasoning under Section 2.[156] The FTC found that, had Rambus fully complied with the disclosure policy, JEDEC would have either (1) excluded Rambus' technologies from the DRAM standards; or (2) demanded RAND assurances *ex ante*.[157] Critically, the FTC did not determine whether one of these alternatives was more likely.[158] As the FTC itself found evidence supporting the possibility that JEDEC would have included Rambus' technology even if it had been fully informed of Rambus' IP,[159] the court's analysis focused on whether preventing JEDEC from obtaining a RAND assurance could constitute anti-competitive conduct under Section 2.[160] This scenario, however, was protected by the rule in *Discon*.[161] Applying this doctrine, the court reiterated that a lawful monopolist's deceptive or fraudulent raising of prices does not alone harm competition in the monopolized market.[162] The FTC explicitly recognized the reasonable possibility that JEDEC would have standardized Rambus' technologies even if Rambus disclosed its patent interests.[163] Thus, '... JEDEC lost only an opportunity to secure a RAND commitment from Rambus,' which loss is not a harm to competition from alternative technologies in the relevant markets.[164] The court also dismissed arguments that non-disclosure to an SSO can result in excess royalties which would distort competition in the downstream market, stating that this consequence was 'surely true' in *Discon*, but 'perhaps on a smaller scale.'[165] Accordingly, the court held that the FTC failed to show that Rambus' conduct was exclusionary conduct constituting unlawful monopolization prohibited by Section 2.[166]

The DC Circuit doubts violation of standalone Section 5

The DC Circuit also expressed 'serious' reservations regarding the strength of the FTC record to support a standalone Section 5 action suggested by Commissioner Leibowitz in his concurring opinion.[167] Calling the agency's findings '... murky on both the relevant margins' of what disclosure JEDEC required and what Rambus

155 *Ibid* 463.
156 *Ibid* 462.
157 *Ibid* 463.
158 *Ibid* 463.
159 Remedy Opinion, *In re Rambus Inc* (n 147 above) 12.
160 *Rambus v FTC* (n 150 above) 464.
161 *NYNEX Corp v Discon, Inc*, 525 US 128 (1998) (hereinafter '*Discon*').
162 *Discon* (n 161 above) 464–6.
163 *Ibid* 466.
164 *Ibid* 466.
165 *Ibid* 466.
166 *Ibid* 467.
167 *Ibid* 467.

failed to disclose.[168] For example, it was unclear whether JEDEC participants were required to disclose work on pending applications such as unfiled amendments.[169] Although some participants viewed JEDEC's policies this broadly, the court found that these interpretations '... significantly stretch[ed]' the policies' language.[170] Moreover, the court took issue with the FTC's reliance on unwritten disclosure expectations in this case because JEDEC's disclosure policies '... suffered from "a staggering lack of defining details."'[171] In the court's view, '... the more vague and muddled a particular expectation of disclosure, the more difficult it should be for the Commission to ascribe competitive harm to its breach.'[172]

The court also criticized the FTC's conclusion that Rambus engaged in deceptive conduct relating to its on-chip PLL/DLL and dual-edge clocking technologies, which JEDEC adopted more than two years after Rambus had last participated in the working group meetings. Criticizing the FTC's heavy reliance on a preliminary survey simply gauging participant interest in certain technologies and not requiring any disclosure of IP interests, the court found that the FTC '... has taken an aggressive interpretation of rather weak evidence.'[173] The court found it unlikely that any of the JEDEC participants placed themselves under the broad and early duty to disclose required by the FTC's interpretation.[174]

Limits of Section 2 case highlight opportunities for Section 5

Although the FTC failed in its Section 2 case against Rambus and the DC Circuit seemed to discourage the FTC from pursuing a standalone Section 5 case upon remand, the Rambus scenario still suggests a potential role for Section 5. In particular, the DC Circuit's reservations were directed more towards the sufficiency of the FTC's fact-finding as opposed to an indictment of the standalone Section 5 action itself.[175] The lack of sufficiency was as much a product of the type of evidence relied upon by the FTC as the inherent ambiguity of the JEDEC disclosure policies.

For example, the court criticized the FTC's heavy reliance on testimony by SSO participants that they interpreted JEDEC's disclosure policies to include disclosure of plans to amend pending applications.[176] Not only do these hold-up victims have an incentive to testify to such an understanding *ex post*, JEDEC's policies were inherently unclear as to the scope of required disclosure. JEDEC's policies were unclear

---

[168] *Ibid* 467.
[169] *Ibid* 467–8.
[170] *Ibid* 468.
[171] *Ibid* 468 (citing the Federal Circuit's evaluation of the JEDEC policies in *Rambus v Infineon* (n 133 above) 1102 and the FTC's own Liability Opinion 52).
[172] *Discon* (n 161 above) 469.
[173] *Ibid* 469.
[174] *Ibid* 469.
[175] *Rambus v FTC* (n 150 above) 467.
[176] *Ibid* 467–8.

until at least the clarifications made in October 1993, after Rambus had already begun participating substantially in the SDRAM proceedings.[177] Additionally, the SSO participants were sophisticated commercial entities, and could have been expected to identify the risk of pending applications and amendments thereto. It was not until 2000 that the USPTO required mandatory publication of filed US Patent Applications (subject to limited exceptions) 18 months after filing, a change in the law motivated by the need for additional transparency in the patenting process.[178] The JEDEC policy clarifications of 1993 suggest that this recognition to mitigate risk from unpublished and pending applications was within the reasonable capability of the JEDEC participants. At the very least, if imposing such a responsibility on the JEDEC participants in *Rambus* is hindsight bias, then at least future SSO participants should be held to a higher standard of risk mitigation through a clear disclosure policy.

Thus, *Rambus* suggests at least the following three limiting principles to a standalone Section 5 action in the context of patent hold-up by an SSO participant in violation of the SSO disclosure policies: (1) Section 5 should not be used to prevent hold-up from activities that could reasonably have been anticipated or mitigated by the SSO through appropriate policies at the time the standard was set; (2) the greater the ambiguity in an SSO policy, the more difficult it should be for the FTC to establish a breach of the policy;[179] and (3) ambiguity in the SSO policy should be initially construed against the SSO participants being held up as they likely had the opportunity and requisite knowledge to take steps to mitigate this risk *ex ante*.[180] A fourth potential limiting principle is also suggested by *Rambus*—the FTC should be more inclined to use Section 5 to enforce disclosure rules than licensing rules. *Rambus* clearly showed that SSO disclosure obligations could be so ambiguous as to discourage FTC action. Licensing rules, especially general RAND commitments, are substantially more subjective because compliance is not binary (as in the case of disclosure rules) and is instead gauged within an amorphous spectrum of 'reasonableness.'

---

[177] Text to n 87–88 above.

[178] See United States Patent and Trademark Office, Changes to Implement Eighteen-Month Publication of Patent Applications, 65 Fed Reg 57024 (10 October 2000).

[179] Compare *Rambus v FTC* (n 150 above) 469. The DC Circuit asserted that,

> . . . the more vague and muddled a particular expectation of disclosure, the more difficult it should be for the Commission to ascribe competitive harm to its breach.

This statement appears to combine the inquiry of breach and the assessment of the competitive harm from the breach. The proposed limiting principle attempts to keep the inquiries separate, while still placing the burden on the SSO to establish clear disclosure policies before resort to Section 5 is warranted.

[180] This principle is consistent with the Dissenting Opinion of Chairman Majoras in *N-Data*, which disagreed with the FTC majority's consumer protection theory of liability because of IEEE participants' sophistication, as compared with the unsophisticated consumer victims of *Orkin*. See text to n 225–230 below. (*Orkin Exterminating Co v FTC*, 849 F2d 1354 (11th Cir 1988), involved a court of appeals' affirmance of an FTC holding that an exterminator's breach of over 200,000 contracts with its consumer clients constituted an unfair trade practice).

Regarding these limits, *Rambus* suggests several objective indicia supporting a *prima facie* case of breach of a disclosure duty if the disclosure rules are strengthened to include pending applications and claim amendments. One objective indicium of breach of disclosure duty would be to seek evidence regarding the breadth of actual disclosures of other participants. This is likely a more accurate proxy of SSO participants' understanding of their disclosure obligations than *ex post* testimony. This would be less subjective than the FTC's approach of interviewing participants after the fact, as they have a large incentive to 'recall' a broader understanding of the disclosure policies after an attempted hold-up. Additionally, such a rule would be self-enforcing, as it encourages broader disclosure by participants. Although there is a cost to SSO participants in providing such disclosure, this incremental cost of compliance is likely offset by the business advantages at the heart of SSO participation. These costs might also be minimal as this broader disclosure obligation is commensurate in scope with a patent applicant's existing obligations to disclose *related* co-pending applications to the USPTO during prosecution under *McKesson*.[181] Thus, disclosure rules can reasonably accommodate disclosure of at least some inchoate patent interests.

A second objective indicium would be to locate evidence that alternatives to the adopted patented technology were considered by the SSO. This can be shown in records of presentations made by members advocating certain concepts or technologies for inclusion into the standard. Alternatively or in addition, this information should be reflected in or attached to the official meeting minutes kept by the SSO. Considered alternatives might also be reflected on voting ballots.

A third objective indicium would be to identify when new claims were introduced by the party accused of non-disclosure. If introduced after the adoption of the standard or the participant-patentee's knowledge of material aspects of the standard setting process, then a presumption could be raised that those new claims were introduced to explicitly capture the standard.

Alternatively, in the current regime of publicly available file histories for published US applications,[182] an SSO could reduce costs to participants by only requiring disclosure of patent application numbers and leave monitoring to the individual participants or a designated watchdog in the SSO. However, this leaves participants open to hold-up from unpublished applications (for example, an application subject to a non-publication request and accompanying forfeiture of the right to file internationally). In this case of evasion, potential breach or opportunistic behaviour

[181] *McKesson Information Solutions v Bridge Medical*, 487 F3d 897 (Fed Cir 2007) (holding a patent unenforceable due to inequitable conduct for failure to provide the USPTO examiner with material information relating to the prosecution of a related patent application).

[182] See eg, PublicPAIR (Patent Application Information Retrieval) <http://portal.uspto.gov/external/portal/pair> accessed 15 September 2010.

might still be indicated by the timing of introduced amendments (reflected in the file history, which would be publicly available upon issuance of the patent).

Moving forward, SSOs are likely to improve and clarify their disclosure policies in the wake of *Rambus*. Although this may reduce the likelihood that Section 5 is invoked to prevent similar attempts at hold-up by an SSO participant, Section 5 clearly remains available to protect these hold-up victims in this context, subject to the limiting principles described above. Even the clearest disclosure policies, however, cannot prevent attempted hold-up by a party *outside* of the original standard setting process. As discussed below, the promise of Section 5 to address this scenario was demonstrated by the FTC's next application of Section 5 in standard setting.

C. Non-Participant N-Data: Section 5 Proves to be Adept and Appropriate

The FTC's recent challenge of conduct by Negotiated Data Solutions LLC (N-Data) provides an example of unfair commercial conduct squarely within the ambit of Section 5 and outside the bounds of the antitrust laws. As the FTC did not allege that N-Data violated the antitrust laws, this case has become a lightning rod of controversy regarding the scope of Section 5. As discussed below, the N-Data case provides several waypoints identifying conduct in the standard setting context *by a non-SSO participant* sufficient to trigger a standalone Section 5 violation.

Factual background

In January 2008, the FTC issued a Complaint against N-Data alleging violation of both prongs of Section 5, unfair competition and unfair acts or practices, based upon N-Data's assertion of patents obtained from National Semiconductor (National).[183] In particular, the Complaint alleges that N-Data and its predecessor in interest, Vertical Networks, Inc. (Vertical), engaged in a prohibited course of conduct when it sought to break a licensing commitment made by National to a standard setting organization, the Institute of Electrical and Electronics Engineers (IEEE).[184]

The technology and patents at issue related to a local area network data protocol commonly known as Ethernet.[185] In 1993, an IEEE working group including National employees began work on a new standard, 'Fast Ethernet', which would remain compatible with the original Ethernet standard.[186] Autonegotiation technology

[183] Complaint dated 23 January 2008, *In re Negotiated Data Solutions, LLC*, FTC File No. 051 0094 <http://www.ftc.gov/os/caselist/0510094/080122complaint.pdf> accessed 15 September 2010, 1 (hereinafter 'Complaint').

[184] Complaint (n 183 above) 1.

[185] *Ibid* 2.

[186] *Ibid* 2.

permits this compatibility.[187] In 1994, National proposed that the new standard incorporate a National autonegotiation technology, 'NWay'.[188] The working group considered alternatives to NWay and also considered adopting a Fast Ethernet standard without autonegotiation.[189] As National had pending patent applications on NWay, National offered to license NWay to any requesting party if NWay was incorporated into the standard.[190] National subsequently memorialized this offer in a letter, offering to license NWay to any requesting party on a non-discriminatory basis.[191] The licence would be paid-up and royalty-free after payment of a one off fee of $1,000.[192]

Subsequently, the IEEE adopted a Fast Ethernet standard including the NWay autonegotiation technology.[193] After the IEEE published the Fast Ethernet standard in 1995, manufacturers incorporated the new standard into hundreds of millions of digital devices including computers, routers, switches, modems, and other digital communications equipment until, by 2001, there were no commercially viable alternative autonegotiation technologies for Ethernet.[194] Back in 1998, however, National assigned the issued NWay patents to Vertical.[195] National provided Vertical with a copy of the 1994 letter, and Vertical acknowledged that it had been informed that the patents might be 'encumbered' by National's actions within the IEEE standardization proceedings.[196] The final assignment from National to Vertical was explicitly made subject to any existing licences, which included patents potentially '... encumbered under standards such as an IEEE standard.'[197]

In 2002, Vertical sent a letter to the IEEE asserting that one or more of its patents 'may be applicable' to the Ethernet standard and offering to grant any party a non-exclusive licence on a RAND basis.[198] Vertical did not mention National's previous licence offer.[199] Instead, Vertical's letter was explicitly intended to supersede any of National's previous assurances.[200] Vertical also sent notice letters to approximately 64 companies demanding licensing fees on a per unit basis, which fees were viewed

---

187 *Ibid* 2.
188 *Ibid* 2.
189 *Ibid* 3.
190 *Ibid* 2–3.
191 *Ibid* 3.
192 *Ibid* 3.
193 *Ibid* 3. The FTC Complaint averred that '... various IEEE members were aware of and relied upon National's one thousand dollar licensing commitment when they voted to include NWay as the autodetection technology in the 802.3 standard.'
194 *Ibid* 4.
195 *Ibid* 4–5.
196 *Ibid* 5.
197 *Ibid* 5.
198 *Ibid* 5.
199 *Ibid* 5.
200 *Ibid* 5.

by the FTC as 'a substantial increase' over National's 1994 assurance.[201] When several companies tendered $1,000 to Vertical in attempts to accept the original National offer, Vertical rejected these payments and pursued litigation against several companies.[202] As a result, several companies entered into licensing agreements with Vertical in excess of $1,000 per company.[203] In 1993, Vertical assigned its patents to N-Data and ceased operations.[204] N-Data continued sending notice letters and asserting these NWay patents in infringement suits.[205]

The FTC's proposed consent order

The FTC accused N-Data of exploiting the incorporation of NWay into the Ethernet standard by reneging on a known commitment made by its predecessor in interest.[206] The FTC found N-Data was engaged in patent hold-up—given the large installed base on Ethernet equipment, incompatibility of NWay with alternative autonegotiation technologies and significant costs of abandoning autonegotiation altogether, the companies being pursued by N-Data were locked into using NWay.[207] Thus, the value N-Data sought to extract from the IEEE members was created by the opportunistic nature of its conduct rather than the value of the patents being asserted.[208]

Acknowledging that N-Data's actions were not a violation of the Sherman Act because N-Data had not made any commitment itself, N-Data's conduct still threatened to raise prices for an entire industry and to 'subvert' the IEEE standard setting process in a way that endangered the viability of standard setting generally.[209] The FTC's proposed complaint alleged that N-Data violated Section 5 alone by (1) engaging in an unfair method of competition and (2) engaging in an unfair act or practice.[210]

---

201 *Ibid* 5.

202 *Ibid* 5.

203 *Ibid* 6.

204 *Ibid* 6. The FTC averred that N-Data was also familiar with the 1994 National assurance letter, as a principal of N-Data had represented Vertical in its negotiations to obtain the patents from National. Further, Mr Loudermilk, who owned N-Data, was Vertical's outside patent counsel. FTC, Analysis of Proposed Consent Order to Aid Public Comment, *In re Negotiated Data Solutions, LLC*, FTC File No. 051 0094 (23 January 2008) <http://www.ftc.gov/os/caselist/0510094/080122analysis.pdf> accessed 15 September 2010, 3–4 (hereinafter 'Analysis of Proposed Consent Order to Aid Public Comment').

205 Analysis of Proposed Consent Order to Aid Public Comment (n 204 above) 4.

206 *Ibid* 4.

207 *Ibid* 8; Complaint (n 183 above) 6.

208 Analysis of Proposed Consent Order to Aid Public Comment (n 204 above) 7.

209 *Ibid* 4.

210 *Ibid* 4.

Unfair competition: an 'unusually important role' for Section 5

The Commission found 'little doubt' that N-Data's conduct was an unfair method of competition within the meaning of Section 5.[211] Regarding this claim, the FTC first established that its proposed action complied with the defined limits of Section 5.[212] For one, the FTC found that the form of patent hold-up at issue was inherently coercive and oppressive, as required by *OAG* and *Ethyl*.[213] The FTC also had little doubt about the competitive harm caused by N-Data's conduct, as it threatened to undermine the standard setting process '. . . to the detriment of all consumers.'[214] In particular, N-Data's conduct had an adverse impact on prices for autonegotiation technology.[215] More generally, N-Data's conduct threatened to reduce the value of standard setting by raising the possibility of 'opportunistic lawsuits.'[216] Consequently, firms would be less likely to rely on new or existing standards.[217] Further, new standards might be adversely affected because SSOs would unreasonably seek to avoid incorporating any patented technologies for fear of an N-Data-like hold-up.[218]

The FTC found that N-Data's conduct was 'particularly appropriate' for Section 5 review because National's prior licensing commitment was made to an industry-wide organization, thereby involving numerous, injured third parties lacking privity with patentees and having mixed incentives to pass on royalties.[219] If N-Data's conduct became widespread, no SSO could rely on the good faith assurances of its members.[220] Instead, member companies would exit the business and sell their portfolios to non-practising entities or 'trolls'.[221] Where contractual remedies might prove ineffective, the FTC found that Section 5 intervention may serve 'an unusually important role'[222] and Congress created the FTC 'precisely' to challenge this kind of conduct.[223]

---

[211] Statement of the Federal Trade Commission, *In re Negotiated Data Solutions, LLC* (23 January 2008) <http://www.ftc.gov/os/caselist/0510094/080122statement.pdf> accessed 15 September 2010, 1 (hereinafter 'Statement of the Federal Trade Commission').

[212] Analysis of Proposed Consent Order to Aid Public Comment (n 204 above) 5.

[213] *Ibid* 5.

[214] Statement of the Federal Trade Commission (n 211 above) 2–3.

[215] Analysis of Proposed Consent Order to Aid Public Comment (n 204 above) 5.

[216] *Ibid* 5.

[217] *Ibid* 6.

[218] *Ibid* 6.

[219] *Ibid* 5.

[220] Statement of the Federal Trade Commission (n 211 above) 1.

[221] *Ibid* 1.

[222] Analysis of Proposed Consent Order to Aid Public Comment (n 204 above) 6.

[223] Statement of the Federal Trade Commission (n 211 above) 1.

Unfair act

Regarding the unfair act claim, the FTC relied heavily on the three-pronged test of limiting principles in *Orkin*.[224] Accordingly, the FTC analysed whether N-Data's conduct caused (1) substantial consumer injury, (2) which injury was not outweighed by countervailing benefits, and (3) not an injury consumers could have reasonably avoided themselves.[225] To satisfy the substantial consumer injury prong, the FTC looked to the millions of dollars in excess royalty payments and legal fees spent by companies practising the Ethernet standard, which would likely be passed to ultimate consumers through higher costs.[226] The FTC also found no countervailing benefit to businesses or consumers created by N-Data's conduct.[227] Lastly, the FTC found that the IEEE members had no way to anticipate N-Data's later repudiation of the price commitment *ex ante*, and had no way to avoid the financial injury threatened by N-Data except for expensive litigation.[228] Essentially, as N-Data had engaged in patent hold-up, the third prong of *Orkin* was satisfied.[229]

The accompanying Proposed Consent Order prohibited N-Data from enforcing its patents unless it first offered to license them on the terms initially offered by National, or a paid-up royalty-free licence of $1000 to the use of NWay to implement an IEEE standard.[230]

Opposition and dissenting opinions in *N-Data*

Although N-Data had already agreed to be bound by the FTC's consent order upon its issue,[231] N-Data submitted a statement during the period for public comment noting its continued disagreement with both the facts and legal theories in the FTC's case.[232] N-Data argued that the FTC's application of Section 5 in the absence of an antitrust violation would give the FTC '. . . almost boundless precedent for finding Section 5 violations.'[233] With regard to the unfair competition claim, N-Data argued that the FTC applied the limiting principles so broadly as to make them

224 *Orkin Exterminating Co v FTC*, 849 F 2d 1354, 1364–6 (11th Cir 1988) (hereinafter '*Orkin*') (approving of the FTC's three-part test for finding 'unfairness', based on (1) substantial consumer injury (2) that consumers could not reasonably have avoided, (3) that is not outweighed by the benefits of the practice in question).

225 Analysis of Proposed Consent Order to Aid Public Comment (n 204 above) 7.

226 *Ibid* 8.

227 *Ibid* 8.

228 *Ibid* 8.

229 *Ibid* 8.

230 *Ibid* 9.

231 *Ibid* 1.

232 AD Melamed and AJ Ewalt, Public Comments on behalf of Negotiated Data Solutions, *In re Negotiated Data Solutions, LLC*, FTC File No. 051 0094 (14 April 2008) <http://www.ftc.gov/os/comments/negotiateddatasol/534241-00011.pdf> accessed 15 September 2010 (hereinafter 'Public Comments').

233 Public Comments (n 232 above) 5.

nearly meaningless.[234] For example, N-Data asserted that the IEEE members were not helpless because N-Data would have to bring an infringement action during which these firms could argue that N-Data was bound by National's letter.[235] Further, citing *Discon,*[236] N-Data argued that a mere price or royalty increase was not an injury to competition.[237] N-Data warned that the FTC's decision provides Section 5 with an '. . . uncertain and almost boundless' reach, thereby exposing firms to potential 'hindsight punishment' likely to deter pro-competitive conduct.[238] With regard to the finding of unfair acts, N-Data argued that the IEEE could have avoided harm by adopting rules making assurance letters irrevocable or rejecting Vertical's 2002 letter outright.[239]

Adding to the controversy in *N-Data,* two Commissioners issued dissenting statements to the FTC's majority opinion. Chairman Majoras warned that finding liability only under Section 5 is '. . . not advisable as a matter of policy or prosecutorial discretion.'[240] In her dissent, Chairman Majoras found that the majority's assertions of coercive and oppressive conduct having an adverse impact on prices were impossible to prove on the available evidence.[241] Finding liability for evasion of a contractual price in the absence of a Sherman Act violation would be a 'slippery slope' without meaningful limiting principles.[242] Chairman Majoras also took issue with the majority's inclusion of sophisticated computer manufacturers as 'consumers' qualifying for the FTC's protection.[243] Echoing N-Data's points, the Chairman found that the IEEE could have acted sooner to protect its members from adverse changes to commitment letters or it could have rejected Vertical's letter.[244] Commissioner Kovacic expressed concerns regarding the spillover effects in private state or federal litigation and the alternative bases for liability pleaded by the FTC.[245] In particular, the FTC's application of Section 5 could affect the application of derivative state unfair competition statutes allowing treble damages.[246]

---

234 *Ibid* 4.
235 *Ibid* 4.
236 *Discon* (n 161 above).
237 Public Comments (n 232 above) 5.
238 *Ibid* 5.
239 *Ibid* 7.
240 Dissenting Statement of Chairman Majoras, *In re Negotiated Data Solutions LLC,* File No. 0510094 (23 January 2008) <http://www.ftc.gov/os/caselist/0510094/080122majoras.pdf> accessed 15 September 2010, 2 (hereinafter 'Dissenting Statement of Chairman Majoras').
241 Dissenting Statement of Chairman Majoras (n 240 above) 4.
242 *Ibid* 4.
243 *Ibid* 5.
244 *Ibid* 6.
245 Dissenting Statement of Commissioner Kovacic, *In re Negotiated Data Solutions LLC,* File No. 0510094 (23 January 2008) <http://www.ftc.gov/os/caselist/0510094/080122kovacic.pdf> accessed 15 September 2010, 1–2 (hereinafter 'Dissenting Statement of Commissioner Kovacic').
246 Dissenting Statement of Commissioner Kovacic (n 245 above) 2.

Commissioner Kovacic also took issue with the majority's failure to integrate the two theories of liability.[247]

Objective limiting principles suggested by N-Data

As a preliminary matter, the application of Section 5 adheres to the judicial limiting principles discussed previously. In particular, the infringement threats brought by N-Data were outside the scope of Section 2, but still oppressive to the standard's participants and adopters. As N-Data was not a party to the original IEEE process for Fast Ethernet, there was no way for the IEEE participants or later adopters to anticipate or even reasonably mitigate the risk of a later party threatening the industry as a whole. Thus, the outcome of *N-Data* falls within the general limits of standalone Section 5 liability.

Moreover, in *N-Data*, the FTC expressly noted that merely breaching a prior commitment in the standard setting context was not enough to constitute an unfair act or practice under Section 5.[248] Instead, the standard setting context, the materiality of National's licensing commitment to NWay's adoption, and the inability of the IEEE members to exercise their own countermeasures were all critical to the finding of liability.[249] Additionally, a mere breach of a previous licensing commitment was unlikely to constitute an unfair method of competition under Section 5.[250] Instead, N-Data was liable because the commitment being breached was in the context of standard setting and threatened to undermine the process or render it anti-competitive.[251] Thus, mere breach of an obligation is insufficient. There must be some anti-competitive effect. In the standard setting context, this effect is satisfied when the activity undermines the standard setting process.

Additionally, although other equitable doctrines such as waiver or implied licence may be utilized by a defendant as an affirmative defence to infringement, the SSO context also encourages an interventionist role by the FTC. In particular, the incorporation of the patented technology into the standard creates (1) a backdoor shortcut to infringement analysis (ie, if patent claim reads on a standard and a product is compatible with the standard, infringement is established); and (2) a large group (sometimes an entire industry) of past/present/future defendants. Public policy would encourage FTC intervention to save present and future defendants from wasteful expenditures defending against litigation that should never have been brought in the first place.

However, the dissents of Chairman Majoras and Commissioner Kovacic do substantially undercut the viability of the consumer protection theory advanced by the

[247] *Ibid* 2.
[248] Analysis of Proposed Consent Order to Aid Public Comment (n 204 above) 9.
[249] *Ibid* 9.
[250] *Ibid* 6.
[251] *Ibid* 6.

FTC's Proposed Consent Order. Specifically, to preserve any distinction between the two prongs of Section 5 (between unfair competition and unfair acts), 'consumers' should likely be restricted to the more conservative interpretation argued by Chairman Majoras. Preserving this distinction would also address some of the concerns of Commissioner Kovacic. For the consumer protection theory to apply, it appears that this theory would hinge on the ability of SSO participants and standards adopters to pass on higher royalties to end-consumers. For example, a greater ability to pass on higher royalties suggests a greater potential harm to end-consumers intended to be protected even under a strict reading of the second prong of Section 5. Thus, in this context, any liability under the second prong should be based on some evidence that the cost from the higher *ex post* royalties are at least highly likely to be passed on to end-consumers by the hold-up targets.

N-Data illustrates a principle from *Rambus* regarding the certainty of the obligation and 'breach.' In particular, the $1000 paid-up licensing obligation from National was clear and unambiguous. If IEEE had made such offers irrevocable earlier, there would be no question of the basis for Section 5 liability. Thus, in view of *N-Data*, Section 5 can be clearly applied to licensing obligations as long as they are objective and irrevocable.

N-Data also involves an application of another *Rambus* principle holding SSOs principally responsible for preventing or mitigating the risk of patent hold-up. According to this principle, standalone Section 5 actions should not be bought to protect SSO participants or standard adopters from hold-up that the SSO could have prevented through better contracting/policies. *N-Data* provides a stark example of the limits of an SSO's policies. In short, SSO policies cannot easily bind a non-party to the standard setting process. Due to the freely assignable nature of patent rights and a lack of transparency regarding patent licensing generally, there was no way for the IEEE to reasonably avoid later hold-up by N-Data. For example, while recording of assignments in the USPTO confers a concrete benefit to a patent owner, there is no requirement or significant incentive to publicly disclose most patent licences. In the absence of a concrete doctrine that allows for an obligation to run with the patent or sufficient public information on licences and the patents 'encumbered' thereby, the FTC has a definite role in enforcing Section 5 to prevent or restrict wasteful infringement actions brought by subsequent owners of patents subject to an earlier SSO agreement. Thus, the third-party nature of N-Data itself bolsters the argument for FTC intervention in this case.

D. Limiting Principles for Section 5 as Applied to Patent Hold-up in Standard Setting

As illustrated by the previous analysis of *Rambus* and *N-Data*, Section 5 may alone be a particularly adept tool to enforce fair competition involving technological standard setting. In turn, application of Section 5 in this context would be subject to limiting principles described above depending on the party (SSO participant or otherwise) engaging in the patent hold-up. Synthesizing the analysis of *Rambus* and *N-Data* reveals potential general principles limiting the application of Section 5 alone to find liability for patent hold-up in standard setting:

(1) Section 5 should not be used to protect SSO participants from hold-up from sources or activities that could reasonably have been anticipated by SSO participants and could have easily been prevented through SSO policies.
(2) Section 5 should be used primarily to enforce clear SSO disclosure obligations,[252] although limited application is justified to enforce objective and irrevocable SSO licensing obligations[253] where the SSO could not have reasonably have anticipated or mitigated risks associated with breach of the licensing obligation. Section 5 should not be used to enforce an alleged breach of a general RAND commitment.
(3) Even absent evidence of additional costs being imposed on end-consumers, standard setting hold-up should be subject to liability under the unfair methods of competition theory of Section 5, where evidence demonstrates harm undermining the integrity of the standard setting process.

E. Section 5 May be the Most Effective Instrument to Combat Welfare-Destructive Hold-ups

The preceding sections outline objective limits for the application of Section 5 in the standard setting context to prevent patent hold-ups. These cases illustrate the core benefit to standalone Section 5 actions in the standard setting context—to preserve the integrity and efficiencies of the standard setting process from actions beyond the traditional reach of the Sherman Act and also beyond the ability of SSOs to mitigate harm themselves.

The FTC has a unique role in protecting harm to the consumer where private incentives of SSO participants fail to provide adequate protection. Absent FTC involvement, the costs of hold-up may remain hidden because hold-up targets have little incentive to complain when they can simply pass on the costs to consumers.[254] Additionally, there is often no venue for resolving complaints within the SSO.[255] Moreover, in the case of third-party hold-up, resort to an SSO procedure to resolve a complaint would be futile.

252 For example, clearer in scope than *Rambus*.
253 For example, the $1000 National licence offer in *N-Data*, if it were irrevocable when made.
254 *IP 2 REPORT* (n 41 above) 40.
255 *Ibid* 40.

Furthermore, the availability to private litigants of such doctrines as patent misuse (an equitable doctrine available to an alleged infringer as an affirmative defence),[256] waiver, or equitable estoppel[257] does not make private litigation so attractive that Section 5 remedies are rendered superfluous as a means to combat attempted hold-ups. Engaging in private patent litigation is inordinately expensive. For example, the AIPLA Economic Report from 2007 listed the average litigation expense to be $461,000 for cases with less than $1 million at risk only through the end of discovery.[258] For cases going to trial, the total expense for that class of cases was $767,000.[259] If over $25 million was at risk, the cost of litigation through trial skyrocketed to an average of nearly $5.5 million.[260] In contrast, the FTC may be able to protect multiple defendants or even an entire industry from patent-hold up without each of the sued defendants having to spend money litigating separately. In short, FTC enforcement of Section 5 thus avoids wasteful expenditures by a wide range of industry participants for defending against nuisance patent litigation.

Conclusion

Properly constrained by limiting principles, Section 5 of the FTC Act may be the most effective vehicle to combat anti-competitive patent hold-ups that arise in the wake of standard setting activity. The standard antitrust laws may not adequately reach a variety of hold-up situations (particularly in light of the case law developed in the *Rambus* matter), and limitations on private litigation suggest it is an inadequate means to deal with hold-up problems. Accordingly, although principled application of Section 5 to other business problems will have to be developed over time, it is already apparent that Section 5 may be particularly well-tailored to deal with competitive abuses that occur in the standard setting context.

---

256 See *Morton Salt Co v GS Suppiger Co*, 314 US 488 (1942); *CR Bard v M3 Systems*, 157 F3d 1340, 1372 (Fed Cir 1998),

> Patent misuse is viewed as a broader wrong than antitrust violation because of the economic power that may be derived from the patentee's right to exclude. Thus misuse may arise when the conditions of antitrust violation are not met.

See also 35 USC 271(d) (limiting patent misuse).

257 See *Qualcomm Inc v Broadcom Corp*, 548 F3d 1004 (Fed Cir 2008) (limiting scope of the remedy of patent unenforceability for patents not disclosed to an SSO, in breach of a duty to disclose).

258 For example, the AIPLA Economic Report from 2009 listed the average litigation expense to be $498,000 for cases with less than $1 million at risk only through the end of discovery (AIPLA Report of the Economic Survey 2009, I-128). For cases going to trial, the total expense for that class of cases was $967,000 (*ibid* I-128). If over $25 million was at risk, the cost of litigation through trial skyrocketed to an average of $6.25 million (*ibid* I-129).

259 *Ibid* I-90.

260 *Ibid* I-91.

15

STANDARD SETTING ANALYSIS UNDER US LAW

Michael A Carrier

Introduction

Standards are common platforms that allow products to work together. Standards play an important role in the economy, covering products as far ranging as ATM machines, telephone networks, and DVDs. Standards are especially needed in *network effects markets* (such as telephone or e-mail systems), in which users benefit from an increase in the number of other users in the system.

Even though standards are vital, antitrust traditionally viewed the process of setting standards with suspicion. Standard setting organizations (SSOs) tend to be composed of industry rivals discussing sensitive information such as price. As Adam Smith worried—'People of the same trade seldom meet together even for merriment or diversion, but the conversation ends in a conspiracy against the public, or in some contrivance to raise prices.'[1]

Despite antitrust's concern, competitors have good reason to engage in such discussions. Before the selection of a standard, a SSO often can choose from an array of alternative technologies.[2] In contrast, after a standard is chosen and the industry has invested in a particular technology, flexibility is severely restricted. If the selected technology is patented, the owner could impose excessive licensing terms that reflect not just the value of the patent but also the significant costs of switching to a new technology. The patentee, in other words, could *hold up* the standard's implementation.

---

[1] Adam Smith, *An Inquiry into the Nature and Causes of the Wealth of the Nations* (1776) 145.
[2] In certain settings, there may be only one superior technical option.

This threat of hold-up explains why SSOs have required members to provide certain information before the standard's selection. Some SSOs have mandated that participants disclose patents that could be implicated by the standard. Many have required members to agree to license their IP on reasonable and non-discriminatory (RAND) terms. One SSO has compelled participants to specify the maximum royalties it would impose. Although the US antitrust enforcement agencies have recently softened their position, antitrust law traditionally has been suspicious of such price-related activities. And this suspicion deters SSO members from sharing information that could prevent hold-up.[3]

In this chapter, I first describe the various types of standard setting. Second, I offer a brief synthesis of US antitrust treatment of SSOs. Third, I set forth the potential anti-competitive concerns of standard setting. And fourth, I articulate the pro-competitive benefits that SSOs (and their IP rules) could offer.

A. Standard Setting

Types

There are several types of standards and processes for setting them. Governments often enact performance standards, which address product quality, health, and safety.[4] Safety standards, for example, define features that products must possess to be sold on the market or to obtain approval from a standard setting body.[5] Governments also have enacted other standards that they require market participants to adopt, such as standards for high-definition television (HDTV) and for interconnection between telephone networks.[6] Government-set standards have become less important in recent years and have suffered weaknesses such as failures to abandon inefficient standards and susceptibility to undue influence by entities with an interest in the outcome.[7]

---

[3] For example, firms in the computer hardware and software industries have explained that their standards-related discussions of licensing terms have been inhibited by antitrust concerns. Federal Trade Commission, *To Promote Innovation:—The Proper Balance of Competition and Patent Law and Policy* (October 2003) <http://www.ftc.gov/os/2003/10/innovationrpt.pdf> accessed 15 September 2010, Ch 3, 3.

[4] Letter from TO Barnett, Assistant Attorney General, Dept of Justice, to RA Skitol (30 October 2006) <http://www.usdoj.gov/atr/public/busreview/219380.pdf> accessed 15 September 2010, 7 ('VITA letter').

[5] JJ Anton and DA Yao, 'Standard setting Consortia, Antitrust, and High-Technology Industries' (1995) 64 Antitrust LJ 247, 247.

[6] H Hovenkamp, MD Janis and MA Lemley, *IP and Antitrust: An Analysis of Antitrust Principles Applied to Intellectual Property Law* (2002) 35-5 to 35-6.

[7] *Ibid* 35-14 to 35-15; MA Lemley, 'Intellectual Property Rights and Standard-Setting Organizations' (2002) 90 California L Rev 1889, 1900 ('Lemley, SSOs').

A second type involves *de facto* standard setting, which occurs when one firm dominates the market.[8] In such a case (exemplified by the QWERTY keyboard layout and Microsoft Windows operating system), the firm's position as market leader allows it to select the standard and (where the standard is protected by IP) force rivals to obtain licences.[9] Because of the benefits of adopting the same product as others, de facto standards are particularly likely in network effects markets.[10]

This chapter focuses on the third type, standards voluntarily set by private industry groups known as SSOs. Hundreds of SSOs operate throughout the world, in varying degrees of formality, size, and scope.[11] Formal SSOs include Underwriter's Laboratory, the Institute of Electric and Electronic Engineers, and the National Institute of Standards and Technology. In the United States, SSOs often fall under the umbrella of the American National Standards Institute, which requires due-process principles and open participation.[12] Firms also have created informal consortia, such as the World Wide Web consortium, to exercise greater control and obtain faster implementation. While SSOs are voluntary in nature, they are sometimes incorporated into state laws or municipal ordinances that require compliance.[13]

Many companies participate in a range of SSOs. One of the central aspects of SSOs is their openness, with the standard available to all members and, often, outsiders as well. Standards are used throughout various industries and facilitate the interaction of users and providers.[14] They have typically been set by engineers, though lawyers have played a larger role in recent years. Because standards bodies do not have enforcement authority, disputes (including numerous antitrust claims) have been adjudicated by courts.[15]

An important category of standards set by SSOs, and the focus of this chapter, consists of *interface (or interoperability) standards.* These standards, vital to the economy, allow products made by different firms to work together.[16] The plethora of

---

[8] RM Webb, 'There is a Better Way: It's Time To Overhaul the Model for Participation in Private Standard-Setting' (2004) 12 J Intellectual Property L 163, 168.

[9] JM Mueller, 'Patent Misuse Through the Capture of Industry Standards' (2002) 17 Berkeley Technology LJ 623, 633.

[10] Lemley, SSOs (n 7 above) 1899.

[11] US Department of Justice and Federal Trade Commission, *Antitrust Enforcement and Intellectual Property Rights: Promoting Innovation and Competition* (2007), <http://www.justice.gov/atr/public/hearings/ip/222655.pdf> accessed 15 September 2010, Ch 2, 33 fn 5 ('*Antitrust Enforcement and IP Rights*').

[12] See generally C Shapiro, 'Setting Compatibility Standards: Cooperation or Collusion?' in R Dreyfuss et al (eds) *Expanding the Boundaries of Intellectual Property: Innovation Policy for the Knowledge Society* (Oxford: OUP, 2001) 81, 84.

[13] DJ Gifford, 'Developing Models for a Coherent Treatment of Standard-Setting Issues under the Patent, Copyright, and Antitrust Laws' (2003) 43 IDEA The J of L and Technology 331, 333 (2003).

[14] *Ibid.*

[15] Shapiro (n 12 above) 85.

[16] Lemley, SSOs (n 7 above) 1892–93.

communications technologies today—such as laptop computers, personal digital assistants, cell phones, pagers, and other devices—would not be possible without interface standards.[17] In addition, nearly every aspect of using a computer or navigating the Internet is based on interface standards.[18] For example, the application programming interfaces defining compatibility with the Windows operating system offer a standard that ensures that applications need not become their own 'mini-operating systems'.[19] Interface standards also are beneficial in clearing patent thickets and fostering cumulative innovation, with one generation's patented invention building on those of previous generations.[20]

Another benefit of interface standards is their ability to create new markets by avoiding *standards wars*, which occur when substitute products with incompatible designs are introduced in the marketplace.[21] In the period before one format emerges victorious, some consumers delay their purchases because they do not wish to be stuck with the losing format. Just to pick one example, 60 per cent of consumers indicated they would not purchase HD DVD or Blu-ray high-definition DVD players/recorders '. . . until there was just one format.'[22] Standards wars are particularly likely in network effects markets, in which there are significant benefits to customers having compatible products.[23]

IP rules

SSOs have adopted an array of IP-related rules designed to limit a patentee's imposition of licensing terms that could hold up the standard's implementation. *Disclosure rules* inform SSO members of IP held by participants.[24] The rules most typically cover patents and sometimes apply to patent applications and other IP rights. Disclosure rules allow members to make informed decisions but could increase costs for IP owners, even causing them to leave the organization.[25] Supplementing disclosure rules, some SSOs have adopted *search rules* that require members to search for relevant IP rights.[26]

SSOs also typically implement *licensing rules* that specify the terms under which the IP would be licensed. The most popular such rule requires members to license their

---

[17] Mueller (n 9 above) 633.

[18] L Seltzer, 'The Standards Industry: Corporate Consortia Are Supplanting Traditional Rule-Making Bodies' (15 April 2001) Internet World, 50.

[19] Lemley, SSOs (n 7 above) 1896; Gifford (n 13 above) 337.

[20] S Scotchmer, 'Standing on the Shoulders of Giants: Cumulative Research and the Patent Law' (1991) 5 J Economic Perspectives 29, 29.

[21] *Antitrust Enforcement and IP Rights* (n 11 above) Ch 2, 34 fn 6.

[22] M Panaritis, 'HD DVD Ready To Bow out of War with Blu-Ray' (18 February 2008) Philadelphia Inquirer, A1.

[23] *Antitrust Enforcement and IP Rights* (n 11 above) Ch 2, 34 fn 6.

[24] *Ibid* 42.

[25] *Ibid* 43.

[26] Lemley, SSOs (n 7 above) 1905.

patents on a '*reasonable and nondiscriminatory*' (RAND) basis.[27] 'Reasonable' royalties do not specify an exact amount, leaving the parties to argue over the term's meaning. In particular, patentees whose patents have been incorporated into the standard tend to have higher conceptions of what constitutes a reasonable royalty than licensees.[28] Because of this uncertainty, some SSOs have required members to specify their most restrictive licensing terms, such as the maximum royalty they would charge.[29]

A few SSOs, such as those developing standards for the Internet, require members to commit to *royalty-free licensing*.[30] Such licensing dispenses with the difficulties of determining reasonable royalties and considers the views of open-source developers that might not otherwise use the standards. But the natural concern with such licensing is a potential reduction in innovation incentives.[31]

Patent hold-up

One of the dangers facing standard setting is the likelihood of patent hold-up. Before a standard's adoption, an SSO often can choose from an array of alternative technologies. At that time, a patentee has no more leverage than any other potential supplier and is unable to command more than a competitive price.[32]

After the SSO chooses a standard, however, the owner of the selected technology may gain significant power. If the technology is patented, the owner could impose royalties so high that members are effectively prevented from using the standard. In many cases, the royalties are passed on to consumers, who are forced to pay higher prices.[33]

Nor can SSO members, faced with demands for excessively high royalties, migrate easily to a different technology. After a standard is selected, industry participants begin designing, testing, and producing goods that conform to the standard.

---

27 *Ibid* 1906.

28 PJ Weiser, 'Making the World Safe for Standard Setting' (2007), <http://ssrn.com/abstract=1003432> accessed 15 September 2010, 26 (citation omitted); see also *Nokia Corp v Qualcomm, Inc*, No. CIV A 06-509-JJF, 2006 WL 2521328 (D Del 29 August, 2006); Lemley, SSOs (n 7 above) 1965; JS Miller, 'Standard Setting, Patents, and Access Lock-in: RAND Licensing and the Theory of the Firm' (2007) 40 Indiana L Rev 351, 358.

29 See VITA letter (n 4 above).

30 *Antitrust Enforcement and IP Rights* (n 11 above) Ch 2, 47.

31 DJ Weitzner, Testimony, US Dept of Justice and FTC, 'Joint Hearings on Competition and Intellectual Property Law and Policy in the Knowledge-Based Economy: Standards and Intellectual Property—Licensing Terms' (18 April 2002) <http://www.ftc.gov/opp/intellect/020418weitzner.shtm> accessed 15 September 2010, 2.

32 DP Majoras, Chairman, Federal Trade Commission, 'Recognizing the Procompetitive Potential of Royalty Discussions in Standard Setting' (23 September 2005) <http://www.ftc.gov/speeches/majoras/050923stanford.pdf> accessed 15 September 2010, 2.

33 *Ibid* 3; J Farrell et al, 'Standard Setting, Patents, and Hold-Up' (2007) 74 Antitrust LJ 603, 645.

That, as former FTC Chairman[34] Deborah Platt Majoras explained, is '. . . the whole idea of engaging in standard setting.'[35] But these efforts, in learning about a particular technology and investing in equipment and complementary products, typically do not have value if the user switches to an alternative technology.[36] In addition, the industry typically will incur the costs of selecting a new standard. As a result of all these costs, the industry will be locked into the chosen standard.

Hold-up is not just a theoretical concern. Of many examples that could be offered, I will describe three. The first is provided by the Union Oil Company of California (Unocal). In the 1990s, the California Air Resources Board sought to reduce pollution by adopting a standard for reformulated gasoline.[37] After refiners invested $4 billion to comply with the standard, Unocal revealed a 'clean fuels' patent covering the standard and demanded royalties.[38] Given their significant investments, the leading California refiners did not have a choice but to invest an additional $1.5 billion to comply with the standard.[39] Unocal's enforcement of its patents would have cost consumers roughly 5 cents per gallon, or more than $500 million annually.[40]

In the second example, computer memory developer Rambus refused to disclose patents and applications and engaged in other deceptive conduct in a semiconductor engineering SSO. As a result, manufacturers faced significant hold-up. Abandoning the technology might have been easy early in the standard's development but, less than a decade later, became 'near[ly] impossible'. As the FTC's expert in the case explained, lock-in 'grows over time', as manufacturers make sunk investments in complementary goods before 'deploy[ing] the standardized product in volume'.[41]

Lock-in arises from the accumulation of investments at each stage. For example, in the computer memory setting, it involves (1) designing chips and products conforming to the standard, (2) testing and verifying the designs, (3) building, testing, and qualifying prototypes, and (4) ramping up production on a commercial scale. As a result of these investments, manufacturers could spend hundreds of millions of dollars if forced to switch to a different technology. In the Rambus case, for example,

[34] The FTC Act adopts the term 'Chairman', and this title has been used in connection with all such officials. 15 USC § 41.

[35] Majoras (n 32 above) 3.

[36] DG Swanson and WJ Baumol, 'Reasonable and Nondiscriminatory (RAND) Royalties, Standards Selection, and Control of Market Power' (2005) 73 Antitrust LJ 1, 9; Farrell et al (n 33 above) 612 fn35.

[37] Farrell et al (n 33 above) 619.

[38] Mueller (n 9 above) 625.

[39] Farrell et al (n 33 above) 620–1.

[40] *Re Union Oil Co of California*, No 9305, (FTC 25 November, 2003), <http://www.ftc.gov/os/2003/11/031126unionoil.pdf> accessed 15 September 2010, para 63, *rev'd* (FTC 7 July 2004); Mueller (n 9 above) 627.

[41] *Re Rambus*, No. 9302, 2006 WL 2330117, 100 (2 August 2006).

Cisco stated that the cost of redesigning and requalifying its products could have exceeded $1 billion.[42]

A third instance is provided by the World Wide Web Consortium, which develops technology standards for the Web, allowing users to link to and distribute documents across the world.[43] Hold-up led this group to suffer significant delay, inefficient resource allocation, and excessive patentee control over design.[44] The head of the SSO's technology activities pointed to five design efforts, making up ten per cent of its activity, that suffered from patent hold-up.[45] These efforts included standards for a Platform for Privacy Preferences (expressing Web sites' privacy preferences in a standard format) and Synchronized Multimedia Integration Language (allowing authors to write interactive multimedia presentations).[46]

Hold-up and other concerning activities have surfaced in the SSO antitrust cases that the agencies have brought and the courts have considered. Many of the most important cases fall into two categories. The first involves SSO members manipulating the standard setting process. In the second, participants engaged in deceptive conduct such as refusing to disclose patents. In addition to these categories, three recent actions are worthy of attention. One illustrated the danger of SSOs reducing the price paid to patentees. Another sanctioned an SSO rule requiring members to declare their most restrictive licensing terms before standard selection. And a third challenged an alleged refusal to comply with RAND obligations.

B. US Case Law

Manipulation

The first set of cases involves the manipulation of the standard setting process. The case of *Allied Tube & Conduit Corp v Indian Head, Inc* involved a fire protection association's code that specified the types of allowable electrical conduit (hollow tubing that carries electrical wires through a building's walls and floors) and was adopted by many state and local governments. Manufacturers of steel conduit packed an SSO meeting to defeat a rival's attempt to certify plastic conduit. The manufacturers

[42] *Ibid* 100, 108.

[43] Weitzner, Testimony (n 31 above) 1.

[44] D Weitzner, 'Supplemental Comments' (6 November 2002) <http://www.w3.org/2002/11/15-doj-ftc-ipr-weitzner-suppl.html> accessed 15 September 2010.

[45] *Ibid.* The five include Platform for Privacy Preferences (P3P), XML Linking (Xlink), Scalable Vector Graphics (SVG), Synchronized Multimedia Integration Language (SMIL), and Voice Extensible Markup Language (VoiceXML).

[46] W3C, 'The Platform for Privacy Preferences 1.0 (P3P1.0) Specification' (16 April 2002) <http://www.w3.org/TR/P3P/> accessed 15 September 2010; W3C, 'Synchronized Multimedia Integration Language (SMIL 2.0)' (2nd ed, 7 January 2005) <http://www.w3.org/TR/2005/REC-SMIL2-20050107/> accessed 15 September 2010.

recruited more than 200 persons to attend the meeting to vote against the proposal, telling them how to vote and communicating with them through walkie-talkies and hand signals. The steel manufacturers were successful in rejecting the proposal that would have allowed plastic conduit by four votes. The Supreme Court concluded that this activity violated the antitrust laws.[47]

At the heart of the second case, *American Society of Mechanical Engineers v Hydrolevel Corp*, was an SSO's Boiler and Pressure Vessel Code, adopted by 46 states. The code governed 'low-water fuel cutoffs', which blocked the flow of fuel to water boilers when the water level fell so low that it could no longer moderate the boiler's temperature. The firm dominating the market for low-water fuel cutoffs benefited as its vice president, in his capacity as vice chairman of the relevant SSO subcommittee, interpreted the code so that its rival's product was deemed unsafe, thus discouraging customers from buying the product. The Court sought to '. . . ensure that standard setting organizations will act with care when they permit their agents to speak for them' and concluded that the conduct presented an antitrust offence.[48]

Deception

The second set of cases involves a participant's deceptive conduct. In *Re Dell*, the FTC examined the standard for VL-bus, which transferred instructions between a computer's central processing unit and its peripherals (such as hard disk drives and modems). The Video Electronics Standards Association required members to disclose IP they possessed that conflicted with the proposed standard. A Dell representative certified that the proposal did not infringe any of its trademarks, copyrights, or patents. Based on these assurances, the SSO adopted the standard. In the succeeding eight months, the standard was widely adopted, being included in more than 1.4 million computers. At that point, Dell informed SSO members that their implementation of the standard violated its 'exclusive rights'. The FTC entered into a consent agreement that prevented Dell from enforcing the patent against those implementing the standard.[49]

The FTC also challenged the misrepresentation of the Union Oil Company of California (Unocal) before the California Air Resources Board, which developed standards for low-emissions, reformulated gasoline. Unocal participated in the rulemaking proceedings, asserting that its research data (on which the agency relied) was 'nonproprietary' and failing to disclose its relevant patents and applications. California refiners spent billions of dollars modifying their refineries to comply with the new standards. Unocal then claimed that the standards infringed its patents and sought to collect royalties that would have cost consumers more than $500 million annually.

[47] 486 US 492, 496–97 (1988).
[48] 456 US 556, 559–63, 577–8 (1982).
[49] *Re Dell*, 121 FTC 616, 617, 623–4 (1996).

The parties ultimately settled, with Unocal agreeing not to enforce patents related to the standards.[50]

A final FTC challenge occurred in *Re Rambus*. Rambus developed computer memory technologies known as DRAM (dynamic random access memory), which processes information and is used in computers, printers, and cameras. Rambus participated in the Joint Electron Device Engineering Council (JEDEC), a semiconductor engineering SSO made up of DRAM manufacturers and purchasers, as well as producers of complementary products. Although the SSO's disclosure policy was not clear, the FTC found that Rambus '... engaged in representations, omissions, and practices likely to mislead JEDEC members', which '... significantly contributed to its acquisition of monopoly power'.[51]

In 2008, the DC Circuit, focusing on causation, reversed this conclusion. It found that JEDEC might have adopted Rambus's technology even absent any deception. The FTC had stated that '... there was insufficient evidence that JEDEC would have standardized other technologies had it known the full scope of Rambus's intellectual property'.[52] The court also found that any '... loss of an opportunity to seek favorable licensing terms' did not, without more, constitute 'antitrust harm'.[53]

Potential price reduction harms

A different type of concern is presented by a potential SSO conspiracy to depress the price paid to suppliers. The case of *Sony Electronics, Inc v Soundview Technologies, Inc* involved standards for the 'V-chip' technology in television sets that allows parents to block violent or sexually explicit programming.[54]

The plaintiff, Soundview, alleged that members of an industry association agreed to reduce the price for licences for a Soundview patent needed for the standard. An SSO subcommittee found that Soundview possessed the patent most likely to be infringed in the standard's operation. Although Soundview informed the SSO that it would license its patent '... on a non-exclusive, non-discriminatory basis', the organization allegedly did not respond to this offer.[55]

---

50 *Re Union Oil Co of California*, No 9305 (FTC 25 November 2003), <http://www.ftc.gov/os/2003/11/031126unionoil.pdf> accessed 15 September 2010, para 9, 10, 26, 31, 61, 63, *rev'd*, No. 9305 (FTC 7 July 2004); FTC Statement, *Re Union Oil Co of California*, Docket No 9305 (10 June 2005), <http://www.ftc.gov/os/adjpro/d9305/050802statement.pdf> accessed 15 September 2010.

51 *Re Rambus*, No 9302, 2006 WL 2330117, at 5–6, 8, 50, 67, 104, 118 (2 August 2006).

52 *Rambus Inc v FTC*, 522 F3d 456, 464 (DC Cir 2008).

53 *Ibid* 466–7. For a different analysis, see *Broadcom Corp v Qualcomm Inc*, 501 F3d 297, 314 (3d Cir 2007) (denying a motion to dismiss since a false promise of RAND licensing could:

> ... harm[ ] the competitive process by obscuring the costs of including proprietary technology in a standard and increasing the likelihood that patent rights will confer monopoly power on the patent holder.

54 157 F Supp 2d 180 (D Conn 2001) ('*Sony Electronics*').

55 *Sony Electronics* (n 54 above) 181–2.

The SSO then apparently agreed on a uniform price for a licence for Soundview's technology of five cents per television set. The court found that this adequately alleged *monopsony* harm, by which a powerful buyer reduces the price of goods it purchases. Although SSO members did not purchase fewer television sets as a result of the alleged conspiracy to reduce price, the price reductions could have harmed innovation. The court initially relied on the presence of these 'complex questions' in refusing to dismiss Soundview's case.[56] Although the court later granted summary judgment against Soundview (because SSO members' technology did not infringe Soundview's patent), the case nonetheless provides the most direct illustration of potentially anti-competitive monopsony harms.[57]

Announcement of maximum royalties

A related development involved the announcement by SSO members, before a standard's adoption, of their most restrictive licensing terms. The SSO VMEbus International Trade Association (VITA) developed computer architecture standards enabling engineers to design systems used in ultrasound machines, semiconductor manufacturing equipment, and weapons systems radar. VITA's experiment with RAND licensing was unsuccessful, with members demanding excessively high royalties that delayed or blocked standards.[58] As a result, VITA revised its patent policy and sought approval in the form of a business review letter, by which a party asks the Department of Justice's Antitrust Division to review proposed conduct and state its enforcement intentions.[59]

One element of the VITA policy required members to make a 'good faith and reasonable inquiry' into its patents and disclose any patents or applications that they believed could become essential to the specification being developed. Members also were required to specify the maximum royalty rate (in dollars or percentage of sales price) and the most restrictive non-price licensing terms they would request.[60]

Thomas Barnett, the head of the Antitrust Division, began his analysis by explaining that the Division analyses most standard setting activities under the Rule of Reason.[61] He also pointed to the difficulties of hold-up, explaining that SSO members '... can choose among multiple substitute technological solutions' early in the process. In contrast, after a standard is developed, '... it can be extremely expensive or even impossible' to substitute technologies, often requiring '... the entire standard setting

---

[56] *Ibid* 183–6, 190.

[57] *Sony Electronics, Inc v Soundview Technologies, Inc*, 281 F Supp 2d 399, 402 (D Conn 2003).

[58] VITA letter (n 4 above) 1–4.

[59] 28 CFR § 50.6. The letter addressed VITA and its standards development subcommittee, VSO.

[60] VITA letter (n 4 above) 5–6.

[61] The analysis would differ if the standard setting process was '... used as a sham to cloak naked price-fixing or bid rigging.' VITA letter (n 4 above) 8.

process... to be repeated'. Barnett applauded VITA's new policy, concluding that it would allow the group to evaluate technologies not only on technical merit but also on licensing terms. He also did not worry about licensees forcing a price reduction since the policy prohibited the joint negotiation or discussion of licensing terms. Barnett concluded that the Division '... has no present intention to take antitrust enforcement action' against VITA's new policies.[62]

RAND refusal

The final development constitutes the agencies' most aggressive prosecution of SSO activity to date. In January 2008, the FTC announced a complaint against, and settlement with, Negotiated Data Solutions (N-Data). N-Data licensed patents used in equipment employing Ethernet, a networking standard '... used by almost every American consumer who owns a computer.' This technology enables devices in a local area network to '... automatically configure themselves to optimize their communication.' N-Data's predecessor had committed to license its technology for a one-off royalty of $1,000 per licensee. But N-Data later demanded royalties 'far in excess of that commitment.'[63]

By a vote of three to two, the Federal Trade Commission challenged N-Data's action. It did not allege a violation of the Sherman Act, but instead claimed an unfair method of competition and unfair act or practice under Section 5 of the Federal Trade Commission Act. The majority asserted that N-Data's behaviour harmed consumers and businesses and explained that its exercise of its 'unique' authority was needed to '...preserv[e] a free and dynamic marketplace'.[64]

Deborah Platt Majoras, the former Chairman of the FTC, dissented, worrying that the majority did not '... identif[y] a meaningful limiting principle' for determining an unfair method of competition. She also found no allegation of improper or exclusionary conduct and questioned whether N-Data had market power.[65] Then-Commissioner (and later Chairman) William Kovacic also dissented, stating that the FTC's challenge would have an unacknowledged effect on state enforcement and that the majority's failure to distinguish between its two theories of liability—unfair

62 VITA letter (n 4 above) 8–10.

63 FTC, 'FTC Challenges Patent Holder's Refusal to Meet Commitment to License Patents Covering "Ethernet" Standard Used in Virtually All Personal Computers in US' (23 January 2008) <http://www.ftc.gov/opa/2008/01/ethernet.shtm> accessed 15 September 2010; FTC, Statement, *Re Negotiated Data Solutions LLC*, File No. 0510094 <http://www.ftc.gov/os/caselist/0510094/080122statement.pdf> accessed 15 September 2010.

64 *Ibid.*

65 Dissenting Statement of Chairman Majoras, *Re Negotiated Data Solutions LLC*, File No 0510094 <http://www.ftc.gov/os/caselist/0510094/080122majoras.pdf> accessed 15 September 2010.

methods of competition and unfair acts or practices—masked weaknesses in its challenge.[66]

Having sketched some of the guideposts in standard setting analysis, I next set forth some factors relevant to determining how such activity should be analysed. SSOs potentially threaten anti-competitive effects such as allowing firms to boycott rivals and reducing the price paid to suppliers. On the other hand, they provide significant pro-competitive justifications in increasing consumer choice and preventing hold-up.

C. Anti-Competitive Concerns

Though SSOs do not usually result in significant anti-competitive effects, there is a range of activity that could potentially cause antitrust concern. First, SSOs could restrict product diversity and consumer choice by creating an unnecessary standard and excluding certain products. As a leading treatise explains—'It is not clear . . . why the world would need consistent rules for steak knives, or coffee mugs, or couches.'[67]

Second, SSOs could increase prices. Such a result could flow from, for example, product safety restrictions or licensing board certifications.[68]

For our purposes, these two concerns are not central. In particular, the prevalence of interface standards that promote interoperability (and are the focus of this chapter) will almost always outweigh these concerns. The absence of interoperability standards will often lead to fewer products in the marketplace, diminishing any concerns of reduced diversity. In addition, unnecessary standardization does not appear to be a significant problem, and SSOs seem more likely to reduce price and make products available than increase price.

A third concern is that SSOs could increase collusion in the downstream market for goods sold to consumers. By bringing together rivals and aggregating information, the organizations provide a ready-made setting for collusion. Standardized products also facilitate collusion by making it easier to monitor rivals' activity.[69] To be clear, just because SSOs agree on standards does not mean they will agree to set prices for the products they sell. But the antitrust agencies should be sensitive to this important

[66] Dissenting Statement of Commissioner Kovacic, *Re Negotiated Data Solutions LLC*, File No 0510094 <http://www.ftc.gov/os/caselist/0510094/080122kovacic.pdf> accessed 15 September 2010.

[67] Hovenkamp, Janis and Lemley (n 6 above) 35-9.

[68] *Ibid* 35-8 to 35-9.

[69] *Ibid* 35-8.

concern and challenge any such behaviour. As former Chairman Majoras correctly observed—'... summary condemnation is almost certainly warranted' when,

> ... manufacturing rivals cross over the line from discussing the price of technology they will 'buy' if they choose a particular standard and start discussing—and fixing—the price of the products they sell.[70]

A fourth concern, arising in SSOs in which membership is restricted to a subset of an industry, involves a concerted refusal to deal with competitors, or boycott. Many of these *closed* SSOs have legitimate reasons for their exclusion, such as preserving neutrality, reducing concerns of capture, limiting free-riding, or failing to meet reasonable requirements.[71] But in certain cases, organizations could be forced to admit outsiders. Such a remedy should be limited to cases in which SSOs (1) exclude rivals, (2) offer members a significant market advantage that they could not otherwise obtain, and (3) lack legitimate business reasons for their exclusion.[72]

Fifth, SSOs could cause monopsony harm. Monopsony, the lesser-known mirror image to monopoly, occurs when a powerful buyer, by limiting its purchases, reduces the price it pays to suppliers.[73] This concern could apply to SSOs since licensing rules and royalty discussions could allow SSO members (the buyers) to force patent owners (the suppliers) to reduce the royalties they can charge.[74]

In settings in which SSO members have the power to reduce licence prices below competitive levels, monopsony threatens two harms. First, it could have adverse effects on allocative efficiency, reducing the quantity and price of the input below competitive levels. Second, it could have effects on dynamic efficiency, reducing innovation. For several reasons, however, these effects are not likely.[75] For starters, the danger of monopsony is less in the SSO setting because:

(1) SSOs consist of not only licensees but also *licensors*, reducing the likelihood of excessive price reductions;[76]
(2) SSOs deal with an array of potential suppliers since members do not know ex ante the standard that will be selected;[77]

---

[70] Majoras (n 32 above) 10. Relatedly, if the organization is only a sham for cartel-type activities, per se condemnation would be appropriate.

[71] Hovenkamp, Janis and Lemley (n 6 above) 35-21 to 35-22.

[72] *Ibid* 35-23.

[73] M Jacobson and GJ Dorman, 'Joint Purchasing, Monopsony, and Antitrust' (1991) 36 Antitrust Bulletin 1, 5.

[74] Technically, the activity in SSOs would resemble an *oligopsony* because it involves multiple buyers. *Ibid* 6.

[75] For an argument that there is a greater risk of oligopsony than I discuss, see J Gregory Sidak, 'Patent Holdup and Oligopsonistic Collusion in Standard-Setting Organizations' (2009) <http://ssrn.com/abstract=1081997> accessed 15 September 2010.

[76] G Ohana et al, 'Disclosure and Negotiation of Licensing Terms Prior to Adoption of Industry Standards: Preventing Another Patent Ambush?' (2003) 24 European Competition L Rev 644, 654.

[77] MA Lemley, 'Ten Things to do about Patent Holdup of Standards (and one not to)' (2007) 48 Boston College L Rev 149, 162 ('Lemley, Patent Holdup').

(3) unilateral announcements of licensing terms do not resemble joint monopsony action depressing royalties; and
(4) a party owning a patent essential to a standard's implementation has far more bargaining power than the typical seller confronting a monopsony.

In addition, the primary harm linked to monopsony, reduced purchases, is absent in this setting. For reasons I have explained elsewhere,[78] because of the *upward-sloped supply curve* (which results from buyers purchasing the lowest-cost items first and higher-cost goods later), buyers' reduced purchases in non-SSO settings can artificially depress price.[79] In contrast, in the SSO context, because nearly all the costs attributed to supplying patented items flow from the initial costs of creation, the supply curve for patented goods is flat.[80] As a result, any reduction in purchases does not affect price.

In the SSO setting, even if buyers reduce a supplier's price, they tend not to reduce the number of items purchased. In fact, disclosure and negotiations before a standard's selection are designed to *increase* its implementation, which would *expand* the number of licences.[81] This is directly contrary to monopsony's goal.[82]

The primary concern of monopsony—the welfare loss of fewer inputs purchased—thus does not apply to IP. Nonetheless, there may be a dynamic concern in that a reduction in price paid to patentees could lower innovation incentives. While this could conceivably occur, innovators substantially benefit from having their patents reflected in standards widely adopted in an industry. As a result, any effects of depressed licensing rates tend to be significantly outweighed by more numerous licences.

D. Pro-Competitive Justifications

SSOs

Despite their potential anti-competitive effects, SSOs more typically offer an array of powerful pro-competitive justifications.[83] First, they can improve product quality by

[78] Carrier, *Innovation* (see note introducing the author at the beginning of this chapter) 338–9.

[79] JM Jacobson and GJ Dorman, 'Monopsony Revisited: A Comment on Blair & Harrison' (1992) 37 Antitrust Bulletin 151, 156.

[80] H Greene, 'Buyer Price-Fixing and Intellectual Property' (2008) Duke L & Technology Rev <http://intprop.law.duke.edu/documents/2008/Greene%20CLE%20Article.pdf> accessed 15 September 2010, para 42. The presence of non-trivial transaction costs makes the conclusion modestly more nuanced.

[81] Ohana et al (n 76 above) 654.

[82] Preventing hold-up also may align with patentee manufacturers' preference for first-mover advantages in arriving quickly on the market rather than high royalties.

[83] See generally MA Carrier, 'Why Antitrust Should Defer to the Intellectual Property Rules of Standard-Setting Organizations: A Commentary on Teece and Sherry' (2003) 87 Minnesota L Rev 2019.

collecting information related to safety and performance records.[84] Second, interoperability standards enable firms to use a common platform and can reduce price and enhance competition in the marketplace. Third, the standards contribute to a greater realization of network effects and prevent buyers from being stranded in a product that loses the standards war.[85] Fourth, they prevent hold-up and create markets that might not otherwise exist.

Further strengthening the pro-competitive benefits, the industries in which SSOs have developed are those with the greatest potential for bottlenecks, patent thickets, and thwarted innovation. A study conducted by Professor Mark Lemley demonstrated that SSOs have concentrated '. . . in precisely those industries where the unconstrained enforcement of patents could be most damaging to innovation', namely, software, Internet, telecommunications, and semiconductors.[86] The danger of hold-up is more pronounced in these industries since a single product may require licences for hundreds or thousands of patents.[87]

Just as ominous, the industries are marked by cumulative innovation, with one generation's patented invention built on those of previous generations.[88] The clearing of patent thickets and fostering of cumulative innovation and new markets through SSOs offers strong benefits for competition and innovation. Significant to begin with, the pro-competitive benefits of SSOs are magnified even further in removing the potentially explosive landmines of the patent system.

IP rules

Not only are the SSOs themselves pro-competitive, but their IP rules are as well. The rules are designed to address the issue of hold-up. As stated before, an SSO often can choose among an array of alternative technologies before a standard's selection. But after the standard has been chosen and the organization has invested in technologies to implement it, it has less flexibility to switch. At that point, participants can charge higher-than-expected royalties knowing that the members are not easily able to abandon the technology. As the example of the VITA SSO revealed, even promises to license patents on a reasonable and non-discriminatory basis will not prevent hold-up when the details are left vague and are the subject of dispute after the standard has been adopted. The cases involving failure to disclose demonstrate even more strongly the dangers of hold-up. For that reason, search, disclosure, and licensing rules are pro-competitive.

[84] Hovenkamp, Janis and Lemley (n 6 above) 35-10.
[85] Shapiro (n 12 above) 88.
[86] Lemley, SSOs (n 7 above) 1954.
[87] *Ibid* 1953; Lemley, Patent Holdup (n 77 above) 150.
[88] Scotchmer (n 20 above) 29.

Search rules merely require SSO members to search for IP that might read on a standard. Most SSOs do not impose this obligation because of the burden it would impose. But even when they do, such an obligation would not lead to anti-competitive effects. Even if a broad obligation to search dissuades potential patentees from entering the organization, that does not constitute anti-competitive effect.

Disclosure rules inform SSO members deciding on a standard of the IP that would be implicated by the selection of certain standards. Disclosure rules generally differ from information-sharing arrangements that antitrust courts have condemned since the information divulged prevents the strategic hiding and ex post exploitation of IP. The American National Standards Institute recommends that all accredited standards bodies require early disclosure, which provides notice, gives an opportunity to evaluate the technology, and offers time to negotiate licence terms.[89]

Licensing rules are even more essential in avoiding the hold-up problem. These rules offer a pro-competitive justification by circumventing a potential bottleneck and contributing to the creation of a product that might not otherwise exist. RAND licensing aims to address hold-up by requiring patentees to agree, before the standard's selection, to reasonable licensing.

Because the requirement is vague, some SSOs have required participants to provide more specific terms, such as the maximum royalties they would charge. A simple announcement of terms under which patents would be licensed does not constitute an antitrust violation since there is no agreement between rivals. And even discussion of the terms in the SSO should be evaluated under the Rule of Reason. As Professor Mark Lemley explains, '. . . [T]he parties are going to have to have these conversations individually or collectively anyway' and it is preferable to have them before the standard's adoption.[90]

In its Technology Transfer Guidelines, the European Commission recognized the benefits of SSO members negotiating licensing terms, stating that parties are '. . . free to negotiate and fix royalties for the technology package and each technology's share of the royalties either before or after the standard is set.' It explained that such agreement '. . . is inherent in the establishment of the standard . . . and cannot in itself be considered restrictive of competition.' And it highlighted the efficiencies from agreeing to royalties before the standard's adoption to 'avoid the choice of the standard conferring a significant degree of market power.'[91]

[89] American National Standards Institute, *Guidelines for Implementation of the ANSI Patent Policy* 5 (2003) <http://standards.ieee.org/board/pat/607patagen_5-1_pt2.pdf> accessed 15 September 2010; see generally Weiser (n 28 above) 22.

[90] Lemley, Patent Holdup (n 77 above) 162.

[91] Guidelines on the Application of Article 81 of the EC Treaty to Technology Transfer Agreements (EC) [2004] OJ C 101/2 para 225.

The IP rules of SSOs bear some resemblance to other types of activity that have received substantial antitrust deference in the US. In one case, the Supreme Court upheld blanket licences that allowed licensees to perform any of the millions of copyrighted musical works in the package, thereby reducing transaction costs and creating 'a different product'.[92] In a second category, the antitrust agencies have upheld nearly all cross-licence agreements and patent pools they have examined because they resolve bottlenecks among owners of blocking patents that could have otherwise prevented the use of products.[93]

Conclusion

As this chapter has revealed, SSOs sometimes threaten anti-competitive effects. They could involve deception as occurred in the Dell and Unocal cases. They could manipulate the process to hurt competitors. They could threaten a boycott, excluding rivals without justification. They could exercise buying power to depress prices and reduce innovation incentives. And they could engage in downstream collusion, raising the price paid by consumers.

But in most cases, standards increase competition and offer consumers a greater range of products. They increase network effects and allow the creation of products that might not otherwise be available. In markets that coalesce around one product, jointly-set standards are superior to the alternative of a dominant firm's de facto standard. And these benefits are even more potent given that the industries in which SSOs have developed are those with the greatest potential for bottlenecks and thwarted innovation.

The IP rules of SSOs also are important in reducing hold-up. A patentee's power may exponentially increase after the organization has adopted a standard and its members have invested in technologies to implement it. Requiring members to disclose relevant patents provides essential information to the SSO. Nailing down licensing terms increases this information and reduces surprise attributed to hold-up.

In conclusion, while SSO activity could lead to several types of anti-competitive concern, US courts and the antitrust agencies have good reason to defer to most SSO activity.

---

92 *Broadcast Music, Inc v Columbia Broadcasting System, Inc*, 441 US 1, 5, 20–2 (1979).

93 Letter from JI Klein, Assistant Attorney General, Department of Justice to GR Beeney, Sullivan and Cromwell (16 December 1998), <http://www.usdoj.gov/atr/public/busreview/2121.htm> accessed 15 September 2010 (DVD patent pool); Letter from JI Klein, Acting Assistant Attorney General, Department of Justice to GR Beeney, Sullivan & Cromwell (26 June 1997) <http://www.justice.gov/atr/public/busreview/215742.htm> accessed 15 September 2010 (MPEG pool).

16

EUROPEAN COMPETITION LAW, NON-PRACTISING ENTITIES, AND FRAND COMMITMENTS

Philippe Chappatte and Paul Walter

Introduction

The benefits of standards are well recognized for both undertakings and consumers. However, the last decade has seen an increasing number of non-practising entities ('NPEs') who acquire patents that are essential to a standard with a view to obtaining substantial royalties that are inconsistent with commitments to license such patents on fair, reasonable, and non-discriminatory terms. This chapter considers some of the special features that allow such NPEs to pursue these strategies and discusses how European competition law may help to address demands for unreasonable royalties. Finally, this chapter discusses some amendments that could be made to the IPR policies of standard setting organisations to bring greater transparency to licensing terms for standards to help prevent excessive royalty demands from NPEs and other essential patent holders.

A. Standards

Benefits of standardization

The former European Commissioner for Competition, Mrs Neelie Kroes, has noted that,

> . . . standards are good because they create the level playing field on which all can compete. More than that, good standard-setting helps consumers, boosts competitiveness and can spur market growth.[1]

[1] Neelie Kroes European Commissioner for Competition Policy, 'Setting the standards high', Address at Harvard Club of Belgium (15 October 2009, Brussels), SPEECH/09/475.

Standardization allows consumers to use standardized products throughout many different countries regardless of the manufacturer. Standardization also reduces risks of purchasing technology that might become obsolete if subject to inter-technology competition.[2] Consumers also benefit from increased competition between suppliers and reduced switching costs normally associated with standardization.

There are also numerous benefits for the members of standards setting organizations ('SSOs'). For example, by pooling their respective technologies and selecting the best, firms are likely to create a better technology collectively than they would be able to achieve independently. Furthermore, standardization expands the customer base for products complying with the standard and accelerates adoption of new technology and equipment, leading to greater revenues. This increases firms' incentives to invest in research and development aimed at improving the standard or at introducing new products and services.

Market power

It is however well recognized that standardization raises issues of market power for essential patent owners. Once a patented technology is incorporated as an essential part of a standard, it is not possible to manufacture standard compliant products without infringing the IPR covering that technology. The market power of the essential patent holders will typically increase as a result of standardization.[3] For example, if a number of alternative technologies compete prior to the adoption of a standard, none of them could command more than the competitive price. However, once a standard has been chosen, the industry will start developing and producing goods that comply with the standard. Once significant sunk costs have been incurred, the industry becomes effectively locked-in to that standard, since these investments would be lost in the event of a switch to a different technology. The owner of an essential patent in theory has the ability to (i) block companies from producing any products compliant with the standard and (ii) demand royalties for its patent that are significantly higher than the royalties it could have demanded had the technology not been included in the standard, or before the standard was adopted and competition eliminated (known as the 'hold-up' problem).

[2] For example, in relation to high-definition DVD formats, Toshiba's HD DVD technology has effectively been marginalized as a result of the success of Sony's Blu-Ray technology. In years past, Betamax technology became obsolete as a result of the success of VHS.

[3] *Broadcom Corporation v Qualcomm Incorporated*, Case No 06-4292, United States Court of Appeal (3d Cir 2007), III.A.2.b:

> Although a patent confers a lawful monopoly over the claimed invention . . . its value is limited when alternative technologies exist . . . That value becomes significantly enhanced, however, after the patent is incorporated in a standard . . . Firms may become locked in to a standard requiring the use of a competitor's patented technology. The patent holder's IPRs, if unconstrained, may permit it to demand supracompetitive royalties.

FRAND Commitments

To address these market power issues, many SSOs have adopted IPR policies which require IPR owners to (i) disclose all patents essential to a standard; and (ii) commit to licensing their standard-essential patents on fair, reasonable, and non-discriminatory ('FRAND') terms.[4] By requiring firms to give FRAND commitments before industry participants are locked-in, SSOs extract a commitment from the essential patent owners that they will moderate their royalty claims and act in a reasonable and non-discriminatory manner. In committing to license on FRAND terms, the patent owner accepts the obligation to moderate its royalty demands in exchange for the benefits that will accrue from the increased licensing opportunities that are created by the inclusion of its technology in standard compliant products.

B. Non-Practising Entities

Historically, patents declared essential to ICT standards were generally owned by companies that were active in both R&D and the supply of products or services. Few patent owners charged royalties for the use of their patents declared essential to the second generation mobile standard, GSM, for example. This century has seen a paradigm shift in patent strategies, with an increasing number of companies actively seeking royalty payments for the use of their patents relating to standards. FRAND commitments have proven to be generally effective in constraining the royalties charged by the vast majority of such companies to levels which strike a reasonable balance between the interests of licensors and licensees.

However, the last decade has seen the development of a new business model in the ICT sector which seeks to exploit the market power obtained through standardization. Businesses are being set up to acquire alleged essential patents with a view to obtaining substantial royalties. Such businesses are generally non-practising entities (or NPEs) that do not supply goods or services. These businesses typically do not contribute to the market with innovation or product development but instead wait until the industry is locked into a standard through investment in standard specific infrastructure to tax the industry with substantial royalty demands.[5]

---

[4] All three European SSOs have committed themselves to ensure that standards, including any IPRs they might contain, can be used by market operators on fair, reasonable, and non-discriminatory conditions—see General Guidelines for the Co-operation between CEN, CENELEC and ETSI and the European Commission and the European Free Trade Association, (28 March 2003) [2003] OJ C91/5.

[5] It should be noted that not all NPEs engage in this behaviour. There are a number of NPEs that contribute positively to the welfare of standardization activities such as research universities and companies that actively invest in R&D but choose an intellectual property licensing approach as their business model rather than a manufacturing or product sales approach. See D Geradin, A Layne-Farrar and A Jorge Padilla, 'Elves or Trolls? The Role of Non-Practicing Patent Owners in

The ability of these businesses to act in this manner is due to a number of reasons. Typically, they are not members of SSOs and have therefore not given commitments to license on FRAND terms. The patents are often acquired from companies in financial trouble that regard their patents as surplus to requirements.[6]

Furthermore, unlike 'traditional' essential patent owners that are typically involved in the development of standards, these NPEs are often not engaged in standards related R&D. These businesses often acquire alleged essential patent portfolios from third parties after adoption of the standard to which the patents relate. They can therefore exploit the hold-up position created by standard setting in order to pursue excessive royalty practices. NPEs are also often not constrained by the threat that SSOs may be reluctant to accept their technology for incorporation into future standards.

The key advantage, however, enjoyed by these businesses is the fact that they do not supply products and services and thus do not require cross-licences that might otherwise constrain a licensor's behaviour.[7] NPEs are able to pursue injunctive relief against licensees in litigation proceedings in the knowledge that they are not exposed to the risk of similar actions being brought against them.[8] A notable example of this bargaining power was shown in the litigation between NTP, an NPE holding certain wireless e-mail patents, and Research In Motion (RIM), the manufacturer of the Blackberry wireless communication device. NTP filed a claim against RIM after the Blackberry manufacturer had not met demands to pay royalties for the wireless e-mail patents. NTP's action was successful and the Court awarded damages of approximately $23 million.[9] Subsequently, the district court increased the damages to around $53 million and issued a permanent injunction preventing RIM from manufacturing, importing, using, or selling any of the accused Blackberry devices.[10] Although the injunction was stayed pending appeal, the threat of an injunction ceasing all Blackberry services led to RIM settling with NTP for a sum of $612.5 million in full and final settlement of all claims against RIM.[11]

---

the Innovation Economy' (May 2008) TILEC Discussion Paper No 2008-018 <http://papers.ssrn.com/sol3/papers.cfm?abstract_id=1136086> accessed 15 September 2010, 1.

[6] M Macdonald, 'Beware of the Troll' (26 September 2005) The Lawyer <http://www.thelawyer.com/beware-of-the-troll/116783.article> accessed 15 September 2010.

[7] D Chisum, 'Reforming Patent Law Reform' (2005) John Marshall Rev Intellectual Property L 336.

[8] See, for example, Federal Trade Commission, 'To Promote Innovation: The Proper Balance of Competition and Patent Law and Policy' (2003) <http://www.ftc.gov/os/2003/10/innovationrpt.pdf> accessed 15 September 2010, 31 and 38–41. Injunctions can greatly magnify the bargaining power of the holder of an essential patent; the threat to obtain an injunction can enable a patent holder to negotiate royalties far in excess of the economic value of the patent holder's contribution—a 'classic hold-up' situation. See J Farrell, J Hayes, C Shapiro, and T Sullivan, 'Standard Setting, Patents and Hold-Up' (2007) 74 (3) Antitrust LJ 638.

[9] NTP, Inc v Research In Motion, Ltd 418 F3d 1282, 1291 (Fed Cir 2005).

[10] *Ibid.*

[11] Press Release, 'Research In Motion and NTP Sign Definitive Settlement Agreement to End Litigation' (3 March 2006).

Where complex and detailed standards have been developed cooperatively by many companies, the risk of hold-up by NPEs is greater. For example, there are thousands of patents declared essential to the third generation mobile standard, UMTS (the ETSI database suggests that over 50 companies have collectively declared over 10,000 patents and patent applications essential to UMTS). The standardization of such technologies means that implementers of the standard are unable to 'design around' truly essential patents. Therefore, even though an NPE may only have a handful of alleged essential patents among many thousands of other patents declared essential to a standard, they will have a very strong bargaining position if they can threaten a manufacturer with an injunction against the sale or use of all standard compliant products. It has been estimated that patents covering industry standards are 13 times more likely to be litigated than comparable non-standard related patents.[12]

Traditionally, NPEs have focused on bringing suits in the United States due to factors such as the prospect of large damages awards, the absence of a 'loser pays' costs system, wide discovery orders, and the presumption of patent validity. However, there has been a recent string of Federal Circuit Court of Appeals and US Supreme Court opinions that has seen the US become less attractive to NPE litigants. Notably, the Supreme Court judgment in *eBay*[13] has made it much more difficult for NPEs to obtain injunctions in patent infringement cases.[14] It has been reported that, in the year following the Supreme Court's judgment in *eBay*, NPE's found themselves unable to obtain injunctions in the US despite establishing patent infringement, whereas injunctions continued to be issued in favour of practising patent holders.[15]

NPEs are therefore increasingly incentivized to bring their actions in European courts where permanent injunctions are still readily available. Although the damages awarded may not be as high as those granted in certain US jurisdictions, the potential availability of permanent injunctions very much strengthens the NPE's bargaining position in licensing negotiations. There has recently been a number of high profile

---

[12] Mark Lemley, 'Ten Things to do about Patent Hold-up of Standards (and one not to)', (2007) 48 Boston College L Rev.149.

[13] *eBay Inc v MercExchange, LLC*, 547 US 388 (2006).

[14] See also: *KSR Intl Co v Teleflex, Inc*, 550 US 398 (2007), which has made it easier to establish 'obviousness' and thereby invalidate the claimant's patents on that basis; *Re Seagate Tech, LLC*, 497 F3d 1360 (Fed Cir 2007), which has made it more difficult to establish 'willful' infringement in order to be eligible for enhanced damages; *Lucent Techs v Gateway, Inc*, 2009 US App LEXIS 20325 (Federal Circuit 11 September 2009), which has arguably made it more difficult to obtain high damages awards if the patented invention is but a small/insignificant feature of the accused products; *Quanta Computer, Inc v LG Electronics, Inc*, 553 US (2008), which has made it easier for many downstream manufacturers to prove patent exhaustion; and *MedImmune, Inc v Genentech, Inc*, 549 US 118 (2007), which has made it easier for a would-be target to file a declaratory judgment action and thereby select a more favourable venue.

[15] 'Patent Trolls Look to Europe' Executive Counsel, Intellectual Property (September/October 2008).<http://www.executivecounsel.info/v5/i5/patent-trolls-look-to-europe/>accessed15September 2010.

cases which suggest that the Netherlands and Germany are increasingly becoming favoured hunting grounds for NPEs in respect of standards related patents.[16]

C. How Competition Law Can Help Defend Against NPE Litigation

In addition to the usual defences that might be used against infringement actions in relation to standards-related patents (eg invalidity and non-infringement), a defence based on the European competition rules may prove an effective constraint to limit the royalties charged by NPEs to a reasonable level.

Article 101 TFEU requires FRAND licensing

It is commonly accepted that where there are potentially competing technologies prior to a standard being adopted, collective standardization reduces or eliminates inter-technology competition.[17] For example, standard setting in the telecoms industry often involves thousands of decisions taken collectively by committees as detailed specifications are developed. Many of these decisions involve a choice between competing technical contributions from different participating innovators who are, in parallel with the standardization process, running R&D programmes to support the standardization process and filing patent applications to protect their innovations. Once a technical contribution is selected for inclusion in the standard, the competitive constraints that were exerted by the competing technologies will be eliminated.

Where standardization involves such a collaborative process that restricts competition, Article 101 TFEU will be infringed unless the conditions for exemption under Article 101(3) TFEU are satisfied, ie the standard contributes to economic progress and allows a fair share of the benefits to pass to consumers (eg reasonably priced downstream products). Most importantly, the standard setting arrangements must not afford the possibility of eliminating competition in the downstream product markets and therefore the patents declared essential to the standard must be licensed on FRAND terms. This principle is clearly set out by the European Commission in the current version of the Guidelines on the applicability of Article 81 of the EC Treaty to horizontal cooperation agreements ('Horizontal Cooperation Guidelines')[18]

---

[16] Actions have been brought by NPEs such as IPCom and High Point in the Netherlands and Germany against companies practising the third generation mobile standard, UMTS.

[17] See for example, *Broadcom Corporation v Qualcomm Incorporated*, No 06-4292 (3d Cir 2007): 'A standard, by definition, eliminates alternative technologies.' See also *RIM v Motorola*, Civil Action No 3:08-CV-0284-G consolidated with 3:08-CV-0317-G, US District Court, Northern District of Texas, Dallas Division, Order denying Motorola's action to stay, dismiss or transfer, 11 December 2008.

[18] 'To avoid elimination of competition in the relevant market(s), access to the standard must be possible for third parties on fair, reasonable and non-discriminatory terms', para 174 of the Horizontal

and has been applied by the Commission in practice.[19] These principles have also been generally applied, for example, in the context of patent pool agreements.

On 4 May 2010, the European Commission launched a public consultation in relation to its review of the rules applicable to horizontal cooperation agreements. A substantial feature of the review is a new draft version of the Horizontal Cooperation Guidelines.[20] The new draft includes significant revisions to the chapter on standardization, '... drawing on recent case related experience in the field [and aimed] at ensuring that standards are set in such a way that the specific benefits of standard-setting are realized and passed on to European consumers.'[21] The Draft Horizontal Cooperation Guidelines support the position that Article 101 TFEU will normally require that patents declared essential to the standard must be licensed on FRAND terms,

> ... the standard-setting organisation's rules must seek to avoid the misuse of the standardisation process through hold-ups and the charging of abusive royalty rates by IPR holders. These objectives should be ensured in standard-setting organisations through rules which are binding on the standard-setting organisation's members ... The IPR policy should also require that all holders of essential IPR in technology which may be adopted as part of a standard provide an irrevocable commitment in writing to license their IPR to all third parties on fair, reasonable and non-discriminatory terms ('FRAND commitment') ... FRAND commitments are intended to prevent IPR holders from making the implementation of a standard difficult by refusing to license or by requesting unfair or unreasonable fees (in other words excessive fees) after the industry has been locked-in to the standard and/or charging discriminatory fees.[22]

Article 102 TFEU requires FRAND licensing

Article 102 TFEU applies to the unilateral conduct of a dominant firm that holds a position of market power. It is widely recognized that,

> ... standardisation can confer market power on a company whose patent has been incorporated in a standard when it is commercially indispensable to comply with the standard in question, and when there is a lock-in to the standard.[23]

As a result of the market power that is gained or enhanced through the incorporation of patented technology into a standard, a company acquires '... a special responsibility

---

Cooperation Guidelines. See also para 152 of the Guidelines on the application of Article 81 of the EC Treaty to technology transfer agreements ('Technology Transfer Guidelines').

[19] See, eg European Commission press release 'Antitrust: Commission welcomes IPCom's public FRAND declaration' (10 December 2009) MEMO/09/549.

[20] See, 'Draft Communication from the Commission, Guidelines on the applicability of Article 101 of the Treaty on the Functioning of the European Union to horizontal co-operation agreements' (2010) SEC(2010) 528/2 ('Draft Horizontal Cooperation Guidelines').

[21] European Commission press release, 'Competition: Commission consults on review of rules applicable to horizontal cooperation agreements' (4 May 2010), IP/10/489.

[22] Draft Horizontal Cooperation Guidelines (n 20 above) paras 280, 282, and 283.

[23] C Madero Villarejo and N Banasevic (DG Competition), 'Standards and Market Power' (May 2008) Global Competition Policy 3.

not to allow its conduct to impair genuine undistorted competition'.[24] Once a patent becomes essential, access must be provided for downstream companies to be able to manufacture products that comply with an industry standard. The FRAND commitment not to exploit the additional market power that may arise from inclusion in a standard reflects the requirements of Article 102 TFEU, which expressly includes the following examples of abuse—'directly or indirectly imposing unfair purchase or selling prices or unfair trading conditions' ('fair and reasonable' components of FRAND) and 'applying dissimilar conditions to equivalent transactions with other trading parties, thereby placing them at a competitive disadvantage' ('non-discriminatory' component of FRAND). The Draft Horizontal Cooperation Guidelines specifically recognize that an '... abuse of the market power gained by virtue of IPR being included in a standard constitutes an infringement of Article 102'.[25]

Articles 101 and 102 TFEU require FRAND commitments to travel

Where the European competition rules apply (as will be the case for most EU-wide ICT standards), it is clear that all patents declared essential to a standard must be made available to all interested parties on FRAND terms. As a matter of competition law, the obligation to license on FRAND terms should therefore travel with the patent. If an NPE buys a patent that is declared essential to a standard and is, or should be, aware of the FRAND commitment given by the previous owner, it should be bound by that commitment.

Support for this position has been provided by the European Commission in its public statements regarding IPCom. Nokia lodged a formal complaint with the Commission against IPCom in January 2009 alleging that IPCom has failed to license on FRAND terms a portfolio of patents relating to wireless networks acquired from Robert Bosch.[26] In December 2009, the Commission announced that, following discussions with IPCom, it welcomed the public declaration by the company that it is prepared to grant irrevocable licences to the relevant essential patents on FRAND terms and conditions.[27] The Commission noted in its press release that,

> The transfer of FRAND commitments after the sale of standard-essential patents is important from a competition law perspective... The unrestricted access to the underlying proprietary technology on FRAND terms for all third parties safeguards the pro-competitive economic effects of standard setting. Such effects could be

[24] Case 322/81 *NV Nederlandsche Banden-Industrie Michelin v Commission* [1983] ECR 3461 para 57.

[25] Draft Horizontal Cooperation Guidelines (n 20 above) para 284.

[26] The complaint arose from initial proceedings filed by Nokia against Robert Bosch in December 2006 in Germany seeking a declaration that it was obliged to honour its agreement to grant licences to Nokia on FRAND terms. See 'Nokia Files Formal Complaint against IPCom in Telecom Patents Dispute' (7 January 2009) MLex Market Intelligence.

[27] See European Commission press release, 'Antitrust: Commission welcomes IPCom's public FRAND declaration' (n 19 above). Nokia has now withdrawn its complaint to the European Commission.

> eliminated if, as a result of a transfer of patents essential to a standard, the FRAND commitment would no longer apply.[28]

The Commission concluded that when patents declared essential to a standard are transferred from one owner to another, so should any relevant FRAND commitments. Therefore, the fact that an NPE is not a member of the relevant SSO and/or did not provide a commitment to license on FRAND terms should not be determinative on the issue of reasonable royalties. The NPE should be required to offer FRAND terms and conditions in order to adhere to its obligations under the European competition rules. The Commission has provided further support for this position in the Draft Horizontal Cooperation Guidelines,

> To ensure the effectiveness of the FRAND commitment, there should also be a requirement on all IPR holders who provide such a commitment to take all necessary measures to ensure that any undertaking to which the IPR owner transfers its IPR (including the right to license that IPR) is bound by that commitment.[29]

In order to determine the maximum FRAND royalty that might be charged by such an NPE, the royalty rate applied by the previous owner in compliance with its FRAND obligation may be a relevant benchmark.[30] Any attempt to exploit *ex post* market power by charging royalty rates higher than those that were imposed by the previous owner of the patent should in general be constrained as not being fair. Support for this approach can be found in the *N-Data* case. In that case, N-Data (an NPE) obtained certain patents essential to an Ethernet standard adopted by the IEEE, after the industry had become committed to the standard, and after the original patent-holder had committed to licensing the technology for a one-off fee of $1,000 per licensee. The FTC's complaint alleged that N-Data obtained the patents in full knowledge of the licensing commitment made by the previous owner. Nevertheless, N-Data reportedly refused to comply with the commitment and instead demanded royalties far in excess of $1,000 per licensee. The FTC's complaint alleged that N-Data in effect exploited the lock-in created by the industry's commitment to the standard in order to demand higher royalties than the industry otherwise would have paid for the technologies. In response to the FTC's investigation, N-Data agreed to a consent order which required it to change its licensing terms to bring them in line with the original patent-holder's commitment.[31]

---

[28] European Commission press release, 'Antitrust: Commission welcomes IPCom's public FRAND declaration' (n 19 above).

[29] Draft Horizontal Cooperation Guidelines (n 20 above) para 286.

[30] See, for example, *Nokia v IPCom* [2009] EWHC 2742 (Pat) at paras 35 and 36. Nokia has claimed in its action against IPCom that the latter has breached Article 102 TFEU by demanding royalties for its essential patents which are significantly higher than the royalties charged by Bosch (the previous owner of the patent portfolio) to Samsung for the same patents.

[31] Federal Trade Commission press release, 'FTC Challenges Patent Holder's Refusal to Meet Commitment to License Patents Covering "Ethernet" Standard Used in Virtually All Personal Computers in U.S.'(23 January 2008).

D. Improvements to the FRAND System

Although the European competition rules provide some protection against unreasonable royalty demands from NPEs, the risk of litigation remains given the potential for disagreement over the definition of what is a fair and reasonable rate for a particular essential patent portfolio. The frequency of such disputes could be reduced in the future if further clarity were added to the definition of FRAND in respect of a particular standard prior to its adoption (*ex ante*).

There also needs to be a transparent and predictable system to determine royalty rates for essential patents in order to achieve the key objectives of standardization such as spurring market growth and encouraging new entry. A new manufacturer which has no patents to cross-license needs to be able to estimate roughly for the purposes of its business plan what level of royalties will be payable to essential patent owners if it wants to manufacture standardized products. Such an assessment is likely to be difficult in relation to complex standards potentially covered by thousands of declared essential patents owned by many companies. A system should therefore be implemented which would allow all potential users of a standard to estimate the total royalty stack and assess how that total should be allocated as between the many patent holders.

This issue has been the subject of some discussion by the European Commission, which is increasingly active in the field of standards in Europe.[32] The European Commission Directorate General for Industry and Enterprise recently issued a White Paper, which noted the importance of modernizing standardization policy in the European Union in order to exploit fully the potential of standard setting.[33] The White Paper suggests that SSOs should consider implementing systems requiring the,

> . . . declaration of the most restrictive licensing terms, possibly including the (maximum) royalty rates before adoption of a standard as a potential route to providing more predictability and transparency.[34]

[32] The Commission has called for 'a stronger role for standardisation in support of innovation' and commented that standardization is 'increasingly important, in particular in the ICT domain'. See Final Communication from the Commission to the Council, the European Parliament and the European Economic and Social Committee, 'Towards an Increased Contribution from Standardisation to Innovation in Europe' (2008) COM(2008) 133 and Draft Position Paper, 'Standardisation, Competition and Intellectual Property Rights' (2008) ICT/SC No 55.

[33] Commission White Paper, 'Modernising ICT Standardisation in the EU—The Way Forward' (2009) COM(2009) 324 final.

[34] *Ibid* para 2.4.

Support for such *ex ante* declarations has also been provided by the former European Commissioner for Competition, Mrs Neelie Kroes, who has noted that the benefits of such a system could be significant since,

> ... competition on price can happen before the standard is set, and the choice of technology in the standard can be based on a full understanding of the price and quality trade-offs that businesses and individuals make every day.[35]

Although at first sight, such *ex ante* declarations may appear attractive, they are unlikely to provide an effective solution to the hold-up problem or to lead to fair and reasonable royalty rates, in particular in relation to complex dynamic standards, such as UMTS, and in fact may lead to further problems.

Ex ante auction issues

An important distinction should be made between *ex ante* declarations and an effective *ex ante* auction. An *ex ante* auction is a competitive process where the owners have incentives to lower the cost of their respective technologies to obtain the rewards that would be expected to flow from the inclusion of their technology in the standard. Such an auction can be expected to result in a truly competitive rate.[36] For an *ex ante* auction to be effective in a standard setting context, it is suggested that at least the following prerequisites need to be fulfilled:

(1) the alternative technologies that are candidates for inclusion in a standard must be well developed, ie known and stable; and
(2) the essential patent landscape must be sufficiently transparent in order to identify which companies own the patents that are essential to each alternative technology.

Neither of these prerequisites can be met in a typical complex dynamic standard setting process. For example, telecommunication standards such as the third generation mobile standard, UMTS, are not developed through the use of 'building blocks' of patented 'off the shelf' technologies. The standardization models used by SSOs such as ETSI and 3GPP are based on an ongoing dynamic process involving the setting of requirements and the identification of a multitude of technical problems which are then solved by substantial R&D efforts conducted by a large number of different contributors as the standardization process develops over a number of months and years.

---

[35] N Kroes, European Commissioner for Competition Policy, 'Setting the standards high' (n 1 above).

[36] See J Farrell, J Hayes, C Shapiro and T Sullivan (n 8 above). See also D Swanson and W Baumol, 'Reasonable and Non-discriminatory (RAND) Royalties, Standards Selection, and Control of Market Power' (2005) 73 Antitrust L J 1, 7ff.

In this dynamic environment, important on-going R&D is carried out in parallel with the standardization process. Many licensors and licensees will typically not have a clear picture of the strength or depth of the relevant essential patent portfolios relating to a given standard for a number of years after the first release of the relevant standard. The number of essential patents granted to SSO members will typically remain at a relatively low level in the early years of the standardization process.[37] The essential patent landscape in relation to a standard should become clearer after a number of years when the bulk of the patent applications relating to the technology as developed on first release have been processed.[38] In circumstances where, prior to the adoption of the standard, it is uncertain which patent applications will be granted, the scope of the granted patents, and which patents will read on the standard and who will own all or even the bulk of the essential patents, it would be impossible to hold an effective *ex ante* auction for the inclusion of such technologies.

***Ex ante* declarations issues**

Where it is unfeasible to hold an effective *ex ante* auction, there is a risk that a system of unilateral *ex ante* declarations would give rise to excessive cumulative royalty costs. A request for a declaration of most restrictive licensing terms could solicit excessive claims as companies seek to maximize their *ex post* negotiating positions. The absence of a clear patent landscape *ex ante* showing the relative importance of the essential patent owners is likely to provide incentives for excessive royalty claims. Such declarations would not reflect any competitive process and could not therefore be equated to a negotiated rate achieved in arm's length negotiations prior to the adoption of a standard.

Furthermore, in circumstances where there are a large number of potential essential patent holders, *ex ante* royalty declarations by such a high number of companies would be likely to result in an excessive aggregated royalty rate for the standard, absent an effective auction.[39]

---

[37] For example, at the time of the adoption of the UMTS standard in December 1999, the ETSI database contained less than 200 declarations of WCDMA related patents that were potentially essential to the standard (WCDMA is the radio access technology for UMTS). As at December 2009, this number of declarations of WCDMA related patents had grown to over 10,000.

[38] The European Patent Office has reported that in 2008, on average, a patent was granted around 43 months after an application was filed. See EPO Annual Report (2008) <http://www.epo.org/about-us/office/annual-reports/2008/business-report/patent-process.html> accessed 15 September 2010.

[39] It is established economic theory that if multiple owners of complementary products, which are essential inputs, set prices for these inputs independently, the combined price will exceed the rate that would be charged by a single owner of all the products concerned (Cournot effects). This problem is greater the larger the number of complementary product owners, the stronger the degree of complementarity between these and the larger the mark-up. The importance of the Cournot effect in standard setting has been underlined by recent economic literature. See, eg C Shapiro, 'Navigating the Patent Thicket: Cross Licenses, Patent Pools and Standard Setting' (May 2000) Competition Policy Centre,

These concerns are not theoretical. For example, operators in the mobile telecoms industry sought to obtain an indication from essential patent owners as to what they would charge for licences to use patents declared essential for fourth generation mobile systems. In July 2007, NGMN[40] announced that it had launched an initiative to deliver more transparency and predictability towards IPR costs associated with next generation mobile technologies. The aim of the initiative was to reveal indicative cumulative IPR costs for various standards in order to provide an early opportunity for technology customers to consider the IPR cost of potential next generation technologies in their decision-making process.[41] The initiative required each participant to provide to an independent third party its main proposed licensing terms and conditions. Although the aggregate amount claimed is confidential, it is well known in the industry that the aggregate royalties claimed by the relevant IPR owners was wholly excessive and would be a substantial cost burden on both handset manufacturers and operators who wish to introduce higher quality services to consumers. The fact that the NGMN IPR Initiative was unsuccessful in achieving its aims of fair and reasonable royalty rates for LTE has been publicly recognized.[42]

Defining FRAND prior to formal adoption of a standard

An alternative solution for increasing the transparency of royalty demands for standards could involve SSOs defining the meaning of FRAND on a standard-by-standard basis prior to the official adoption of the standard.

Currently, SSOs' roles are limited as regards the definition and enforcement of FRAND commitments. SSOs typically are not mandated to negotiate with essential IPR owners that refuse to license their declared essential IPR on FRAND terms. Often the only practical course of action open to SSOs in circumstances where FRAND commitments are not given or honoured is to change a standard's specification to avoid infringing essential IPR which will not be made available on FRAND terms.[43] This is unlikely to provide a solution once operators have invested billions of euros in equipment conforming to the standard as first released.

University of California, Working Paper No CPC00-11 <http://129.3.20.41/eps/le/papers/0303/0303005.pdf> accessed 15 September 2010.

[40] NGMN is an organization that was set up in September 2006 by seven leading mobile operators (China Mobile, KPN, NTT DoCoMo, Orange, Sprint Nextel, T-Mobile, and Vodafone) in order to develop a common vision for mobile networks and technology. NGMN is run by the network operators' members.

[41] 'NGMN announces new sponsors and unprecedented agreement on innovative IPR regime', 17 July 2007.

[42] See for example, France Telecom Orange's response to the questionnaire attached to the White Paper 'Modernising ICT standardisation in the EU—The Way Forward' (n 33 above),

> The legitimate objective of transparency and predictability cannot be guaranteed, on the contrary, such declaration could even lead to unreasonable royalty aggregations that could mislead the whole industry about the commercial viability of a standard implementation: example of the Trusted Third Party royalty aggregation process defined and run by the NGMN.

[43] See, eg Art 8.1 of ETSI's IPR Policy.

If SSOs (through consensus among their members—the membership of an SSO is likely to include in principle the bulk of the patent owners and their potential licensees) were able to define FRAND and require all members to agree to such a definition in order to participate in the standard setting activities, these principles could then be applied *ex post* in order to provide a benchmark for the appropriate rates to be charged for essential patents (including by any subsequent owners of an essential patent)—the competition rules referred to above may be invoked to underpin the FRAND commitment.

As ETSI demonstrated in the context of the UMTS standard, an SSO can establish consensus as to how FRAND should be defined in relation to a particular standard. As part of the provisional decision to select WCDMA technology for inclusion in the UMTS standard, ETSI members established the UMTS IPR Working Group to provide guidelines for the FRAND licensing of UMTS essential patents. The UMTS IPR Working Group represented '... 40 of the world's leading wireless companies actively contributing in global standardisation under the auspices of ETSI (Europe), ARIB (Japan), TIA (U.S.A.) and ITU'. The Working Group conducted an industry consultation exercise in which senior executives of ETSI members were asked for their views on methods to achieve FRAND royalties for the UMTS standard.[44] Following industry consultation, the UMTS IPR Working Group published the IPR Report which issued recommendations for the licensing of UMTS essential patents.[45]

The IPR Report recommended that the '... [Maximum cumulative royalty rate] should be set in single figures to meet the perceived view of what is "affordable".'[46] The recommendation of a single-digit cumulative royalty rate was intended to reflect an equitable balance between the interests of essential patent holders, purchasers of infrastructure equipment and mobile handsets, and end-consumers.[47] The IPR Report also recommended that the maximum cumulative royalty rate should be divided among the holders of essential patents according to the principle of equality of essential patents,

> The first step in this obligatory consideration is the recognition that there is only one criterion for the weighting of essential patents and that is their essentiality, and it is this principle which justifies each patent attracting the same standard royalty rate.[48]

[44] The UMTS IPR Working Group held meetings with the senior executives of the following participating ETSI members: AirTouch, Alcatel, Bosch, Ericsson, Fujitsu, Golden Bridge Technology, GSM Association, InterDigital, Lucent Technologies, Matsushita, Mitsubishi, Motorola, NEC, Nokia, Nortel, NTT DoCoMo, Oki, Siemens, Sipro Lab Telecom, and Sony.

[45] ETSI, 'Third Generation Mobile Communications: The way forward for IPR' (January 1999) (the 'IPR Report').

[46] See IPR Report (n 45 above) 23. Although the industry consultation indicated a permissible cumulative range of between 5–15 per cent depending on the product, there was a mutual understanding that the mean should be in single figures.

[47] The IPR Report (n 45 above) states that the goal of the UMTS IPR Working Group was to '... ensure an equitable balance is maintained between rewarding the patent holders and, at the same time, maintaining the commercial viability of their usage.'

[48] IPR Report (n 45 above) 20.

This experience suggests that it should be possible for SSO members to reach a consensus on an appropriate definition of FRAND on a standard-by-standard basis. The key objective for SSO members should be to achieve a cumulative royalty rate that represents a fair balance so that, on the one hand, sufficient royalties are available for each essential patent holder to achieve a reasonable rate of return on its research and development investment, while on the other hand, ensuring that implementation costs for the standard are sufficiently low to allow a competitive downstream industry to the ultimate benefit of end-consumers. Where this balance lies will depend on the specific facts of a standard and it is unlikely to be the case that a single FRAND aggregate rate will be appropriate for all ICT standards.

Some commentators have suggested that such measures should be prohibited under competition law since it effectively amounts to a buyers' cartel.[49] However, an agreement between licensors and licensees who are members of an SSO should not be considered as a buyers' cartel if a consensus can be reached. The European Commission's Technology Transfer Guidelines recognize that the development of such *ex ante* licensing frameworks (with maximum cumulative royalty rates) may be efficient and pro-competitive,

> Undertakings setting up a technology pool that is compatible with Article 81 [now Article 101 TFEU], and any industry standard that it may support, are normally free to negotiate and fix royalties for the technology package and each technology's share of the royalties either before or after the standard is set. Such agreement is inherent in the establishment of the standard or pool and cannot in itself be considered restrictive of competition and may in certain circumstances lead to more efficient outcomes. In certain circumstances it may be more efficient if the royalties are agreed before the standard is chosen and not after the standard is decided upon, to avoid that the choice of the standard confers a significant degree of market power on one or more essential technologies.[50]

The Draft Horizontal Cooperation Guidelines also recognize the benefits of increased transparency and state that,

> . . . it is important that parties involved in the selection of a standard be fully informed not only as to the available technical options and the associated IPR, but also as to the likely cost of that IPR.[51]

However, the Draft Horizontal Cooperation Guidelines only consider the application of the competition rules in this regard to systems of unilateral *ex ante* declarations, which as noted above are in many cases unlikely to lead to fair and reasonable royalty rates.

[49] See for example, JG Sidak, 'Patent Holdup and Oligopsonistic Collusion in Standard-Setting Organizations' (2009) 5 (1) J Competition L & Economics123.

[50] Guidelines on the application of Article 81 of the EC Treaty to technology transfer agreements, 2004 OJ C 101/2, para 225.

[51] Draft Horizontal Cooperation Guidelines (n 20 above) para 287.

The Draft Horizontal Cooperation Guidelines suggest that rules allowing for the 'joint negotiation or discussion of licensing terms in particular royalty rates' may lead to a restriction of competition within Article 101(1) TFEU.[52] However, it is clear that such rules can also have pro-competitive effects and yet the Draft does not provide any guidance as to how the Commission will assess such rules under Article 101(3) TFEU. Disputes about the definition of FRAND royalties could be avoided if the current review is used by the Commission to provide more guidance to SSOs as to what measures might be collectively adopted by their members in order to bring greater precision to the definition of FRAND terms.

Conclusion

The benefits of standards are well recognized for both undertakings and consumers. However, the increasing number of non-practising entities seeking to demand substantial royalties in respect of essential patents threatens to undermine the economic benefits of standardization. Although the European competition rules provide some protection against unreasonable royalty demands, the risk of litigation remains given the potential for disagreement over the definition of FRAND terms and conditions. We believe that the frequency of such disputes could be reduced in the future if SSOs were encouraged to define more specifically 'fair and reasonable' on a standard-by-standard basis prior to the official adoption of the standard. The European Commission should therefore use the opportunity presented by the review of the Horizontal Cooperation Guidelines to provide SSOs with further guidance as to how *ex ante* joint discussions aimed at defining the meaning of FRAND for a particular standard would be assessed under Article 101 TFEU.

[52] Draft Horizontal Cooperation Guidelines (n 20 above) para 287.

17

TAKING CONTRACTS SERIOUSLY: THE MEANING OF THE VOLUNTARY COMMITMENT TO LICENSE ESSENTIAL PATENTS ON 'FAIR AND REASONABLE' TERMS

*Roger G Brooks and Damien Geradin**

Introduction

The literature addressing the meaning of a commitment made by holders of patents 'essential' to a standard to licence such patents on 'fair, reasonable, and nondiscriminatory' (FRAND) terms and conditions is now substantial. While reaching quite different conclusions, a number of authors have addressed this as a question of economic theory—what limitations (if any) on the freedom of the parties negotiating a licence to essential patents will best ensure efficient outcomes?

On the basis of such analyses, authors have variously argued that, in order to satisfy a 'fair and reasonable' commitment, a patent holder,

- must charge no more than the incremental value of his invention over the next best technical alternative;[1]

* The authors represent Qualcomm Inc. in a variety of matters involving FRAND-related issues. The views expressed in this paper cannot however be attributed to Qualcomm Inc. or any of the authors' other clients.

[1] MA Lemley and C Shapiro, 'Patent Holdup and Royalty Stacking' (2007) 85 Texas L Rev 1991, 1996; M Dolmans, 'Standards, IP, and Competition: How to Avoid False FRANDs' (2008) *Fordham IP Law Institute, Standard Setting—The Interplay With IP and Competition Laws* 12–3; J Temple Lang, 'Licensing, Antitrust and Innovation under European Competition Law' (April 13–14, 2007) Fordham IP Property Conference 2, 5–6 (arguing that '. . .the royalty for the technically better technology would tend to be driven down to a level based on its relative value, that is, how much better it was than the second-best').

- must not negotiate for a royalty-free cross-licence as part of the consideration for a licence;[2]
- must set his royalty rate based on a mathematical proportion of all patents essential to the practice of a standard;[3]
- must set his royalty rate in such a way as to prevent cumulative royalties on the standardized product from exceeding a low percentage of the total sale price of that product;[4]
- must not raise requested royalty rates after the standard has been adopted, or after the relevant market has grown to maturity;[5]
- is not entitled to seek injunctive relief against a standard implementer should they fail to agree on licence terms.[6]

The types of economic arguments relied on by these authors to justify these restrictive regimes may well be useful in debating public policy and the proper application of national competition law—although one of the present authors and others have elsewhere critiqued the merits of many of these calls for what is essentially government intervention in the private licensing process.[7] But in this paper we step back to

[2] Dolmans (n 1 above) 14 (arguing that it would be 'unreasonable for a licensor to insist that licensees provide cross-licenses of (or non-assertion of patent clauses covering) their technology royalty-free or on below-competitive rates').

[3] P Chappatte, 'FRAND Commitments—The Case for Antitrust Intervention' (2009) 5 European Competition J 319, 340; Temple Lang (n 1 above) 6 (arguing that 'fair and reasonable' requires proportionality).

[4] Lemley and Shapiro (n 1 above) 2010–1.

[5] Chappatte (n 3 above) 336; C Shapiro and H Varian, *Information Rules: A Strategic Guide to the Network Economy* (Harvard Business School Press, 1999) 241:

> Reasonable *should* mean the royalties that the patent holder could obtain in open, up-front competition with other technologies, not the royalties that the patent holder can extract once other participants are effectively locked in to use technology covered by the patent.

DG Swanson and WJ Baumol, 'Reasonable and Nondiscriminatory (RAND) Royalties, Standards Selection, and Control of Market Power' (2005) 73 Antitrust LJ 1, 10:

> If the primary goal of obtaining RAND licensing commitments is to prevent IP holders from setting royalties that exercise market power created by standardization, then the concept of a 'reasonable' royalty for purposes of RAND licensing must be defined and implemented by reference to *ex ante* competition.

[6] J Farrell et al, 'Standard Setting, Patents, and Hold-Up' (2007) 74 Antitrust LJ 603, 638,

> ...[A] patent holder that has made a commitment to license on a FRAND basis should not be able to get (or threaten) an injunction against use of the technology to comply with the standard.

Temple Lang (n 1 above) 10 (arguing that 'a licensor that is subject to a FRAND obligation has no right to seek an injunction against a company that is willing to take a licence on FRAND terms').

[7] Damien Geradin has published a series of articles in combination with other authors, including Anne Layne-Farrar, Jorge Padilla, and Miguel Rato, which criticize the various arguments raised in the papers cited in the preceding footnotes on the grounds that these papers were not supported by legal and economic analysis, but instead merely reflected the policy preferences of their authors. See, eg, D Geradin and M Rato, 'FRAND Commitment and EC Competition Law: A Reply to Philippe Chappatte' (2009) European Competition Journal (forthcoming 2010); D Geradin, 'Pricing Abuses by

ask a different question—what do these arguments and proposed regimes have to do with the contract that is the source of the FRAND obligation?

This paper is divided into five parts. Part A reviews the basic fact that a FRAND commitment is the result of a voluntary contract between essential patent holders and a standards setting organization such as ETSI, with the important corollary that the meaning of that commitment must be determined through the legal methods of contractual interpretation. Using a FRAND undertaking to ETSI as an example, it identifies the main categories of information potentially relevant to contract construction, including for instance the contract language itself, and the 'negotiation history' of the ETSI IPR Policy. Part B shows that none of these categories of information support any of the restrictive limitations listed at the opening of this introduction. On the contrary, 'fair and reasonable' are on their face flexible terms the specific content of which is substantially left to the negotiation between the parties. Our research also shows that all attempts made subsequent to the ETSI IPR Policy's adoption to alter the balance of interests between essential patent holders and implementers by changing the meaning of FRAND have been rejected by the ETSI membership. Part C addresses issues regarding the judicial enforcement of a FRAND undertaking. First, we demonstrate that, when it is alleged that a patentee has failed to offer 'fair and reasonable' terms, the role of a court is not to determine what 'fair and reasonable' terms would be, but whether the terms offered, taking into account all of the specific circumstances between the parties and prevailing market conditions, fall outside the range of reasonableness contemplated by the FRAND commitment. Second, we conclude that a licensee should not be able to collaterally attack the enforceability of a licence based on a prior FRAND commitment. Third, we note that what is 'fair and reasonable' after full adjudication of infringement and validity may be higher than what would have been 'fair and reasonable' in the context of pre-litigation negotiations. Part D demonstrates that the assertion of some authors that the making of a FRAND commitment should be construed as a waiver of the patent-holder's right to seek injunctive relief is contradicted rather than supported by the history of the adoption of the ETSI IPR policy. Part E invites further research into the 'intent of the parties' with respect to the 'non-discriminatory' component of FRAND, observing that considerable discussion of this term at the time of the ETSI IPR Policy's development centred on fears of 'protectionist' discrimination based on the nationality of companies.

Essential Patent Holders in a Standard-Setting Context: A View from Europe' (2009) 76 Antitrust LJ 329; D Geradin et al, 'Competing Away Market Power? An Economic Assessment of Ex Ante Auctions in Standard Setting' (2008) 4 European Competition J 443; D Geradin et al, 'The Complements Problem Within Standard Setting: Assessing the Evidence on Royalty Stacking' (2008) 14 Boston U J Science & Technology L 144; V DeNicolo et al, 'Revisiting Injunctive Relief: Interpreting *eBay* in High-Tech Industries with Non-Practicing Patent Holders' (2008) 4 J Competition L and Economics 571; D Geradin and M Rato, 'Can Standard-Setting Lead to Exploitative Abuse? A Dissonant View on Patent Hold-up, Royalty-Stacking and the Meaning of FRAND' (2007) 3 European Competition LJ 101.

A. The Contractual Basis of FRAND Obligations

National (or EU) competition laws may impose their own limitations on patent owners, but a FRAND obligation is *solely* the result of a *voluntary* contract entered into by the patent owner on an identifiable date.[8] We will take ETSI and its IPR Policy as typical of major standard setting organizations (SSOs), and note that it is crystal clear in the ETSI IPR Policy that an obligation to license a patent on FRAND terms arises not by automatic operation of ETSI policy, but (at the earliest) only if and when the patent owner agrees, in writing, to license on FRAND terms.[9]

The contract and its interpretation

If a FRAND undertaking is a contract, then there are legally proper methods for determining what that contract means, and they do not include lengthy flights of economic theory. On the contrary, both the civil law and common law traditions of contract interpretation and enforcement fundamentally look to discern and give effect to *the intent of the parties.*[10]

In that context, we note that the 'parties' to a FRAND undertaking are the patent owner and ETSI, while the 'parties' that developed and agreed upon the underlying ETSI IPR Policy were the diverse set of industry participants that make up the ETSI

[8] J Miller, 'Standard Setting, Patents, and Access Lock-In: RAND Licensing and the Theory of the Firm'(2007) 40 Indiana L Rev 351 ('The RAND promise, embedded in SSO bylaws to which participants agree, is primarily a matter of contract law.'); M Lemley, 'Intellectual Property Rights and Standard-Setting Organizations' (2002) 90 California L Rev 1889, 1909 ('SSO IP rules have legal significance only to the extent they are enforceable. Because the IP policies are at base agreements by members of the SSO to abide by certain rules regarding IP ownership, their enforceability is initially a question of contract law.'). ETSI's IPR policy clearly recognizes and takes into account the strictly voluntary nature of any FRAND undertaking: Section 8 of the ETSI IPR Policy contains a mechanism for dealing with the 'non-availability of licences' including a member's refusal to license on FRAND terms. Where an IPR owner informs ETSI of such a refusal prior to the publication of a standard, the General Assembly first tries to find a 'viable alternative technology'. If none exists, and the IPR owner refuses to reconsider its position, there is a procedure for ETSI to decide whether ETSI '... should pursue development of the concerned parts of the STANDARD or a TECHNICAL SPECIFICATION based on the non-available technology and should look for alternative solutions.' (§ 8.1.3).

[9] Section 6.1 of ETSI's IPR policy provides that when essential IPR is disclosed, ETSI will request—but not oblige—the owner of the IPR to undertake in writing that it is prepared to grant irrevocable licences on FRAND terms and conditions, and as such to waive its right to refuse to offer a licence to those seeking one. Even if the owner of an essential IPR decides not to make a FRAND commitment, it does not necessarily follow that the IPR in question will be excluded from the standard. For instance, under Article 8.1 of ETSI's IPR policy, in such case ETSI's General Assembly will examine whether alternate technical solutions exist. Where it concludes that this is not the case, the Director General may request the owner of the IPR to reconsider. However, the latter is under no obligation to agree to license.

[10] AL Corbin, *Corbin on Contracts* (1952) § 538. The intent of the parties also plays a fundamental role in the interpretation of contracts in civil law systems. See, for instance, Article 1156 of the French Civil Code whereby the judge must search for the intent of the parties when the contract was concluded or modified. Under this provision, the 'subjective intent' of the parties ('what they really meant') is more important that the literal language of the contract itself.

membership—not academic economists or competition authorities. As a result, there is no reason at all to suppose that the 'founding fathers' of ETSI settled on IPR policies that are functionally interchangeable with EU competition law, as some authors more or less suggest.[11] Nor is there any reason to suppose that the agreement they reached did or was intended to implement idealized economic theory.

We propose, then, unlike others, to take the FRAND obligation seriously as a contract. Using a FRAND undertaking to ETSI as the example, we will ask when the contract was formed, and what the parties actually agreed to.[12]

The contract and the evidence of the intent of the parties

A particular FRAND obligation comes into existence as the last step in a lengthy history. The relevant terms of the ETSI IPR Policy were fixed by vote of the ETSI membership in 1994.[13] However, the adoption of the ETSI IPR Policy did not create any FRAND commitment; it merely set out the terms under which ETSI may (if it follows its rules) consider member-owned IPR for inclusion in standards. No contract is formed, no FRAND commitment is created, until a patent holder submits a

[11] M Dolmans, 'EC competition law and IP licensing in a standard-setting context: Are hold-up, royalty stacking and patent ambush serious concerns and, if so, what are possible responses?' *ABA Section of Antitrust Law, IP Committee, Brown Bag on Standards and IP, Brussels* (22 June 2007) ('Article 82 [now Article 102 TFEU] obligations are substantially similar to the contractual obligations under FRAND commitments.').

[12] Some commentators have raised the spectre that the contractual nature of a FRAND commitment could enable such a commitment to be evaded after a commitment is made by selling the patent to a third party that has not made such a commitment. However, despite decades of SSO operation in reliance on contractual FRAND commitments, the only two instances we are aware of in which a purchaser of patents has claimed not to be bound by a prior FRAND (or similar) commitment is the position taken but more recently abandoned by IPCom in connection with patents purchased from Bosch. See 'Antitrust: Commission welcomes IPCom's public FRAND declaration' MEMO/09/549 (10 December 2009) <http://europa.eu/rapid/pressReleasesAction.do?reference=MEMO/09/549&format=HTML&aged=0&language=EN> accessed 15 September 2010 (European Commission welcoming,

> the public declaration by German IP licensing company IPCom, following discussions with the Commission, that it is ready to take over Bosch's previous commitment to grant irrevocable licenses on fair, reasonable and non-discriminatory (FRAND) terms to patents held by IPCom which are essential for various standards set by the European Telecommunications Standard Institute (ETSI) and Universal Mobile Telecommunications System (UMTS),

and N-Data's attempt to ignore a prior owner's agreement to license certain essential patents for $1000.

We suggest that an argument can be made that, given the online publication of FRAND declarations by major SSOs and the sophistication of participants in such industries, a purchaser of a patent which has been made subject to a FRAND declaration takes with either actual or constructive notice of that declaration, can be presumed to have negotiated a price taking that 'encumbrance' into account, and should therefore be equitably estopped from asserting the patent in a manner inconsistent with that undertaking. This was essentially the result reached by the FTC in the N-Data case (*Re Negotiated Data Solutions LLC*, Order and Decision, No. C-4234 (22 September 2008) <http://www.ftc.gov/os/caselist/0510094/080923ndsdo.pdf> accessed 15 September 2010). Of course, the details of such an approach must be worked out within the legal doctrines of particular jurisdictions.

[13] A current version of the ETSI IPR Policy is available at <http://www.etsi.org/WebSite/document/Legal/ETSI_IPR-Policy.pdf> accessed 15 September 2010.

written agreement to license identified patents (whether identified individually or categorically) on FRAND terms. Certainly it is this written agreement or 'undertaking' that is the contract, but since such undertakings commonly repeat or refer to the 'fair, reasonable, and nondiscriminatory' terminology of the ETSI IPR Policy, and are written against the background of that policy, such an undertaking cannot be construed as a free-standing document, but must be construed (as it was written) with reference to the IPR Policy.

However, such an 'undertaking' in many cases was not made until a decade or more after the ETSI IPR Policy was crafted and adopted. ETSI pronouncements and actions relating to the IPR Policy during those interim years may also cast light on the understanding of the parties at the time of contract formation, and so cannot be ignored. Such evidence will include ETSI policy documents, as well as the record of revisions to or 'interpretations' of the IPR Policy that have been proposed but *rejected* by the ETSI membership over time.

In sum, we identify four main categories of information potentially relevant to contract construction—(i) the contract language itself; (ii) information as to the preexisting 'understanding of the industry' as to what a FRAND undertaking to an SSO meant, at the time the FRAND concept was incorporated into the ETSI IPR Policy; (iii) information concerning the actual deliberation and debate by the ETSI members at the time the policy was adopted; and (iv) subsequent comment and action relating to the meaning of FRAND by ETSI.

The specific language of a particular declaration made by a patent holder would of course also be relevant. However, since this class of evidence of intent would by its nature pertain only to individual declarations, we will not give it any further consideration in this discussion of general principles.

It is indeed possible that economic theory might make additional contributions by enabling us to better *understand* the course of the contract negotiations, or the contemporaneous industry practices, but nothing in either the civil law or common law tradition could permit economic theory to *substitute* for or overrule evidence of the actual intent of the parties. Further, if one did wish to use economic theory to predict or better understand the IPR Policy compromises actually reached by the ETSI members, one would need to look to game theory models that take into account the institutional interests and bargaining power of the ETSI member organizations, and we have not seen that complicated game attempted.

B. The Contractual Meaning of 'Fair and Reasonable'

In this part, we rely on the main categories of information potentially relevant to contract construction identified above to determine the meaning of the 'fair and reasonable' terms that are part of FRAND.

The plain language

The starting point of any contract interpretation must be the language of the contract itself. The terms 'fair and reasonable' are on their face terms implying wide latitude; they are permissive words to which there is even conceptually no one right answer.[14] For example, in connection with the sale of a relatively illiquid property such as a house or a tract of real estate, negotiations between the seller and one or more potential buyers could result in a considerable range of prices (perhaps differing depending on the urgencies of the parties), any one of which the outside observer would have to concede to be at least 'fair' or 'reasonable'. The same is surely true of prices and terms for patent rights.

But we can say more. When a patent holder commits to license on 'fair, reasonable, and nondiscriminatory' terms in response to and pursuant to Section 6.1 of the ETSI IPR Policy, it is appropriate that, when considering the 'plain meaning', we look to what was before the declarant—the 'plain meaning' of FRAND as it appears in context within the IPR Policy.

The ETSI IPR Policy states as its 'Policy Objectives' the following (emphasis added),

> 3.1 It is ETSI's objective to create STANDARDS and TECHNICAL SPECIFICATIONS that are based on solutions which best meet the technical objectives of the European telecommunications sector, as defined by the General Assembly. In order to further this objective *the ETSI IPR POLICY seeks to reduce the risk to ETSI, MEMBERS, and others applying ETSI STANDARDS and TECHNICAL SPECIFICATIONS, that investment in the preparation, adoption and application of STANDARDS could be wasted as a result of an ESSENTIAL IPR for a STANDARD or TECHNICAL SPECIFICATION being unavailable.* In achieving this objective, the ETSI IPR POLICY seeks a balance between the needs of standardization for public use in the field of telecommunications and the rights of the owners of IPRs.
>
> 3.2 *IPR holders* whether members of ETSI and their AFFILIATES or third parties, *should be adequately and fairly rewarded for the use of their IPRs* in the implementation of STANDARDS and TECHNICAL SPECIFICATIONS.
>
> 3.3 ETSI shall take reasonable measures to ensure, as far as possible, that its activities which relate to the preparation, adoption and application of STANDARDS and TECHNICAL SPECIFICATIONS, *enable STANDARDS and TECHNICAL SPECIFICATIONS to be available to potential users* in accordance with the general principles of standardization.

The above language makes clear that the rationale behind the FRAND commitment—and the 'fair and reasonable' terms that are part of it—is twofold: (i) to ensure dissemination of the essential IPR contained in a standard, thereby allowing it to

---

[14] See for instance Case C-336/07 *Kabel Deutschland* [2008] ECR I-10889 para 46 (discussing the notion of 'reasonable' 'must carry' obligations that may be imposed by EU Member States upon cable operators on the basis of the Universal Service Directive).

remain *available* for adoption by members of the industry, whilst at the same time (ii) making certain that holders of those IPR are able to *reap adequate rewards* from their innovations.[15]

In this respect, the ETSI IPR Policy's reference to the terms 'fair and reasonable' is not accidental. Because the circumstances surrounding the negotiation of licensing agreements can widely differ,[16] only flexible terms such as 'fair and reasonable', the precise content of which is left to negotiation between the parties, can ensure the widest availability of the technology embodied in the standard in the widest possible variety of circumstances, without unduly diminishing the innovation incentives that patent law was designed to create.[17] Thus, as pointed out by the European Commission in its Communication on 'Intellectual property rights and standardization' that was issued while the ETSI IPR Policy was being negotiated, beyond the broad goal that essential technology be *available*,

> . . . it is not feasible or appropriate to be more specific as to what constitutes 'fairness' and 'reasonableness' since these are subjective factors determined by the circumstances surrounding the negotiation.[18]

---

[15] The latter goal lies at the heart of the patent system, and is as essential to the success of the standards enterprise as the former. As the United States Supreme Court observed in *Verizon Comm v Trinko* 540 US 398 (2004),

> The mere possession of monopoly power, and the concomitant charging of monopoly prices, is not only not unlawful; it is an important element of the free-market system. The opportunity to charge monopoly prices—at least for a short period—is what attracts 'business acumen' in the first place; it induces risk taking that produces innovation and economic growth. *Ibid* 407.

[16] Not all standard implementers seeking to obtain a licence from a given essential patent holder will be similarly situated. Generally, a range of variables will traditionally be negotiated between licensors and licensees, all of which may be of appreciable value, such as cross-licensing, volume of licensed products, exhaustion of patent rights, technology transfer, technical support, upfront fees, jurisdiction, scope of licence (eg, products, territory, have made rights, etc), possible product purchases, the formation of broader business relationships and cooperation, etc. Granting a licence cannot be confused with selling a product at a standard price (which would be the royalty). Because licensors and licensees seek to exchange a potentially diverse assortment of 'value' (the royalties being just one possible element of consideration), any interpretation of a FRAND commitment as 'dictating or specifying a particular licensing result' would prove a Procrustean bed.

[17] In this respect, FRAND is very much akin to a general clause, albeit a contractual one.

> General clauses or standards ('Generalklauseln', 'clauses générales') are legal rules which are not precisely formulated, terms and concepts which in fact do not even have a clear core. They are often applied in varying degrees in various legal systems to a rather wide range of contract cases when certain issues arise such as abuse of rights, unfairness, good faith, fairness of duty or loyalty or honesty, duty of care, and other such contract terms not lending themselves readily to clear or permanent definition.

S Grundmann and D Mazeaud (eds), *General Clauses and Standards In European Contract Law—Comparative Law, EC Law and Contract Law Codification*' (Kluwer Law International, 2006).

[18] Communication of the Commission 'Intellectual Property Rights and Standardization' (27 October 1992) COM (92) 445 final, para 4.3.3.

By contrast, the above extracts of the ETSI IPR Policy do not contain any language hinting at *any* of the very specific and restrictive limitations listed at the opening of this paper, which other authors attempt to read into 'fair and reasonable'.[19]

'Fair and reasonable' in the standards context prior to the ETSI IPR policy

As will be set out below, ETSI very explicitly picked up the 'FRAND' concept from the pre-existent 'RAND' policy of the International Standards Organisation (ISO), and similarly worded policies were at that time already in place at other internationally important SSOs. It is reasonable, then, to suppose that the understanding of technology industry companies as to what 'fair and reasonable' meant in this context was informed by the usage in those other SSOs.

One could review that context at length, but we will limit ourselves here to only a few illustrations. For instance, an ISO document circulated by the ISO/IEC Secretariat in 1999 stated that, even by that date, '. . . ISO has no guidelines as to what constitutes "reasonable" since each patent holder sets its own fee which is based upon commercial considerations at the time.'[20] Similarly, the patent policy of the International Telegraph and Telephone Consultative Committee (CCITT) (now known as ITU-T) in place in 1994 aimed to ensure that patentees '. . . would be willing to negotiate licences with other parties on a non-discriminatory basis on reasonable terms and conditions', but emphasized that the '. . . detailed arrangements arising from patents (licensing, royalties, etc) are being left to the parties concerned, as these arrangements might differ from case to case.'[21] Finally, in a 1992 letter to ETSI, the American National Standards Institute (ANSI) noted that '. . . under the ISO/IEC and ANSI policies licensors remain free to negotiate such license terms as they may deem appropriate so long as such licenses are fair and non-discriminatory'.[22]

---

[19] For instance, nothing can be read in such extracts as suggesting that FRAND imposes any specific and concrete obligations on the owner of standard essential patents with regard to the actual level of royalties or any other terms and conditions provided for in licensing agreements. Nokia's Vice President for Intellectual Property Rights, Dr Ilkka Rahnasto, makes a similar observation. He explains that '. . .the [FRAND] rule leaves the determination of exact terms for the parties to decide. This case-by-case determination allows parties to a particular licensing transaction to find their own interpretation of "fair and reasonable".' I. Rahnasto, *Intellectual Property Rights, External Effects and Anti-trust Law* (Oxford: OUP, 2003) 4.105. He further adds: 'In connection with standardization, the term "fair and reasonable" is usually understood as a reference to the economic reality. Generally, a licence is fair and reasonable if the terms would be acceptable in arm's-length-negotiations.' *Ibid* 6.34. 'Fair and reasonable' licensing terms would therefore consist of those terms determined through fair, bilateral negotiations between individual IPR owner and standard implementer in accordance with the market conditions prevailing at the time of such negotiations.

[20] 'Issues Relating to Patents—SC17's Patent Policy'(21 September 1999) ISO/IEC JTC1/SC17 N 1585.

[21] ETSI/GA15(93)18 (enclosure to letter dated 29 October 1992 from CCITT to Eurobit). (We use the ETSI nomenclature to identify ETSI documents. Thus, this document was 'Permanent Document' number 18, submitted at the 15th ETSI General Assembly meeting in 1993; 'Temporary Documents' are given a 'TD' designation.)

[22] ETSI/GA12(92)TD3 4.

Here again, what our initial research has *not* found is any indication, by the time ETSI adopted its current FRAND policy in 1994, that 'fair and reasonable' in the context of the ISO—or other SSOs—had ever been held by the ISO or by any court to imply *any* of the detailed restrictions recently hypothesized by various authors.

'Negotiation history' of the ETSI IPR policy

Key points in the negotiation and adoption of the ETSI IPR Policy itself may also shed light on what ETSI members understand that they are agreeing to when they make a FRAND undertaking.

ETSI as an organization was established in 1988, by the European Conference of Postal and Telecommunications Administrations ('CEPT'). From an early stage ETSI looked to the ISO IPR policy as the baseline. For example, a document submitted by the ETSI Technical Assembly Chairman in 1991 proposed that 'The licens[or] is required to grant licences on fair and reasonable non discriminatory terms as for the ISO policy.'[23] Similarly, the ETSI Director submitted the ETSI Annual Report to the 12th ETSI General Assembly in 1992, which stated that ETSI was '... developing a policy, based on that of the International Standards Organisation (ISO) and the International Electrotechnical Commission (IEC)'.[24]

However, in other respects ETSI's draft policy aimed at what the ETSI Technical Assembly Chairman believed would be an 'advance' over the ISO IPR policy.[25] This proposed package of heightened restrictions on IPR owners included what became referred to as an 'automatic licensing' or 'licensing by default' provision, a requirement of advance declaration of maximum royalty rates, a rule precluding required cross-licences, and a mandatory arbitration requirement.[26]

Commencing at the 12th ETSI General Assembly meeting in April 1992, fierce debate broke out over these proposed heightened restrictions.

At the March 1993 15th ETSI General Assembly, an IPR Policy and Undertaking including some of the novel provisions noted above was approved over heated opposition including threats by some participants to withdraw from ETSI.[27] However, following the approval, even louder opposition broke out. Several important IPR owners objected strongly to the 'automatic licensing' provision, and the Computer and Business Equipment Manufacturers' Association ('CBEMA') filed a complaint

[23] 12 TA TD 7 4 (attached to GA 11 TD 20).

[24] ETSI/GA 12 TD 15 6.

[25] 12 TA TD 7 3.

[26] ETSI/GA12(92)3.

[27] For example, IBM called the proposal 'a source of deep divisions within the ETSI membership' and stated that for 'many members it is the company's strategic assets and policies which are at stake'. ETSI/GA15(93)26. IBM said 'IBM has to evaluate now its future involvement in ETSI'. ETSI/GA15(93)6. Other companies said words to the same effect. ETSI/GA15(93)23.

with the European Commission asserting that novel aspects of the policy (including the requirement of advance disclosure of royalty rates) were anti-competitive. Important participants threatened to withdraw from ETSI if the policy was implemented;[28] so serious was the dissention among the membership that the ETSI Technical Assembly Chairman warned that '. . . other entities with simpler rules may have ambitions to take over ETSI work and ETSI could be out of business in five or ten years'.[29] On 22 July 1994, the ETSI General Assembly voted to '. . . abandon the IPR Undertaking as adopted by the General Assembly meeting during its meeting on 18 March 1993'.[30] The 1993 draft IPR Policy and Undertaking was never actually implemented by ETSI, and following the July 1994 vote ETSI was again without an IPR policy.

Finally, at the 21st ETSI General Assembly in November 1994, the ETSI membership approved an IPR policy from which the heightened restrictions described above had been removed.[31] The 1994 policy remains in effect today, with minor changes.

What this history documents is that not merely was FRAND a concept borrowed in its inception from prior use by the ISO, but that the ETSI membership did not pour *new* meaning into FRAND, as all attempts to do so were rejected. Thus, anyone who wishes to argue some restrictive or idiosyncratic meaning for an ETSI FRAND undertaking, whether based on economic argument or idiosyncrasies of EU or French law, should thus face a substantial burden of proof as a matter of contract interpretation.

Post-1994 ETSI comment on the FRAND undertaking

Post-adoption ETSI commentary and actions establish that the ETSI membership has consistently rejected subsequent efforts to alter the balance of interests between IPR owners and licencees by changing the meaning of FRAND.

In 2003, a number of ETSI members promoted an effort to make FRAND less flexible and discretionary by defining or giving examples of practices that would violate FRAND. The ETSI General Assembly authorized the creation of an Ad Hoc Group ('AHG') to consider and report on such proposals.[32] During this process, multiple participants in the ETSI AHG noted their understanding that the meaning of FRAND was a matter of global consensus, not an ETSI question. A representative of Microsoft observed that '... FRAND is a standard principle throughout all SDOs',

[28] EJ Iversen, 'Standardization and Intellectual Property Rights: ETSI's controversial search for new IPR-procedures' (September 1999) (Paper presented at the first IEEE conference on standardization and innovation in information technology, Aachen, Germany) ('ETSI received between 12–14 letters from parties. . . who threatened to pull out of ETSI if it implemented the 1993 Policy.').

[29] ETSI/GA20(94)22 Rev.1 4.

[30] ETSI/GA20(94)20; ETSI/GA20(94)22 Rev.1.

[31] ETSI/GA21(94)3; ETSI/GA21(94)39 Rev.2 17–18.

[32] ETSI/GA42(03)20.

while a representative of Motorola asserted that the '... FRAND term is identical in ITU policy, Japan SDO, US SDO... and this ["FRAND"] is the standard way to express it'.[33]

But even if FRAND had historically been a global concept, other AHG participants argued that ETSI should nevertheless endorse new specific restrictions under the FRAND umbrella for its own purposes. Proposals included prohibitions on licences that require a royalty-free cross-licence, prohibitions on requiring 'grantbacks' of rights to improvements, and prohibitions on licensing for certain regions of the globe at rates different from those charged for other regions. But none of these restrictions ever were agreed to, whether by the AHG or by the ETSI General Assembly. Instead, the AHG reported to the ETSI General Assembly that 'The ETSI IPR Policy does not define FRAND', and that 'The ad hoc group was unable to define FRAND conditions'.[34] Further, it reported that 'holders of big IPR portfolios' 'saw no sense in... attempts' 'to indirectly define FRAND conditions by giving several examples of bad practices'.[35] The AHG provided with its report an 'Annex A' that contained a list of supposed 'bad practices' that had been proposed by those members who advocated additional restrictions, while noting that these had *not* been agreed to by the AHG. The ETSI GA, while accepting the report itself, went farther and deleted this Annex A entirely.[36]

In 2006, another effort to tighten the permissive nature of 'fair and reasonable' was made within ETSI, with Nokia and two other manufacturers advocating that ETSI should '... make changes to the [ETSI] IPR regime and practices' by 'introduc[ing] the principles of AGGREGATED REASONABLE TERMS and PROPORTIONALITY into the FRAND definition'.[37] The proposal was once again intensely controversial within ETSI, and was not adopted by the General Assembly.[38]

Thus, any party contemplating making a FRAND commitment that looks to the ETSI record to understand what such a commitment would mean will find the ETSI membership *declining* to approve restrictions or interpretations identical or analogous to many of those advocated today by the proponents of the restrictive FRAND regimes.

[33] ETSI/GA/IPR02(03)05 3.

[34] ETSI/GA42(03)20 8.

[35] *Ibid* 9.

[36] ETSI/GA42(03)20 Rev.1; ETSI/GA42(03)34 4–5.

[37] ETSI/GA/IPRR01(06)08 2–3. For a discussion of this proposal and the negative impact it would have had, see D Geradin, 'Standardization and Technological Innovation: Some Reflections on Ex-ante Licensing, FRAND, and the Proper Means to Reward Innovators' (2006) 29 World Competition 511.

[38] ETSI GA/IPRR06(06)24 Rev.1 14.

Most recently, ETSI's 'Guide on IPRs', published in 2007, once again specifically disclaims any notion that ETSI does or intends to impose any more specific (and therefore more restrictive) definition of FRAND terms and conditions, stating instead that '. . . such commercial terms are a matter for discussion between the IPR holder and the potential licensee, outside of ETSI' (§2.2), and 'Specific licensing terms and negotiations are commercial issues between the companies and shall not be addressed within ETSI' (§4.1).

C. Enforcing FRAND Contractual Commitments

What is the role of the court?

To say that 'fair and reasonable' are contract terms that leave wide latitude is not to say that they are meaningless or unenforceable terms. In circumstances where an essential patent holder and a standard implementer[39] are unable to agree on licensing terms, the standard implementer, once accused of infringement, may simply wait and assert defensively that the IP owner has failed to satisfy its obligation to offer fair and reasonable terms, or possibly (depending on the procedures available in a given jurisdiction) could seek a determination through a breach of contract action that FRAND terms have not been offered, and an order requiring compliance with that obligation.[40] If the essential IP owner has in fact failed to offer fair and reasonable terms consistent with its FRAND commitment, then this will represent a breach of contract.

In such cases, however, a court radically misunderstands the FRAND commitment that the IP owner has made, and misunderstands the court's own role, if it seeks to answer the question 'What is the reasonable royalty for this IPR?' As discussed in Section B, the term 'fair and reasonable' was intended to convey substantial latitude and discretion, and excludes the possibility that there is just one 'right' answer.

In agreeing to license on FRAND terms, the IP owner has not agreed to constrain its licensing terms more tightly than the 'range of reasonableness'. Thus, if an offer has been made and refused, then the only contractual question to be adjudicated is whether the terms offered, taking into account all of the specific circumstances

[39] The intended beneficiaries of a FRAND declaration appear to be any parties who wish to perform actions identified in Paragraph 6 of the ETSI IPR Policy with respect to a standard-compliant product. This includes those who wish to '. . . manufacture, including the right to make or have made customized components and sub-systems to the licensee's own design for use in manufacture'. The ability of intended third party beneficiaries of a contract to enforce their rights under that contract is well recognized within the Common Law Tradition, while Civil Law jurisdictions provide comparable enforcement rights under (in the case of France, for example) the doctrine of 'stipulation pour autrui'. French Civil Code Art. 1121.

[40] Geradin and Rato 'Can Standard-Setting Lead to Exploitative Abuse?' (n 7 above) 119.

between the parties and prevailing market conditions, fall outside the *range* of reasonableness contemplated by the FRAND commitment.

This type of analysis is not foreign to courts. Under US patent law, for example, after a jury has awarded 'reasonable royalty' damages, the appeals court does *not* seek to second guess that decision and substitute its own view of what is 'most reasonable'. Rather, the appeals court engages in a deferential review, asking only whether the jury's award falls outside the range of what could be considered reasonable.[41] Similarly, European courts[42] use a 'going rate' or benchmarking method to identify a range of reasonable royalty rates that can serve as the basis for the calculation of damages after a finding of patent infringement, and the trial court enjoys significant judicial discretion in its appraisal. Where a decision awarding damages is appealed, the task of the appeals court is not to determine *ex novo* what the 'reasonable rate' and resulting damage award is, but only to examine whether the lower court exceeded its considerable discretion in awarding reasonable damages.[43]

In the case of FRAND licensing, the initial discretion as to what is 'reasonable' is entrusted to the negotiating parties or, in the absence of agreement, to the IP owner. If the would-be licencee 'appeals' to a court, that court's task is comparable to that of the appeals court in the US and European patent systems. And, as the party advancing the proposition that specific offered terms fall outside the range of reasonableness and thus do not satisfy the FRAND commitment, one would expect that the burden of proof would rest with the potential implementer.

In order to determine whether offered terms and conditions pass this 'range of reasonableness test', there is no reason that courts should not make use of analytical tools already existing in the law. For instance, while the question of what is 'reasonable'

[41] *Micro Chem, Inc v Lextron, Inc* 317 F3d 1387, 1394 (Fed Cir 2003) (a lower Federal court case) (discussing the differences between the experts' opinions regarding royalty rates and affirming jury's determination as reasonable); *Rite-Hite Corp v Kelley Co, Inc* 56 F3d 1538, 1554–5 (Fed Cir 1995) (noting range of possible royalties and affirming lower court's determination of royalty rate as reasonable); *Monsanto Co v Ralph* 382 F3d 1374, 1383 (Fed Cir 2004) (giving deference to jury's determination of royalty rate based on expert testimony regarding *Georgia-Pacific* factors).

[42] It must, however, be noted that in Europe, in contrast with the United States, injunctive relief is considered the primary remedy for patent infringement, over and above monetary compensation. Moreover, any damages awarded must only be compensatory in nature and may not have a punitive character. For these reasons, many cases are settled out of court after a finding of patent infringement and the existing case law on the calculation of damages is therefore very sparse.

[43] Cour de Cassation (Ch Comm) (France), *Sté Ets Delaplace et Sté Sicma c. Sté Van Der Lely*, 19 February 1991, [1991] Annales de la Propriété Industrielle, 4 (noting that the lower court had correctly exercised its judicial discretion in determining the royalty rate serving as the basis for damages after a finding of patent infringement); T Sampson, 'The "Adjusted Future Free Income Ratio": A New Methodology for Determining IPR Royalty Rates?' [2007] 371 European Intellectual Property Rev 377. See also the example given of a 'notional' royalty rate set by the UK Court of Appeal in a copyright case: *Nigel Christopher Blayney (t/a Aardvark Jewellery) v Clogeau St Davids Gold Mines* [2003] FSR 19.

continues to be a flexible inquiry, the much-cited *Georgia-Pacific*[44] case identifies 15 specific factors that US courts routinely consider, and the factors from the *Georgia-Pacific* list have been invoked as useful in other jurisdictions. Interestingly, in one discussion paper created by the ETSI General Assembly Ad Hoc Group in 1993, the reporters (themselves representatives of RIM, not a US corporation) wrote that 'If one were to read the important "*Georgia-Pacific*" case cited in United States law as a method to determine a "reasonable royalty", it can readily be seen to be a test that closely parallels the concept of "fair, reasonable, and non-discriminatory" license obligations.'[45]

Of course other important jurisdictions use different language, but we believe that they fundamentally agree that, when a court must determine a royalty rate, it may and should consider the wide range of information that would be relevant to a business decision-maker confronting the same question.[46] Similarly, we believe that non-US jurisdictions can also find within their own structures examples of the type of deferential review that is appropriate where a court is tasked not to decide what the 'right' answer is, but to decide whether terms offered fall entirely outside the *range* of possibility contemplated by the word 'reasonable'.[47]

Not all of the *Georgia-Pacific* factors will necessarily be relevant to the question of whether proffered licence terms are within the range of reasonableness, and peculiarities of a particular industry or standardized industries in general may properly enter into the equation. Nevertheless, a court may well find that the *Georgia-Pacific* list provides a useful framework or starting point for the inquiry.[48] Notably, royalties received under prior and existing licences for the very patents being litigated often represent the most influential factor in determining 'reasonableness' under the *Georgia-Pacific* framework, and should arguably have the same role in the context of FRAND litigation.

---

44 *Georgia-Pacific Corp v US Plywood Corp.* 318 F Supp 1116 (SDNY 1970) (a lower Federal court).

45 ETSI GA/IPR02(03)05 1.

46 *General Tire & Rubber Co v Firestone Tyre & Rubber Co* [1975] FSR 273 (HL):

> . . . [E]vidence may consist of the practice, as regards royalty, in the relevant trade or in analogous trades; perhaps of expert opinion expressed in publications or in the witness box; possibly of the profitability of the invention; and any other factor on which the judge can decide the measure of loss.

Cofrinex v Helary, Paris Court of Appeal 12 July 1977; Sec. 139 Para. 2 German Patent Act (royalty rate of a hypothetical licence agreement must be determined in the light of all relevant circumstances).

47 *Flint v Lovell* [1935] 1 KB 354 (CA) (Greer, LJ) (explaining that an award of damages is reversible only if 'the amount awarded [is] an entirely erroneous estimate of the damage to which the plaintiff is entitled').

48 At least one US court has adopted the *Georgia-Pacific* factors to assess the reasonableness of a licensing offer challenged on FRAND grounds: *ESS Tech, Inc v PC-Tel, Inc*, No. C-99-20292, 2001 WL 1891713, 3–6 (ND Cal 28 November 2001) (a lower Federal court).

FRAND commitments and subsequent licences

If a would-be licensee refuses offered terms and objects that those terms do not satisfy the patent owner's agreement to offer FRAND terms and conditions, then the court must undertake the analysis discussed above. However, *after* the parties have negotiated and executed a licence agreement, a complaint by the licensee that the terms of that licence are not FRAND presents very different issues.

While the doctrinal description will differ in different jurisdictions, the point is not complicated—it cannot be proper for a party, aware of rights it is entitled to, to claim under an existing contract (here, the FRAND commitment), to negotiate and sign a licence, enjoy the benefit of that licence for as long as it pleases, and then collaterally attack the licence as unenforceable (and perhaps claim past damages) on the theory that the licence terms violated the preceding contractual commitment. Within the common law tradition, this is a result of the doctrine of integration,[49] or alternatively of the rule that, even in the absence of complete integration, a collateral contract may not be used to contradict the terms of a subsequent agreement.[50]

An extremely important economic truth underlies this principle. It is widely understood that uncertainty itself imposes an economic cost; accordingly, businesses often use the 'stabilizing force of contracts'[51] to reduce or eliminate unpredictability. For this reason, companies commonly negotiate long-term licence agreements at fixed royalty rates, giving the two parties predictability as to revenues and costs, respectively. As the US Supreme Court has explained, 'Markets are not perfect, and one of the reasons that parties enter into... contracts is precisely to hedge against the volatility that market imperfections produce.'[52] Private parties are of course free to negotiate short-term licence agreements, or agreements under which the royalty rate is subject to frequent re-negotiation, or periodic modification based on some external criteria. But they don't do this, precisely because predictability is extremely important to many aspects of the conduct of a business, including, eg, decisions about investments in research and development. As a result, uncertainty relating to '... contract sanctity can have a chilling effect on investments and a seller's willingness to enter

[49] *Restatement (Second) of Contracts* (1981) § 213:

> (1) A binding integrated agreement discharges prior agreements to the extent that it is inconsistent with them. (2) A binding completely integrated agreement discharges prior agreements to the extent that they are within its scope. (3) An integrated agreement that is not binding or that is voidable and avoided does not discharge a prior agreement. But an integrated agreement, even though not binding, may be effective to render inoperative a term which would have been part of the agreement if it had not been integrated.

[50] RA Lord, 11 *Williston on Contracts* (4th edn, Minnesota: West Group, 1999) § 33:26 ('[E]vidence of a collateral agreement is not barred by the parol evidence rule if such evidence does not contradict the written contract.').

[51] *NRG Power Marketing v Main Public Utilities* 558 US __, 130 S Ct 693, 696 (2010).

[52] *Morgan Stanley Capital Group, Inc v Public Utility District 1 of Snohomush County* 128 S Ct 2733, 2746 (2008).

into long-term contracts and this, in turn, can harm customers in the long run.'[53] Yet, a rule that would permit a licensee to collaterally attack a licence agreement—potentially years after the fact—on the theory that its terms violate a prior FRAND commitment, would make it *impossible* for licensing parties to negotiate for long-term predictability.

What is 'fair and reasonable' will be higher after adjudication of infringement and validity

US courts and commentators routinely recognize that a 'reasonable royalty' will be higher after a patent has been held valid and infringed in court than it was before that adjudication.[54] Providing empirical and theoretical support for this judicial view, Lemley and Shapiro have demonstrated that nearly half of patents litigated to a final determination in the US are held invalid, while a significant number of those held valid are held to be not infringed.[55] They report in a later paper that average 'reasonable royalty' damage awards set rates more than *double* estimated average negotiated patent royalties, and conclude that this difference is at least in part attributable to the uncertainty surrounding the strength and value of untested patents.[56]

Shapiro points out that, in light of these facts, what is 'fair and reasonable' in the context of an offer to license patents that have not been tested in litigation should be something lower than would be awarded after adjudication of infringement and validity, because of the uncertain strength of the patents.[57] But the reverse is equally true—after a patent has been tested and the uncertainty eliminated, then what is 'fair and reasonable' no longer needs to include any 'uncertainty discount', and should be substantially *higher* than would have been the case pre-litigation.

---

[53] *Ibid* 2749 (quoting *Market Based Rates*, para 6, 72 Fed Reg 33906-33907).

[54] *Maxwell v J Baker, Inc* 86 F3d 1098, 1109–10 (Fed Cir 1996) (a lower Federal court),

> [T]hat an infringer had to be ordered by a court to pay damages, rather than agreeing to a reasonable royalty, is also relevant' to 'an amount sufficient to adequately compensate the patentee for the infringement';

Stickle v Heublein, Inc 716 F2d 1550, 1563 (Fed Cir 1983); *Endress & Hauser, Inc v Hawk Measurement Sys Pty Ltd* 892 F Supp 1123, 1130 (SD Ind 1995),

> Although courts employ the 'willing licensor/willing licensee' model as the basis for determining a reasonable royalty, they do so with the understanding that a 'reasonable' royalty after infringement is likely to be higher than that arrived at between truly willing patent owners and licensees.

VE O'Brien, 'Economics & Key Patent Damages Cases' (2000) 9 University of Baltimore Intellectual Property LJ 1, 19 and 20, fn 70 (observing that 'the hypothetical negotiation already has a built-in bias toward a royalty rate that is higher than those observed in practice' and that the Federal Circuit often sustains awards 'based on a royalty several times that observed in the real world').

[55] MA Lemley and C Shapiro, 'Probabilistic Patents' [2005] J of Economic Perspectives 75, 80; Lemley and Shapiro (n 1 above) 2019–20.

[56] Lemley and Shapiro (n 1 above) 2032–33.

[57] J Farrell et al (n 6 above) 637 fn 134 (citing and discussing Shapiro).

This 'that was then, this is now' aspect of FRAND is not only theoretically correct, it stands as a critically important deterrent to excessive litigation. Lemley and Shapiro have also noted that, in the ordinary licensing context, the risk of injunction and complete exclusion from the market motivates prospective infringers to obtain a licence instead of litigating.[58] However, if an infringer of essential patents is entitled to the *same* terms after unsuccessful litigation as he was entitled to before, then this incentive disappears; the infringer will have strong incentives to litigate even a weak case in the hopes of 'getting lucky' with an invalidity or non-infringement ruling, and will face no downside risk beyond attorneys' fees. The Former Chief Judge of the Federal Circuit has noted exactly this incentive problem in the context of *Georgia-Pacific* royalty determinations, explaining that an infringer who, after unsuccessful litigation, 'could count on paying only the normal, routine royalty non-infringers might have paid. . . would be in a "heads-I-win, tails-you-lose" position'.[59] Thus, a static definition of 'fair and reasonable' unaffected by litigation would expose FRAND declarants to a much greater risk of non-meritorious litigation than that faced by parties unconstrained by FRAND. It is unlikely that any standards setting organization intended, by requiring FRAND declarations, to create this perverse incentive to attack rather than to pay for the intellectual property of its members.

D. FRAND Agreement Containing No Agreement Not to Seek Injunction

A number of authors have argued that by making a FRAND commitment, an essential patent holder waives its right to seek injunctive relief in case of infringement (the 'waiver theory').[60] One of the present authors has shown elsewhere that there is no support for such assertions in either statute nor case law.[61] But we would emphasize here the contractual starting point—the idea that there has been a waiver of all injunction rights can find no basis in the language of the IPR policies of ETSI and other major SSOs. In the case of ETSI, at least, there is evidence of the intent of the parties, and it points just the other way. This is so because injunction waiver language was in at least one version of the 'aggressive' ETSI policy debated and (temporarily)

---

[58] Lemley and Shapiro 'Probabilistic Patents' (n 55 above) 75, 80.

[59] *Panduit Corp v Stahlin Bros Fibre Works* 575 F2d 1152, 1158 (6th Cir 1978) (a lower Federal court) (Markey, J) (quoting *Troxel Mfg Co v Schwinn Bicycle Co* 465 F2d 1253, 1257 (6th Cir 1972)).

[60] M Dolmans, 'Standards for Standards' (2002) 26 Fordham Intl LJ 163; Miller (n 8 above) 351.

[61] Geradin and Rato 'Can Standard-Setting Lead to Exploitative Abuse?' (n 7 above) 116 et seq. Lemley explicitly admits that he is 'aware of no cases treating this issue', stating that it is his 'policy preference' that an IPR owner's commitment to an SSO be construed as itself implying the grant of a licence, with the result that the IP owner is precluded form seeking an injunction for patent infringement. Lemley 'Intellectual Property Rights and Standard-Setting Organizations' (n 8 above) 1926.

adopted in 1993,[62] but was entirely removed from the replacement policy adopted in November of 1994.

Given the base-line fact that a patent owner's right to injunctive relief against infringement is recognized under EU law,[63] international trade law,[64] and US federal law,[65] clear evidence should be required to establish that a patentee by making a FRAND undertaking has intentionally and voluntarily waived that right, as is required in both civil[66] and common law legal systems.[67] Given the history of ETSI's policy, it does not appear that this showing could possibly be made in regard to undertakings made to ETSI.

This is not to say that a FRAND undertaking has no impact on a party's ability to obtain an injunction; it does. If a patentee has failed to comply with its undertaking to offer a licence on FRAND terms, then no injunction should be available. But conversely, if it is determined that the patentee did comply with its obligation to offer FRAND terms, and the infringer continues to practice without taking a licence, then we can find no contract-based argument that the patentee has waived its right to enjoin that ongoing infringement.

E. The Other Half of FRAND

We have focused in this paper on the 'fair and reasonable' component of FRAND, because the meaning of 'fair and reasonable' has attracted far more controversy than the meaning of 'non-discriminatory', and because space does not permit a thorough examination of the contractual meaning of the entirety of FRAND in a single article. Yet 'ND' remains something of a cipher. No serious commentator suggests that it means that all licensees must be given 'the same' terms,[68] but nor is there any consensus as to what it *does* mean—why it is included in the policy at all. Does it mean anything more, less or different than meeting the requirements of various competition laws, such as the Robinson Patman Act in the United States or Article 102(c) of the Treaty on the Functioning of the European Union?

---

[62] ETSI/GA15 TD 25, sec 13 ('The Signatory hereby undertakes not to seek an injunction against a Party in respect of any Essential IPR in respect of [enumerated situations]').

[63] Directive 2004/48/EC of the European Parliament and of the Council of 29 April 2004 on the enforcement of intellectual property rights [2004] OJ L157, Article 9.

[64] Agreement on Trade-related Aspects of Intellectual Property Rights (TRIPS agreement), Article 41(1), the Annex 1C to the Marrakech Agreement Establishing the World Trade Organisation (signed in Marrakech, Morocco on 15 April 1994).

[65] Title 35 of the US Code Annotated, Chapter 29, Section 283 (US Patent Act).

[66] Belgian Supreme Court, 19 September 1997, *Arr Cass* 1997, 840 and French Supreme Court, 10 May 2000, Case No 97-13907.

[67] *Schoon v Troy Corp*, No 1677-N, 2006 Del Ch LEXIS 123, *7 (Del Ch 27 June 2006) (a state case).

[68] Given that many licensees will be differently situated (for the reasons discussed above in n 16), such an approach would in fact lead to discrimination.

We urge that a 'contract interpretation' investigation of the meaning of 'ND' in SSO IPR policies is much needed. And interestingly, while it is 'FR' that gets all the attention today, there was actually more discussion of the meaning of 'ND' in the course of the ETSI debates that resulted in the adoption of the ETSI IPR Policy. While a systematic study remains to be done, we remark that much of that discussion addressed fears of 'protectionist' discrimination based on the nationality of companies, or based on membership in the EU.[69] These fears have not materialized as real-world issues, and it may be that the 'meaning of ND' has become somewhat mysterious precisely because the requirement has been so successful that the fears it primarily targeted have become forgotten, and thus invisible. But, with regard to ND, a careful 'intent of the parties' analysis remains to be done.

Conclusion

The effort to conflate a contractual FRAND commitment with either idealized economic theory or the competition law of any jurisdiction is ill-conceived. In short, a FRAND commitment and the limitations that competition law may impose on intellectual property rights are simply two separate things, and intellectual clarity requires that each be considered in its own right, and according to the analytical methods appropriate to it.

Our research shows that, if a FRAND commitment is taken seriously as a contract—as it should be—then efforts to look to FRAND as a source of cumulative royalty caps, particular formulas for calculating or apportioning royalties, or limitations on remedies against unlicensed infringers are not only without basis, but are contradicted by the ordinary methods of contract interpretation.

---

[69] For example, at the July 1992 ETSI GA meeting, the Technical Assembly Chairman explained certain features of the Technical Barriers to Trade Agreement,

> There are two relevant articles. The first is Article 2. This is the one that gives a legal obligation on the European Community. It requires that an entity importing into the European Community is given no less favourable treatment than an entity established in the Community. However, Article 2 covers technical regulations and the obligations fall on Governments and the Community. It does not apply to voluntary standards or impose obligations on private standards bodies or individuals. However Article 4 obliges Governments and the Community to make best endeavours so that private standards making bodies also meet the requirements of Article 2.

ETSI/IPR/GA(92) 4. Earlier, ETSI had approved an agreement with the Standards Institution of Israel agreeing that the,

> application of international standards free of discrimination is essential for the development of global telecommunications and removal of barriers to trade. Both agree to pursue Intellectual Property Right policies in respect of telecommunications standards that support this objective.

ETSI/GA 11(91) 8; ETSI/GA 11(91) 25 Rev.1.

Part V

LICENSING

18

HARDCORE RESTRICTIONS IN TECHNOLOGY TRANSFER AGREEMENTS UNDER REGULATION (EC) NO 772/2004

Stefan Enchelmaier

Introduction

The present Block Exemption Regulation for technology transfer agreements[1] has not, as far as one can see, been the subject of any judgment by the European Court of Justice.[2] In this, the Regulation is no different from its predecessor, Regulation 240/96,[3] and in fact from most other block exemption Regulations adopted over the decades. Nevertheless, the absence of authoritative guidance from the Court is particularly regrettable in the present context. Regulation 772/2004 operates in the new administrative environment created by Regulation 1/2003. This Regulation has, among others, abolished the possibility for undertakings to notify agreements to the Commission. Self-assessment was already a choice within the context of the old notification regime established by Regulation 17 of 1962—undertakings were never under a duty to notify, only under a duty not to infringe Article 85 EEC (later Article 81 EC, now Article 101 TFEU).[4] Under the present regime, however, those wishing to enter into a technology transfer agreement must work out for themselves whether

---

[1] Commission Regulation (EC) No 772/2004 of 27 April 2004 on the application of Article 81(3) of the Treaty [now Article 101(3) TFEU] to categories of technology transfer agreements [2004] OJ L123/11. Together with the Regulation, the Commission published its Guidelines on the application of Article 81 of the EC Treaty to technology transfer agreements [2004] OJ C101/2 (hereinafter 'Guidelines').

[2] At least there is no result if '772/2004' is entered in the search form on the Court's website: <http://curia.europa.eu/jurisp/cgi-bin/form.pl?lang=en> accessed 16 September 2010.

[3] [1996] OJ L31/2.

[4] V Korah, *An introductory guide to EC competition law and practice* (7th edn) (Oxford: Hart Publishers, 2000) 209 para 6.4 (not repeated in later editions).

their agreement comes within the terms of Regulation 772/2004.[5] The Commission's Guidelines go some way towards enabling parties to do so, but they are short of detail particularly when it comes to a central question for the drafters of technology transfer agreements, namely which clauses must under no circumstances be contained in an agreement, so-called 'hardcore restrictions'.

This chapter will analyse the provisions of Article 4(1) and (2) of Regulation 772/2004 one by one, explore some preliminary questions necessary for their understanding, and place the results in the wider context of the relationship between competition law and the law governing the European Internal Market.

A. Hardcore Restrictions: Article 4, Regulation 772/2004

The scope and meaning of Article 4 do not disclose themselves easily to the reader. The bulk of the text, that is, paragraphs 1(c) and 2(b), is formulated in a sort of double negative—there is a general prohibition of market and customer allocation, but this is followed by a whole host of exceptions. The same is true, if to a lesser extent, of paragraph 1(b) and (d), and of paragraph 2(a) and (c); only the prohibition in paragraph 1(a) is in no way qualified. Quite what is prohibited as a hardcore restriction is revealed, therefore, only once the exceptions are subtracted from the general prohibitions.

Another problem is the relative dearth of explanations given by the Commission. There is none in the recitals, although readers familiar with Community law would not expect to find in the recitals more than very broad policy statements and summaries of the operational provisions of the Regulation. More unfortunately, there is very little in the Commission's Guidelines, too, and much of the literature on the subject is confined to paraphrasing the meagre commentary in the Guidelines.[6]

[5] Findings of inapplicability and informal guidance letters can be discounted for present purposes, compare Commission Notice on informal guidance relating to novel questions concerning Articles 81 and 82 of the EC Treaty [now Articles 101 and 102 TFEU] that arise in individual cases (guidance letters) [2004] OJ C101/78.

[6] Without any claim to completeness, the following may be mentioned: S Anderman, 'The new EC competition law framework for technology transfer and IP licensing' in J Drexl (ed), *Research Handbook on Intellectual Property and Competition Law* (Cheltenham: Edward Elgar Publishing, 2008), 107–38; S Anderman and J Kallaugher, *Technology Transfer and the New EC Competition Rules: Intellectual Property Licensing after Modernisation* (Oxford: OUP, 2006); B Backhaus, 'Kartellrechtliche Grenzen der Vertragsgestaltung im Rahmen einer Zusammenarbeit von Unternehmen bei der Entwicklung von Arzneimitteln' [2005] Gewerblicher Rechtsschutz und Urheberrecht: Internationaler Teil 359–69; B Bird and A Totougni, 'The new EC technology transfer Regulation: two years on' (2006) 28 Eur Intellectual Property Rev 292; J-M Coume, 'IP rights and EU competition law: can your IP licensing agreement benefit from safe harbour?' [2007] ECLR 23; F Dethmers and P Posthuma de Boer, 'Ten years on: vertical agreements under Article 81' [2009] ECLR 424–39; M Dolmans and A Piilola, 'The new technology transfer block exemption—a welcome reform, after all' (2004) 27 World Competition 351; M Feil, 'The new block exemption Regulation on technology transfer agreements in light of the US antitrust regime on the licensing of intellectual property' [2005] Intl Rev Intellectual Property and

In the following we shall, therefore, seek to define the terms first of the prohibitions, then of the exceptions, and finally ask what remains impermissible once the exceptions have given contour to the rules. When this is done, the question can be asked whether the prohibitions amount to a coherent policy, and how they fit into the scheme of the internal market, the term used in Article 101 TFEU (formerly the 'common' market in Article 81 EC/85 EEC) in accordance with Article 26(2) TFEU (ex Article 14(2) EC).

Finally, Regulation 772/2004 implements Article 101(3) TFEU,[7] but only to the extent stipulated in the Regulation. If any of the conditions for the block exemption to apply are not fulfilled, the situation falls to be assessed under Article 101(3). The Commission's yardstick for this assessment is surprisingly simple:[8] Article 101 TFEU is unlikely to be infringed where there are four or more independently controlled technologies, in addition to the technologies controlled by the parties to the agreement, that may be substitutable for the licensed technology at a comparable cost to the user. This applies, however, only outside the area of hardcore restrictions.[9] This is another reason why it is important for undertakings to know what is and what is not permissible under Article 4.

Competition L 31; F Fine, 'The EU's new antitrust rules for technology licensing: a turbulent harbour for licensors' (2004) 29 ELR 766; F Fine, *The EC competition law on technology licensing* (London: Sweet & Maxwell, 2006); D Garrod and A Rosen, 'The new safe harbour for technology transfer agreements' [2004] Competition L J 31; M Hansen and O Shah, 'The new EU technology transfer regime—out of the straightjacket into the safe harbour?' [2004] ECLR 465; F-E Hufnagel, 'Die neue Gruppenfreistellungsverordnung Technologietransfer—kein Lizenzvertrag ohne Kartellrecht?' [2004] *Mitteilungen der deutschen Patentanwälte* 297; P-M Louis, 'Le nouveau règlement d'exemption par catégorie des accords de transfer de technologie: une modernisation et une simplification' [2004] Cahiers de Droit Européen 377; M Lubitz, 'Die neue Technologietransfer-Gruppenfreistellungsverordnung' [2004] Europäische Zeitschrift für Wirtschaftsrecht 652; L Peeperkorn, 'IP licences and competition rules: striking the right balance' (2003) 26 World Competition 527; C Ritter, 'The new technology transfer block exemption under EC competition law' (2004) 31 Legal Issues of Economic Integration 161; V Schumacher and C Schmid, 'Die neue Gruppenfreistellungsverordnung für Technologietransfer-Vereinbarungen' [2006] Gewerblicher Rechtsschutz und Urheberrecht 1; S Topping, 'The new EU technology transfer block exemption and guidelines' [2004] World Intellectual Property Rev 26; P Treacy and T Heide, 'The new EC technology transfer block exemption Regulation' [2004] Eur Intellectual Property Rev 414; M Turner and D Callaghan, 'Software licensing under the competition law spotlight—the new technology transfer block exemption' (2004) 20 Computer L and Security Rep 273; H Ullrich, 'Technologietransfer und Kartellrecht: Der Bedeutungsverlust der EG-Gruppenfreistellungsverordnung' (2007) *VPP-Rundbrief* 51.

[7] See the opening words of Article 2 Reg 772/2004: 'Pursuant to Article 81(3)... it is hereby declared...'

[8] Ullrich (n 6 above) 59 (text to fn 84); Schumacher and Schmid (n 6 above) 10 sub 3: 'rule of thumb'; Ritter (n 6 above) 175 (text to fn 62).

[9] Guidelines (n 1 above) para 131.

Competing and non-competing undertakings, Article 4(1) and (2)

The distinction between competing and non-competing undertakings is the first to be considered for a proper understanding of Article 4. The first paragraph of that provision applies to competing undertakings, the second to non-competing ones.

The licensing agreements encompassed by the Regulation enable the licensee to make 'contract products' with the technology that is the subject-matter of the agreement, Article 1(1)(f) Regulation 772/2004; these products can be either goods or services, Article 1(1)(e). The technology will, therefore, be an intangible input to these products. Two markets will hence be affected by the agreement: the market of the technology involved, and of the product(s) made with it. The competitive relationship between the parties to the agreement will for that reason have to be gauged with regard to both markets. That assessment under Article 1(1)(j)(i) and (ii) in principle follows the traditional method of defining the relevant market. Within the framework of the Regulation (not under Article 101(3) TFEU, though), the assessment has a number of peculiar features. The Regulation must be easily applicable by undertakings, or else it would offer too little legal certainty. Likewise, the Regulation must take a generalizing perspective; too much differentiation would, again, not be conducive to legal certainty. This concern with '. . . adequate legal security for undertakings' is specifically highlighted in Recital 4 of the Regulation.

For these reasons, the market share thresholds in Article 3(1) and (2) refer to the presence of the licensed technology on the product market(s) affected by the agreement, even if the companies are competitors on the technology market only. These market shares are easier to ascertain than those for the relevant technology, which is once removed from the product markets. Connected with this is the second peculiarity of the assessment of the parties' competitive relationship. For the purposes of the Regulation (again, not for a full assessment under Article 101 TFEU), only actual competition on the technology markets counts, whereas on the product markets, actual or potential competition is taken into consideration as usual. The potential of an undertaking to enter a technology market is even further removed from the readily measured product markets; it is also difficult for one party to an agreement to assess how far the other has progressed with its efforts to develop its own technology.[10]

From this follows a third peculiarity. The parties to an agreement may be non-competitors because one undertaking produces goods or services with a technology for which it will not or must not (as a mere licensee) grant licences: it is, therefore, not a competitor on the technology market(s). The other party, by contrast, will not produce with its technology, but instead recoups its development costs by granting licences: as a consequence, it is not a competitor on the product market(s). In this

[10] Fine (n 6 above) 777.

situation, the potential competition on the technology market of the first party, in case it is the proprietor of technology it is using, does not enter the assessment. To allow such parties the same restrictions as genuine non-competitors might, however, have unwelcome consequences on the markets affected, as the parties have an incentive to agree the most restrictive allowable terms.[11] This is reinforced by the provision of Article 4(3) which retains for the parties the more generous rules of Article 4(2) if they become competitors *after* the agreement is entered into (not *because*—the perspective is *ex ante*)[12] and as long as it is not materially altered. In this situation, the onset of competition between the parties can depend on the mere whim of one of them.

It can only be mentioned in passing here that the Commission considers as non-competitors also undertakings in a unilateral or mutual 'blocking position' with regard to the other's technology; this follows from the requirement in Article 1(1)(j)(i) that the parties '. . . license out competing technologies *without infringing each others' intellectual property rights*' (emphasis added). Nevertheless, the Commission views parties' submissions to that effect with a certain suspicion. It requires 'objective evidence' that, in particular, the technologies are not in reality substitutes for each other.[13] This suspicion is only understandable: the most natural conclusion from two undertakings working in closely related technological fields, even working on the same solutions to a given problem, is that they are indeed competitors. A blocking position, if it really exists, would in other words be a symptom of competition rather than a hindrance to it.[14]

Reciprocal and non-reciprocal agreements

A second distinction that occurs in Article 4 is that between reciprocal and non-reciprocal agreements. A reciprocal agreement requires both mutual licences and competing technologies, Article 1(1)(c). A non-reciprocal agreement, by contrast, has only one party granting the other a licence, or mutual licences for technologies that do not compete, Article 1(1)(d).[15] From this it follows that the distinction is relevant only if the undertakings are competitors; as a consequence, it does not occur in Article 4(2): the licences there in issue are, by definition, all non-reciprocal.

Article 4(1) treats non-reciprocal agreements more leniently than reciprocal ones—in a non-reciprocal agreement, the licensee's own technology remains unfettered; with it, the licensee is free to compete in any way he likes with the licensor. As far as products, quantities, territories, customer groups, and any other parameter of

[11] Ullrich (n 6 above) 61 sub (ii).
[12] Fine (n 6 above) 773 (text to fn 44); Garrod and Rosen (n 6 above) 36, last para.
[13] Guidelines (n 1 above) para 32.
[14] Similarly, Ullrich (n 6 above) 62 sub (iii).
[15] See, for a discussion of problems with that distinction in practice, Bird and Totougni (n 6 above) 295 sub (5).

competition are concerned, any demand that cannot be satisfied with the licensed technology because of restrictions in the agreement can turn to products made with the licensee's own technology. The licensee is an additional source of goods or services made with the licensed technology; otherwise, competition can go on as before.

Nevertheless, there is a risk that the licensee's 'in-house' competition with his proprietary technology against his licensed technology will not be as vigorous as if he had only products made with his own technology to offer on the market. The contractual ties and the mutual advantage derived from the agreement as its *sine qua non* will create a certain reticence to compete aggressively, regardless of whether the licensing conditions as such give incentives to this effect. This is comparable to the 'spill-over' effect that both cooperative and concentrative joint ventures may have and which is assessed under Article 101(3) TFEU, either directly or by virtue of Article 2(4), (5) Regulation 139/2004.

In a reciprocal agreement this effect is compounded—it gives each party to the agreement a say over how its competitor (the licensee) uses the licensed technology, and imposes the converse restrictions on the licensor. Both parties, in the use of both technologies, thus have an incentive to take the other side's interest into consideration when devising their own business strategy.

Price restrictions, Article 4(1)(a), Article 4(2)(a)

Article 4(1)(a) prohibits the restriction, in an agreement between competitors, of a party's ability to determine its prices when selling products to third parties. For non-competing undertakings, paragraph 2(a) allows maximum sale prices and price recommendations as long as the parties are genuinely free not to adhere to the recommendation. This prohibition is as wide as its model in Article 101(1)(a) TFEU. It encompasses restrictions on either or both parties, and it is not limited to 'contract products' as defined above. It follows from Recital 19 and the definition of 'technology transfer agreement' in Article 1(1)(b), that the wording is no editing mistake.[16] The reason for the inclusion of non-contract products is not obvious, in particular as Recital 19 makes any stipulations for more than one level of trade (such as the sale of products—contract products or others) subject to the provisions of the competition rules applicable to supply and distribution agreements.[17]

Another glance at Article 101 shows that the proviso in Article 4(2)(a) translates the 'indirect' fixing of prices that is prohibited by the Treaty article as pressure or incentives to observe price recommendations as if they were fixed or minimum prices;

[16] See also Guidelines (n 1 above) para 79: '. . . fixing of prices for products sold to third parties, *including* the products incorporating the licensed technology' (emphasis added).

[17] Fuchs reaches the same conclusion in Immenga andMestmäcker (eds), *Wettbewerbsrecht, EG/Teil 1* (4th ed) (2007), TT-VO para 222 (p 1753).

the Court has qualified such arrangements as concerted practices.[18] Under Article 4(1)(a), the Commission views a calculation of royalties on the basis of individual product sales (so-called 'running royalties') in certain circumstances as indirect price fixing.[19] This is unconvincing—the increase in the licence costs for a larger production run may prompt the licensee to produce less, but it does not make its prices more foreseeable or fix them at a particular amount. All that can be said is that each individual product will be more expensive by the amount *pro rata* of the licence fee.[20] Maximum prices and genuine price recommendations allow consumers to compare prices within and between brands more easily. Such prices also facilitate uniform sales promotions by the various distributors of one company.

The more generous standard in Article 4(2)(a) cannot, however, be transposed to Article 4(1)(a). If the undertakings offer competing products, there is no reason to allow them to coordinate their prices in any way, either directly, or indirectly through recommendations. If the undertakings are competitors only on the technology markets, that is, if their products do not compete, there is no economic or technological advantage (Article 101(3) TFEU) to be had from making any agreements about prices anyway.

Limitations of output, Article 4(1)(b)

The general prohibition against limitations of output again refers to any products, whether or not they were made with the licensed technology. Again (see the preceding point) it is not quite clear why non-contract products should specifically be mentioned. If the restriction of output of non-contract products is strictly necessary for the implementation of the technology transfer agreement, it will, as an 'ancillary restraint',[21] share in the exemption of the agreement. If it is not, it will not benefit from the block exemption anyway, without the need to spell this out explicitly. Even as a clarification in this respect Article 4(1)(b) is, therefore, of limited value only.

Be that as it may, a 'direct' limitation can take the shape of (absolute) quota or (relative) market share clauses. An 'indirect' limitation would be present, for instance, where intermediate goods were limited. The passage in Article 1(1)(b) about the 'sale

[18] See, eg, Case T-213/00 *CMA CGM v Commission* [2003] ECR II-913 para 175.

[19] Guidelines (n 1 above) para 80; equally critical, but with the contention that the Commission is too generous, are Bird and Totougni (n 6 above) 294 sub (4); for a less critical view, see Backhaus (n 6 above) 366 sub 4.

[20] Also critical of the Commission's view is V Korah, 'Legislative Comment: Draft block exemption for technology transfer' (2004) 25 ECLR 247, 255 f,

> . . . [Running royalties] serve many useful functions. They may settle opposing views as to the value of a licence at an early stage when it is not known how valuable it will become; they are a useful way of allocating risk; they also enable the licensee to pay from its earnings on the sale of the protected product.

Fuchs (above n 17) para 224, is of the same opinion.

[21] Thus the ECJ's constant jurisprudence since Case 42/84 *Remia and Nutricia v Commission* [1985] ECR 2545 paras 19, 20.

and purchase of products' would seem to allow such limitations. Another example would be the 'running royalties' discussed above, or royalties (running or not) that rise once certain production thresholds are crossed. This might make further production too expensive for the market to absorb.

The exceptions from this general prohibition are limitations on the output of contract products imposed on the licensee in a non-reciprocal agreement, and on only one party in a reciprocal agreement. If the agreement is non-reciprocal and the parties are actual competitors on the product markets, the licensee can still use the other technology at his disposal (his own technology, or one licensed to him by a third party) to cover any unsatisfied demand for products made with the licensed technology. Consumers will, by definition in Article 1(1)(j)(ii), consider these products as interchangeable. The limitation of output is therefore acceptable. The same is true if the non-reciprocal nature of the agreement follows from the fact that the licensed technologies do not compete, ie cannot be used for the production of competing products—limitations with regard to one market will in that case not influence competition on another. If the parties are competitors on the technology market (only actual competition counts, see above), they may still be potential competitors on the product markets. In that case, the limitation will also be acceptable, although its benefits, while still qualifying as technological progress in the sense of Article 101(3) TFEU, will be less marked.

If, finally, the parties are competitors only on the technology market, the limitation is nevertheless permissible. In this situation the above-mentioned factors are absent that make the limitation acceptable in the other two paradigms. Nonetheless, an overall reduction in the quantities of products will not necessarily ensue.[22] Quite on the contrary—there is no reason why the licensor should seek to achieve by means of a licensing agreement what he might as well bring about by a simple unilateral decision, particularly as such a strategy would in no way be threatened by a licensee who is not even a potential competitor on the product market.

These exceptions leave two output restrictions prohibited—first, restrictions of output of contract products imposed on the licensor in a non-reciprocal agreement; secondly, restrictions of output of contract products imposed on both parties to a reciprocal agreement. Restrictions of the first type would remove the only alternative source of supply for both consumers in, and for parallel traders into, a particular territory, if the agreement also contains restrictions as under Article 4(1)(c)(iii) and (iv). A provision in accordance with cl (iii), the agreement of a 'sole licence', leaves the licensor free to produce in the particular territory. Often, however, such a licence will be granted precisely because the licensor cannot or will not produce sufficient quantities in that territory. Cl (iv) takes this one step further and removes not only the

[22] Thus the Commission, Guidelines (n 1 above) para 83.

presence of the licensor, but also his active and even passive sales into the territory assigned exclusively (as defined in Article 1(1)(l)) to the licensee. The same applies to the alternative in cl (iv) of the assignment of an exclusive customer group as defined in Article 1(1)(m).

Restrictions of the second type (bilateral output restrictions) would allow the parties to create artificial scarcity in their respective 'home' markets (that is, where they first market the contract products). This would invariably leave insufficient quantities for parallel trade, so that other markets would also be affected.

B. The General Prohibition of the Allocation of Markets or Customers, Article 4(1)(c), 4(2)(b), and its Exceptions

For the purpose of Article 4(1)(c) and (2)(b), 'markets' are defined always, using the categories of product and geographical markets. 'Customers' can, again as usual, be participants on the demand side on any level of trade, ranging from industrial and commercial clients—these are envisaged, for instance, in Article 4(1)(c)(vii)—to end-users, typically private consumers, Article 4(2)(b)(v).

First exception: licensee's production restrictions, Article 4(1)(c)(i)

This exception allows the limitation of production, not as in Article 4(b) in quantitative terms, but with regard to the characteristics of the contract products. The licensee's production with the licensed technology (its own technology must always remain freely useable, Article 4(d)) may be confined to certain technical fields or product markets. The distinction between 'fields of use' and 'product markets' is immaterial: each can comprise several of the other.[23] Moreover, the limitation along either field(s) of use or product market(s) has the same legal consequences. What is required in any case, however, is a positive description of the fields or markets allowed to the licensee: every other production is prohibited. This (among others) distinguishes Article 4(c)(i) from (ii), where every production is allowed unless specifically prohibited. 'To produce'[24] with the licensed technology in this context means '... to produce in order to sell'.[25] This follows from the reference to 'markets'—these are real or virtual places of economic exchange; most licensees would not be able to afford the licence

[23] The Commission in its Guidelines (n 1 above) para 179 gives the example of a moulding technique that can be used to make plastic bottles and plastic glasses (same field of use, different product markets) and of an engine technology used in four- and six-cylinder engines (same product market, different fields of use); for another example (maize seed variety licensed for animal food/for human foodstuffs), see Anderman, 'The new EC competition law framework for technology transfer and IP licensing' (n 6 above) 119.

[24] The French version is the same (*produir*), the German has 'to use' (*gebrauchen*).

[25] *Contra,* but without any reasons given, Schumacher and Schmid (n 6 above) 7 (text to fn 61) f.; apparently of the same opinion is Louis (n 6 above) 389.

fee for long were they not allowed to offer their goods or services to third parties against consideration.

The exception does not distinguish between reciprocal and non-reciprocal agreements. The wording 'licensee(s)' indicates with the inclusion of the plural that reciprocal agreements are caught as well as non-reciprocal agreements. The block exemption is only ever available to agreements between two undertakings, Article 2, first paragraph—the plural must, therefore, mean that each of the two can be both licensor and licensee. A restriction as in the present exception does not exclude the respective licensor from the field or market, but only introduces the licensee as an additional competitor with the licensed technology (*ex hypothesi,* he already is a competitor in his own right with his other technology). This exception is, hence, only a consequence of every proprietor's right to determine how far he wants to alienate his right: if he can assign it partially or totally, he must *a fortiori* be allowed to transfer it to a more limited extent of his choosing.

In all, this exception leaves as a hardcore restriction any limitations placed on the use of (each) licensor's own technology with regard to any field of use or product market. Such restrictions, however, are also, although not specifically, blacklisted in lit. (d).

Second exception: either party's production restrictions, Article 4(1)(c)(ii)

Compared to the restrictions in the previous exception, this one adds 'one or more exclusive territories'. Also, the obligation here is a negative one, namely to keep out of specific fields of use, product markets, or exclusive territories.[26] From the wording, it is not immediately clear whether 'reserved for the other party' refers only to the exclusive territory, or also to the fields of use and product markets. In paragraph (1)(c)(iv) and in paragraph (2)(b)(i), the phrase qualifies 'exclusive territory' and 'exclusive customer group'; similarly, when other licensees are meant, the Regulation uses 'allocated', namely in paragraph (1)(c)(v) and in paragraph (2)(b)(ii). The same follows from the definition of an 'exclusive territory' in Article 1(1)(l). 'Reserved for the other party', hence, applies only to the exclusive territory. Nevertheless, there is no reason to allow such a restriction if the one party is not active in the field of use or the product market which the other party must keep out of.[27]

Here again,[28] 'to produce within a territory' refers to production as much as to subsequent sales in the designated territory. This follows from no. (iv), with which no. (ii) is mutually complementary, both read jointly with the definitions of 'exclusive territory'

[26] Conversely, restrictions on production outside those perimeters are not permissible: Dolmans and Piilola (n 6 above) 353 (text to fn 11).

[27] The last point also implies that the better view is that the restriction must be symmetrical in the sense that one party must only be forbidden a market etc on which the other is an actual competitor. Otherwise, there would be no benefit to the consumer through an increase of technological choice on the market(s) affected.

[28] See above on Article 4(c)(i).

and 'exclusive customer group' in Article 1(1)(l) and (m). The latter definition expressly speaks of the 'selling' of contract products to the exclusive group; the phrase '. . . produce for a particular customer' (obviously not for free!) in the former definition also shows that sales are meant primarily.[29] As a consequence of such a restriction, items for sale in the exclusive territory by the other party must be imported from outside that territory, with the added logistical costs and other burdens attendant on such imports.

In that way, particular geographical and product markets (the latter defined by the needs of specific customer groups) can be assigned to either party. This enables that party to work the licensed technology on that market with reduced interference by the other party using the same technology. This reasoning, however, applies only if the technology is protected in the exclusive territory. Otherwise, the protection of the intellectual property would be extended for no cogent reason. Such differences in protection cannot occur, however, in relation to know-how: if it is public in one Member State, it is public in all.

If the agreement is non-reciprocal because the licensed technology does not compete with the one the licensee already has, the result will be an addition to the licensee's technological capabilities. This adds the licensee as a new competitor on the markets of the products made with that technology, to the benefit of competition on that market. In view of his fairly low market share (Article 3), the licensee also has an incentive to reap the full potential of the licensed technology in the assigned market. He is not in a position of strength on that market such that he could let the licensed technology lie idle for 'strategic' purposes. The incentive to use the technology as actively as possible would be lessened if the licensee faced local competition from the licensor, who is already an established and experienced user of the technology. If, on the other hand, the agreement is non-reciprocal because it is non-mutual (ie only one party grants the other a licence to a competing technology), each party's other technology is left unrestricted, and must remain so by virtue of Article 4(1)(d). The licensed technology, in other words, competes with the licensee's proprietary technology. Even if output restrictions are imposed on the licensee's production, or if parallel imports of products made with the licensed technology into or from the allowed territory turn out too cumbersome, the licensee remains free to satisfy any demand with his own technology.

All this leaves as impermissible hardcore restrictions limitations in a reciprocal agreement of either or both parties' production to certain fields of use, product markets, or exclusive territories. If both parties are subject to such restrictions, the agreement allows each licensor control over the market penetration of his technology, and enables him to seal Member States' markets off against each other. As a consequence,

[29] *Contra,* Fuchs (above n 17) 1760 para 249 who gives no reasons beyond the wording of the provision.

products would have to be brought in by parallel traders if they can find sufficient quantities. One of the parties may, however, lawfully be made subject to output restrictions, Article 4(b). This might not be too serious a worry because each remains free to exploit his own technology, Article 4(d). Nevertheless, a reciprocal agreement creates a mutuality of interests likely to dissuade the parties from aggressively competing against each other with their respective proprietary technologies. The same applies, albeit to a lesser extent, if only one party to a reciprocal agreement is subject to such restrictions.

It is not amenable to any precise calculation whether such a restriction might not as well be viewed as similarly acceptable as the permissible restriction under Article 4(c)(i). The Commission's view that it is not, is arguable and, at any rate, legally binding within the scope of the Regulation.

Third exception: sole licences, Article 4(1)(c)(iii)

The third exception allows so-called 'sole' licences, both in a reciprocal and non-reciprocal agreement. Such a licence excludes other licensees (although by no means also their products!) from the territory assigned to the sole licensee; if the licence is not delimited by territory but by field of use etc, Article 4(c)(i) applies. The licensor's production and marketing, however, are not subject to any restrictions,[30] unless these are specifically stipulated and in conformity with Article 4(c) (ii) and (iv). The Regulation highlights this freedom of the licensor by speaking of a 'particular' rather than an 'exclusive' territory. In an exclusive territory, pursuant to the definition in Article 1(1)(l), only one undertaking (here, the licensee) would be allowed to produce the contract products. In a reciprocal agreement, mutual sole licences are problematic (but accepted by the Regulation) because they facilitate the sealing-off of markets as discussed above.[31]

Fourth exception: sales into exclusive territories or to exclusive customer groups, Article 4(1)(c)(iv), 4(2)(b)(i)

Cl (iv) allows restrictions in a non-reciprocal agreement on either or both parties' sales into the exclusive territory, or to the exclusive customer group, reserved for the other party. The sales that may be prohibited can be active sales, that is, business solicited by a party; they can also be passive sales, where the undertaking merely receives unsolicited orders. By contrast with cl (iii), this allows for so-called 'exclusive' licences. The reference to 'exclusive territories' imports into cl (iv) the prohibition of producing in that territory, but this adds little: 'to produce', as has been explained above, means 'to produce in order to sell'. Cl (iv), thus, applies to sales from an establishment outside the exclusive territory, cls (i) and (ii) to sales from within.

[30] Anderman (n 6 above) 125.
[31] Concurring, Treacy and Heide (n 6 above) 417 after fn 41.

Similarly, with regard to the prohibition of active sales to the exclusive customer group of the other party, cl (iv) merely repeats what is already stipulated in Article 1(1)(m), namely that active sales may lawfully be prohibited. Article 4(2)(b)(i) avoids this duplication by referring only to passive sales; the permission to prohibit active sales is implied by the term 'exclusive territory'.

The prohibition of even passive sales, imposed on both parties, leads to a complete insulation of one party on the markets affected from competition with the licensed technology by the other party; we can leave aside the exception for alternative sources of supply in Article 1(1)(l) *in fine*. Nonetheless, such an arrangement is acceptable within the market share thresholds set by Article 3. The parties are not allowed under any of the exceptions from Article 4(1)(c) to quell parallel trade in goods that each party has sold in its reserved territory or to its reserved customer group. Such trade can be engaged in by exclusive customers or by (other) parallel traders to whom the exclusive customers have sold some of the contract products. Under Article 4(1)(b), in the context of a non-reciprocal agreement such as in issue under subsection (c)(iv), at least the licensor's output of contract products must not be capped. Additionally, in consequence of Article 4(1)(d) the licensee remains free, using his own technology, to compete with the licensor regarding any territory and any customer group.

Within the scope of application of Article 4(2), the licensee has no competing technology (at least in the typical case, see above on the assessment under the Regulation of whether there is a competitive relationship between the parties). To compensate for this, Article 4(2)(b)(iii) (by the proviso) and (c) allow licensees certain active and passive sales on a scale that is unlikely substantially to undermine the licensor's allocation of territories and customer groups among his licensees.

This leaves as prohibited in any case restrictions on active and/or passive sales in a reciprocal agreement. Such restrictions would reduce the role of the competing technology as an unencumbered counterweight to the licensed technology. The close and mutual ties between the parties would be a strong incentive for them to carve up the markets affected by the agreement.

Fifth exception: delimitation of licensees' territories, Article 4(1)(c)(v), 4(2)(b)(ii)

The exception in Article 4(1)(c)(v) allows restrictions, in a non-reciprocal agreement, of active sales by one (the 'first') licensee into the exclusive territory or to the exclusive customer group allocated to *another* ('second') licensee, provided the second licensee was not a competitor of the licensor when he entered into his licensing agreement with the licensor. Under cl (v), therefore, three parties are envisaged. Only the 'first', however, is from the outset a competitor of the licensor's, so that *their* agreement falls to be assessed under Article 4(1)(c)(v). If both licensees are novices, Article 4(2)(b)(ii) applies.

In this way, and particularly if combined with an exclusive licence pursuant to cl (iv), the licensor can allow his licensees to penetrate their respective markets without 'free-riding' from other licensees.[32] Free-riding occurs where one undertaking exploits the marketing efforts of another, without engaging to a comparable extent in costly and time-consuming advertising, distribution, and customer services. As the agreement between licensor and one licensee does not bind third parties such as the other licensees, the licensor must specifically impose restrictions in its licensing agreements to regulate licensees' conduct vis-à-vis each other.

As cl (v) indicates, passive sales do not constitute free-riding: by contrast with the licensor, other licensees are always allowed to supply customers who turn to them without prompting. If the licensee in a particular market considers that such sales deprive him of the fruits of his licence, he must persuade the licensor to impose output restrictions under lit. (b) on the other licensee(s). By reducing the available quantities, this will also have an indirect effect on the possibility of parallel trade between the markets affected; no other measures to reduce such trade are permissible. Because the agreement is non-reciprocal, the competitive pressure by the other technology remains.

This leaves two prohibitions impermissible under Article 4(1)(c)(v)—first, a prohibition of *active* sales by the licensee if the other licensee to be protected thereby was a competitor of the licensor when he concluded his own licence; secondly, the prohibition of *passive* sales imposed by the licensor on the licensee if the other licensee is new to the market, ie was no competitor when he entered into *his* licence with the licensor. The second prohibition has just been discussed: the licensee cannot expect more than protection against free-riding by the licensee-competitor of the licensor's. In the first case, where the other licensee, too, was all along in competition with the licensor, the other licensee does not even need the limited protection against passive sales. As he has his own technology, he has additional possibilities to counter any marketing offensives by fellow licensees.

Article 4(2)(b)(ii) follows similar considerations. Here, however, the protection of one licensee against passive sales (and *a fortiori* active sales, too) by another licensee lasts for only two years. The protection of the licensee is thus more intensive, while it lasts, than that under paragraph (1)(c)(v). The reason is that under Article 4(2), none of the licensees are competitors of the licensor and, hence, of one another: if two of them were competitors on the same market as their licensor, they would of necessity compete against each other on that same market. They therefore lack any experience with this or a similar technology before they enter the market, and hence deserve complete protection in the start-up phase of two years. Once this time is up, restrictions of active sales remain permissible, but passive sales by other licensees must not

[32] Ritter (n 6 above) 177 sub (1).

be prohibited anymore. There are no exceptions from this exception, and hence no restriction that remains impermissible under Article 4(2)(b)(ii).

Sixth exception: own-use ('captive use') restrictions, Article 4(1)(c)(vi), 4(2)(b)(iii)

The obligation under Article 4(1)(c)(vi), (2)(b)(iii) on a licensee to produce the contract products only for its own use prevents the licensee from supplying the licensor's or other licensees' competitors. In that respect, the provision complements Article 4(1)(c)(iv) and (v) and (2)(b)(i) and (ii). This obligation is permissible as part of a reciprocal as well as of a non-reciprocal agreement. Impermissible, however, remain restrictions on active or passive sales as spare parts for the licensee's own products. This guarantees the spread of the licensed technology (and hence technological progress in the sense of Article 101(3) TFEU) at least through the medium of the licensee's own products. Output limitations according to Article 4(1)(b) can ensure that the contract products are indeed only used as 'spare parts' for the licensee's own products, leaving only as little as possible for use in competitors' products.

Output restrictions are not mentioned in Article 4(2). If, however, a licensee competing with the licensor may be subject to such restrictions, then this must also be true of a licensee who would not have been able to enter the market but for the licence.[33] The benefit to competition even of a licence with an output limit will be greater in the case of a non-competitor. Such a licensee could not simply have increased his output of products made with his own technology. In territories or with customer groups not currently supplied by the licensor, even the limited licence introduces an entirely new alternative choice. This would have been less the case where the new offer merely rounded off the programme of an already established competitor. Furthermore, if licensor and licensee are not competitors because the licensee will not grant licences in his own technology, and the licensor will not produce with his,[34] the situation would be even closer to the one for which Article 4(1)(b) allows certain output restrictions, namely a situation of competition between licensor and licensee. Confirmation of this comes from Article 4(2)(iv): a permission to produce the contract products only for one particular customer limits output indirectly, that is, not by an absolute number but by the factual demand of that specific customer.

Seventh exception: single-customer restrictions, Article 4(1)(c)(vii), 4(2)(b)(iv)

Also permissible is the limitation, in a non-reciprocal agreement, of the licensee's production of contract products to what is required as an alternative source of supply for a particular customer. This, in effect, equals the allocation to that licensee of a

[33] Concurring, Ritter (n 6 above) 179, last para.

[34] See above on the distinction between competing and non-competing undertakings.

customer group with one member only. Because of this narrow demand base, the licensee has no reason to neglect his own technology in case he is a competitor of the licensor. If he is not, for the reasons explained in the preceding paragraph there will still be a benefit to competition even of a licence thus circumscribed.

This leaves as an impermissible restriction the same limitation in a reciprocal agreement. The reason is unclear (at least the Commission has never explained it): the limitation in cl (vi) confines the licensee to a similarly narrow customer base, but is not limited to non-reciprocal agreements. Under cl (vi), the licensee may produce as much as he thinks he can sell, plus spare parts. Under cl (vii), by contrast, this decision is in the hands of the customer. Additionally, that customer has other sources of supply. His demand of the licensee's products may, therefore, be smaller from the outset. Nevertheless, this does not amount to a relevant difference as long as there are no quantitative limits on the output of the production of the particular customer. If the licensee thinks that tooling up for a licence with such limitations is not worth his while, he can always decline the licence and concentrate on his own technology instead, or seek a removal of the restriction. The same is true if the licensee is not in competition with the licensor.

Parties to a reciprocal agreement who contemplate including a limitation as in cl (vii) thus have an uncomfortable choice. They can either follow the letter of the Regulation, abandon the limitation (or modify the agreement so that it is non-reciprocal), and be on the safe side. Alternatively, they can trust that their agreement will be individually exempt under Article 101(3) TFEU, seeing that in all material respects it is similar to one containing a restriction permissible under cl (vi).

Eighth exception: restrictions regarding distribution systems in agreements between non-competitors, 4(2)(b)(v), (vi), (c)

Three exceptions from the general prohibition of market and customer sharing relate to distribution systems, two of these—Article 4(2)(b)(vi) and (c)—relate specifically to selective distribution systems.

Article 4(2)(b)(v) allows restrictions of sales to end-users by wholesaler-licensees. This comprises any sales, active or passive.[35] The allocation of functions among distributors, that is, the creation of a hierarchy in accordance with each distributor's level of trade, is not normally in breach of Article 101(1) TFEU, particularly (but not only) in selective distribution systems.[36] This restriction is meant to protect the investments required of retailer-licensees. Because of the smaller scale and more

[35] See above the section on the Fifth exception: delimitation of licensees' territories, Article 4(1)(c)(v), 4(2)(b)(ii).

[36] Case 26/76 *Metro-SB-Großmärkte v Commission (No. 1)* [1977] ECR 1875 paras 28 ff (hereinafter '*Metro (No. 1)*'); Fuchs (n 17 above) 1722 para 301 aptly speaks of a prohibition of 'leapfrogging'.

localized catchment area of their operations, retailer-licensees cannot compete on an equal footing with the wholesalers. This refers to choice, delivery time, and price. Wholesalers normally stock the licensor's full range of products in some quantities and on large, often green-field, premises organized for optimum logistics. Retailers must make do with the smaller premises they find locally, typically in town centres, which allows them to stock only smaller quantities and sometimes only samples of specific products. These constraints reduce the efficiency of their businesses and, in turn, raise their prices. This is exacerbated by the fact that within the ambit of the Regulation, every licensee has to carry out some 'production' in application of the licensed technology;[37] 'pure' distribution systems do not come within the Regulation's scope, as Recital 19 emphasizes.

There are no (retro-)exceptions from cl (v). The licensor does, of course, remain at liberty to allow his wholesalers to do business direct with end-users. To some extent, the rise of online sales has blurred the distinction between the levels of trade anyway: some retailers maintain shops with a limited turnover for walk-in customers, and run large out-of-town warehouses for their extensive internet business. Wholesalers' licences will often contain territorial limitations, too, but these are also becoming less meaningful in the internet environment.[38]

The other two exceptions to be discussed here serve to protect a licensor's selective distribution system. In such a system, applicants have to fulfil objective, non-discriminatory, pre-established criteria in order to be admitted as distributors, see Article 1(1)(k). If these criteria are purely *qualitative*, ie relate to the presentation of the sales premises, training of personnel, and after-sales services, the system is compatible with Article 101(1) TFEU. By contrast, the selective distribution system is *quantitative* if it stipulates a minimum turnover for each distributor; an obligation to carry the full range of the licensor's products or sales promotions for specific brands, and similar obligations which an individual distributor will only ever be able to fulfil with regard to one or few producers; minimum distances between outlets, particularly in scarce (downtown) locations, and other objective limitations on the potential number of distributors (including, eg, licensing requirements).[39] A distribution system with these characteristics requires, but is usually eligible for, exemption under Article 101(3) TFEU—it limits the number of potential distributors and thus exposes each one admitted to fewer intra-brand competitors; by the same token, however, it increases each distributor's incentives to concentrate on the development

[37] See Article 2, first para *in fine:* '... agreements ... permitting the production of contract products'.

[38] See the references and discussion in Dethmers and Posthuma de Boers (n 6 above) 434.

[39] Combinations of quantitative selection at the wholesale level with qualitative selection at the retail level and vice versa, or of both types of selection on one and the same level, are possible and in principle permissible.

of the brand and to compete more vigorously with distributors of other brands (so-called 'inter-brand' competition).

Against this backdrop, Article 4(2)(b)(vi) allows restrictions of sales to unauthorized distributors by members of a selective distribution system. Strictly speaking, cl (vi) is tautological (it says the same thing twice), because the definition of a 'selective distribution system' in Article 1(1)(k) already envisages an undertaking by the licensees not to sell contract products to unauthorized distributors.

In contrast with cl (v), this restriction is independent of the level of trade at which the member is operating. If any such sales were permissible, they would undermine the choice of criteria for the selection, and the selection itself, both of which are the licensor's prerogative. Sales to outsiders would also mean that the benefits of the system—care for the brand by the selected distributors in competition with other brands and in the interest of consumers—are no longer guaranteed, while the system's inherent limitation of the number of distributors remains in place. The distribution system would, hence, breach Article 101(1) TFEU and lose its eligibility for exemption under Article 101(3) TFEU. Because, however, in view of its benefits the decision in favour of a selective rather than a non-selective distribution system is a legitimate one, the licensor need not tolerate any actions that call this decision into question. As a consequence, there are no retro-exceptions from cl (vi).

Apart from this, cross-supplies between members at any level of the selective distribution system remain allowed; only those at the wholesale level will normally be prohibited, in keeping with cl (v), from selling to end-users.

In the same vein, Article 4(2)(c) prohibits all restrictions on sales by selected retailers to end-users. All such sales are permissible, whether active or passive. Due to the necessarily small quantities involved, the typical lack of logistical capacity, and the spatial distance of such a retailer from the markets of most other licensees and their customers, retailer-licensees will not be able seriously to undermine the marketing efforts of other licensees. Their activities will only ever amount to pinpricks.

For the same reason, lit. (c) comprehensively governs which sales are allowable to selected retailer-licensees. Lit. (b) is to that extent overridden by lit. (c): passive sales cannot be prohibited within the exceptions from lit. (b) if at the same time under lit. (c) all sales, including active ones, by qualifying retailers to end-users must remain permissible. Lit (c) is, in this respect, *lex specialis* to lit. (b).[40] One may doubt this because it would mean that lit. (b)(vi) ceased to apply to these retailers. As has been noted above, however, it is part of the definition in Article 1(1)(k) of a selective distribution system that its members undertake not to sell the contract products to

40 Concurring, Fuchs (n 17 above) 1773 para 304.

unauthorized distributors. This obligation, therefore, remains unaffected if lit. (c) takes precedence over lit. (b) in its entirety.

This rank of lit. (c) is highlighted by its promotion to the same level as lit. (b). From its connection with selective distribution systems, the rule in lit. (c) could alternatively have been placed as cl (vii) in lit. (b). The prominence instead attached to the rule makes good substantive sense as well: qualitative selection, not least because of the investment in premises and personnel that it requires of distributors, shifts competition away from the parameter of price, and more towards service and other aspects.[41] Territorial and customer restrictions on top of that (with the exception of those imposed on wholesalers, see above) would not add any further benefits for consumers, but would only further reduce inter-brand competition as well as competition between the members of the selective distribution system (so-called 'intra-brand' competition).

The qualification in lit. (c) ('without prejudice') is only seemingly an exception. The sales premises are usually taken into consideration in the selection of distributors at retail level, be it in respect of the premises' location, or their presentation or suitability for pre- and after-sales services (see above). Any operation, therefore, out of an establishment that was not inspected and approved of by or on behalf of the licensor takes the retailer out of the selective distribution system, if only with regard to this particular outlet. Just as by virtue of Article 4(2)(b)(vi), members of a selective distribution system may be prohibited from supplying unauthorized distributors, they must not, themselves, act as unauthorized distributors.

Restrictions on the exploitation and development of the licensee's own technology, Article 4(1)(d)

Article 4(1)(d) contains two prohibitions. The first safeguards the licensee's ability to exploit its own technology. 'To exploit' in this context means to grant licences or to produce with the technology; out and out assignments of the rights in the protected invention are outside the scope of the Regulation,[42] as are encumbrances (the use of the patent etc as security for credit). The blacklisting of such clauses is meant to maintain a plurality of technologies. Restrictions of this type make no contribution to economic and technological progress, contrary to Article 101(3) TFEU. The wording of the provision, however, is too wide: in a reciprocal agreement, restrictions on the licensor in accordance with lit. (c)(ii) and (iii) are permissible.

Article 5(2) is substantially identical with Article 4(1)(d), but applies if the parties are not competitors. The classification in that case of such clauses as non-core

[41] *Metro (No. 1)* (n 36 above) paras 20–1.

[42] See the definition of 'technology transfer agreement' in Article 1(1)(b), after the semicolon: 'assignments . . . where part of the risk associated with the exploitation of the technology remains with the assignor'; see also Louis (n 6 above) 381.

restrictions may be problematic if the parties are non-competitors merely because the licensee will not grant licences to its own competing technology, and the licensor does not want to produce contract products with his technology (in which case they are merely potential competitors on the technology market, and not competitors on the product market). Nevertheless, as soon as that situation changes, the clause becomes a core restriction and brings the whole agreement down with it; it does not benefit from Article 4(3) which, as an exception from the rule in Article 4(1), applies narrowly only within Article 4. The parties will relax the clause in time in order to prevent this outcome.

The second prohibition is directed against restrictions on research and development. These restrictions, too, produce no benefits for consumers. The qualification ('unless') does not amount to an exception—it merely protects the intellectual property rights of the licensor.[43] He need not surrender these rights gratuitously (or at all) if he does not wish to do so, and likewise does not have to tolerate the destruction of his asset by the licensee's divulging the secret information (Article 1(1)(i)(i)). The requirement of proportionality (see the wording 'indispensable') prevents any abuse of this right.

C. Technology Transfer in Context: Competition Law and the Internal Market

The above analysis of Article 4(1) and (2) has largely confirmed the first impression: the exceptions have almost swallowed the rule. Time and again, we have seen that ultimately, two factors safeguard competition on geographical and product markets that have lawfully been shared out by the parties to technology transfer agreements—the rule in Article 4(1)(d), itself limited by preceding exceptions, that the parties must remain at liberty to exploit their own technologies in competition with each other;[44] and parallel trade as the bedrock of the internal market. Parallel traders are, as it were, competitors of last resort, when the parties have seized all lawful opportunities to allocate markets or customers. Parallel traders have this position thanks to the Court of Justice's jurisprudence on the exhaustion of rights. To put it very succinctly, products embodying intellectual property rights can in principle be marketed *everywhere* in the internal market once they have been marketed *anywhere* in that market.[45]

[43] Know-how is an intellectual property right for the purposes of the Regulation (only), Article 1(1)(g).

[44] By contrast, Dolmans and Piilola (n 6 above) 355, cite Article 4(1)(d) to argue that the Commission could have treated reciprocal agreements more leniently.

[45] For a recent overview, see Enchelmaier, 'The inexhaustible question: free movement of goods and intellectual property in the European Court of Justice's case law, 2002–2006' (2007) 38 Intl Rev Intellectual Property and Competition L 353.

Parallel trade is only possible where there are quantities of goods (services lend themselves less to parallel trade) that have not been absorbed by the market where they were first commercialized. A proprietor's only lawful lever against parallel trade, therefore, is a limitation of output, both (or alternatively) his own and his licensees'. The Court has recently appeared to veer away from this constant jurisprudence.[46] The judgments in question—*Adalat* and *Glaxo*—could, on the one hand, be read as isolated aberrations, limited to trade in pharmaceuticals and at any rate, not transferable to the context of licensing. On the other hand, the Court's case law on the free movement of goods has recently shown an alarming disregard for conceptual coherence; instances of this are the judgments in *Moped Trailers*,[47] *Mickelsson and Roos*,[48] and *Copad v Dior*.[49]

It seems that anything is possible right now, and ad hoc solutions to individual cases reign, more intuitively found than rationally deducted from tried and tested principles. If this were to become the norm, restrictions in technology licensing agreements could be used by the parties to carve up the internal market and to stifle competition in exactly those markets that were supposed to receive their special dynamics from the spread of technology. But all is not lost—Regulation 240/96, despite all its weaknesses in other respects, was much more concerned to prevent market sharing. The Commission's 'new economic approach' led to the present, over-differentiated and ultimately counter-productive jungle that is Article 4 of Regulation 772/2004. It is to a large extent in the hands of the Commission to revert to a regime similarly stringent to the one that preceded Regulation 772/2004. Parties to anti-competitive agreements will certainly be emboldened to challenge the Commission's policy more often than was the case with past block exemption Regulations, be it in the national courts or (where possible) in the European Courts. Nevertheless, it remains to be seen whether (and it must be hoped that) the Court will still concede the Commission a substantial leeway in the implementation of those tasks entrusted to it like, in particular, competition policy.

Conclusion

Article 4 of Regulation 772/2004 is overly complex, thus failing to provide legal certainty;[50] it is also in substance rather generous to market and customer sharing by parties to technology transfer agreements. In this way, the Regulation must to an extent rely on parallel trade to ensure that markets remain open. Just at a time when

46 See Heinemann in this volume, Ch 13; for another critical view, see Ullrich (n 6 above) 63.
47 Case C-110/05 *Commission v Italy* [2009] ECR I-519.
48 Case C-142/05 *Åklagaren v Percy Mickelsson and Joakim Roos* [2009] ECR I-4273.
49 Case C-59/08 *Copad SA v Christian Dior couture SA* [2009] ECR I-3421.
50 Concurring, Fine (n 6 above) 771, 786; Schumacher and Schmid (n 6 above) 10 sub V.

the Commission cannot, as it could before Regulation 1/2003 entered into force, systematically control individual agreements anymore, the Court has cast grave doubts over the continuance of its long-standing jurisprudence on parallel trade. This is a worrying, but not a hopeless situation for competition law as it grapples to define its relation with technology transfer.

19

CANARIES IN THE COAL MINE: HAS NEO-CLASSICAL ECONOMICS LOST GROUND AT THE INTERSECTION OF IP LICENSING AND ANTITRUST LAW IN THE UNITED STATES?

Kevin Arquit

Introduction

In the wake of the 'Great Recession' of 2008,[1] there is a not-so-quiet war going on. On one side stand stalwart supporters of the neo-classical 'Chicago-Harvard School' of law and economics, who claim both that the market is by-and-large self-correcting and that market participants act 'rationally'.[2] On the other is a creeping counter-reformation made up of various theories like 'behavioural' economics, which posit cognitive biases that permanently preclude a pure rational choice model.[3]

[1] I use the term 'Great Recession' to refer to the economic crisis of 2007 through 2010 for convenience's sake, despite the fact it has been applied to every recent recession. That said, given the scope and scale of the crisis, it seems unlikely to have a recessionary rival any time soon (as opposed to a depressionary one).

[2] See eg J Cassidy, 'After the Blowup' *New Yorker* (New York 11 January 2010) 28 (noting general lack of ideological movement by Robert Lucas, Eugene Fama and John Cochrane, among others). It is somewhat inaccurate to refer to adherents of classic neo-classical economics as belonging solely to the 'Chicago School,' referring to the University of Chicago. There used to be clear borders between the conservative Chicago School and the slightly more moderate Harvard School. Those have largely been erased in the modern era of antitrust jurisprudence and scholarship. See WE Kovacic, 'The Intellectual DNA of Modern US Competition Law for Dominant Firm Conduct: The Chicago/Harvard Double Helix' (2007) 11 Columbia Business L Rev 1; RA Posner, 'The Chicago School of Antitrust Analysis' (1979) 127 U Pennsylvania L Rev 925 (describing convergence of Chicago and Harvard Schools) (hereinafter Posner, '*Chicago School*'). Accordingly, throughout this paper, I refer to the Chicago-Harvard School of antitrust.

[3] See eg *ibid* (noting interventionist positions of, for instance, Richard Thaler and Gary Becker); J Cassidy, 'Economics: Which Way for Obama?' *New York Review of Books* (New York 12 June 2008);

Behavioural economics (and similar theories)[4] believe that the driving assumption of neo-classical macro-economics—that market participants, buyers, and sellers alike, act rationally to maximize wealth after gathering as much information as possible[5]—is fundamentally outmoded. Defenders of the Chicago-Harvard School essentially respond that the model is not perfect, nor is it perfectly predictive—but it remains the best of any alternative. It provides a consistent framework on which to construct effective economic and fiscal policy, and it is the most effective in fostering near- and long-term economic growth and innovation. It is, much like democracy as a form of government, the worst system—except for all the others.

This debate is raging throughout government, business, and academia, and implicates a wide array of academic disciplines, legal fields, and areas of public policy. For instance, prominent economists are actively campaigning in New York Times best-selling books for the formal inclusion of behavioural economics and psychology in any economic model deployed to prevent future catastrophes like the Great Recession.[6] Supporters of a new, consolidated Consumer Financial Protection Agency proceed from the starting premise that consumers of financial products simply cannot understand what they are buying,[7] which echoes the behavioural economists and their support for what Thaler calls 'libertarian paternalism' (and most just call paternalism).[8] In securities regulation generally, President Obama's pick to head

<http://www.nybooks.com/articles/archives/2008/jun/12/economics-which-way-for-obama/> accessed 16 September 2010 (noting the contrasting approaches adopted by leading presidential candidates).

4 Throughout this article, I refer generally to behavioural economics as the main nemesis of orthodox Chicago-Harvard School neo-classical economics informed by rational choice theory. There are other theories and economic propositions that qualify as 'post' Chicago School. However, they all fundamentally differ from neoclassical economics in their rejection of the argument that individuals and firms act rationally in making economic decisions. See eg JT Rosch, Commissioner, Federal Trade Commission, 'Remarks Before the Conference on the Regulation of Consumer Financial Products, Managing Irrationality: Some Observations On Behavioral Economics And The Creation Of The Consumer Financial Protection Agency' (6 January 2010) <http://www.ftc.gov/speeches/rosch/100106financial-products.pdf> accessed 16 September 2010; JT Rosch, Commissioner, Federal Trade Commission, 'Remarks Before the Bates White Antitrust Conference, Antitrust Law Enforcement: What To Do About The Current Economics Cacophony' (1 June 2009) <http://www.ftc.gov/speeches/rosch/090601bateswhite.pdf> accessed 16 September 2010, 7–12.

5 See eg Posner, *Chicago School* (n 2 above) 928 (describing possible core assumptions of Chicago School).

6 See eg GA Akerloff and RJ Shiller, *Animal Spirits: How Human Psychology Drives the Economy, and Why It Matters for Global Capitalism* (Oxford: Princeton University Press, 2009); D Ariely, *Predictably Irrational: The Hidden Forces That Shape Our Decisions* (London: Harper, 2009).

7 See eg Regulatory Restructuring: Enhancing Consumer Financial Products Regulation: Hearing Before the House Financial Services Committee, 111th Congress (2009) (testimony of Professor Elizabeth Warren) <http://www.house.gov/apps/list/hearing/financialsvcs_dem/warren_testimony.pdf> accessed 16 September 2010.

8 CR Sunstein and RH Thaler, 'Libertarian Paternalism Is Not An Oxymoron' (2003) 70 U Chicago L Rev 1159.

the Securities and Exchange Commission, Mary L Schapiro, has moved dramatically and swiftly to recalibrate the Commission toward stricter regulation. Schapiro has, for example, reversed the Bush administration policy requiring the consent of the commissioners (most of whom were opponents of government regulation) before staff could move to resolve major cases, and has aggressively recruited top aides who evince a pro-regulatory bent.[9] Professor Thaler, one of the most prominent behavioural finance scholars, was a frequent advisor to the Obama campaign. Cass Sunstein, Thaler's co-author in a recent book boosting behavioural economics as a basis for government policy,[10] heads the administration's Office of Information and Regulatory Affairs.

A number of prominent academics and jurists have also reversed course on their past faith in the market.[11] Perhaps most notable among them is Judge Richard Posner whose 1976 classic *Antitrust Law* extolled the virtues of the Chicago-Harvard School of law and economics.[12] Judge Posner openly embraced behavioural economics in a notable break with his colleague on the Seventh Circuit, Judge Frank Easterbrook, in a case recently decided by the Supreme Court.[13] In a decision concerning fees paid to mutual fund managers, Judge Posner, dissenting from the court's refusal to hear the case en banc, starkly criticized the majority's faith in the market and adherence to neo-classical orthodoxy, arguing that Judge Easterbrook's economic analysis

[9] See eg S Labaton, 'SEC Chief Pursues Tougher Enforcement' *New York Times* (New York 22 February 2009) B1; S Labaton, 'SEC Nominee Offers Plan for Tighter Regulation' *New York Times* (New York 15 January 2009) B3; ML Schapiro, Chairman, United States Securities and Exchange Commission, 'Remarks Before the 37th Annual Securities Regulation Institute, Embracing The Change' (20 January 2010) <http://www.sec.gov/news/speech/2010/spch012010mls.htm> accessed 16 September 2010; ML Schapiro, Chairman, United States Securities and Exchange Commission, 'Remarks Before the SEC-CFTC Joint Meetings on Regulation Harmonization' (2 September 2009) <http://www.sec.gov/news/speech/2009/spch090209mls.htm> accessed 16 September 2010; ML Schapiro, Chairman, United States Securities and Exchange Commission, 'Remarks Before the United States Senate Committee on Banking, Housing and Urban Affairs, Testimony Concerning Regulation Of Systemic Risk' (23 July 2009) <http://www.sec.gov/news/testimony/2009/ts072309mls.htm> accessed 16 September 2010.

[10] CR Sunstein and RH Thaler, *Nudge: Improving Decisions About Health, Wealth & Happiness* (New Haven:Yale University Press, 2008).

[11] The list of individuals wracked by self-doubt after the 'Great Recession' includes such weighty personalities as Alan Greenspan. See EL Andrews, 'Greenspan Concedes Flaws In Deregulatory Approach' *New York Times* (New York 24 October 2008)B1. When asked by Representative Henry Waxman, Democrat from California, whether his ideology '. . . pushed [him] to make decisions that [he] wished he had not made' Greenspan responded: 'Yes, I've found a flaw. I don't know how significant or permanent it is. But I've been very distressed by that fact.' *Ibid.*

[12] Compare RA Posner, *Antitrust Law* (2nd edn, Chicago: University of Chicago Press, 2001) with RA Posner, *A Failure of Capitalism: The Crisis of '08 and the Descent into Depression* (Harvard: Harvard University Press, 2009) (hereinafter '*Failure*'); see also Cassidy (n 2 above).

[13] Judge Posner has criticized the proposed Consumer Financial Protection Agency as overly paternalistic, but noted in a prominent op-ed on the subject that '. . . [b]ehavioral economists are right to point to the limitations of human cognition.' RA Posner, 'Treating Financial Consumers as Consenting Adults' *Wall Street Journal* (New York 22 July 2009).

(which assumed that competition in the marketplace for investment advisors would suffice to tamp down excessive fees),

> . . . is ripe for reexamination on the basis of growing indications that executive compensation in large publicly traded firms often is excessive because of the feeble incentives of boards of directors to police compensation.[14]

Fortunately or unfortunately (depending on one's economic perspective) the Supreme Court left the Seventh Circuit commotion unresolved, finding that '. . . [t]he debate between the Seventh Circuit panel and the dissent from the denial of rehearing regarding today's mutual fund market is a matter for Congress, not the courts.'[15]

One of the primary battlegrounds for this dispute is in antitrust law and enforcement policy.[16] It is manifest in recent academic literature and in the public comments of the new Obama administration's regulators. For instance, in the context of merger review, the Department of Justice's new top antitrust official has been a vocal advocate for theories of antitrust harm that diverge markedly from the Chicago-Harvard School. Assistant Attorney General Christine Varney would put significant stock in vertical effects, which under neo-classical economic theory are only rarely a threat to competition,[17] and 'innovation markets', which would, for instance, permit an agency to block a merger between two firms that do not directly compete with one another on the off chance that they may in the future.[18] Similarly, in a rejection of the Chicago-Harvard School's approach to single-firm conduct, antitrust regulators summarily rejected the Bush administration's policy position on § 2 liability in favour of a more aggressive stance. And, notably, the recently revised Horizontal

---

[14] *Jones v Harris Associates* 537 F3d 728, 730 (US Ct of Apps (7th Cir) 2008) (denial of en banc rehearing) certiorari granted No 08-586 (9 Mar 2009) (Posner, J, dissenting), vacated and remanded 130 S Ct 1418 (2010).

[15] 559 US at 17. Many observers believed that the case could have served as an indication of things to come in terms of the economic mood of the judiciary. A Liptak, 'Supreme Court to Hear Case on Executive Pay' *New York Times* (New York 18 August 2009) A10.

[16] See eg ME Stucke, 'Money, Is That What I Want? Competition Policy & the Role of Behavioural Economics' (2009) University of Tennessee Legal Studies Research Paper No 75 <http://ssrn.com/abstract=1419751> accessed 16 September 2010; Rosch (n 4 above) 7–12; JT Rosch, Commissioner, Federal Trade Commission, 'Remarks Before the New York Bar Association Annual Dinner, Implications of the Financial Meltdown for the FTC' (29 January 2009) <http://ftc.gov/speeches/rosch/090129financialcrisisnybarspeech.pdf> accessed 16 September 2010 (hereinafter '*Financial Meltdown*'); ME Stucke, 'Behavioral Economists at the Gate: Antitrust in the Twenty-First Century' (2007) 38 Loyola U Chicago LJ 513.

[17] See Posner, *Chicago School* (n 2 above) 937–38.

[18] See eg CA Varney, Assistant Attorney General, Dept of Justice, 'Vigorous Antitrust Enforcement in this Challenging Era, Remarks Prepared for the US Chamber of Commerce' (12 May 2009) <http://www.justice.gov/atr/public/speeches/245777.htm> accessed 16 September 2010 (hereinafter '*Vigorous Antitrust Enforcement*'); CA Varney, 'Remarks before the PLI 36th Annual Antitrust Institute, Vertical Merger Enforcement Challenges at the FTC' (17 July 1995) <http://www.ftc.gov/speeches/varney/varta.shtm> accessed 16 September 2010; see also S Gates and T Srimushnam, 'A New Direction?' *The Deal Magazine* (New York 5 February 2009) <http://www.thedeal.com/newsweekly/community/a-new-direction.php> accessed 16 September 2010.

Merger Guidelines, jointly promulgated by the Department of Justice and Federal Trade Commission, retreat from the relative transparency and clarity of the past incarnation in favour of a more open-ended approach to unilateral effects analysis that likely reflects a pro-regulatory, anti-Chicago-Harvard School, shift in merger review.[19]

There is one area in antitrust law where the assault on neo-classical economics and rational choice theory is not as obvious. That is in the field of intellectual property ('IP') licensing, which, while having undergone significant development over the last four or so decades, remains relatively fixed in the Chicago-Harvard School firmament. That said, one can identify a number of potential developments in IP licensing that could act as the proverbial canaries in the coal mine, indicating that this area of law and policy too is giving ground to the behavioural economists and the other Chicago-Harvard School insurgents. Also notable, these developments could prefigure a recalibration of the innate tension between antitrust and IP. In past decades, the two have settled into a relative equilibrium with antitrust law used largely to restrain IP-holders with true market power that use various stratagems—most often tying or exclusive dealing—to extend the IP grant beyond its proper bounds.

Interestingly, if the shift away from the Chicago-Harvard School were to entail a greater emphasis on regulation and fear of market failure due to irrational behaviour by buyers and sellers or information asymmetry, one could expect to see antitrust law used more aggressively to *limit* the IP 'monopoly', even when its exercise is economically rational and welfare-maximizing. However, if the shift focuses more on the limits of so-called 'static' efficiency relative to 'dynamic' efficiency (the former being concerned with price and output, the latter with broader innovation gains that change the market dynamics entirely), the same rejection of the Chicago-Harvard School could lead to greater focus on *protecting* IP rights over antitrust. For instance, as will be discussed in greater detail, an emphasis on dynamic efficiency could be seen in how courts and commentators approach payments in patent disputes in the generic pharmaceutical industry. Crucially, this particular dilemma only serves to highlight the limits of the post-Chicago-Harvard School economic theories in providing predictability and consistency in antitrust enforcement.

This chapter will provide both a primer on the intersection of IP licensing law and antitrust in the United States and a brief roadmap of where post-Chicago-Harvard School economics could appear in this legal landscape. Specifically, part A of this chapter briefly describes the rise of behavioural economics and a few of its more notable appearances in recent years. Part B provides a short survey of IP licensing law since it first developed out of the early patent misuse cases of the 1910s–30s. Part C sketches the current state of antitrust law as applied to IP licensing. And Part D

[19] See Varney, *Vigorous Antitrust Enforcement* (n 18 above) 11; DOJ and FTC, *Horizontal Merger Guidelines* § 6 (19 August 2010).

showcases those developments that could suggest that the assault on neo-classical economics has reached IP licensing law as well.

A. Cognitive Dissonance: The Rise of Behavioural Economics in Antitrust

In its initial incarnation, modern antitrust law was animated less by economic theory and more by populist concern with large industrial combines and voting 'trusts' that had begun to wield not just vast amounts of economic power in American society but ever-increasing political sway.[20] However, as the study of economics began to percolate through legal circles—notably at the University of Chicago (from whence the Chicago School gets its name)—antitrust theorists realized that too much antitrust regulation could be equally as pernicious as too little. Accordingly, neo-classical price theory entered the mix, injecting terms like 'deadweight loss' into the legal discussion and abiding concerns over reduced output and increased price into the case law. As the populist era faded into the twentieth century, antitrust law became increasingly focused on economics as it, for instance, rejected fear of monopoly-qua-monopoly and realized that the monopoly profit incentive, properly harnessed, is itself an engine of competition.[21]

In cases such as *GTE Sylvania*,[22] this focus on economic theory and deployment of neo-classical economics manifested itself alongside the adoption of the related rational choice theory as a governing idea.[23] Rational choice theory, at its core, is the idea that, while individuals may not always act 'rationally'—that is, in their best interest, consistently, and after some deliberation—most will, and, in particular, the 'Firm' (corporations, partnerships, etc) will do so.[24] Accordingly, it provides a predictive tool that is of great use in antitrust theory because it allows courts and scholars to identify the kind of business conduct that will not self-correct and that requires regulatory or judicial intervention. A good example of this is the growing unpopularity of predatory pricing theory as a basis for judicial intervention. Early on in antitrust cases, predatory pricing—that is, reducing one's price below cost and taking a

---

[20] See generally Posner, *Antitrust Law* (n 12 above) 33–6.

[21] See generally Posner, *Chicago School* (n 2 above) (describing entry of Chicago School into mainstream of antitrust economic theory).

[22] See *United States v Arnold, Schwinn & Co* 388 US 365, 87 S Ct 1856 (1967) overruled by *Continental TV, Inc v GTE Sylvania* 433 US 36, 58–9, 97 S Ct 2549, 2561–2 (1977) (overruling per se treatment of agreements limiting sales of franchisors' television sets to franchised locations).

[23] Compare *Continental TV, Inc v GTE Sylvania* 433 US 36, 56–9, 97 S Ct 2549, 2560–2 (1977) ('Economists also have argued that manufacturers have an *economic interest* in maintaining as much intrabrand competition as is consistent with the efficient distribution of their products.') (emphasis added).

[24] RA Posner, 'Rational Choice, Behavioral Economics, & the Law' (1998) 50 Stanford L Rev 1551, 1551 (defining rational choice as '. . . choosing the best means to the chooser's ends') (hereinafter '*Rational Choice*').

loss to drive competitors out of the market to allow for later monopoly pricing—was often alleged and frequently successful. However, decades of economic inquiry, informed by rational choice theory, have shown that predatory pricing is both rare and rarely successful.[25]

Fundamentally, rational choice theory adopts three assumptions (applicable to individuals and groups of individuals alike). First, people will act rationally; that is, they will act deliberatively after considering the costs and benefits of a particular action. Second, they will seek to maximize utility, which, in the context of a business decision, means they will act to achieve the best bottom line possible. And, third, they will base that decision on the information available to them and will seek to obtain all possible relevant information before deciding (the last characteristic '. . . does not imply omniscience', merely informed choice to the extent reasonable and rational).[26] One can then model outcomes based on these assumptions (which are admittedly unrealistic when applied to individuals as individuals, but acquire greater predictive force when applied to many individuals over time).[27]

For years, this Chicago-Harvard School approach to antitrust—incorporating neoclassical price theory, informed in turn by rational choice theory—has ruled the roost in the courts and at the regulatory agencies. In recent years, however, a series of prominent legal and economic scholars have called for a revision of the Chicago-Harvard School orthodoxy to reflect so-called 'behavioural economics', which—at its core—rejects the assumptions listed above.

According to proponents of behavioural economics, actual humans do not deliberate based on stable preferences using optimal information to maximize their utility. Rather, they exercise 'bounded' decision making that is not always self-interested, fully informed or based on proper deliberation.[28] There is no good definition of 'behavioural economics', and experiments exploring or applying it span both micro- and macro-economics. Many incorporate elements of psychology and sociology. The common theme running through all modern behavioural economics, however, is the notion that humans will not always act in their best interests as economically rational creatures. In particular, behavioural economists and those that would apply behavioural economics to law and economics argue that individuals harbour cognitive

[25] That conclusion was famously reached as long ago as 1958 with John McGee's groundbreaking article in which he found little historical evidence that Standard Oil had achieved a monopoly on oil refining through predatory pricing despite the conventional wisdom that said it had. McGee was one of the first theorists to show that predatory pricing is highly irrational. It is expensive, and of dubious utility because, among other things, competing firms can shut down temporarily to weather the period of below cost pricing; see J McGee, 'Predatory Price Cutting: The Standard Oil (NJ) Case' (1958) 1 J L and Economics 137.

[26] Posner, *Rational Choice* (n 24 above) 1553.

[27] Posner, *Rational Choice* (n 24 above) 1556–7.

[28] See generally C Jolls, CR Sunstein and R Thaler, 'A Behavioral Approach to Law & Economics' (1998) 50 Stanford L Rev 1471.

'quirks' that effectively short-cut rational thought and decision making.[29] Perhaps more significant, many behavioural economists believe these quirks or heuristics are consistent over large numbers of people and over time.

The history of behavioural economics began with psychology experiments in the early 1960s in which psychologists started developing their own models that purported to show consistent deviations from the rational choice model in human cognition and decision making. Psychologists Amos Tversky and Daniel Kahneman, two pioneers in the field, published several influential and widely-cited works. Among these was a 1974 article in *Science* in which Tversky and Kahneman suggested that individuals rely on heuristic principles to form probability judgments that lead to systematic and predictable errors.[30] Through their work, Tversky and Kahneman demonstrated the phenomenon known as 'anchoring' in which people tend to project an estimate by beginning from an initial value and adjusting up or down to produce a final answer. The experiments by Tversky and Kahneman showed that people had a tendency to adjust insufficiently.[31]

Behavioural economists have also noted the importance of a phenomenon known as loss aversion—the '. . . disparity between the strong aversion to losses relative to a reference point and the weaker desire for gains of equivalent magnitude.'[32] Under traditional consumer theory, the fact an individual possesses something should have no effect on a consumer's preference among different commodity bundles; however, the principles of reference-dependence and loss aversion suggest that decisions are affected by whether a transaction is framed as a potential loss or a potential gain. In one experiment, half of the subjects were given coffee mugs and asked at what price they would sell their mugs. The other half of the group was given no mug but asked how much they would pay to acquire one. The results showed that the median selling price was $5.79 while the median purchasing price was $2.25.

[29] See Posner, *Rational Choice* (n 24 above) 1553.

[30] A Tversky and D Kahneman, 'Judgment Under Uncertainty: Heuristics and Biases' (1974) 185 Science 1124, 1128.

[31] In one study, subjects were asked to estimate the percentage of African countries in the United Nations. A 'wheel of fortune' spun in the presence of the subjects produced the initial number—between 0 and 100—provided to the subjects. The subjects were then asked: (i) to answer whether that initial number was higher or lower than the percentage of African countries in the UN and (ii) to then actually provide an estimate of the percentage of African countries in the UN. The arbitrary numbers given to the various experimental groups had a significant effect on the final estimates. For the group provided with 10 as its initial starting figure, the median estimate of the percentage of African countries in the UN was 25; for another group provided with 65 as the initially starting point, its median estimate was 45. *Ibid.*

[32] CF Camerer and G Loewenstein, 'Behavioral Economics: Past, Present, Future' in CF Camerer, G Loewenstein and M Rabin (eds), *Advances in Behavioral Economics* (Oxford: Princeton University Press, 2004) 3, 4; see D Kahneman and A Tversky, 'Prospect Theory: An Analysis of Decision Under Risk' (1979) 47 Econometrica 263 (critiquing expected utility theory as a model of decision making under risk and offered an alternative model—prospect theory—that ostensibly better-accounted for the phenomena demonstrated by people making choices among risky prospects).

Behavioural economists have used loss aversion to explain '... asymmetries in demand elasticities after price increases and decreases.'[33]

'Anchoring' and 'loss aversion' are only two examples of the 'heuristics' that behavioural economists posit that, they claim, cause individuals to consistently deviate from the expected outcomes of the Chicago-Harvard School. And while these heuristics provide an interesting glimpse into psychological idiosyncrasies, they provide little by way of a consistent governing principle for competition law that provides predictable and stable policy and enforcement outcomes. This becomes clear when one looks at how behavioural economics conceivably could influence the intersection of competition law and IP licensing.

B. Brief History of the US Antitrust Law's Treatment of IP Licensing

As with antitrust generally, the law governing the intersection of IP and antitrust is relatively young. Modern US antitrust law dates back to the last years of the nineteenth century and cases reconciling IP with antitrust date back only to the first decade or two of the twentieth. The current law surrounding IP licensing had its initial origins in the patent misuse cases of the 1910s, which usually involved a patentee insisting that its licensees exclusively purchase or use unpatented products produced by the patentee. Notably, there is some ambiguity as to how the patent misuse doctrine relates to the antitrust laws. Some courts have gone as far as to hold that patent misuse is totally grounded in antitrust rather than intellectual property principles, and the common reading is that the two are concerned with the same type of competitive harm (ie, unreasonably extending the ability of the patentee to limit output or raise prices on unpatented products).[34]

Early in the 1910s, the Supreme Court initially approved the practice of tying a patent licence to the purchase of an unpatented product in a case where the producer of

---

[33] Camerer and Loewenstein (n 32 above) 16.

[34] Indeed, one could argue that the rejection of the patent misuse doctrine as something separate from antitrust (that is, something meant to deter the extension of the patent monopoly beyond the patent's claims without regard for the effect on competition) can be associated with Chicago School. See *USM Corp v SPS Tech, Inc* 694 F2d 505, 512 (US Ct of Apps (7th Cir), 1982) ('If misuse claims are not tested by conventional antitrust principles, by what principles shall they be tested?'). This proposition was memorialized in a 1988 amendment to the Patent Act that requires proof of market power before a tie can constitute patent misuse, 35 USC § 271(d)(5) (2006). As discussed below, modern patent misuse doctrine still appears to reach conduct that would not be subject to sanction on antitrust principles, though the two still appear to largely dovetail. Were patent misuse to move away from protecting competition as an animating principle and toward a nebulous policy concern with the 'integrity' of the patent system, this could prefigure a shift in licensing policy away from the Chicago-Harvard School.

a rotary mimeograph included a licence restriction that the licensee purchase stencils, paper, ink, and other products only from the manufacturer.[35] But, only five years later, the Supreme Court shifted course dramatically in *Motion Picture Patents Co v Universal Film Mfg Co*.[36] There, the Court condemned a licensing provision requiring operators of motion picture projectors to screen film only produced by the manufacturer.[37] The case neatly showcased two significant themes in the antitrust treatment of IP licensing, which feature prominently in the following discussion. The first is the cyclical tug-of-war between the primacy of IP vis-à-vis antitrust, and vice versa.[38] The second is the recurrent factual scenario presented in the case: the use of IP licensing by the owner to exert market power beyond the relevant IP (often by tying or exclusive dealing arrangements).[39] How the courts and regulatory agencies treat this fact pattern (ie whether they are more solicitous to the IP right or to the antitrust law's concern with competitive harm) can be important when assessing any potential ideological shift from Chicago-Harvard School price and rational choice theory to theories like behavioural economics.

Motion Picture through the 1960s

Following *Motion Picture*, the courts made increasing use of the misuse doctrine. In a series of cases in the 1930s and 1940s, the Supreme Court struck down several restrictive licensing arrangements.[40] The *Mercoid* case took the doctrine farthest afield from a purely economic (ie neo-classical) focus on competition. There, Mid-Continent sued Mercoid, a producer of an *unpatented* switch used in a heating distribution system, which was subject to its combination patent, for contributory infringement. The suit itself was found to violate the patent misuse doctrine. Notably, the Court paid lip service to the antitrust laws, but really grounded its analysis on the '. . . integrity of the patent system.'[41]

Post-*Mercoid*, the patent misuse doctrine has waned in importance with several Federal Circuit decisions limiting its scope (and confining it largely to the contours of

---

[35] *Henry v AB Dick* 224 US 1, 11, 32 S Ct 364, 365 (1912) overruled in part by *Motion Picture Patents Co v Universal Film Mfg Co* 243 US 502, 37 S Ct 416 (1917).

[36] 243 US 502, 37 S Ct 416 (1917).

[37] *Ibid* 518–9, 421.

[38] For a brief and cogent survey of the oscillating primacy of either IP or antitrust in the relevant case law, see MA Lemley 'A New Balance Between IP and Antitrust' (2007) 13 Southwestern J L and Trade in the Americas 237.

[39] That *Motion Picture* is actually a patent misuse case also highlights the reality that patents are the most common type of IP bumping up against the antitrust laws.

[40] *Carbice Corp v American Patents Development Corp* 283 US 27, 34–5, 51 S Ct 334, 336 (1931) (holding licensing arrangement requiring licensees to purchase unpatented dry ice from patent-holder of dry ice container violated patent misuse doctrine); *Mercoid Corp v Mid-Continent Investment Co* 320 US 661, 661, 64 S Ct 268, 268 (1944).

[41] *Mercoid Corp v Mid-Continent Investment Co* 320 US 661, 668, 64 S Ct 268, 272 (1944) ('If such an expansion of the patent monopoly could be effected by contract, the integrity of the patent system would be seriously compromised.').

antitrust law). However, it provides the frame on which much of the law surrounding the intersection of IP licensing and antitrust lies. It also showcases the tensions that reside at this intersection—between the rational choice/Chicago-Harvard School approach to the problem (which would pay little mind to non-economic concerns about the 'integrity' of the patent system and would probably permit what would otherwise be deemed 'misuse' that did not pose a threat of competitive harm) and the possibly resurgent approach of the Supreme Court in the 1960s, concerned as it was with ill-defined concepts like the 'integrity' of the patent system.

The 'Nine No-Nos' to the IP Guidelines: IP wanes and then waxes

The *Mercoid* hostility to IP rights continued into the 1970s with the formulation of the so-called 'Nine No-Nos', a set of practices initially identified by US Department of Justice Deputy Assistant Attorney General Bruce Wilson, who may have even intended that they be treated under the 'per se' rule of strict antitrust liability.[42] Although the per se approach was never adopted in practice by the Department of Justice ('DOJ'), the practices listed as 'No-Nos' belie a serious lack of confidence at the time that the protection of IP (absent manifest abuse) could be actually *pro*-competitive.[43] This runs counter to the Chicago-Harvard School's position that the robust protection of IP can be a net positive. The Nine No-Nos also failed to address the key insight of neo-classical antitrust theory, which is that anti-competitive harm cannot occur in the absence of market power.

The pendulum swung again, however, in the 1980s, as the US enforcement agencies themselves tried to pedal away from the per se or quasi-per se approach envisioned in the Nine No-Nos.[44] Indeed, the Reagan administration actually introduced legislation—the Omnibus Intellectual Property Rights Improvement Act of 1987[45]—that would have explicitly limited patent misuse claims to antitrust violations and would

---

[42] Wilson stated that these restraints are likely '. . . in virtually all cases [to] lead to antitrust trouble because of their adverse effect on competition.' B Wilson, Deputy Assistant Attorney General, United States Dept of Justice, 'Remarks Before the American Patent Lawyers Association, Law On Licensing Practices: Myth Or Reality? Or Straight Talk From "Alice in Wonderland"' (21 January 1975). He also noted that '. . . the validity of licensing practices other than these nine is to be tested under the rule of reason.' *Ibid.*

[43] The Nine No-Nos were: (1) royalties unrelated to sales of the patented product; (2) tie-outs (restrictions on the licensee unrelated to the patented product); (3) normal ties (requiring the purchase of an unpatented product in exchange for the licence); (4) mandatory package licensing (which is really just a tie); (5) exclusive grantbacks (ie a guaranteed licence to the *licensor* of any improvements developed through the licence); (6) licensee control over subsequent licences; (7) restraints on sales of unpatented products made with a patented process; (8) restraints on resale; and (9) minimum resale price maintenance. See R Gilbert and C Shapiro, 'Antitrust Issues in the Licensing of Intellectual Property: The Nine No-Nos Meet the Nineties' (1997) Brookings Papers: Microeconomics <http://faculty.haas.berkeley.edu/SHAPIRO/ninenono.pdf> accessed 16 September 2010, 285.

[44] See eg CF Rule, 'The Administration's Views: Antitrust Analysis After the Nine No-No's' (1986) 55 Antitrust LJ 365, 365 ('I look at such speeches simply as a way to atone for the sins of the Division's "forefathers" and their nine no-no's.').

[45] See S 539, 100th Congress, title III (1987); H R 1155, 100th Congress, title III (1987).

have applied a rule of reason analysis to any IP restrictions related to technology licensing. Underlying the shift at the agencies was an increasing recognition that technology patent licensing, and especially many of the practices condemned under the Nine No-Nos, can be pro-competitive. For instance, grantbacks can be used to reduce costs associated with bargaining down the road.[46] Ties are, of course, an essential feature of any intellectual property 'pool', where various owners agree to license multiple pieces of IP on blanket terms. These pools are, in turn, essential to defeat the 'patent thicket'.[47]

The backlash against the Nine No-Nos culminated in 1995 with the release of the Intellectual Property Guidelines authored jointly by the DOJ and the Federal Trade Commission ('FTC'),[48] alluded to above, which explicitly embraced the Chicago-Harvard School consumer welfare-based approach to antitrust.[49] More specifically, the Guidelines adopted three broad animating principles, which are, by their terms, solicitous to the IP right and not suspicious of the supposed IP 'monopoly'. These are: (1) IP is comparable to any other form of property; (2) the DOJ and FTC will not presume that IP rights create antitrust market power; and (3) IP licensing '... allows firms to combine complementary factors of production and is generally pro-competitive.'[50]

The Guidelines are limited to conduct that '... harms competition among entities that would have been actual or likely potential competitors in a relevant market in the absence' of the arrangement.[51] That is, generally speaking, the Guidelines are concerned with conduct that serves to limit competition in a market other than the one in which the product covered by an IP right competes. In other words, '... [t]he Agencies will not require the owner of intellectual property to create competition in its own technology.'[52]

In identifying these markets, the Guidelines take an approach that reflects the particular saliency of IP licensing in the technology context. That is, the Guidelines purport to define relevant product markets as 'goods markets', 'technology markets',

[46] See Rule (n 44 above) 370.

[47] Where blocking patents can prevent a product covered by multiple patents—most notably in the hi-tech and standard-setting context—from being brought to market. Carl Shapiro defines the patent thicket as '... an overlapping set of patent rights requiring that those seeking to commercialize new technology obtain licenses from multiple patentees.' C Shapiro, 'Navigating the Patent Thicket: Cross Licenses, Patent Pools, and Standard Setting' (2000) 1 Innovation Policy and the Economy 119.

[48] See DOJ and FTC, *Antitrust Guidelines for the Licensing of Intellectual Property* (1995) (hereinafter 'IP Guidelines' or 'Guidelines'), reprinted in 4 Trade Regulation Rep (CCH) ¶ 13, 132. Note that the Guidelines apply equally to IP protected by patent, copyright, trade secret or know-how arrangements, but not trademark (which does not generally implicate licensing issues).

[49] Guidelines (n 48 above) § 1.0 ('The intellectual property laws and the antitrust laws share the common purpose of promoting innovation and enhancing consumer welfare.')

[50] *Ibid* § 2.0.

[51] *Ibid* § 3.1.

[52] *Ibid* § 3.1. This is also the statutory law of patent misuse. No patent owner can be denied relief for infringement or held guilty of misuse or illegal extension of the patent right for refusing to license a lawfully acquired patent. 35 USC § 271(d)(4) (2006).

and, possibly, 'innovation markets'. Goods markets are the various final, intermediate, and upstream markets implicated by the IP at issue (and can include the IP itself).[53] Technology markets consist of the intellectual property licensed and its close substitutes—ie, the technologies or goods that can act as a competitive constraint on the licensed IP.[54] Technology markets can also be used to analyse the competitive effects of licensing arrangements where the IP is marketed separately from the underlying products. Finally, 'innovation' markets come into play when the licensing arrangement might impact a firm's incentives to develop new or improved products or technology. For instance, a competitor's R&D efforts may act as a constraint on a rival IP owner's incentive to limit money spent on its own R&D. Were a licensing arrangement to impede that constraint by, for instance, limiting a rival's R&D expenditures, an innovation market could be implicated.[55] This framework is then used to analyse a particular licensing arrangement under the rule of reason.[56]

Crucially, the Guidelines establish a 'safety zone' whereby a licensing arrangement that is not facially anti-competitive (eg, price-fixing) and only affects 20 per cent or less of some relevant market will not raise antitrust concerns.[57] As noted, in the vast majority of cases that fall outside the safe harbour, the Guidelines make clear that the agencies will apply a rule of reason approach in assessing any competitive harm.[58]

The Guidelines represent the victory of the Chicago-Harvard School. That is, they assume that so long as the IP holder is not using his intellectual property to exert pricing, output, or innovation influence on a market that is not covered by the IP, it should be free to deploy its IP as it likes and any licensing action (ie, either the grant or refusal of a licence) should be presumed pro-competitive. Indeed, this first principle of the antitrust treatment of IP licensing has also taken root overseas in, for instance, the European Technology Transfer Block Exemption and the Japanese Licensing Guidelines.[59] The current state of antitrust law as applied to IP licensing reflects this consensus.

C. Current State of the Law and Recent Developments

The law on the antitrust treatment of IP licensing practices generally breaks down into two camps: allegedly anti-competitive conduct arising from the unilateral or

[53] Guidelines (n 48 above) § 3.2.1.

[54] *Ibid* § 3.2.2.

[55] *Ibid* § 3.2.3.

[56] *Ibid* § 4.

[57] *Ibid* § 4.3.

[58] *Ibid* § 4.3 ('The Agencies will analyze arrangements falling outside the safety zone based on the considerations outlined in parts 3–5.').

[59] GF Masoudi, Deputy Assistant Attorney General, Dept of Justice, 'Remarks at the Digital Americas 2006 Meeting on Intellectual Property and Competition: Four Principles for Encouraging Innovation' § B (11 April 2006).

concerted refusal to issue a licence, or allegedly anti-competitive conduct arising from a particular licensing arrangement. Typical claims under the former tend to be treated much like refusals to deal outside the IP context. Claims under the latter usually stem from some restriction in the licence on conduct outside use of the protected IP. Typical claims under the latter include challenges to various flavours of exclusive licence, price maintenance, conduct related to payment of royalties, territorial or customer restrictions, restrictions related to the 'field of use' of the IP, tying arrangements and 'package' licensing, agreements not to challenge the validity of the licensed IP and patent 'pools'. Courts have also had occasion to consider whether group boycotts *not* to license IP from a competitor and the settlement of infringement actions can violate the antitrust laws. Each of these issues is discussed briefly below, with notable recent developments highlighted.

Unilateral refusals to deal

As a general rule, just as there is rarely a duty to deal with a competitor in antitrust law writ large,[60] there is no duty to license one's intellectual property. The IP Guidelines are clear on this point. They explicitly point out that antitrust laws do not '. . . impose on the intellectual property owner an obligation to license the use of that property to others.'[61] This protection of the IP right is entirely consistent with a neo-classical approach to both micro- and macroeconomic theory. It follows from the idea that, without more, legitimate monopolies are pro-competitive. Indeed, it reflects the concern with dynamic versus static efficiency noted above. The lack of a duty to license fits neatly with the notion that the IP right should be protected to drive innovative forces that will result in products and services that compete with the protected IP and will prompt firms to create products worthy of IP protection. This strong presumption that a firm will not be required to license its IP, even at the expense of static efficiency, is nicely illustrated by the general rule that a firm may apply for and obtain IP with the intent *not* to use the invention.[62]

The three foundational cases on unilateral refusals to license IP suggest the limited areas where IP holders should be circumspect when denying a rival a licence. In *Data General Corp v Grumman Systems*, the First Circuit created a rebuttable presumption that unilateral refusals to license are lawful.[63] As is frequently true, the case involved

[60] See *Verizon Communications Inc v Law Offices of Trinko* 540 US 398, 411, 124 S Ct 872, 881 (2004) ('Finally, we do not believe that traditional antitrust principles justify adding the present case to the few existing exceptions from the proposition that there is no duty to aid competitors.').

[61] Guidelines (n 48 above) § 2.2; see also 35 USC § 271(d)(5) (2006) (denying patent misuse relief for refusal to license patents).

[62] *Hartford-Empire Co v United States* 323 US 386, 432-33, 65 S Ct 373, 395–6 (1945).

[63] 36 F3d 1147, 1187 (US Ct of Apps (1st Cir), 1994) (hereinafter '*Data General*'),

> . . . [W]e hold that while exclusionary conduct can include a monopolist's unilateral refusal to license a copyright, an author's desire to exclude others from use of its copyrighted work is a presumptively valid business justification for any immediate harm to consumers.

the 'aftermarket' for services for computers manufactured by one of the litigants, in this case Data General Corp.[64] Grumman argued that Data General was unlawfully refusing to license IP necessary for its diagnostic tools used in servicing Data General's computers.[65] The court there refused to find exclusionary conduct after analysing the point under the rebuttable presumption noted above.[66] The Ninth Circuit created a similar presumption in *Image Technical Services v Eastman Kodak Co*, but modified it slightly to emphasize that the presumption would not rest on 'formalistic distinctions', but would be based on actual market reality and would focus the fact-finder on '... the primary interest of both intellectual property and antitrust laws: public interest.'[67]

In *Re Independent Service Organizations Antitrust Litigation*, however, the Federal Circuit rejected the First and Ninth Circuit's middle-road approach, holding that,

> ... [i]n the absence of any indication of illegal tying, fraud in the Patent and Trademark Office, or sham litigation, the patent holder may enforce the statutory right to exclude others from making, using, or selling the claimed invention free from liability under the antitrust laws.[68]

The Federal Circuit also rejected arguments that a unilateral refusal to license could state an antitrust claim where it has the effect of influencing a market other than that covered by the relevant IP.[69] Because the right to exclude is derivative of the patent, the court reasoned, exercise of that right cannot be an antitrust violation even if it does spill over to secondary markets.[70] The key distinction between *Data General Image Technical*, on the one hand, and *Independent Service Organizations*, on the other, is the latter case's narrowing of the potential zone of liability to only those cases where the '... refusal to sell or license [the] patented invention' causes an anti-competitive effect that extends beyond the *statutory patent grant*, as opposed to beyond the market in which the patented device competes.[71] In other words, if the patent grant creates an aftermarket for service of the patented device, a refusal to license that patent, even if results in an aftermarket monopoly, will not be enough to implicate the antitrust laws.

Despite this split in authority, the weight of the cases permits owners of IP to refuse to license their patents or copyrights with broad discretion. As the Guidelines

---

64 *Data General* (n 63 above) 1152.

65 *Ibid* 1153–5.

66 *Ibid* 1189.

67 125 F3d 1195, 1218 (US Ct of Apps (9th Cir), 1997).

68 *Re Independent Service Organizations Antitrust Litigation* 203 F3d 1322, 1327 (US Ct of Apps (Fed Cir), 2000) (hereinafter '*Independent Service Organizations*').

69 *Ibid* 1327.

70 *Ibid* 1327.

71 *Ibid* 1327–8; see also *Intergraph Corp v Intel Corp* 195 F3d 1346, 1358-59 (US Ct of Apps (Fed Cir), 1999) (finding, absent any evidence of intent to monopolize and no duty to deal, that refusal to provide technical assistance and other special benefits not antitrust violation).

conspicuously emphasized and the Federal Circuit reiterated in *Independent Service Organizations*, there is '. . . no reported case in which a court ha[s] imposed antitrust liability for a unilateral refusal to sell or license'.[72]

Europe, however, goes the other way. It has identified a set of factors under which it will impose an obligation on the owner of IP to license it to a rival. In 2007, the Court of First Instance affirmed the European Commission's 2004 decision holding that Microsoft had abused its dominant position in the market for PC operating systems by, among other things, refusing to supply its competitors with interoperability information to permit them to interoperate with Microsoft's workgroup servers.[73] The Court of First Instance held that in certain 'exceptional circumstances' an owner of intellectual property can be forced to disclose proprietary information to permit interoperability.[74] These circumstances are: (1) when the refusal to disclose relates to a product or service that is indispensable to the exercise of an activity in a neighbouring market; (2) where the refusal to license excludes any effective competition in that market; and (3) where the refusal to disclose prevents the appearance of a new product for which there is potential consumer demand.[75] Thus, Europe has effectively adopted the exact opposite of the Federal Circuit's approach in *Independent Service Organizations*, and will impose a clear duty to deal when a refusal to license precludes all competition in a neighbouring market. In practice, this effectively means that a manufacturer who competes with third party servicers in an aftermarket for the maintenance of its equipment would *have* to arm its competitors with technical data to permit them to compete.[76]

It is also noteworthy that the American mistrust of duties to deal with rivals does not necessarily apply in the context of merger review. The reviewing agencies will not infrequently impose an affirmative duty to deal on rivals as a condition of approving a transaction. For instance, in the recent Ticketmaster/Live Nation deal, approved in 2010 by the DOJ, one of the conditions of clearance was that Ticketmaster license ticketing software and divest certain ticketing assets to a smaller insurgent competitor to permit it to compete head-to-head with Ticketmaster.[77]

As discussed in more detail below, any shift toward the imposition of a duty to deal with rivals in the IP context would represent strong evidence that neo-classical economics is on the outs in this field of law.

---

72 *Independent Service Organizations* (n 68 above) 1326.

73 Case T-201/04 *Microsoft Corp v Commission* [2007] ECR II-3601 (hereinafter '*Microsoft*').

74 *Ibid* 6.

75 *Ibid.*

76 Compare Case C-418/01 *IMS Health Co v NDC Health Co* [2004] ECR I-5039.

77 See Press Release, Dept of Justice, 'Justice Department Requires Ticketmaster Entertainment Inc. to Make Significant Changes to its Merger With Live Nation Inc' (25 January 2010) <http://www.justice.gov/atr/public/press_releases/2010/254540.htm> accessed 16 September 2010.

Concerted refusals to license

Agreements among competitors, however, that serve to restrict licensing activity are an entirely separate creature, and are much more suspect under the antitrust laws. After all, a pure concerted refusal to license IP strongly resembles a group boycott in the non-licensing context, which is subject to per se illegality under § 1 of the Sherman Act.[78] Nevertheless, one can imagine a situation where a concerted refusal to license technology—especially in the arena of high technology with its greater concentration of component and blocking patents—could be seen as pro-competitive, or at least efficiency enhancing. One could also potentially make an argument that concerted refusal to license arrangements among co-owned or component patents enhances dynamic efficiency by protecting the monopoly profits earned through the IP grant and providing an incentive for firms to make the investment to produce these products, products that they would otherwise not manufacture out of fear of blocking or component patents. Additionally, one could see an argument for pro-competitive efficiencies in concerted refusals to license in the standard-setting context where a standard body could insist on certain quality control provisions or information sharing restrictions, to name just a few, for an entity seeking to license the standardized technology and constituent IP.

Exclusivity and exclusive dealing

As with concerted refusals to deal, the treatment of exclusive dealing and exclusivity arrangements in the licensing context is generally consistent with non-IP antitrust jurisprudence and regulatory policy. In arrangements where, for instance, the licensor agrees not to license the technology to anyone else (for instance, in the United States, Apple only licensed the iPhone to AT&T), courts will apply the rule of reason.[79] The Guidelines take the same approach, and even acknowledge that because IP can be stolen more easily than other types of property, some exclusivity agreements that would be illegal in other contexts are permissible.[80] Recently, the issue of exclusivity has come to the fore in cases challenging agreements whereby branded pharmaceutical manufacturers pay potential generic entrants to stay out of the market (these payments are discussed in more detail below). These cases raise interesting questions about the economic framework for antitrust and for neo-classical economic theory (some adherents of which would argue that such payments are competitively neutral because they serve, for instance, to provide a greater incentive for competing but non-generic drug manufacturers to enter the market).

[78] See eg *Primetime 24 Joint Venture v Nat'l Broadcasting, Co, Inc* 219 F3d 92, 102–3 (US Ct of Apps (2d Cir), 2000) ('A concerted refusal to license copyrighted programming to Primetime in order to prevent competition from it is a boycott that, if proven, violates the Sherman Act.'); *Clapper v Original Tractor Cab Co* 165 F Supp 565 (US Dist Ct (SD Ind), 1958) aff'd in part & rev'd in part on other grounds 270 F2d 616 (US Ct of Apps (7th Cir), 1959).

[79] See eg *County Materials Corp v Allan Block Corp* 502 F3d 730 (US Ct of Apps (7th Cir), 2007).

[80] *Ibid*; Guidelines (n 48 above) § 4.1.2.

Price/use maintenance, territorial limitations, and boycotts

The most disputed topic under this subheading is maximum and minimum resale price maintenance, which prompts in the United States some of the most heated ideological battles in antitrust and IP. In a marked divergence from prevailing antitrust law at the time,[81] the 1926 case, *United States v General Electric Co*, permitted a patentee to set the price at which products covered by the relevant IP and produced by a licensee are sold.[82] This broad holding, however, had been whittled down in significant ways over the years. For instance, the Supreme Court refused to apply *General Electric* in a case where IP holders cross-license their patents to enable the marketing of a new product and set the resale price of that product.[83] An IP holder can also clearly not use a licence to set the price of unpatented products, and the Guidelines state that '... [i]t has been held per se illegal for a licensor of an intellectual property right in a product to fix a licensee's *resale* price of that product.'[84]

However, the recent Supreme Court decision *Leegin Creative Leather Products, Inc v PSKS, Inc* overturned the per se rule against minimum resale price maintenance on the books since the *Dr Miles* decision in 1911.[85] This decision may, in the context of IP licensing, prefigure a shift *toward* neo-classical economic theory and the Chicago-Harvard School. In the context of cross-licensing arrangements and patent pools, *Leegin* could be used to argue that the pro-competitive justifications of bringing a new product to market (through the combination of separately owned patents) outweighs any supra-competitive pricing resulting from price maintenance by the licensees. Indeed, such an approach would be more consistent with the law on royalty pricing, which effectively holds that the owner of licensed IP may charge whatever royalty the market will bear.[86] In the context of an IP holder with market power, minimum resale price can be maintained by proxy by charging a high royalty rate.

Similar to price maintenance are use and quantity restrictions. As a general rule, an IP owner may limit the number of licensed products produced and sold.[87] An IP holder may also limit the customers to which the licensed products are sold. Further, an IP owner may generally limit the 'field of use' of the licensed IP, but again may not use a field of use restriction to extend the IP grant beyond its scope or to achieve a

[81] See *Dr Miles Medical Co v John D Park & Sons Co* 220 US 373, 31 S Ct 376 (1911) (holding minimum resale price maintenance per se illegal) overruled by *Leegin Creative Leather Products v PSKS, Inc* 551 US 877, 127 S Ct 2705 (2007).

[82] 272 US 476, 47 S Ct 192 (1926).

[83] *United States v Line Material Co* 333 US 287, 68 S Ct 550 (1948).

[84] Guidelines (n 48 above) § 5.2 (emphasis in original); see also *United States v Univis Lens Co* 316 US 241, 62 S Ct 1088 (1942); *Ethyl Gasoline Corp v United States* 309 US 436, 60 S Ct 618 (1940).

[85] 551 US 877, 127 S Ct 2705 (2007); *Dr Miles Medical Co v John D Park & Sons Co* 220 US 373, 31 S Ct 376 (1911).

[86] See eg *Stewart v Abend* 495 US 207, 110 S Ct 1750 (1990).

[87] See eg *United States v EI du Pont de Nemours & Co* 118 F Supp 41 (US Dist Ct (D Del), 1953) aff'd on other grounds 351 US 377, 76 S Ct 994 (1956).

horizontal market allocation.[88] Territorial restrictions are also generally permissible. They only pose concerns when they are pretextual arrangements to further a more sinister scheme, like horizontal market division.[89]

Finally, in a twist on the concerted refusal to license situation, in at least one recent case, an agreement among competitors *not* to negotiate with the IP holder for a licence was successfully challenged.[90]

Royalties

The antitrust treatment of royalty payments under IP licences is one of the more active areas of law in recent years, and is particularly relevant in the standard-setting context. Note that one of the key insights of the Chicago-Harvard School is that monopoly itself is not an evil; rather the unreasonable creation or maintenance of monopoly should be the target of antitrust laws. Accordingly, monopolistic pricing—lawfully imposed—should not be subjected to a penalty (and, indeed, should be encouraged as it creates the incentive for vigorous competition). As a corollary, the owner of IP should be permitted to exploit it to the hilt and, as a general matter, the law permits an IP holder to set whatever royalty it wants.

There are a few exceptions under the current state of the law. It can be problematic to base the royalty formula on something separate from the products or services created under the licence.[91] Recently, the question of discriminatory royalties has come to the fore in connection with the *Broadcom Corp v Qualcomm Corp* decision in the Third Circuit reversing the dismissal of Broadcom's claim that Qualcomm committed an antitrust violation by breaching its commitment to license its W-CDMA mobile telephone technology on 'fair, reasonable, and non-discriminatory' (or 'FRAND') terms.[92] One of the allegations in that complaint, which the court allowed to proceed, was that Qualcomm charged more to companies to license the W-CDMA

[88] See *Gen Talking Pictures Corp v West Elec Co* 304 US 175, 58 S Ct 849 aff'd on rehearing 305 US 124, 59 S Ct 116 (1938).

[89] See eg *Timken Roller Bearings Co v United States* 341 US 593, 598-99, 71 S Ct 971, 974–5 (1951) overruled in part on other grounds by *Copperweld Corp v Independence Tube Corp* 467 US 752, 104 S Ct 2731 (1984).

[90] *Golden Bridge Technology v Nokia, Inc* 416 F Supp 2d 525 (US Dist Ct (ED Tex), 2006).

[91] See eg *United States v Microsoft Corp* 56 F3d 1448 (US Ct of Apps (DC Cir), 1995) (challenging Microsoft's imposition of a royalty on each computer shipped irrespective of whether Microsoft's operating system was installed). But, in a recent district court case in Illinois, the court refused to find liability where the licence provided for payment of a royalty in situations where the patented product could be used (ie, payment was not necessarily tethered to *actual use*). The court refused to find a claim of patent misuse because licensees could effectively use the stock trading software at issue to get the best price but then use other software to make the trade. Were the royalty provision written differently, licensees would then be able to totally avoid payment. *Trading Technologies International Inc v eSpeed, Inc* No 04 C 5312, 2008 US Dist LEXIS 4627, at *11 (US Dist Ct (ND Ill), 18 January 2008).

[92] See 501 F3d 297, 304 (US Ct of Apps (3d Cir), 2007) (hereinafter '*Broadcom v Qualcomm*').

technology if they did not use Qualcomm's chipsets.[93] Likewise, it can be patent misuse for an IP holder to demand royalty payments beyond the expiration of the IP right.[94]

Finally, courts will entertain claims that licensees effectively conspired to fix the price of the patented product through the imposition of royalties. In *Schoenbaum v E.I. DuPont de Nemours & Co*, for instance, the district court denied the defendants' motion to dismiss a case against a royalty arrangement.[95] Monsanto licensed its genetically modified seeds to DuPont and Pioneer Hi-Bred International.[96] The plaintiffs alleged that DuPont and Pioneer conspired to collect a fixed technology fee from farmers for payment to Monsanto in exchange for Monsanto sharing a portion of its supra-competitive profits.[97] The plaintiffs alleged, successfully, that the arrangement served to fix the final prices of the genetically modified seeds produced with the Monsanto technology.[98]

Tying

Tying is arguably the classic claim related to IP licensing, and perhaps the one most often alleged. It closely resembles the patent misuse cases described earlier where the sale of a licence is conditioned on purchase of some good or service that is not covered by the relevant IP. Tying in the licensing context should usually not be an economic concern, especially to adherents of the Chicago-Harvard School.[99] Most courts require plaintiffs to meet four elements to succeed on a per se tying claim: (1) the existence of two separate products; (2) sale conditioned on purchase of something else; (3) market power sufficient to restrain the tied market; and (4) an effect on a 'not insubstantial amount' of interstate commerce.

Notably, in line with the discussion above about the Chicago-Harvard School and economics-centric approach adopted by modern antitrust authorities, one of the more significant cases of recent years involves an IP licensing tie. In *Illinois Tool Works v Independent Ink, Inc*, the Supreme Court firmly rebuffed earlier cases suggesting that the IP right established market power sufficient to meet the third element above.[100] The allegation there was that one company without market power in the market for printers that print bar codes on packages illegally tied the sale of ink to

[93] *Broadcom v Qualcomm* (n 92 above) 318.

[94] See eg *Zila, Inc v Tinnell* 502 F3d 1014 (US Ct of Apps (9th Cir), 2007) (finding that even clear contractual intent to pay royalty in perpetuity is unenforceable).

[95] 517 F Supp 2d 1125 (US Dist Ct (ED Mo), 2007) vacating judgment in part No 4:05CV01108 ERW, 2007 WL 3331291 (US Dist Ct (ED Mo), 6 Nov 2007) (hereinafter '*Schoenbaum v DuPont*').

[96] *Schoenbaum v DuPont* (n 95 above) 1129.

[97] *Ibid* 1131.

[98] *Ibid* 1144.

[99] See RA Posner 'Transaction Costs and Antitrust Concerns in the Licensing of Intellectual Property' (2005) 4 John Marshall Rev Intellectual Property L 325; Posner, *Chicago School* (n 2 above) 926.

[100] 547 US 28, 126 S Ct 1281 (2006).

its equipment. The Court rejected the claim, finding that competition in the foremarket would serve to constrain pricing in the aftermarket for ink.

A related claim is bundled licensing, which is generally permissible so long as there is no conditioned sale. The classic cases involved block booking of motion pictures (ie, forcing the theatre to agree to exhibit separately copyrighted films).[101] However, simply bundling related patents together without any restriction or requirements regarding use will likely not warrant per se treatment.[102] The recent *Princo* decisions, arising from the same dispute as *Philips Corp*, support this general proposition.[103] There, Sony and Philips jointly agreed to bundle two rival technologies into one patent licence required for the manufacture of CD-R and CD-RW discs.[104] The Federal Circuit held in part that where there existed a reasonable belief that one patent could block another, it was not patent misuse to tie the two in a single licensing package.[105] The panel decision noted that while tying patents in a package licence certainly has anti-competitive potential, it can also be markedly pro-competitive in 'clearing blocking patents, integrating complementary technology, and avoiding litigation'.[106]

Finally, there have been some interesting cases recently dealing with the two-products requirement in connection with franchising. These suggest that bulk licensing of IP in connection with franchising or similar arrangements may not constitute a tie where there is some functional interconnection between the relevant IP.[107]

Patent pools and cross-licensing

The key distinction when assessing the legality of patent pools, which are generally analysed under the rule of reason, is whether the pooled patents include 'blocking' patents or competing patents.[108] If the latter, a patent pool with a set royalty is akin to price fixing.[109] However, with blocking patents, the courts have realized since the

---

[101] See *United States v Loew's, Inc* 371 US 38, 83 S Ct 97 (1962); *United States v Paramount Pictures* 334 US 131, 68 S Ct 915 (1948).

[102] *US Philips Corp v ITC* 424 F3d 1179 (US Ct of Apps (Fed Cir), 2005).

[103] See *Princo Corp v ITC* 563 F3d 1301 (US Ct of Apps (Fed Cir), 2009), aff'd on reh'g, No 2007-1386, slip op (US Ct of Apps (Fed Cir), 30 Aug 2010) (en banc).

[104] *Princo*, No 2007-1386, slip op at 4–6.

[105] *Princo*, 563 F3d at 11.

[106] *Ibid* (quoting *Princo*, 563 F3d at 1308).

[107] See eg *Rick-Mik Enterprises Inc v Equilon Enterprises LLC* 532 F3d 963, 974 (US Ct of Apps (9th Cir), 2008) (finding a single product where franchise agreement required franchisee to purchase credit card processing services from the franchisor).

[108] See *Carpet Seaming Tape Licensing Corp v Best Seam Inc*, 616 F2d 1133, 1142 (US Ct of Apps (9th Cir), 1980) ('A well-recognized legitimate purpose for a pooling agreement is exchange of blocking patents.').

[109] Guidelines (n 48 above) § 5.1 ex. 9:

> In the absence of evidence establishing efficiency-enhancing integration from the joint assignment of patent rights, the Agency may conclude that the joint marketing of competing patent rights constitutes horizontal price fixing and could be challenged as a per se unlawful horizontal restraint of trade.

early days of IP licensing antitrust law that such arrangements are pro-competitive and essential to clear up patent 'thickets'.[110]

The *Princo* case, mentioned earlier, involves several notable issues relevant to the antitrust treatment of patent pools. Most significant is the allegation, that Sony and Philips agreed to include Sony's 'Lagadec' IP in the patent pool to prevent it from emerging as a viable competitor to the standardized technology. The Federal Circuit found that, while such conduct could conceivably represent an antitrust violation, it does not present a viable patent misuse defence in an infringement action.[111]

In addition, patent pools that serve to retard innovation have been challenged under the antitrust laws, as the Guidelines note.[112]

Generic settlements

Generally, when patent litigations settle, conditions imposed on the parties (eg cross-licensing or patent pools) may raise antitrust concerns. These are usually handled under ordinary antitrust theories discussed above.

However, a substantial body of case law has developed from challenges to settlements of litigation arising out of the Hatch-Waxman Act.[113] Under this Act, designed to encourage 'generic' entry in the pharmaceutical industry, a 'brand name' drug manufacturer claiming infringement against a potential generic entrant gets an automatic 30 month stay of FDA approval (and a successful generic entrant gets a period of 180 days exclusivity before other generics can enter the market).[114] When these infringement suits settle, the generic manufacturer usually agrees to certain conditions on its activities but will receive payment to stay out of the market. Both public and private plaintiffs typically challenge these 'pay-to-delay' or 'reverse settlement' payments as antitrust violations.[115]

The federal appeals courts have contentiously split on this issue.[116] The Second and Eleventh Circuits have essentially held that generic settlements will not violate the

---

[110] See *Standard Oil Co v United States* 283 US 163, 51 S Ct 421 (1931). In *Zenith Radio Corp v Hazeltine*, however, the Supreme Court found that an arrangement structured to exclude a competitor from the market by restricting members of the pool from licensing outside of the pool may violate Section 1. 395 US 100, 89 S Ct 1562 (1969).

[111] *Princo* (n 103 above) at 29, 41.

[112] See eg *United States v Automobile Mfrs Assoc* 307 F Supp 617 (US Dist Ct (CD Cal), 1969) appeal dismissed *City of New York v United States* 397 US 248, 30 S Ct 1105 (1970).

[113] Drug Price Competition and Patent Term Restoration Act, Pub L 98-417, 98 Stat 1585 (1984).

[114] *Ibid* § 505(j)(5)(B)(iv).

[115] See eg L Layton, 'FTC Sues in "Pay-for-Delay" Pact' *Washington Post* (Washington 3 February 2009) <http://www.washingtonpost.com/wp-dyn/content/article/2009/02/02/AR2009020202968.html> accessed 16 September 2010.

[116] Though note that the DOJ, in its unusual amicus brief adverse to the FTC arguing that the Supreme Court should not grant certiorari in *Schering-Plough Corp v FTC* 402 F3d 1056, 1068 (US Ct of Apps (11th Cir), 2005), claimed no such split (because the reverse payments at issue in the case

antitrust laws unless there was fraud on the patent office, the suit for infringement is objectively baseless, or the settlement includes restrictions on products not covered by the relevant patents.[117] The Sixth Circuit, by contrast, struck down as per se unlawful a $40 million reverse payment in *Re Cardizem CD Antitrust Litigation*.[118] Recently, in a long-running case regarding a pay-to-delay deal for the popular antibiotic Cipro, the Second Circuit dismissed plaintiffs' claims based on precedent set in the decision in its *Tamoxifen* case. However, in a highly unusual move, the three-judge panel called for en banc review, citing '... the compelling reasons to revisit *Tamoxifen* with the benefit of the full Court's consideration of the difficult questions at issue and the important interests at stake.'[119] In effect, the Second Circuit panel *asked* to be overruled. Although the full court declined in September 2010 to hear the case en banc, the panel's highly unusual request underscores the dynamic state of this area of law. Similarly, the FTC—led by now-Chairman Leibowitz—has explicitly declared war against pay-to-delay settlements, against the weight of prevailing judicial precedent.[120] In a recent FTC staff study, the Commission concluded that pay-to-delay settlements cost consumers as much as $3.5 billion a year, and recommended that Congress pass legislation banning the practice.[121] As discussed below, a further shift in this direction would signal a trend away from the Chicago-Harvard School.[122]

D. Canaries in the Coal Mine: Indicators of a Post-Chicago/Harvard World

As a general matter, the retreat of the Chicago-Harvard School in the context of antitrust and IP licensing would likely have two main themes.

The first is, quite simply, more regulation. The primary corollary of neo-classical economics is that market failures are, with few exceptions, quickly self-correcting. Though certain participants will win and others will lose, vigorous competition and

cited by the FTC restricted the generic's ability to market drugs not subject to the patent-in-suit). See Brief for the United States as Amicus Curiae, *Schering-Plough Corp v FTC* No 05-273 (2006).

[117] *Re Tamoxifen Citrate Antitrust Litig* 429 F3d 370, 396 (US Ct of Apps (2d Cir), 2005), amended by 466 F.3d 187 (US Ct of Apps (2d Cir), 2006); *Schering-Plough Corp v FTC* 402 F3d 1056, 1068 (US Ct of Apps (11th Cir), 2005); see also *Asahi Glass Co v Pentech Pharms* 289 F Supp 2d 986 (US Dist Ct (ND Ill), 2003) (Posner, J, sitting by designation).

[118] 332 F3d 896 (US Ct of Apps (6th Cir), 2003).

[119] *Arkansas Carpenters Health and Welfare Fund v Bayer AG* 604 F3d 98 (US Ct of Apps (2d Cir), 2010), reh'g denied, No 05-2851-CV L, 05-2852-CV CON, 05-2863-CV-CON, slip op (US Ct of Apps (2d Cir), 7 Sep 2010).

[120] See eg *Pay-For-Delay: How Drug Company Pay-Offs Cost Consumers Billions* (FTC Staff Study 2010) <http://www.ftc.gov/os/2010/01/100112payfordelayrpt.pdf> accessed 16 September 2010.

[121] *Ibid.*

[122] Presumably, the Chicago-Harvard School would generally argue that so long as the settlement does not restrict products not covered by the IP right, settlement in these cases would be competitively neutral or positive in that the reverse payment and survival of the patent right still create an incentive for either entry by generics or entry of competing products.

sane and limited government oversight will, the Chicago-Harvard School posits, maintain a general equilibrium and cyclic but progressive growth.[123] Given the events of the past two years, including the failure of Lehman Bros and the subsequent multi-hundred-plus billion dollar government bailout of the global financial sector, the self-correcting market is under significant assault from many quarters, many unexpected (though the jury on whether the free market, properly maintained, is inherently unstable remains well out).[124] Accordingly, one of the primary 'canaries' when assessing ideological shifts in law and economics is a general increase in regulation. One simply needs to compare the vast increase in the regulatory state from the late 1950s to the early 1970s with the dramatic deregulation of the Reagan years to get a sense of what a shift such as that would look like. Neo-classical economics was the guerilla insurgent in the former era and the established paradigm in the latter. In the context of IP and competition law, one could point to certain doctrinal or jurisprudential shifts, such as a greater willingness to impose a duty to deal or to challenge pay-to-delay or reverse settlements, as evincing a greater regulatory bent. These are discussed in more detail below.

The second bell-wether area would be in the relative strength of the IP right versus antitrust regulation and vice versa. Here, again, the ongoing debate is between 'dynamic' and 'static' efficiency. According to many, 'static' efficiency is associated with neo-classical economics. That is, the Chicago-Harvard School's focus on consumer welfare leads it to an undue concern for short-term pricing versus potential long-term incentives for innovation and evolution in a given market. This position is disputed—many in the Chicago or Harvard Schools would argue that monopoly lawfully achieved is the ultimate driver of dynamic efficiency. By contrast, 'dynamic' efficiency would be associated with zealous defenders of the IP right versus antitrust regulation. What is ironic about this interplay is that one could point to 'dynamic' efficiency to justify all manner of potential policy choices, many of which are mutually inconsistent. And this irony and inconsistency just goes to showcase the doctrinal failings (ie, unpredictability, instability, lack of internal consistency) of the theories competing with the Chicago-Harvard School.

More regulation

The Chicago-Harvard School adopts rational choice theory precisely because it describes an ordered, self-correcting market. The profit motive creates competition,

---

[123] Compare JT Rosch 'The Redemption of a Republican' (FTC Watch 2009) <http://www.ftc.gov/speeches/rosch/090601redemption.pdf> accessed 16 September 2010,

> I have questioned the basic tenets of orthodox Chicago School law and economics . . . that antitrust law is concerned with maximizing societal welfare; that markets are generally perfect; that, if imperfect, they can and will correct themselves; that, accordingly, rational businesspeople will not engage in predatory conduct (because it is not profit-maximizing since markets will correct themselves).

[124] See eg Posner, *Failure* (n 12 above); Cassidy (n 2 above) 28 (quoting Judge Posner as saying 'So probably the term "Chicago School" should be retired.').

which in turn allocates resources and effort efficiently—which in turn allows consumer welfare to flourish by elevating the best products and talent and punishing the lazy and/or incompetent. By arguing that markets do not and cannot self-correct because buyers and sellers are inherently precluded from making decisions in their self-interest, non-Chicago-Harvard School theorists will reject certain assumptions in the context of IP licensing.

A good test case here would be the reverse payments discussed above. A Chicago-Harvard School economist would likely argue that these reverse payments, when limited to the patented technology, protect the patent right and thus the incentive for other firms to expend resources on developing new and better products. Similarly, the reverse payment could provide an additional incentive for generics manufacturers to enter the market. That is, it provides generics manufacturers an added cushion. If the patent turns out to have strong claims, the generics manufacturer can settle the infringement suit. If the patent is weak, the generics manufacturer can continue the fight and enter the market.

New pay-to-delay or reverse settlement payments cases could also showcase a greater emphasis at the regulatory agencies on inchoate concerns—which are totally alien to neo-classical economic theory—for 'fairness' or equity. Even if payments to generics manufacturers that have the effect of keeping them out of the market are competitively positive or neutral (ie, they do serve to create an incentive for later entry or for the development of competing but not bioequivalent and therefore patentable pharmaceuticals), the added cost for individuals purchasing the branded drugs during the temporary period of supra-competitive pricing is simply 'unfair' and therefore should be regulated.

Precisely these types of nebulous fairness concerns were cited, in the merger review context, by the current FTC Chairman, Jon Leibowitz, and his colleague J Thomas Rosch in their concurring opinions in the 2008 *Ovation Pharmaceuticals* case.[125] For many years, there was only one drug indicated to treat a serious heart condition (called patent ductus arteriosus), which is life-threatening for babies born prematurely.[126] That drug, Indocin, was produced by Merck, which had a broad product portfolio.[127] In 2005, the rights to Indocin were purchased by Ovation, a smaller pharmaceutical company that focused only on producing that one drug.[128] Shortly thereafter,

[125] J Leibowitz, Concurring Statement of Commissioner Leibowitz, *FTC v Ovation Pharmaceuticals, Inc* (16 December 2008) <http://www.ftc.gov/os/caselist/0810156/081216ovationleibowitzstmt.pdf> accessed 16 September 2010 (hereinafter '*Concurring Statement*'); JT Rosch, Concurring Statement of Commissioner Rosch, *FTC v Ovation Pharmaceuticals, Inc* (16 December 2008) <http://www.ftc.gov/os/caselist/0810156/081216ovationroschstmt.pdf> accessed 16 September 2010.

[126] Complaint for Permanent Injunction and Other Equitable Relief, Including Disgorgement of Unlawful Monopoly Profits, *FTC v Ovation Pharmaceuticals* (16 December 2008) (hereinafter '*Complaint, Ovation Pharmaceuticals*') *available at* <http://www.ftc.gov/os/caselist/0810156/081216ovationcmpt.pdf.> accessed 16 September 2010

[127] *Complaint, Ovation Pharmaceuticals* (n 126 above).

[128] *Ibid.*

a potential competing drug, Neoprofen, entered the approval pipeline. In 2006, Ovation purchased the rights to Neoprofen—and proceeded to raise the price by more than 1000 per cent.[129] The FTC rightly sought to disgorge profits and unwind the purchase of Neoprofen, because the acquisition was effectively the creation of a total monopoly.[130] The then-Commissioner Leibowitz, however, included the following statement in his concurring opinion,

> Ovation's profiteering on the backs of critically ill premature babies is not only immoral, it is illegal. Moreover, the company's behavior is a stark reminder of why America desperately needs health care reform and why vigorous antitrust enforcement is as relevant today as it was when the agency was created almost one hundred years ago in 1914.[131]

These concerns over fairness, morality, and health policy have no place in neo-classical antitrust policy. Similar verbiage in cases challenging pay-to-delay or reverse settlement payments would suggest a shift away from neo-classical economics in the IP and antitrust context.

Another good example would be the standard setting cases. Currently, there are two competing judicial approaches to the problem. In *Broadcom v Qualcomm*, the Third Circuit allowed a claim to go forward challenging Qualcomm's breach of its alleged promise to license patents essential for mobile communications technology on FRAND terms.[132] The court even went so far as to suggest that an *Aspen Skiing* unilateral refusal to deal claim could work here.[133] By contrast, the DC Circuit reversed the FTC on a very similar claim in *Rambus, Inc v FTC*, in which the FTC alleged that Rambus failed to adequately disclose intellectual property in the standard setting process, allowing it to lock in its patents and extract supra-competitive royalties.[134] There, the Court keyed into the causation issue, expressing significant doubt that, but for the alleged non-disclosure, a non-proprietary or FRAND-licensed technology would have been incorporated.[135] The key pivot point here is that the neo-classical economists would argue that, even when one's IP is locked into a standard (be it through deceit, as was alleged in *Broadcom*, or otherwise), the markets will still self-correct and courts and agencies should remain hands off. By contrast, sceptics would argue that it is relatively easy, and relatively common, for firms to attempt to have their IP integrated into a standard, which can then be exploited to garner

[129] *Ibid.*
[130] *Ibid.*
[131] Leibowitz, *Concurring Statement* (n 125 above) 1.
[132] 501 F3d 297 (US Ct of Apps (3d Cir), 2007).
[133] *Ibid* 316.
[134] 522 F3d 456 (US Ct of Apps (DC Cir), 2008).
[135] *Ibid* 466–7.

supra-competitive royalties. Again, a shift in this direction could prefigure a shift away from neo-classical theory in the realm of IP licensing.[136]

It should also be noted in this context that any shift away from the general approach of the 2007 DOJ/FTC report on antitrust and IP licensing could be a litmus test for any change in approach.[137] It, like the Guidelines, acknowledges the pro-competitive benefits of many licensing arrangements previously thought dangerous (eg, tying or extending royalties beyond the life of the patent).

Indeed, the 2007 report was cited in the now repudiated Bush administration report on § 2 liability, which suggested that a unilateral refusal to license intellectual property can never give rise to a claim under § 2.[138] If the FTC and/or DOJ were to revisit this aspect of the agencies' single-firm theory, it would be a significant departure from orthodox Chicago-Harvard School theory (which holds that compulsory licensing reduces static efficiency by encouraging inefficient manufacturers to enter the market).

Recalibrating the interface

As noted earlier, the antitrust and IP laws have historically cycled in and out of relative prominence. In the early years of the interface, the IP laws arguably won out as the courts excused conduct extending the scope of the patent right beyond its appropriate bounds. Later, antitrust came back with a vengeance, condemning as per se unlawful what would now be considered relatively innocuous conduct.

One possibility would arguably de-emphasize antitrust in favour of certain IP rights. Here, courts and agencies would begin to focus more on what's called 'dynamic' efficiency rather than 'static' efficiency. Static efficiency refers to efficiency in price and output setting (and is generally the bailiwick of antitrust, and especially antitrust informed by neo-classical price theory). However, static efficiency (by imperiling the sanctity of the IP right) can weigh on dynamic efficiency, which refers to the more long-term tendency of brand new technologies to upset the apple cart totally.[139] Dynamic efficiency arguments can challenge Chicago-Harvard School assumptions about rationality because they suggest that firms will not act rationally in price and output decisions in service of future innovation gains. Ironically, this challenge to the

[136] See eg Rosch, *Financial Meltdown* (n 16 above) 4 (citing American enforcers promoting trust in the free market in the context of intellectual property licensing, and standard setting specifically).

[137] DOJ and FTC, *Antitrust Enforcement and Intellectual Property Rights: Promoting Innovation and Competition* (2007) <http://www.ftc.gov/reports/innovation/P040101PromotingInnovationand Competitionrpt0704.pdf> accessed 16 September 2010.

[138] See Department of Justice, *Competition and Monopoly: Single-Firm Conduct Under Section 2 of the Sherman Act* (2008) <http://www.justice.gov/atr/public/reports/236681.pdf> accessed 16 September 2010, 124.

[139] Or 'Apple' cart. As Mark Lemley put it, '. . . [a]sk yourself whether you'd rather pay monopoly price for an iPod or a competitive price for an eight-track player.' Lemley (n 38 above) 248.

Chicago-Harvard School would result in *less* antitrust regulation, not more.[140] A shift in favour of dynamic competition could also manifest itself in any revision of the Horizontal Merger Guidelines or the IP Guidelines to emphasize the potential importance of dynamic technological change in any given market.[141] In the context of merger review, the Bush administration already took a giant leap in that direction in the recent XM/Sirius merger. There, the DOJ and FCC cleared a merger to monopoly in the satellite radio space,[142] reasoning that dynamic competition concerns required the inclusion of non-satellite radio technology as actual or potential competitors with a merged XM/Sirius.[143] A similar analysis in any of the licensing contexts discussed in this chapter could signal a shift in favour of dynamic competition—and away from the pure consumer welfare focus of neo-classical economics—in the IP licensing/antitrust space.

These are just a few examples of areas where behavioural economics and its cousins could make inroads at the intersection of IP and antitrust in the United States. Again, the fundamental marks of a shift away from the primary of neo-classical, Chicago-Harvard School economics in IP licensing/antitrust law would be either an increase in regulation or a recalibration of the balance between the protection of the IP right and antitrust regulations that police abuses of that right. Given the prevailing political and economic winds, a shift away from neo-classical economics seems more likely than not.

---

[140] See eg JG Sidak and DJ Teece, 'Dynamic Efficiency in Antitrust Law' J Competition L and Economics (forthcoming).

[141] *Ibid* 23.

[142] Note that the regulatory agencies did not find the satellite radio space to be a distinct product market for antitrust purposes.

[143] See Press Release, Dept of Justice, 'Statement of the Department of Justice Antitrust Division on its Decision to Close its Investigation of XM Satellite Radio Holdings Inc.'s Merger With Sirius Satellite Radio Inc' (24 March 2008) <http://www.justice.gov/opa/pr/2008/March/08_at_226.html> accessed 16 September 2010,

> ... [T]he evidence did not show that the merger would enable the parties to profitably increase prices to satellite radio customers for several reasons, including... the competitive alternative services available to consumers [and] technological change that is expected to make those alternatives increasingly attractive over time.

20

SHIFTING TOWARDS A DYNAMIC EFFICIENCY TEST?: EVALUATING LICENSING AGREEMENTS UNDER ANTITRUST LAW

Vladimir Bastidas Venegas

Introduction

Although the intersection of patent and antitrust law (hereinafter 'the intersection') has been turned and twisted by judges, practitioners, and academics for over a century, it still remains unclear, giving rise to fierce legal and intellectual debate. This intensity is hardly surprising. Both innovation and competition are essential in market systems. To what an extent these areas of law clash or strive for the same objective is an issue disputed in itself, illustrating the confusion and diversity of opinions amongst commentators. Though recently the focus of the attention recently has been on cases of unilateral conduct, the main battleground concerns the assessment of cooperation through licensing agreements. Licensing agreements are intriguing because of their complex economic effects, rendering assessments under antitrust law very difficult. How much leeway should be given to patent holders and the licensees with respect to preserving the objectives of the patent system of innovation and dissemination, while at the same time, minimizing any negative effects on rivalry and consumer welfare?

In the light of this debate, the objective of this chapter is to examine and analyse the evolution of different standards and tests applied to licensing agreements as proposed by different legal scholars. These will be evaluated from an economic perspective and subsequently used to analyse the current EU competition rules as applied to licensing agreements.

This chapter comprises three parts. First, in Part A, a short summary will be presented of the most important elements as regards innovation and licensing from an economic perspective. Second, in Part B, the different standards and tests will be presented and analysed. Finally, in Part C, an evaluation will be made of the approach expressed by the Commission in Regulation 772/2004 on the application of Article 81(3) of the Treaty (now 101(3) TFEU) to categories of technology transfer agreements (hereinafter the 'TTBER'),[1] as well as in the Guidelines on the application of Article 81 of the EC Treaty (now 101 TFEU) to technology transfer agreements (hereinafter referred to as the 'TT Guidelines').[2]

A. Economic Features of Licensing Agreements

Economic and legal scholarship addressing the economic rationales and welfare effects of licensing agreements has always been relatively scarce. However, following the ideas of New Institutional Economics on transaction costs and the choices between different forms of organization, a body of literature relevant to licensing has slowly formed.[3] This literature examines the costs and incentive structures associated with specific forms of transactions, such as mergers or distribution contracts. The form of the transaction generally affects the possibility of the parties to protect themselves from transactional hazards (costs) as well as the incentives of the parties to make specific investments into the contractual relation, increasing the efficiency of the transaction in question. This particular branch of economic theory is therefore interesting for antitrust law as it explains the rationale of different types of agreements and clauses. This literature points towards several factors that are pertinent for an assessment of licensing under antitrust law.

To begin with, it is important to make a clear distinction between inventions and innovations.[4] An *invention* is that which is protected by a patent right, while an *innovation* constitutes the end-result, the product which also has been manufactured, marketed, and distributed to the end-user.[5] This may consist of one or several

[1] Commission Regulation (EC) 772/2004 on the application of Article 81(3) of the Treaty to categories of technology transfer agreements [2004] OJ L 123/11.

[2] Commission Notice, Guidelines on the application of Article 81 of the EC Treaty to technology transfer agreements [2004] OJ C 101/2.

[3] See *inter alia* OE Williamson, *The Economic Institutions of Capitalism* (1985); JE Oxley, 'Appropriability hazards and the governance in strategic alliances' (1997) 13 J of L, Economics, and Organization 387; E Caves, H Crookell and JP Killing, 'The Imperfect Market for Technology Licenses' (1983) 45 Oxford Bulletin of Economics and Statistics 249; C Bessy and E Brousseau, 'Technology Licensing Contracts—Features and Diversity' (1999) 18 Intl Rev L and Economics 451.

[4] R Eisenberg, 'Patents and the Progress of Science: Exclusive Rights and Experimental Use' (1989) 56 U Chicago L Rev 1017, 1038.

[5] M Carrier, 'Resolving the Patent-Antitrust Paradox Through Tripartite Innovation' (2003) 56 Vanderbilt L Rev 1047, 1054 (hereinafter 'Tripartite Innovation').

particular inventions. This distinction brings to light the fact that the process of innovation normally does not end with the patented invention itself but requires additional inputs that in turn require additional investments either by the patent holder or other parties contributing to the innovation process.[6] This has crucial implications for the assessment of licensing agreements, as any protection of innovation cannot limit itself to the protection of a patent holder's initial incentives for investing in the R&D of the invention covered by the patent. The possibilities of executing a specific type of transaction as required by the innovation process, eg mergers, joint ventures, licensing agreements, must also be protected. Consequently, promoting innovation implies the protection of parties other than the patent holder in order for efficient transactions to take place. An antitrust assessment cannot solely focus on the 'reward' of the patent holder but must give room for the division of profits between the cooperating parties. Unfortunately, many economists and legal scholars still fail to consider the implications of the distinction between inventions and innovations in their evaluations of licensing agreements under antitrust law. Their focus remains on the share of the patent reward captured by the patent holder.

Secondly, innovation is to a high degree cumulative.[7] Notably, this impacts the balance between innovation incentives and competition. In patent law, strong patent rights cannot always be presumed to have only positive effects on innovation as follow-up innovation may also be discouraged.[8] Likewise, if antitrust law protects patent holders too much, this may also have a disincentive effect on the second generation of innovators.[9] A broad view of innovation must consequently consider the effect of incentives/disincentives on follow-up innovation.

Thirdly, it is widely acknowledged that licensing agreements are generally characterized by high transaction costs,[10] perhaps to a larger extent than many other types of contracts. Demand is often uncertain regarding innovation *ex ante*,[11] causing difficulties in designing contracts that permit patent holders to appropriate sufficient value from the patent, and at the same time, provide licensees with adequate incentives

[6] DJ Teece, 'Profiting from technological innovation: Implications for integration, collaboration, licensing and public policy' (1986) 15 Research Policy 285, 288–90.

[7] TM Jorde and DJ Teece, 'Rule of reason of horizontal arrangements: agreements designed to advance innovation and commercialize technology' (1992) 61 Antitrust L J 579, 581; ML Tushman, and P Anderson, 'Technological Discontinuities and Organizational Environments' (1986) 31 Administrative Science Quarterly 439, 441.

[8] S Scotchmer, 'Standing on the Shoulder of Giants: Cumulative Research and the Patent Law' (1991) 5 J of Economic Perspectives 29, 31.

[9] This is in particular the case as regards the use of grant-back clauses. See *inter alia* A Gutterman, *Innovation and Competition Policy: A Comparative Study of the Regulation of Patent Licensing and Collaborative Research & Development in the United States and the European Community* (London: Kluwer Law International, 1997) 316.

[10] Caves et al (n 3 above) 250; Oxley (n 3 above) 388. T Wada, 'Hybrid Modes of Licensing and Interfirm Knowledge Transfer' (PhD Dissertation, University of California, Berkeley 2003) 1.

[11] Bessy and Brousseau (n 3 above) 452.

for investing in and developing the licensed technology. Moreover, the nature of *technology*, a form of information, also creates difficulties in its transfer from one party to another. Problems can arise in that the patented *invention*, or the patent disclosure, is insufficient for an effective transfer of technology between patent holders and licensees.[12] In order for such a transfer to be efficient, the patent holder must include additional information (know-how), which in turn opens up the risks of transactional hazards in the form of lock-in or spill-over effects.[13] Furthermore, it may be difficult to specify the nature of the knowledge in the licensing contracts so as to make them enforceable in court.[14] Know-how may consist of tacit information, which is in the 'heads' of engineers that cannot be codified.[15] When patents are weak, it can be particularly difficult for patent holders to protect their technology, either because licensees may use such technology in an 'unauthorized' way, or because the patent rights are open to challenge. These hazards in combination with the fact that the investments made by the parties may be quite substantial, necessitate that licensing agreements protect both parties in order to induce efficient licensing.[16] Naturally, licensing agreements that effectively protect both parties, also typically provide a great deal of control, attracting attention from antitrust authorities. In addition, since transactional hazards may be greater in contracting between competitors (eg because the risks of spill-over are more significant), the degree of control required may be higher, which in turn increases the scrutiny under antitrust law.[17]

If innovation is to be protected under antitrust law, such rules obviously must take these important factors into consideration. Hence, antitrust law needs to incorporate

---

[12] *Ibid* 461.

[13] Spill-over effects emerge because licensees obtain knowledge permitting the development of follow-up innovation that in the future may undermine the position of licensors, see Wada (n 10 above) 38–41. Lock-in effects may emerge due to sunk costs committed to tooling up and which are specific to the transaction. These investments cannot be used if licensees were to license-in technology from another party. This author has not come across an explicit discussion on the existence of such investments (concerning licensing agreements) in the economic literature. Legal scholars usually discuss investments made by licensees to develop the technology and to adapt manufacturing assets, see *inter alia* SD Anderman and J Kallaugher, *Technology Transfer and the New EU Competition Rules—Intellectual Property Licensing after Modernisation* (Oxford: OUP, 2006) 176–7; V Korah, *Intellectual Property Rights and the EC Competition Rules* (Oxford: Hart, 2006) 100; C Ritter, 'The New Technology Transfer Block Exemption under EC Competition Law' (2004) 31 Legal Issues of Economic Integration 161, 177. However, importantly, for these costs to give rise to a transactional hazard, they must be asset specific (as well as susceptible to opportunistic behaviour), see Williamson (n 3), 30–32.

[14] JP Choi, 'A Dynamic Analysis of Licensing: The Boomerang Effect and Grant-Back Clauses' (2002) 43 Intl Economic Rev 803, 804–5.

[15] DJ Teece, 'Capturing value from knowledge assets: The New Economy, Markets for Know-how, and Intangible Assets' (1998) 40 California Management Rev 55, 63.

[16] Bessy and Brousseau (n 3 above) 460.

[17] Following the ideas of the Chicago School, both European and US law use the distinction between horizontal and vertical agreements in the analyses of licences. Additionally, the safe harbour rules for horizontal licences in the EU are stricter. Articles 3 and 4, TTBER; US Department of Justice and the Federal Trade Commission, 'Antitrust Guidelines for the Licensing of Intellectual Property' (1995), section 3.3.

a broader perspective, or definition, of innovation. Individual licensing agreements and licensing restraints must also be better understood with regard to their function in terms of resolving the transactional hazards to which a particular technology transfer is subject. This perspective suggests the significant need for a greater emphasis on assessing the necessity, or indispensability, of individual clauses rather than focusing on market power and market shares. This is not to say that an antitrust assessment should not carry out a determination of competitive harm in terms of increased prices and restriction of output, but rather, that after identifying *prima facie* competitive harm, the justifiability of individual restraints must focus on a more detailed examination of transactional hazards.

B. Theories and Standards

Theories and standards regarding the intersection may be grouped in three different categories. The first is based on a formal division between the scopes of operation of antitrust and patent law, respectively. Cases falling into the grey zone are settled by giving primacy to either patent or antitrust law. These theories are referred to as theories based on the *scope of the patent*. The second category applies a balancing of welfare effects between, on one hand, short-term competition (*allocative efficiency*) and on the other hand, ex ante innovation (*dynamic efficiency*), referred to here as *economic balancing tests*. The third category, based on an innovation-centred approach, means, that the goal of *dynamic efficiency* is incorporated explicitly in the antitrust assessment. Thus there is no conflict between patent and antitrust law, with the promotion of innovation as the most important goal of antitrust law. The following subsections do not address the legal tests as established by the courts, but rather the theoretical framework as developed by scholars. An exception, however, is made with regard to the inherency doctrine as created by the US courts, making a discussion of case law inevitable.

Patent scope theories

Patent scope theories, or scope-of-the-patent theories, invoke the legal rights inherent in the patent grant in the analysis of potentially anti-competitive behaviour. This analysis takes two forms. On one hand, the patent scope may be viewed purely from a legal perspective and then the rights that fall within or outside the scope are defined. This is referred to as the *inherency doctrine*.[18] The second method uses the rights granted by the patent to determine in economic terms the value of the patent. The value is then used to evaluate the legality of different licensing practices.

[18] The claim is not made here that the 'inherency doctrine' is an accepted term for early patent scope tests. Rather the concept is used here (as a label) simply to distinguish it from later patent scope tests that arguably were more economic in their methodology. The inherency doctrine and the scope of the patent theory seem to be used interchangeably in the legal scholarship.

Licensing practices exceeding the value of the patent are deemed anti-competitive. Accordingly, these theories are referred to here as *patent value* theories.

The inherency doctrine

The inherency doctrine was developed in the US by the courts there when dealing with the first cases involving the patent-antitrust interface. The US Supreme Court developed this formalistic theory in order to determine the powers given under the patent grant and then to asses whether particular licensing practices fell within those rights. Another way of expressing the inherency doctrine is as the *limited licence doctrine*, meaning that a patent right owner is entitled to impose lesser restrictions that are part of the larger right in the patent, that is, the right to exclude use altogether.[19] The use of the inherency doctrine would in particular be a determinant for the application of *patent misuse*.[20]

The main feature of the inherency doctrine is the distinction between those breaches of licensing clauses triggering a patent infringement, and those that only could be enforced through general contract law.[21] Importantly, antitrust law could only be invoked against clauses falling within the scope of general contract law, while clauses *inherent* in the patent right would be immune from antitrust challenge. For instance, an output restriction imposed on licensees is encompassed by the patent holder's right to refuse to license to manufacturers altogether. In an early case, *Bement*, the patent grant was also viewed as covering price restrictions imposed on licensees as the patent holder was free to charge any price (even a monopolistic price) if it would 'reserve' the market to itself.[22] This, which today is hardly uncontroversial, was seen as a completely legitimate practice under patent law and thus immune from antitrust law.

The approach of the inherency doctrine created controversy when the courts expanded the scope of the patent grant in infringement suits directed at third parties.

---

[19] L Kaplow, 'The Patent-Antitrust Intersection: A reappraisal' (1984) 97 Harvard L Rev 1813, 1845.

[20] It appears that the term 'misuse of the patent' stems from *Morton Salt*. RJ Hoerner, 'The Decline (and Fall?) of the Patent Misuse Doctrine in the Federal Circuit' (2001) 69 Antitrust L J 669, 670; *Morton Salt Co v Suppiger Co.*, 314 US 488 (1942). However, the rationale of the *patent misuse doctrine* is attributed to the ruling in *Motion Picture Patents*. Patent misuse prevents the enforcement of the patent. See MA Lemley, 'The Economic Irrationality of the Patent Misuse Doctrine' (1990) 78 California L Rev 1599, 1609; RC Feldman, 'The Insufficiency of Antitrust Analysis for Patent Misuse' (2003) 55 Hastings LJ 399, 409; *Motion Picture Patents Company v Universal Film Manufacturing Company* et al, 243 US 502 (1917).

[21] See these two examples that arrive to the opposite conclusion: GF Munson, 'Control of Patented and Copyrighted Articles after Sale' (1917) 17 Yale LJ 270; FY Gladney, 'The Right to Use a Patented Article—A Study of *Henry v A. B. Dick Co.*' (1916) 1 St Louis L Rev 203. This has been referred to as the distinction between *covenants* and *conditions* in later legal scholarship, see WF Baxter, 'Legal Restrictions on Exploitation of the Patent Monopoly: An Economic Analysis' (1966) 76 Yale LJ 267, 278.

[22] *Bement v National Harrow Co*, 186 US 70 (1902).

A practice developed whereby patent holders imposed tying clauses on licensees, forcing them to limit the use of the patented product with unpatented products supplied by the patentee. These cases, which in reality involved sale contracts of the patented product subject to a tying condition rather than an explicit licence, expanded the possibility for bringing patent infringement suits for breaches of contract that otherwise formally would have been assessed under general contract law. Thus, a sale of the product was not assessed as an ordinary sales contract where restrictions imposed on the purchaser could be found illegal for restraints of trade. Instead, the sale of patented machines subject to conditions on the use of the machine were seen as licensing agreements, thus immunizing the terms and conditions of the sale even where these resulted in restraints of trade. Additionally, patent holders not only sued those purchasers (licensees) that breached the conditions of sale, but also third parties that had sold the unpatented goods to licensees for *contributory infringement*. In the hey-day of the inherence doctrine, this meant that the patent holder could control the sales of unpatented goods by third parties to a certain extent, something that could not easily be read into the rights conferred under the patent grant. This approach was first endorsed in *Button-fastener* by the Sixth Circuit Court of Appeals, and subsequently by the US Supreme Court in *Dick*.[23]

Although this approach clearly is formalistic, there is some form of economic reasoning underpinning the interpretation given by the US Supreme Court. Judge and later Justice Lurton wrote the opinions in both *Button-Fastener* and *Dick*, justifying them by pointing out that it was the competitive superiority of the patented invention that permitted the patent holder to 'control' the market for unpatented goods. Even in cases where the unpatented goods could only be used with the patented machine, purchasers of the machine would only accept the conditions imposed by the patent holder as long as the patented machine gave them a benefit. Thus, it made no difference whether the patent holder chose to be rewarded through the sales of the machine or the sales of unpatented products together with the machine.[24] Although, the approach seems not to have been unwelcomed at the time of the judgment,[25] it would be heavily criticized later by scholars. It is far-reaching and problematic to construe patent law in such a way as to expand the rights given under the patent, the right to exclude, to the use and control of unpatented goods.[26] Inherent in this is the

[23] *Heaton-Peninsular Button-Fastener Co v Eureka Specialty Co et al*, 77 F 288, 35 LRA 728, 25 CCA. 267, (6th Cir 1896); *Henry v AB Dick Company*, 224 US 1 (1912).

[24] *Henry v Dick Company* (n 23 above); *Button-Fastener* (n 23 above).

[25] GH Montague, 'The Supreme Court on Patents' (1912) 21 Yale LJ 583; ES Rogers, 'Restrictions on the Use of Patented Articles' (1912) 10 Michigan L Rev 608.

[26] GS Rich, 'The Relation between Patent Practices and the Anti-Monopoly Laws' (1942) 24 J of the Patent Office Society 85–106; issue no. 3, 159–81; issue no. 4, 241–83; issue no. 5, 328–56; issue no. 6, 422–37; GA Smith, 'Contest over Latitude of Patent Rights' (1922) 8 American Bar Association J 355, 355.

underlying notion that patent holders could be overcompensated by such a practice.[27]

The inherency doctrine up until *Dick* had mainly been a way of granting patent rights and licensing practices immunity. However, after *Dick*, the pendulum started to swing in the direction of favouring competition. Patent and competition concerns were increasingly perceived as being in conflict with one another.[28] A first step was taken by the US Supreme Court in *Standard Sanitary* in 1912[29] in which the Court held that not even a licensing agreement could be used to cartelize a whole industry. This indicated that practices that had previously been defined as rights inherent in the patent grant in special circumstances would be assessed under antitrust law. Thus, there was no longer full immunity as in the previous cases. This view implied that *overcompensation* of the patent holder would be a ground for breaking immunity. A second factor was the introduction of the Clayton Act 1914 in which tying clauses restricting competition were made illegal, irrespective of whether they concerned patented or unpatented products.[30] This legislation was particularly targeted at the situation in *Dick*. Moreover, the Supreme Court overruled *Dick* in *Motion Picture Patents* by hindering suits for contributory infringement when tying the use of unpatented products, thus confirming the approach of the Clayton Act.[31] A third step was taken in *United Shoe*, where the Court held that patent grants could not immunize contracts that in themselves were illegal under state or federal law.[32] This development swung the pendulum heavily in favour of competition. By 1942, the patent holder in *Morton Salt* was barred from bringing an infringement suit as tying clauses were included in the contracts constituting patent misuse even though the case concerned an unauthorized making of the patented invention and not contributory infringement.[33] Although this may give the impression that the competition concerns were weighed against the interests of protecting innovation, these cases also represent the use of the inherency doctrine as a form of limiting the protection under

[27] For instance, Rich claims that by 'monopolizing' unpatented products which would be used for a continuing period of time in comparison with asking for a royalty on the sales of the machine would give the patent holder more profit. Rich (n 26 above) 252. A similar view is expressed by Smith pointing out that the unpatented goods may be more valuable than the patented article, see Smith (n 26 above) 355.

[28] This is described as the shift towards a 'separate spheres' model. WK Tom, and JA Newberg, 'Antitrust and Intellectual Property: From Separates Spheres to Unified Field' (1997) 66 Antitrust L J 167, 170–3.

[29] *Standard Sanitary Mfg v US*, 226 US 20, (1912).

[30] Clayton Act, 15 USC §§ 14.

[31] *Motion Picture Patents Company v Universal Film Manufacturing Company et al*, 243 US 502 (1917).

[32] *United Shoe Machinery Corp et al v United States*, 258 US 451 (1922).

[33] *Morton Salt Co v Suppiger Co*, 314 US 488 (1942).

patent law.[34] The framework for that analysis was thus the same as in the earlier cases at the intersection.

The strongest argument against the inherency doctrine is that the 'test' applied does not involve an explicit assessment of the interest involved, namely incentives for R&D and competition.[35] In other words, each individual case could differ in its effects on R&D and competition, meaning that a rule either prohibiting or permitting the suits of third parties selling unpatented goods would always risk being under- or over-inclusive if based on a statutory construction of 'rights'. Thus, the inherency doctrine suffered from the shortcoming of inflexibility due to the lack of economic reasoning in each individual case.

The main blow to the inherency doctrine and its methodology was delivered by the Supreme Court's ruling in *General Electric* in 1926, marking an evolution towards a patent value test.[36] This case concerned a price restriction imposed on the sales of patented goods. In its analysis of the conditions the patent holder could impose on licensees the Court held that,

> [T]he patentee may grant a license to make, use and vend articles under the specifications of his patent for any royalty or upon any condition the performance of which is reasonably within the reward which the patentee by the grant of the patent is entitled to secure.[37]

The Court inferred from the right to reserve sales to the patentee also a right to specify the *methods* of sale by licensees as long as the conditions were 'normally and reasonably adapted to secure the pecuniary reward for the patentee's monopoly'.[38] It follows that the object of this assessment is also to define that which is within or outside the scope of the patent. However, the methodology for making the assessment is economic, as it requires a determination of the patent's value. This marks a shift from a formalistic analysis of what rights are granted under the patent, to a more economic assessment incorporating the economic purpose of the patent grant.[39]

Patent value theories

The patent value approach has its foundation in *General Electric.* This case gave rise to a number of different tests and methods as developed by legal scholars as to

[34] Importantly, it is not claimed here that the patent misuse doctrine was an application of antitrust law. Rather, the delimitation under patent law took competition aspects into concern. See Tom and Newberg (n 28 above) 172.

[35] TR Powell, 'The Nature of a Patent Right' (1917) 17 Columbia L Rev 663; Baxter (n 21 above) 275–9, 312–4.

[36] *United States v General Electric Co*, 272 US 476, (1926).

[37] *General Electric* (n 36 above) 489.

[38] *Ibid* 490.

[39] FK Beier, 'Patent License Agreements under German and European Antitrust Law' (1972) 3 Intl Rev of Industrial Property and Copyright L 1, 13.

determining which licensing restraints fall within or outside the value of the patent.[40] Of these, two theories are presented here that illustrate the methodology of the patent value approach.

The theory most hostile to patents is *Baxter's comparability test*.[41] In his analysis of the intersection, Baxter departs from the view that patents are given by society as to resolve the public goods problem of knowledge. Patents are a means of inducing inventions by granting the patent holder a reward through a monopoly on the invention. Accordingly, the patent holder is entitled to the reward permitted by the patent grant. However, the reward is limited to what users of the invention are willing to pay for exploiting it.[42] In addition, that reward comes at a price, namely the cost flowing from the patent in terms of distorting resource allocation, which decreases the social value of the invention. Consequently, there is a tension between patent rights and competition. The patent right is described as an island in a sea of prohibition. What is not permitted in the exercise of the patent right must be prohibited.[43] It is the task of the courts to determine this delineation in hard cases by ensuring that patent holders are only allowed to restrict competition as to receiving their reward under the patent system.

Baxter proposes the following test to make this determination,

> [A] patentee is entitled to extract monopoly income by restricting utilization of his invention, notwithstanding that utilization of other goods and services are consequently restricted, provided that in each case he confines the restriction to his invention as narrowly and specifically as the technology of his situation and the practicalities of administration permit.[44]

According to Baxter, the only justification for the patent system is to use the market as a reward mechanism for the invention in question. Consequently, the bargaining outcome between licensors and licensees will set the reward for each individual patented invention. There will be benefits to licensors as well as disadvantages to licensees stemming from the licensing contract. These are to be *comparable* to the value of the invention to licensees and to society. Accordingly, this test is referred to as the *comparability test*.[45]

---

40 See HF Furth, 'Price-Restrictive Patent Licenses Under the Sherman Act' (1958) 71 Harvard L Rev 815, (the competitive superiority test); GR Gibbons, 'Price Fixing in Patent Licenses and the Antitrust Laws' (1965) 51 Virginia L Rev 273, (the value of the patent test); DF Turner, 'The Patent system and Competitive Policy' (1969) 44 U New York L Rev 450; DW Hayter, 'When a License is Worse than a Refusal: a Comparative Competitive Effects Standard to Judge Restrictions in Intellectual Property Licenses' (1988) 11 Berkeley Technology L J 281, (the comparative-competitive-effects-test); S Maurer and S Scotchmer, 'Profit Neutrality in Licensing: The Boundary between Antitrust Law and Patent Law' (2007) 8 American L and Economics Rev 476.

41 Baxter (n 21 above).

42 *Ibid* 273.

43 *Ibid* 312.

44 *Ibid* 313.

45 *Ibid* 313–14.

Two factors are crucial in the assessment under the comparability test. First, any restrictions must be confined to the exploitation of the patented invention and not extend to unpatented items. Antitrust law should not interfere with licensing agreements seeking to extract the full value of the patent. Although Baxter acknowledges that in the process of profit-maximizing the use of the patent, other goods and services may be detrimentally affected, this is acceptable as long as the patent holder does not appropriate more than the value of the patent. This first factor consequently sets the outer limit for the quantity of the reward.

Secondly, antitrust law should also capture those restrictions that potentially could harm *comparability* meaning the competitive process determining the reward of the patent monopoly to its holder, and which supposedly corresponds to the value of the patent to society. Clauses that harm comparability include in particular restrictions aimed at protecting licensees from one another, allowing them to charge a supra-competitive price, as well as any other restrictions that allow a monopolization of the end-product in competition with other substitutable technologies. This finding is based on the notion that without a bargaining process between the two parties that are actively seeking to maximize their own position, the patent holder could be over-compensated and licensees 'bribed' through a share of monopoly profits.[46] Baxter's test therefore could be interpreted as that the mechanism of determining the reward to the patent holder is also the means to restrict monopoly losses.[47] The comparability test, at least in theory, purports to make a balance between reward and incentives to patent holders, and on the other hand, losses from the social cost of the patent monopoly. Baxter's comparability test is also illustrative of the difference between the inherency theory and tests based on patent value. To take an example, an output restriction imposed on licensees would be seen as 'inherent' in the patent right, and thus immune from antitrust challenge irrespective of its effects on allocative efficiency. However, for Baxter such restrictions eliminate comparability since licensees may share the industry profits with the licensor while a simple royalty obligation could secure the patent reward for the patent holder.[48]

One criticism of the *comparability test* is that its view of innovation is restricted to the rewards to the patent holder and thus the initial incentives for R&D leading to the patented invention. Such a view embodies a narrow view of innovation which characterizes both the inherency doctrine and the patent value tests. There are two problems with such an approach. As explained above, a view that only focuses on the incentives of the patent holder also ignores follow-up innovation. Moreover, the narrow view on innovation neglects the fact that licensees are a part of the innovation process contributing additional investments necessary for realizing the commercial

[46] *Ibid* 314.
[47] Kaplow (n 19 above) 1854.
[48] Baxter (n 21 above) 329–39.

value of the innovation. Obviously, Baxter is not concerned with transactional hazards, such as free-riding problems between licensees, but only looks at their interest in having the smallest amount of royalties paid and the least restrictive licensing agreement concluded. Accordingly, the comparability test strikes down any division of profits between the parties, as this is seen as a transfer of parts of the patent reward to licensees, meaning that the allocation of the patent reward is no longer inducing R&D. Additionally, as pointed out by Carrier, the suspicious view on licensing restrictions also hinders the patent holder from maximizing the profits derived from the invention.[49] The wording of the test, requiring that the restrictions be confined 'narrowly and specifically', could induce courts to search for less restrictive alternatives and thus second-guess the actions of the parties. Consequently, the application of this test could risk under-rewarding the patent holder, leading to suboptimal incentives for initial R&D.[50] As a result, this test leans in favour of competition concerns and antitrust law.

A different patent value test leading to a completely different outcome in its application is the *competitive superiority test*, designed by Bowman, who belongs to the Chicago School.[51] Departing from the goal of antitrust law, promoting allocative efficiency, a patent may per se lead to a restriction of output. The trade-off in the antitrust-patent law intersection is that allocative efficiency is increased in the long-term by patents (by producing inventions that consumers would be willing to pay for) while in the short-run allocative efficiency is simultaneously decreased due to the patentee's ability to restrict output.[52] For this reason, Bowman claims that an ordinary antitrust standard is not useful when dealing with patent rights, since these rights are meant to give market power, and in the extreme case monopoly.[53] Importantly, the assessment of patents in the context of licensing agreements should differentiate between profit maximization (which may include a monopolistic price) and the extension of the legal patent monopoly to unpatented products. Only in cases where the latter is established should there be intervention against patent licences. Accordingly, Bowman claims that,

> To evaluate the foregoing examples of patentee-licensee contracts in terms of whether they 'extend monopoly' is to apply a test that assumes the propriety of allowing a patentee to use any method of charging what the traffic will bear if, but only if, the reward to the patentee arising from the conditional use measures the patented product's competitive superiority over substitutes.[54]

[49] M Carrier, 'Unraveling the Patent-Antitrust Paradox' (2002) 150 U Pennsylvania L Rev 761, 795 (hereinafter 'Unraveling'); Kaplow (n 19 above) 1853.

[50] Carrier, Unraveling (n 49 above) 795; JA Ordover, 'Economic Foundations and Considerations in Protecting Industrial and Intellectual Property' (1985) 53 Antitrust LJ 503, 513.

[51] WS Bowman, *Patent and Antitrust Law; A Legal and Economic Appraisal* (Chicago: University of Chicago Press, 1973).

[52] *Ibid* 3.

[53] *Ibid* 53.

[54] *Ibid* X.

It follows that the main task of the courts is to prevent licensors from imposing conditions that do not measure the competitive superiority of the patented invention. This test may give the impression that appraising the competitive superiority of a patent also determines the maximum reward accruable to the patent holder. If that were the case, Baxter's and Bowman's tests would be practically identical. However, *Bowman's superiority test* is merely a method of keeping licensors from extending its monopoly beyond the patent scope, and not a way of measuring the *legitimate* reward of the patent holder in each individual case. Accordingly, the *competitive superiority test* lacks the second step, the comparability method, in Baxter's test. Instead, competitive superiority is inferred from the fact that licensees agree to the licensing contract, meaning that there is no test specifically determining whether a licensing practice 'stays' within the competitive superiority of the patent.[55] Instead, Bowman engages in an analysis of individual restraints (through an evaluation of US case law) where the determination is done. He places particular emphasis on the distinction between horizontal and vertical licences. His detailed analysis of vertical licences indicates particularly that restraints in such agreements generally have a goal of legitimate profit-maximization based on free-riding concerns, price discrimination, and quality control.[56] Bowman's test also tends to allow licences where licensees only would accept the agreement with the purpose of monopolizing the industry and share the resulting profits, as long as other unpatented competing products are unaffected.[57] While such a 'method' promotes licensing practices including a legitimate division of profits between the parties, eg because licensees also share the investments and risks, there is no possibility to impede practices resulting in overcompensation of the patent holder or cartelization.[58] The test thus tips in favour of the primacy of patent law. In addition, the test does not consider follow-up innovation as it only measures the effects on innovation with regard to the initial incentives of the patent holder.

Looking at these two tests, illustrative of the two extremes within the category of patent value theories, they suffer to some extent from the same fundamental deficiencies. Both have a limited view of innovation in the sense that they only focus on the initial investments made by the patent holder, thus ignoring incentives of follow-up innovators. Moreover, Baxter's test neglects the issue of transaction costs and the protection of the licensee investments. Therefore it is likely to be over-inclusive.

---

[55] Ordover (n 50 above) 513.

[56] Bowman (n 51 above) chapters 5 and 6.

[57] This is exactly why Baxter stresses that the bargaining between licensee and licensor must be kept uncorrupted by such monopolization aims. Otherwise the social costs of the restrictions may exceed the value of the patent to society. See Baxter (n 21 above) 314; Ordover (n 50 above) 513. Turner also criticizes such an approach, see DJ Turner, 'Basic Principles in Formulating Antitrust and Misuse Constraints on the Exploitation of Intellectual Property Rights' (1985) 53 Antitrust LJ 485.

[58] Kaplow criticizes the test for being limited to measuring the patent reward and not the monopoly losses. Kaplow (n 19 above) 1851–2.

In this sense, Bowman's test is more lenient, including some arguments and reasoning based on transaction costs. However, inferring a pro-competitive aim from the acceptance of the license by licensees seems to be going too far.

Economic balancing tests

An economic balancing test weighs the detriments of reduced competition, static allocative efficiency, against the benefits of inducing new inventions, dynamic efficiency. Importantly, this approach explicitly recognizes that the creation of new inventions is also a form of competition in the market.

Perhaps the most elaborated test is Kaplow's ratio test, which explicitly balances monopoly losses and the incentives for R&D.[59] Kaplow points out that licensing contracts and individual clauses may either increase or decrease one or the other. A clause may well increase the monopoly loss, but if it also increases the reward to the patentee, the net effect of the agreement may ultimately increase the incentives for further R&D. In such cases, the negative effect caused by the monopoly loss, which of course would be objectionable from an antitrust law point of view, would still lead to increased consumer welfare. However, Kaplow does not calculate the net effect by subtracting the deadweight loss from the reward acquired by the patent holder. Instead, he calculates the ratio of the incremental reward from the individual restraint and the deadweight loss caused by it.[60]

$$\frac{\text{Patent reward}}{\text{Monopoly losses}} = \text{Ratio}$$

Accordingly, the ratio depends on how much the reward is increased/decreased in relation to how much the deadweight loss is increased/decreased by each individual licensing restriction. As a general rule, licensing restrictions with higher ratios should be allowed under this theory. However, this calculation is not sufficient as it does not include a standard of comparison. Therefore, the ratio of an individual restraint must be compared with the optimal ratio. Here it may be assumed that patent law has made the 'correct' balance. Accordingly, the ratio for increasing the patent life by one year may be used as a standard. If the individual ratio is lower than the optimal ratio, the practice should be forbidden. If, on the other hand, the individual ratio is higher, a second comparison must be made with the ratio of shortening the patent life by one year.[61] If the individual ratio is higher, the practice is permitted.

The most important objection to the ratio test is the self-evident difficulty in applying the test in reality. Calculating the optimal ratio and individual ratios of licensing

[59] *Ibid* 1851.

[60] *Ibid* 1831.

[61] *Ibid* 1830. This calculation is made to measure whether the licensing practice costs less (in providing the incremental reward) than the last year of patent life.

agreements requires information that is unavailable in actual cases. This makes the *ratio test* an interesting theoretical construction but not viable as a test for actual application.

Nevertheless, the theoretical underpinnings of the test are still of interest. The main strength of the ratio test is that it makes an explicit trade-off between increasing reward to the patent holder and welfare losses in the product market. This is done in a manner that the inherency doctrine, as well as the *comparability* and the *competitive superiority tests* are not capable of matching. As discussed above, the inherency doctrine is a very inflexible rule tending to label a practice as within or outside the scope of the patent without any specific assessment in the case. Moreover, Baxter's test fails to make a trade-off, since every time licensees' bargaining incentives may be corrupted by the nature of the individual restraint it is assumed that *comparability* does not occur. The test therefore does not weigh the welfare effects of reward and restraint as against each other. As regards Bowman's test, there is no balancing since as soon as a practice increases the patent reward, and such reward is not derived from unpatented products, the practice is allowed, irrespective of the harm caused by monopoly losses. By contrast, the *ratio test* does not rest on presumptions made in the earlier tests but rather explicitly balances deficiencies in the product market as against the benefits for R&D investment in each individual case. Consequently, antitrust law could be used to fine-tune the competition assessment already done in patent law. This is an important departure from those previous theories, explicitly relying and protecting the initial balance established under patent law.

However, the main problem with the ratio test from a theoretical perspective is that it still uses the reward of the patent holder as a means of measuring the incentive mechanism of the patent system. Thus it incorporates a narrow view on innovation. First, it ignores the existence of follow-up innovation. Second, the test suffers also from the same problems as associated with the inherency doctrine and Baxter's comparability test, namely that the risks of licensees are not explicitly considered and protected.

In addition to Kaplow's ratio test, Ordover has also offered a framework based on economic balancing in order to assess the legality of practices in the antitrust-patent intersection.[62] Ordover defines the intersection as the trade-off between static and dynamic efficiency.[63] Thus, instead of presenting patent law as merely an incentive to producing inventions, he views that process as a method of competition, but at the R&D stage, *ex ante*. In contrast, antitrust law aims at promoting competition at the product or service stage incorporating the invention, *ex post*.[64] If there is free entry at the *ex post* stage, firms would risk not being able to appropriate the social

[62] Ordover (n 50 above) 503.
[63] *Ibid* 511.
[64] *Ibid* 510.

value of the invention. If *ex ante* competition is stimulated, eg by allowing the patentees greater latitude in exploiting their rights, this may also lead to increased *ex post* competition, for example through a variety of products developed by different parties.[65] Accordingly, the existence of two temporally connected markets must be conceptualized: an upstream market for ideas and a downstream market for product or services. As the two markets are connected vertically, an antitrust assessment requires an investigation of market conditions, such as entry barriers, market shares, and concentration at both levels.[66]

In making this assessment, *Ordover* suggests that in the analysis (based on the US rule of reason approach), one must ask,

> Would the same quantum of knowledge, information, be created in the absence of the restraint? How would the degree of socially beneficial spillovers be affected in the long run, if the restraint were prohibited or modified?[67]

In answering these questions, Ordover suggests looking at the structure of both the upstream and downstream markets, whether patents are interpreted narrowly or broadly, whether incentives for R&D are created by policies other than patent rights, and at the nature of the R&D in the specific industry, in particular if innovation is drastic or incremental.[68] In general, where parties are only operating at one level, anti-competitive effects are not like to occur.[69]

Several points of Ordover's framework are interesting, distinguishing it from previous approaches. To begin with, antitrust and patent law are not perceived of as having separate fields of operation. Antitrust law should consider the effects of licensing agreements on the upstream R&D market as well as the downstream product market. This stands in contrast to the *patent scope theories*, which presume that the realization of the patent reward at the *ex post* stage always increases competition at the *ex ante* stage and thereby confines antitrust law to operate only at the *ex post* level. Instead, antitrust law applies across the two inter-temporally connected markets. This not only incorporates a balancing exercise, but also opens up the possibility of using antitrust law to fine-tune the initial balance made in patent law, between innovation and product market competition. In fact, Ordover additionally claims that higher importance should be given to the effects on the upstream market, that is, output expansion at the R&D level, when calculating net benefits from patents and licensing practices.[70] This already implies an innovation-centred approach more characterizing for a dynamic efficiency test.

65 *Ibid* 510.
66 *Ibid* 514.
67 *Ibid* 517–18.
68 *Ibid* 518.
69 *Ibid* 516.
70 *Ibid* 517.

Secondly, by detaching itself from patent reward as a measurement method for R&D incentives, the two-market framework opens up for an assessment of follow-up innovation incentives, although this is not explicitly discussed by Ordover. For instance, a grant-back clause has effects on the R&D market for both licensors and licensees, meaning that the framework as such may consider such factors.

Third, although not explicitly stated, the shift from protection of patent reward to an innovation market also opens up for a broader view of innovation, including the activities of licensees. If patent reward no longer is the important factor, this means, in theory, that sharing of profits between licensors and licensees is not automatically condemned, a common feature of patent scope theories and the ratio test. Because of the lack of detail in how Ordover views individual restraints, it is not possible to conclude whether his approach views the activities of licensees positively, but this is at least theoretically possible. In particular, his second question referring to beneficial long-term spill-over effects could be construed to include justifications based on licensing efficiency, and thus the benefits of dissemination of technology.[71]

Dynamic efficiency test

With a dynamic efficiency test the protection of innovation is incorporated as an explicit goal of antitrust law. Moreover dynamic efficiency has primacy over other goals of antitrust law. Thus, static and dynamic competition no longer are opposed to one another.

Carrier is the only scholar to date that has designed a test having an innovation-centred approach, referred to here as a *dynamic efficiency test*.[72] Similar to Ordover, Carrier's starting point is that the intersection is a conflict between static and dynamic efficiency. Carrier sees innovation as the prime contributor to growth and thus, the most important kind of efficiency. He concludes that innovation must be positioned as a goal of antitrust.[73]

Carrier also refutes that an *economic* test would be useful for courts to apply since the values that such tests rely on are often impossible to calculate.[74] He introduces a

---

[71] Ordover also states that,

> For example, if patentees are allowed greater latitude in exploiting their property rights so that a value of the patent (or copyright) is enhanced, ex ante competition into the R&D activity may be stimulated. This should be considered a plus, on both short-run and long-run accounts. On the short-run account, this increases a chance that more than one viable new product (or process) will be developed.

It is difficult to see whether this would fall within the first or second question. In any case it appears to open up for arguments of protecting the licensee. *Ibid* 511.

[72] Carrier, Tripartite Innovation (n 5 above). Carrier has also in earlier articles discussed a previous version of the same test specifically applied to Section 2, Sherman Act. Carrier, Unraveling (n 49 above).

[73] Carrier, Unraveling (n 49 above) 815; Carrier, Tripartite Innovation (n 5 above) 116. He claims that innovation should be the *most important* goal of antitrust.

[74] Carrier, Unraveling (n 49 above) 799–800.

more practical oriented test using the framework given by a *rule of reason approach*. The parties may show that the activity which follows from the transaction complies with a *reasonable necessity* standard. Importantly, this assessment is made at the justification after having established *prima facie* an anti-competitive effect in the market. According to Carrier, 'reasonable' is to ensure that the activity is required for innovation, although not *absolutely* necessary. Such a standard would prevent the courts from applying the *less restrictive clause standard* and second-guessing the behaviour of the parties.[75] The standard is more than simply showing plausibility that the activity will promote innovation. The word 'necessity' is aimed at establishing a link between the activity and the innovation.[76]

The *reasonable necessity* test measures innovation at three different levels. First, activities taken to promote innovation before the patent has been given are considered, referred to as the *product creation stage*.[77] Second, innovation is considered at the post-product level, the *recovery of investment stage*.[78] Third, innovation is considered in the context of multiple patents or a bottleneck scenario.[79] According to Carrier, if reasonable necessity is shown for an activity promoting innovation at any of these three stages, this provides a justification for the activity in question.[80] An essential factor in Carrier's approach is sector-specificity. This entails that factors, such as the importance of patents in inducing innovation, patent strength and the importance of spillovers for innovation, will affect the antitrust assessment of individual licensing practices. For instance, if patents are important in inducing innovation, more leeway for the parties may be given, when cooperating in the product creation stage.

Turning to the application of the first stage of the test, the inquiry is whether the patent-based activity is reasonably necessary for *product creation*. According to Carrier, the test is met when in the absence of the activity in question, the product would not be created, or in the alternative, where the activity makes it *significantly easier* to create the product.[81] Relevant factors in this inquiry are company-specific information, such as the parties' capabilities and market positions, as well as sector-related information including the difficulties and expense of creating the product. The effectiveness of patents as an appropriability mechanism in the sector in question is also assessed.[82] For instance, Carrier gives the example of two small R&D companies that need to collaborate (through agreement or merger) in order to achieve the capability to create a new product.

[75] Carrier, Tripartite Innovation (n 5 above) 121.
[76] *Ibid* 122.
[77] *Ibid* 124.
[78] *Ibid* 131.
[79] *Ibid* 134.
[80] *Ibid* 124.
[81] *Ibid* 124.
[82] *Ibid* 126–7.

The test at stage two focuses on whether the activity in question is appropriate for allowing the patent holder to exploit and develop its technology. Development and commercialization are important in providing the patent holder a means of appropriating the value of the patent.[83] This assessment considers the capabilities of the firm to develop and commercialize, but also the protection of both licensors and licensees against free-riding.[84] In this assessment, the courts are not supposed to second-guess the parties by ascertaining whether the licence constitutes the most efficient method of using the patent or whether the royalty level is correlated to the patent holders R&D investments.[85] For instance, a licensee that offers complementary assets would be a strong indicator that the licensing agreement is reasonably necessary. Sector-specific factors are also to be included. For instance, in industries where imitation is easy, courts must be sensitive to the needs of patent holders to recover initial investments.[86] The approach under the second stage combines two different factors. On one hand, the 'patent reward' is incorporated to a certain extent, although not applying it as a standard for the maximum level of reward. On the other hand, it also takes a broad view of innovation in the sense that licensees are seen as providing an added value to the innovation meriting protection. This distinguishes Carrier's approach from previous tests, as the activities by licensees are incorporated into the innovation process and not merely seen as benefits in current or future competition in the product market.[87]

As concerns the third stage, Carrier suggests that the inquiry of reasonable necessity is aimed at resolving bottlenecks. Carrier points in particular to the problems of *follow-on innovation* and *hold-up* and the alternative solutions available. In his view, bottlenecks may emerge when basic patents are strong and broad, protecting technology, used in downstream markets for other products invoking that basic technology, or where the innovation is cumulative.[88] Moreover, bottlenecks may also be caused because multi-component products use different technology that in turn can be protected by several different patent rights.[89] The emergence of bottlenecks serves as an argument particularly favouring certain licensing arrangements, such as patent pools and strategic alliances. Whenever these arrangements resolve these bottlenecks, such activities should be found to be reasonable and necessary. The balancing in these

---

[83] *Ibid* 132.
[84] *Ibid* 132.
[85] *Ibid* 132–3.
[86] *Ibid* 133.
[87] Naturally, one may ask what the difference really is between referring to dynamic competition and increased product market competition. If dynamic competition brings new products to the market, the benefits would still be in the form of competition through new, different or better products. However, in cases where those products also grant licensors and licensees market power, there is still a benefit in the form of the new, different or better product being released on the market as such.
[88] Carrier, Tripartite Innovation (n 5 above) 135–44. Carrier also points to the anti-commons problem in the agricultural sector.
[89] *Ibid* 144–52.

cases should be done between the effect of licensing restrictions and the detrimental effects in terms of delay on follow-on innovation.[90] The focus at this stage is on promoting the commercialization of products and follow-up innovation. This also distinguishes Carrier's approach from other tests, since incentives to innovation are not equated with protection of the patent reward,[91] corresponding to the broader view of innovation as expressed in the economic scholarship.

Carrier's approach consequently takes into consideration many of the factors that are relevant according to economic theory, in particular the broad view of innovation and transactional hazards. Hence, licensing agreements must be assessed under a different standard than the value of the patent. In practice, Carrier simply follows normal antitrust assessment as a first step, meaning that a competitive harm must be identified *prima facie*. This arguably also removes a great number of those cases that under a patent scope standard would seem suspicious simply because the licensing restraint does not appear to fall within the patent reward. Such an approach is reasonable considering that without an identified competitive harm, it is unnecessary to intervene against licensing practices. The discussion on the legitimacy of the licensing practice thus becomes relevant in the justification stage.

Moreover, Carrier's approach protects innovation irrespective of the stage at which the particular practice takes place, expressing a broad view of innovation. This is also interesting because it gives primacy to innovation over 'ordinary' competition concerns when the activity is reasonable justified. However, one may get the erroneous impression that Carrier thus means that R&D incentives are more important than competition, thus incorporating a *Schumpeterian* view. This is not correct considering that the three stages also protect competition through dissemination, in particular through the resolution of bottlenecks. As strong patents may justify antitrust intervention at the second stage, an 'ordinary' antitrust assessment is also partly incorporated at the justification stage.

The reliance on sector-specific factors, however, is a troublesome element in Carrier's test. While one may sympathize with the aim of facilitating judicial scrutiny in individual cases, too much emphasis is put on such factors. Whether the fact that patents are strong or not in a particular sector will not necessarily mean that a particular licensing practice is aimed at resolving such a problem. It can be argued that such factors will come into play if the transactional hazards with the specific licence are considered. However, they will then be more specific to the case at hand and more connected to the content of the licence agreement. It seems somewhat contradictory to apply a reasonable necessity standard and to subsequently rely on too broad presumptions about the conditions in certain sectors. A more rigorous evaluation of

[90] *Ibid* 152.

[91] As stated above, this view is also hinted in Ordover's framework, but Carrier is more explicit and detailed in his analysis of follow-up innovation.

specific clauses is inevitable, since the effects vary depending on the factual circumstances of the individual case. This does not mean that a court should apply a stricter standard, such as proportionality, indispensability, or an absolute necessity test. Instead, competition authorities and courts should have an increased understanding of the licensing practices concerned, and the problems that such clauses are meant to resolve. Less focus should be placed on the reward aspects (either for the licensor or the licensee).

C. An Evaluation of the Commission's Approach

What is the approach of the Commission in the TTBER and the TT Guidelines in light of the theories discussed above?[92] A preliminary point of interest is the view of the Commission as to intellectual property rights (IPRs) generally. IPRs are seen as promoting *dynamic competition* through improved or new products and processes, while competition law ensures that companies are pressured to innovate.[93] Thus, according to the TT Guidelines, IPRs and competition law are complementary and there is no inherent conflict between these two different bodies of law.[94] The TT Guidelines do not set *dynamic efficiency* as an explicit goal of competition law. They simply state that competition indirectly promotes this goal.[95] Peeperkorn, one of the authors of the TTBER, provides an explanation for this approach. In his view, the claim that an explicit goal of innovation should be included under competition law is based on the assumption that there is a contradiction between innovation and the goal of current competition policy, which is to promote price competition.[96] By contrast, the tendency in economic theory points to the fact that a competitive structure in the market leads to more innovation, as it pressures companies to compete in terms of technology.[97] Thus, product market competition and a strict competition policy serve as a stick to promote dynamic efficiency by pressuring firms

---

92 The focus here is the reasoning by the Commission in the TT Guidelines. The TTBER may be seen as an expression of the underlying reasoning in the TT Guidelines as regards specific restraints.

93 TT Guidelines (n 2 above) para 7.

94 *Ibid* para 7.

95 *Ibid* para 7. Regarding the fourth condition in Article 101(3) TFEU the Commission states,

> ... Ultimately the protection of rivalry and the competitive process is given priority over potentially pro-competitive efficiency gains which could result from restrictive agreements. The last condition of Article [101(3) TFEU] recognises the fact that rivalry between undertakings is an essential driver of economic efficiency, including dynamic efficiencies in the shape of innovation.

Commission Notice, Guidelines on the application of Article 81(3) of the Treaty (Exemption Guidelines) [2004] OJ C 101/08, para 105.

96 L Peeperkorn, 'IP Licenses and Competition Rules: Striking the Right Balance' (2003) 26 World Competition 527, 532–3.

97 Peeperkorn is relying on Arrows' finding that a competitive market structure is prone to give incentives to innovation. This would contravene the claims made by some (Schumpeter) that a concentrated market or monopoly leads to more innovation.

to innovate.[98] If the interpretation given by Peeperkorn is an accurate depiction of the Commission's view, this implies that an eventual balancing between *ex ante incentives* (to conduct R&D which leads to the IPR in question) and *ex post* short-term price competition will favour the latter.

The assessment under Article 101 TFEU is divided into two parts—the finding of a restriction of competition and competitive harm under the prohibition in Article 101(1) TFEU and then the evaluation of possible justifications under the exemption in Article 101(3) TFEU. Such a structure leaves the assessment of the specific innovation concerns primarily to the exemption stage. Following the reform of competition law in the last decade, the approach under Article 101(1) TFEU has shifted to a more Chicago School approach meaning that likely competitive harm in the form of increased prices or restricted output must be established to justify intervention by competition law.[99] Such an approach is incompatible with a patent scope approach, since the determination of patent scope is not used as a tool to define restrictions of competition.[100] Looking at economic theory on licensing, such an approach also makes sense. Irrespective to which degree the parties are restrained in their freedom of action, it is dubious to intervene against practices not impacting the conditions in the market in terms of price, output, or innovation. Moreover, since licensees also contribute to innovation, clauses protecting licensees from licensors or resulting in a division of profits between the parties merit protection if innovation is to be promoted.

When screening the negative effects on competition the Commission further looks at innovation, technology, and product markets.[101] Although such a view may give the impression that the Commission explicitly protects innovation markets, such a protection is limited. As stated in the TT Guidelines, the use of an innovation market is primarily confined to situations where the future product market may be identified, and only in particular circumstances will the Commission explicitly protect competition in innovation by securing that a number of independent research poles

[98] Peeperkorn (n 96 above) 533. This idea also has support in some theoretical literature as well as empirical studies. For an overview, see S Ahn, 'Competition, Innovation and Productivity Growth: A review of theory and evidence' (2002) OECD, Economic Department Working Paper No 317.

[99] TT Guidelines (n 2 above) para 15; Exemption Guidelines (n 95 above) para 24. For the shift in the assessment under Article 101(1), see Anderman and Kallaugher (n 13 above), 53–4.

[100] The Commission's previous practice cannot be viewed as based solely on a patent scope approach. Regulation 240/96 block exempted clauses based on the *nature* of the clause, but sometimes the rationale would include protection of licensee investments and was therefore not based on the level of patent reward. Nonetheless, there are examples in the older case-law that obviously applied patent scope theory. See in particular *Windsurfing* where the Commission and the Court engaged in an exercise of assessing whether the licensing clauses fell within the subject-matter or not before making an assessment under the competition rules. Commission Regulation (EC) No 240/96 of 31 January 1996 on the application of Article 85(3) of the Treaty to certain categories of technology transfer agreements [1996] OJ L 31/2; Case 193/83 *Windsurfing International Inc v Commission*, [1986] ECR 611.

[101] Article 3, TTBER (n 1 above); TT Guidelines (n 2 above) paras 20 and 25.

are kept on the market. One example is the Commission's reasoning regarding grant-back clauses in the TT Guidelines. There it is stated that the stronger the position of the licensor, the higher the negative effects of grant-back obligations because the licensee constitutes an important source of future competition.[102] Obviously, such reasoning is based on the position in the technology market and not the innovation market.

As regards the exemption stage, the requirements of economic and consumer benefit evaluates whether the agreement has any pro-competitive effects. The indispensability requirement establishes a causal link between restriction of competition and the pro-competitive effects. Finally, the elimination of competition test prohibits the agreement in case the negative effects on 'static' competition are too severe. The most important of these requirements is the indispensability requirement since ultimately it is this assessment that evaluates the need for a particular licensing agreement or clause in the specific case.

In this regard, it is important to acknowledge that the Commission recognizes a number of transactional hazards which are resolved by individual clauses. The reasoning by the Commission with regard to the imposition of territorial restrictions and non-compete obligations on the licensee, and the use of no-challenge clauses to protect know-how, are examples which show awareness of the transactional hazards involved.[103] However, there are also examples to the contrary. For instance, as regards the use of minimum royalty obligations combined with running royalties, the TT Guidelines state that the licensor may *normally* demand a lump-sum payment in order to cover specific investments made into the contractual relation.[104] Thus, a less restrictive alternative than minimum royalty obligations could be used. Such a view ignores that licensees may not be willing to pay such lump-sum payments if there is uncertainty of future demand for products using the licensed technology. The lump-sum payment is an asset-specific investment which cannot be recovered if the technology turns out to be a commercial failure.[105] At the same time, the patent holder must secure that a licensee with market power will not behave opportunistically once the transferred technology becomes a success. The minimum royalty obligation thus protects the relation-specific investments made by the licensor which are necessary to make the technology transfer effective. In fact, the clause resolves the transactional

[102] TT Guidelines (n 2 above) para 110.

[103] TT Guidelines (n 2 above) para 174, referring to free-riding concerns between licensees justifying territorial protection. TT Guidelines (n 2 above) para 201, stating that non-compete obligations may be justified to hinder the licensee from using transferred know-how when producing goods under a licence from a third party. TT Guidelines (n 2 above) para 112, stating that know-how, once transferred, cannot be granted back by the licensee. Thus, without a no-challenge clause protecting the transferred know-how, a licensor may choose not to license.

[104] TT Guidelines (n 2 above) para 203.

[105] The statement ignores the transactional hazards for the licensee.

hazard for both parties by reducing the risks caused by the relation-specific investments.

Moreover, the Commission sometimes uses market position to assess indispensability also in cases where market position is not a relevant factor to transactional hazards. For instance, the assessment of the indispensability of sales restrictions between the parties is dependent on whether the licensor or licensee is a weak competitor.[106] While such a view seems to make common sense, the justification for sales restrictions is usually connected to free-riding problems between the parties and the protection of further investments made by either licensors or licensees.[107] The absence or existence of such investments, however, is not considered in the situation and the assessment is solely based on market position. Such a view is in dissonance with a transactional hazards approach, where each restraint must be connected to issues such as free-riding. The same may be said of the example on grant-back clauses described above. Whether a strong grant-back is justified because of the benefits of transferring the technology and because the know-how included in the licence requires indirect protection through the transferred patent is not considered, but only the possible strong market position of licensors.[108]

Another interesting aspect is that the Commission's application of the elimination of competition requirement gives primacy to static competition.[109] The aforementioned examples also imply that the Commission views dissemination as the prime benefit of licensing as it brings new products to the market and thus promotes product market competition. However, when the positive effects on competition due to dissemination are not realized because of the presence of market power, ordinary static competition concerns take over. While such an approach encompasses a limited form of balancing, it may hardly be said to be innovation centred. For instance, in cases of exclusive licensing of competing technology to a licensee with significant market power, the conditions under Article 101(3) TFEU are unlikely to be met when there are entry barriers in the technology market and the licensed technology

[106] TT Guidelines (n 2 above) para 170.

[107] Obviously, the Commission sees the aim of the clause to protect additional investments, but there is not discussion on why such a protection is needed and how a territorial exclusivity provides such protection. An analogy is made in the literature between technology and vertical distribution agreements regarding exclusivity and territorial protection, see eg Ritter (n 13 above) 177; JW Schlicher, 'Some Thought on the Law and Economics of Licensing Biotechnology Patent and Related Property Rights in the United States' (1987) 69 J of the Patent and Trademark Office Society 264, 270.

[108] As discussed above, efficient transfer also requires the transfer of unprotected know-how. Patents are sometimes used in order to facilitate such a transfer because it provides licensors a punishing mechanism if licensees use the know-how in an unauthorized way. At the same time, the transfer of know-how also facilitates spill-over to the licensee since the latter may more easily develop a superior technology in the future. A grant-back clause thus protects licensors from this transactional hazard (*misappropriation*), see Choi (n 14 above); Bessy and Brousseau (n 3 above).

[109] Exemption Guidelines (n 95 above) para 105.

constitutes a real source of competition in the market.[110] Such an assessment neglects the benefits of the licence in terms of *innovation* and increased welfare effects, but only protects static efficiency.

All in all, the approach by the Commission seems to incorporate an economic balancing approach. The framework, however, is unlike those of Kaplow and Ordover, in the absence of a balancing of the benefits of *ex ante* innovation incentives, but rather, the benefits of dissemination against static efficiency. While such a view on one hand includes a broader view of innovation in the sense that licensee activities are also part of dissemination, it neglects to a certain extent the benefits of innovation as such. This is illustrated in particular by the Commission's protection on follow-up innovation incentives regarding grant-back clauses whose benefits are primarily assessed as future competition on technology and product markets.

Conclusion

The tests proposed by legal scholars in the patent-antitrust intersection have shifted from a formalistic approach to economic balancing and thereby integrating the policy aims of innovation into the competition law assessment. Although the Commission's approach does not easily fit in with any of the different models described above, the TTBER and TT Guidelines incorporate some form of balancing test while still emphasizing the protection of product market competition and static allocative efficiency. Such an approach is not necessarily disadvantageous for innovation as it also permits a favourable view and protects the investments made by the licensees, which are crucial in the innovation process. On the other hand, since on balance the aspects of product market competition will weigh more heavily, cooperation which is good for innovation and efficiency in the long run but which requires more rigid control between the parties may still be discouraged because of the competition law regime.

110 TT Guidelines (n 2 above) para 166.

INDEX

Lightning Source UK Ltd.
Milton Keynes UK
UKOW04n1145270314

228940UK00001B/3/P